Lecture Notes in Computer Science 13838

Founding Editors

Gerhard Goos
Juris Hartmanis

T0177978

The series Lecture Notes in Computer Science (LNCS), including its subseries Lecture Notes in Artificial Intelligence (LNAI) and Lecture Notes in Bioinformatics (LNBI), has established itself as a medium for the publication of new developments in computer science and information technology research, teaching, and education.

LNCS enjoys close cooperation with the computer science R & D community, the series counts many renowned academics among its volume editors and paper authors, and collaborates with prestigious societies. Its mission is to serve this international community by providing an invaluable service, mainly focused on the publication of conference and workshop proceedings and postproceedings. LNCS commenced publication in 1973.

Luca Di Gaspero · Paola Festa · Amir Nakib ·
Mario Pavone
Editors

Metaheuristics

14th International Conference, MIC 2022
Syracuse, Italy, July 11–14, 2022
Proceedings

Editors
Luca Di Gaspero [ID]
University of Udine
Udine, Italy

Paola Festa
University of Napoli - Federico II
Napoli, Italy

Amir Nakib [ID]
University of Paris Est
Paris, France

Mario Pavone [ID]
University of Catania
Catania, Italy

ISSN 0302-9743 ISSN 1611-3349 (electronic)
Lecture Notes in Computer Science
ISBN 978-3-031-26503-7 ISBN 978-3-031-26504-4 (eBook)
https://doi.org/10.1007/978-3-031-26504-4

Preface

The Metaheuristics international conference – MIC – is nowadays the biennial main event focusing on progress in the area of Metaheuristics and their applications, whose first edition was established in 1995. MIC 2022 is therefore the 14th edition of a long series, and took place at the Department of Architecture of the University of Catania, Italy. As the community may know, the MIC conference is traditionally held in odd years, and this means that this edition should have taken place in 2021. However, due to the COVID 19 pandemic situation and given that the main feature of MIC is to provide a friendly environment for discussion of on new ideas, collaborations and meeting old and new friends, in accord with the Steering Committee we decided to postpone the conference to 2022 with the hope – and the bet – that it could be held mostly in person. We won our bet, and we were really pleased to welcome a large number of enthusiastic participants! MIC 2022 continues, then, its mission in playing a unique role as an information- and knowledge-sharing forum in the metaheuristics area, enabling attendees to learn and network across the broad range of its offerings.

MIC 2022 was held from July 11–14, 2022 in beautiful Ortigia, the historical island of the city of Syracuse, Sicily, Italy. By hosting the event, the city of Syracuse gave the participants the opportunity to enjoy the riches of its historical and cultural atmosphere, its traditions, and the beauties of its natural resources, such as the sea, the Roman Amphitheatre, the Greek Theatre, and the Ear of Dionysius, which is a limestone cave in the shape of a human ear. Syracuse is also the city where the great mathematician Archimedes was born.

In this edition, we wanted to give greater prominence to issues concerning the integration between metaheuristics and machine learning, proposing reputable Plenary talks by leading scientists who provided a broad view on the state of the art and on the future of this growing and emerging topic. Keynote speakers have always been one of the most important parts of MIC, and our six speakers this year were the following:

- Christian Blum, Spanish National Research Council, Spain
- Kalyanmoy Deb, Michigan State University, USA
- Fred Glover, Entanglement, Inc., USA
- Salvatore Greco, University of Catania, Italy
- Holger H. Hoos, Leiden University, The Netherlands
- Gary Kochenberger, Entanglement, Inc., USA
- El-Ghazali Talbi, University of Lille 1, CNRS, INRIA, France

Around 125 of valid submissions were received of which 72 papers were selected. Among those overall accepted submissions, 34 were of regular papers, 23 of short papers and 15 consisted of oral presentations of recently published works. The best 48 manuscripts, among short and regular papers, were included in this Volume. Each paper was reviewed by a team of 102 PC members and reviewers that we deeply thank for their contribution to MIC 2022's success. As organizers, we are very happy and proud to have had 150 participants overall of whom more than 110 were present. In total,

MIC 2022 offered to the participants 8 parallel sessions of oral presentations, whose topics ranged from the foundation of metaheuristic techniques to their application in the solution of combinatorial problems arising in many real-world domains. A number of satellite social events also took place, with the purpose of giving the conference delegates the opportunity to finally interact in person after this long pandemic time.

Finally, we would like to thank and recognize the enormous efforts of all members of the organizing committee, included the local members, who made MIC 2022 possible by donating their time, expertise, and enthusiasm. Without their hard work and dedication, a successful event like MIC 2022 would not have been possible. Of course, we would not have been able to organize an event on such a high scientific level without the excellent work done by all program committee members, the financial manager and conference secretariat, as well as the distinguished plenary speakers, to whom we express and confirm our great appreciation. The main thanks are addressed to all authors who supported MIC 2022 by sending their important and excellent contributions. A last special thanks we would like to address to the steering committee members for their helpful advice and support, and mainly for trusting us to organize this wonderful scientific event. Thanks to the success of MIC 2022, the Metaheuristics International Conference still continues to be the premier event for science and technology in metaheuristics, where scientists from all over the world meet to exchange ideas and sharpen their skills.

July 2022   Luca Di Gaspero
                                                                             Paola Festa
                                                                             Amir Nakib
                                                                             Mario Pavone

Organization

Organizing Committee

General Chairs

Luca Di Gaspero	University of Udine, Italy
Paola Festa	University of Naples "Federico II", Italy
Amir Nakib	Université Paris Est Créteil, France
Mario Pavone	University of Catania, Italy

Local Organization

Vincenzo Cutello	University of Catania, Italy
Andrea Schaerf	University of Udine, Italy
Raffaele Cerulli	University of Salerno, Italy
Angelo Cavallaro	University of Catania, Italy
Carolina Crespi	University of Catania, Italy
Attilio Di Natale	University of Catania, Italy
Georgia Fargetta	University of Catania, Italy
Rocco A. Scollo	University of Catania, Italy
Francesco Zito	University of Catania, Italy

MIC Steering Committee

Fred Glover	Entanglement, Inc., USA
Celso Ribeiro	Universidade Federal Fluminense, Brazil
Éric Taillard	University of Applied Sciences of Western Switzerland
Stefan Voss	University of Hamburg, Germany

Keynote Speakers

Christian Blum	Spanish National Research Council, Spain
Kalyanmoy Deb	Michigan State University, USA
Fred Glover	Entanglement, Inc., USA
Salvatore Greco	University of Catania, Italy
Holger H. Hoos	Leiden University, The Netherlands
Gary Kochenberger	Entanglement, Inc., USA
El-Ghazali Talbi	University of Lille 1, CNRS, INRIA, France

Contents

Application of CMSA to the Electric Vehicle Routing Problem with Time Windows, Simultaneous Pickup and Deliveries, and Partial Vehicle Charging

Mehmet Anıl Akbay[1]([✉]) [iD], Can Berk Kalayci[2] [iD], and Christian Blum[1] [iD]

[1] Artificial Intelligence Research Institute (IIIA-CSIC), Campus UAB, Bellaterra, Spain
makbay@iiia.csic.es, christian.blum@csic.es
[2] Department of Industrial Engineering, Pamukkale University, Denizli, Turkey
cbkalayci@pau.edu.tr

Abstract. As a consequence of the growing importance of environmental issues, partially due to a negative impact of transportation activities, the use of environmentally-friendly vehicles in logistics has become one of the prominent concepts in recent years. In this line, this paper addresses a variant of the vehicle routing problem, the electric vehicle routing problem with time windows and simultaneous pickup and deliveries, which are two essential real-life constraints. Moreover, we consider partial recharging of electric vehicles at charging stations. A recent self-adaptive variant of the matheuristic "Construct, Merge, Solve & Adapt" (CMSA) is applied to solve the tackled problem. CMSA combines heuristic elements, such as the probabilistic generation of solutions, with an exact solver that is iteratively applied to sub-instances of the original problem instances. Two constructive heuristics, a Clark & Wright Savings algorithm and a sequential insertion heuristic, are probabilistically applied to generate solutions which are then subsequently merged to form a sub-instance. The numerical results show that CMSA outperforms CPLEX in the context of small problem instances. Moreover, it is shown that CMSA outperforms the heuristic algorithms when large problem instances are concerned.

Keywords: Electric vehicle routing · Time windows · Simultaneous pickup and delivery · Partial recharging · Construct · Merge · Solve & adapt

1 Introduction

In spite of the remarkable advancements in exact solution techniques such as dynamic programming and mathematical programming—think of popular, high-performing tools such as CPLEX[1] and Gurobi[2]—solving complex combinatorial optimization problems using pure exact techniques still may require overly long computation times, especially when considering large-sized problem instances. In those cases in which

---

[1] https://www.ibm.com/analytics/cplex-optimizer.
[2] http://www.gurobi.com/.

© The Author(s), under exclusive license to Springer Nature Switzerland AG 2023
L. Di Gaspero et al. (Eds.): MIC 2022, LNCS 13838, pp. 1–16, 2023.
https://doi.org/10.1007/978-3-031-26504-4_1

exact techniques and solvers fail to deliver good-enough solutions within a reasonable computation time, approximate techniques are used to obtain such solutions in much lower computation times. In order to take advantage of exact solvers also in the context of large-sized problem instances, recent years have seen a sharp increase in the number of hybrid algorithms that combine approximate techniques and exact methods and tools. The resulting algorithms are often called *hybrid metaheuristics* or *matheuristics* [18]. A broad overview on these methods can be found in [6,7]. One of the recent hybrid algorithms in this direction is "Construct, Merge, Solve & Adapt" (CMSA) [5]. CMSA is based on applying an exact solver iteratively to sub-instances of the original problem instance. In other words, the search space is reduced before the exact solver is applied. Search space reduction is achieved in a bottom-up manner by, first, probabilistically generating solutions to the tackled problem and, second, by merging these solutions in order to obtain a sub-instance. The literature provides successful applications of CMSA to a range of combinatorial optimization problems such as the maximum happy vertices problem [25], the route planning for cooperative air-ground robots [3], refueling and maintenance planning of nuclear power plants [13], the prioritized pairwise test data generation problem in software product lines [17] and the minimum positive influence dominating set problem [1].

1.1 Our Contribution

A recent variant of CMSA—called Adapt-CMSA [2]—was introduced in order to reduce the parameter-sensitivity of CMSA. The authors of [2] were able to show that Adapt-CMSA, in contrast to standard CMSA, does not need to be tuned separately for different problem sizes, respectively types. In this paper we propose the application of Adapt-CMSA to a variant of the classical vehicle routing problem (VRP): the electric vehicle routing problem with time windows and simultaneous pickup and deliveries (EVRP-TW-SPD). Classical vehicle routing problems aims to find optimal routing plans for a fleet of vehicles distributing goods from depots to customers, respectively demand points. The objective is generally to minimize the number of vehicles, the total distance traveled, or both, considering the limited capacity of vehicles. However, using electric vehicles instead of those with internal combustion engines requires to consider restrictions such as a limited driving range and en-route battery charging. The problem addressed in this study also considers two important real-life constraints, namely, time windows (TW) and simultaneous pickup and deliveries (SPD). TW constraints force vehicles to visit a customer within a predefined time interval. Moreover, when SPD constraints are considered, each customer has two different demands: (1) the goods to be delivered to the demand point (*delivery demand*), and (2) the goods to be collected from the demand point (*pickup demand*). So, once a vehicle visits a certain customer, both demands must be met simultaneously. This approach usually arises as a reverse logistics practice [11,35]. However, despite the importance of reverse logistics in terms of sustainable operations, to the best of our knowledge SPD has not yet been considered within the scope of electric vehicle routing problems [24]. Finally, while—in the classical EVRP—batteries of electric vehicles are always completely charged when a vehicle makes use of a charging station, in this study we allow partial recharging. Note that this scenario is much more realistic than the one with full charging.

1.2 Organization of the Paper

The remaining part of this paper is organized as follows. Related literature is provided in Sect. 2. Section 3 presents the technical description of the problem together with the mathematical model. Section 4 describes the application of the proposed algorithm to the EVRP-TW-SPD. An experimental evaluation together with computational results are given in Sect. 5. Finally, concluding remarks and outlook to future lines of research are presented in Sect. 6.

2 Related Literature

Since it was first introduced to the literature in [10], the VRP and its variations have become one of the most popular research topics in combinatorial optimization and logistics. Many researchers and practitioners have introduced new, every time more realistic models as well as efficient problem-solving strategies. Toth and Vigo [34] and Golden et al. [19] present a range of extensions of vehicle routing problems as well as recent advances and challenges. Moreover, Montoya et al. [29] provide a broad taxonomy and a classification of publications from the VRP literature. Apart from introducing new problem variations, researchers also focused on developing solution methodologies to efficiently solve existing problem variants. A taxonomic review of metaheuristic solution approaches for VRPs is presented in [14].

Due to the growing interest in the use of alternative fuel technologies in transportation, new VRP variants that take advantage of electric vehicles considering their limited driving range and en-route charging necessities have been introduced to the literature. These studies are either presented under the name of EVRP or, more generally, as green vehicle routing problems. Asghari and Al-e-Hashem [4] and Moghdani et al. [27] provide a systematic review and categorization of green vehicle routing problems. The route optimization of rechargeable vehicles was first considered by Conrad and Filiozzi [9]. They presented a mathematical formulation of the problem, aiming to minimize the number of vehicles and the total cost related to the traveled distance, the service time, and the vehicle recharging cost. After this pioneering work, several researchers presented EVRP variations together with different algorithmic solutions. Erdoğan et al. [15] formulated a green vehicle routing problem in terms of a mixed-integer programming model and proposed two contraction heuristics along with a customized improvement technique to solve large-sized problem instances. Schneider et al. [33] extended the EVRP by including time window constraints into the model. Moreover, they proposed a metaheuristic algorithm based on variable neighborhood search (VNS) and tabu search (TS). Felipe et al. [16] extended the green vehicle routing problem introduced by [15], considering several real-life assumptions such as partial recharging and multiple charging technologies with different charging speeds and costs. They developed a mathematical model and proposed several heuristic solution methods for problem instances of realistic size. In addition, Keskin and Çatay [21] introduced an EVRP with TWs and partial recharging. They proposed an adaptive large neighborhood search (ALNS) based solution approach and analyzed if the partial recharging of batteries improved the obtained solutions when compared to the full charging option.

Furthermore, they also considered a fast charging option in [22] and proposed two different mathematical models as well as a matheuristic. Moreover, Montoya et al. [28] introduced a new model that takes into account the non-linear charging time of batteries. They reported that the time spent for charging batteries is non-linear, and ignoring this fact may cause the generation of infeasible and/or costly solutions. Keskin et al. [23] considered also the fact that an electric vehicle might have to wait in a queue when visiting a charging station. They proposed a two-stage simulation-based heuristic using a large neighborhood search algorithm. Sadati and Çatay [31] recently introduced a multi-depot green vehicle routing problem and developed a mixed-integer linear programming model. They proposed a solution method based on VNS and TS and reported on the computational properties of the algorithm. Duman et al. [12] proposed exact and heuristic algorithms based on branch-and-price-and-cut and on column generation to solve the EVRP with TWs. Huerta-Rojo et al. [20] proposed an algorithm named ACOLS, a combination of ant colony optimization and local search, for the EVRP with TWs and partial recharging.

3   Problem Description

In the following, we present a technical description together with a MILP model of the EVRP-TW-SPD under the partial recharging assumption. Our model is based on the model for the EVRP-TW proposed in [33] and modified by [21] for allowing partial recharging of vehicle batteries. We have extended this model by adding SPD constraints. In order to preserve the comprehensibility, we use the same notation as in the works cited above. Let $V = \{1,...,N\}$ refer to the set of N customers and F to the set of charging stations, respectively. Based on F, a set of dummy charging stations F' is defined that contains multiple copies of each charging station from F in order to allow for multiple visits to any of the charging station. Both 0 and $N+1$ refer to a single depot. Each route starts from 0 and ends at $N+1$. Set $V' = V \cup F'$ includes all customers and dummy charging stations. Henceforth, when a set is marked by a sub-index $0, N+1$ or both, it means that the set includes the respective instance(s) of the depot. Based on this, the following sets can be defined: $F'_0 = F' \cup \{0\}, V'_0 = V' \cup \{0\}, V'_{N+1} = V' \cup \{N+1\}, V'_{0,N+1} = V' \cup \{0\} \cup \{N+1\}$. In accordance with the defined sets and notations, the EVRP-TW-SPD can be defined on a complete, directed graph $G(V'_{0,N+1}, A)$ with the set of arcs $A = \{(i,j)|i,j \in V'_{0,N+1}, i \neq j\}$. Each arc is associated with a distance d_{ij} and travelling time t_{ij}. The constant h represents the battery consumption rate of an electric vehicle per unit distance traveled. Each customer with a delivery demand $q_i > 0$ or a pickup demand $p_i > 0$ (or both) must be served by an electric vehicle from a homogeneous fleet of vehicles that all have the same loading capacity C and battery capacity Q. Moreover, each vertex $i \in V'_{0,N+1}$ has a time window $[e_i, l_i]$ that denotes the earliest and latest possible visiting times allowed. Additionally, each customer $i \in V$ has a service time s_i, which refers to the time necessary for the visit. If a charging station is visited, the electric vehicle's battery is charged with a constant charging rate of $g > 0$.

The problem can be modeled by means of a MILP as follows. First, the model contains a binary decision variable x_{ij} that takes value 1 if arc (i, j) is traversed and 0 otherwise. Next, decision variables τ_i keep track of the service starting time at customer

$i \in V$. Moreover, decision variables y_i and Y_i record the battery state of charge on arrival, respectively departure, at (from) vertex $i \in V'_{0,N+1}$. Furthermore, for each arc $(i,j) \in A$, variable u_{ij} denotes the remaining cargo to be delivered to customers of the route, while v_{ij} denotes the amount of cargo already collected (picked up) at already visited customers. Note that each customers must be visited exactly once. The MILP model can then be stated as follows.

$$\text{Min} \quad \sum_{i \in V'_0, j \in V'_{N+1}} d_{ij} * x_{ij} + \sum_{j \in V'_{N+1}} x_{0j} * M \tag{1}$$

$$\text{s.t.} \quad \sum_{j \in V'_{N+1}, i \neq j} x_{ij} = 1 \qquad\qquad \forall i \in V \quad (2)$$

$$\sum_{j \in V'_{N+1}, i \neq j} x_{ij} \leq 1 \qquad\qquad \forall i \in F' \quad (3)$$

$$\sum_{i \in V'_0, i \neq j} x_{ij} - \sum_{i \in V'_{N+1}, i \neq j} x_{ji} = 0 \qquad\qquad \forall j \in V' \quad (4)$$

$$\tau_i + (t_{ij} + s_i)x_{ij} - l_0(1 - x_{ij}) \leq \tau_j \qquad \forall i \in V_0, j \in V'_{N+1}, i \neq j \quad (5)$$

$$\tau_i + t_{ij}x_{ij} + g(Y_i - y_i) - (l_0 + gQ)(1 - x_{ij}) \leq \tau_j \quad \forall i \in F', \forall j \in V'_{N+1}, i \neq j \quad (6)$$

$$e_j \leq \tau_j \leq l_j \qquad\qquad \forall j \in V'_{0,N+1} \quad (7)$$

$$0 \leq u_{0j} \leq C \qquad\qquad \forall j \in V'_{N+1} \quad (8)$$

$$v_{0j} = 0 \qquad\qquad \forall j \in V'_{N+1} \quad (9)$$

$$\sum_{i \in V'_0, i \neq j} u_{ij} - \sum_{i \in V'_{N+1}, i \neq j} u_{ji} = q_j \qquad\qquad \forall j \in V' \quad (10)$$

$$\sum_{i \in V'_{N+1}, i \neq j} v_{ji} - \sum_{i \in V'_0, i \neq j} v_{ij} = p_j \qquad\qquad \forall j \in V' \quad (11)$$

$$u_{ij} + v_{ij} \leq Cx_{ij} \qquad\qquad \forall i \in V'_0, j \in V'_{N+1}, i \neq j \quad (12)$$

$$0 \leq y_j \leq y_i - (hd_{ij})x_{ij} + Q(1 - x_{ij}) \qquad \forall i \in V, \forall j \in V'_{N+1}, i \neq j \quad (13)$$

$$0 \leq y_j \leq Y_i - (hd_{ij})x_{ij} + Q(1 - x_{ij}) \qquad \forall i \in F'_0, \forall j \in V'_{N+1}, i \neq j \quad (14)$$

$$y_i \leq Y_i \leq Q \qquad\qquad \forall i \in F'_0 \quad (15)$$

$$x_{ij} \in 0,1 \qquad\qquad \forall i \in V'_0, j \in V'_{N+1}, i \neq j \quad (16)$$

In this study, solutions using fewer vehicles—that is, with fewer routes—are preferred over others, even if the total distance traveled is higher than in other routes. Therefore the purely distance-based objective function from [21] is extended by adding an extra cost $M > 0$ for each vehicle used. Note, in this context, that the number of vehicles used in a solution is equal to the variables on outgoing arcs of the depot (0) that have value 1. The objective is to minimize the objective function from (1). Constraints (2) and (3) control the connectivity of customers and charging stations. More precisely, constraints (2) enforce that each customer must be visited by an electric vehicle. Constraints (3), on the other side, allow a charging station to be used, or not. Constraints (4) guarantee the balance of flow. Constraints (5) and (6) calculate arrival and departure times considering service and battery charging times. Constraints (7) enforce that each node can only be visited within the allowed time windows. Sub-tours are also eliminated by constraints (5)–(7). Constraints (8)–(12) guarantee that the delivery and pickup demands of customers are satisfied simultaneously. Finally, constraints (13)–(15) are battery state constraints.

4 Adapt-CMSA for the EVRP-TW-SPD

In this section we will describe the Adapt-CMSA algorithm that we designed for the application to the EVRP-TW-SPD. However, before describing the algorithm we first explain the solution representation in detail.

4.1 Solution Representation

Any solution S produced produced by the algorithm is kept in the form of a binary, square matrix with $|V'_{0,N+1}|$ rows and columns. Position (i, j) of S, denoted by s_{ij}, is hereby defined as follows:

$$s_{ij} := \begin{cases} 1 & \text{if the edge } (i, j) \text{ is a part of the solution} \\ 0 & \text{otherwise} \end{cases} \quad \forall i, j \in V'_{0,N+1} \quad (17)$$

Note that if the arc (i, j) is a part of a solution S, one of the vehicles used in S has passed from node i to node j.

Example. Let vector $\mathbf{I}$ contain the complete set of node indexes of an example problem instance with 3 charging stations and 5 customers. Note that indexes 0 and 9 represent the depot.

$$\mathbf{I} = (\ \underbrace{0,}_{\text{depot}}\ \underbrace{1, 2, 3,}_{\text{charging stations}}\ \underbrace{4, 5, 6, 7, 8,}_{\text{customers}}\ \underbrace{9}_{\text{depot}}\ )$$

A solution S with two vehicles that take routes T_1 and T_2 (where $T_1 = <0\text{-}7\text{-}6\text{-}1\text{-}8\text{-}9>$ and $T_2 = <0\text{-}3\text{-}4\text{-}5\text{-}2\text{-}9>$) is represented as follows:

$$S = \begin{bmatrix} s_{00} & s_{01} & \cdots & s_{09} \\ s_{10} & s_{11} & \cdots & s_{19} \\ \vdots & \vdots & \ddots & \vdots \\ s_{90} & s_{91} & \cdots & s_{99} \end{bmatrix} = \begin{bmatrix} 0&0&0&1&0&0&0&1&0&0 \\ 0&0&0&0&0&0&0&0&1&0 \\ 0&0&0&0&0&0&0&0&0&1 \\ 0&0&0&0&1&0&0&0&0&0 \\ 0&0&0&0&0&1&0&0&0&0 \\ 0&0&1&0&0&0&0&0&0&0 \\ 0&1&0&0&0&0&0&0&0&0 \\ 0&0&0&0&0&0&1&0&0&0 \\ 0&0&0&0&0&0&0&0&0&1 \\ 0&0&0&0&0&0&0&0&0&0 \end{bmatrix}$$

Algorithm 1. Pseudo-code of Adapt-CMSA for the EVRP-TW-SPD

1: **input 1:** values for CMSA parameters t_{prop}, t_{ILP}
2: **input 2:** values for solution construction parameters α^{LB}, α^{UB}, α_{red}, d_{rate}, h_{rate}
3: $S^{bsf} :=$ GenerateGreedySolution()
4: $\alpha_{bsf} := \alpha^{UB}$
5: Initialize(n_a, l_{size}, γ)
6: **while** CPU time limit not reached **do**
7:     $C := S^{bsf}$
8:     **for** $i := 1, \ldots, n_a$ **do**
9:         $S :=$ ProbabilisticSolutionConstruction($S^{bsf}, \alpha_{bsf}, l_{size}, d_{rate}, h_{rate}$)
10:         $C :=$ Merge(C, S)
11:     **end for**
12:     AddRandomEdges(C, γ)
13:     $(S^{cplex}, t_{solve}) :=$ SolveSubinstance(C, t_{ILP}) {This function returns two objects: (1) the obtained solution (S^{cplex}), (2) the required computation time (t_{solve})}
14:     **if** $t_{solve} < t_{prop} \cdot t_{ILP}$ and $\alpha_{bsf} > \alpha^{LB}$ **then** $\alpha_{bsf} := \alpha_{bsf} - \alpha_{red}$ **end if**
15:     **if** $f(S^{cplex}) < f(S^{bsf})$ **then**
16:         $S^{bsf} := S^{cplex}$
17:         Initialize(n_a, l_{size}, γ)
18:     **else**
19:         **if** $f(S^{cplex}) > f(S^{bsf})$ **then**
20:             **if** $n_a = n^{init}$ **then** $\alpha_{bsf} := \min\{\alpha_{bsf} + \frac{\alpha_{red}}{10}, \alpha^{UB}\}$ **else** Initialize(n_a, l_{size}, γ) **end if**
21:         **else**
22:             Increment(n_a, l_{size}, γ)
23:         **end if**
24:     **end if**
25: **end while**
26: **output:** S^{bsf}

4.2 The Adapt-CMSA Algorithm

Algorithm 1 shows the pseudo-code of our Adapt-CMSA algorithm for the EVRP-TW-SPD. First, function GenerateGreedySolution() initializes the best-so-far solution S^{bsf} to a feasible solution obtained utilizing an insertion heuristic (as explained in detail below). Then, parameters α_{bsf}, n_a, l_{size} and γ are initialized in lines 4 and 5. The handling of these parameters will also be outlined in detail below.

At each iteration, Adapt-CMSA builds a sub-instance C of the original problem instance. Just like a solution, a sub-instance C is a two-dimensional matrix whose elements are binary variables c_{ij} for all $i, j \in V'_{0,N+1}$. Hereby, variable c_{ij} refers to arc (i, j). C is initialized at each iteration to the best-so-far solution S^{bsf}, that is, $c_{ij} := s_{ij}^{bsf}$ for all $i, j \in V'_{0,N+1}$. Then, a number of n_a solutions are probabilistically constructed in lines 8–11. The function for the construction of a solution, ProbabilisticSolutionConstruction($S^{bsf}, \alpha_{bsf}, l_{size}, d_{rate}, h_{rate}$), receives—apart from the best-so-far-solution S^{bsf}—four parameters as input. Hereby, parameter α_{bsf} (where $0 \leq \alpha_{bsf} < 1$) is used to bias the construction of new solutions towards the best-so-far solution S^{bsf}. More specifically, the similarity between the constructed solutions and S^{bsf} will increase with a growing value of α_{bsf}. Parameter l_{size} controls the number of considered options at each solution construction step. A higher value of l_{size} results in more diverse solutions which, in turn, leads to a larger sub-instance. Next, $0 \leq h_{rate} \leq 1$ is the probability to choose the C&W savings heuristic as solution construction mechanism. Otherwise, the mechanism of the insertion heuristic is used. Finally, $0 \leq d_{rate} \leq 1$ is the probability of performing a step during solution construction deterministically.

After the construction of a solution S (line 9), the so-called *merge step* is performed in function Merge(C, S). In particular, $c_{ij} := s_{ij}$ for all $i, j \in V'_{0,N+1}$ with $s_{ij} = 1$, that is, every edge that is either used in S^{bsf} or in any of the solutions constructed in the current iteration is an option in the sub-instance. After probabilistically constructing n_a solutions and merging them to form the sub-instance C, the algorithm calls function AddRandomEdges(C, γ) to randomly add $\lceil \gamma \cdot |V'_{0,N+1}| \rceil$ additional edges/options to the current sub-instance. This mechanism augments the size of the sub-instance and increases the possibility of obtaining high-quality solutions when solving the sub-instance with CPLEX, which is precisely done in function SolveSubinstance(C, t_{ILP}); see line 13. Hereby, t_{ILP} is the CPU time limit for the application of CPLEX, which is applied to the ILP model from Sect. 3 after adding the following constraints:

$$x_{ij} = 0 \quad \text{iff } c_{ij} = 0 \quad \forall i, j \in V'_{0,N+1} \tag{18}$$

That is, the available edges in the model are reduced to the ones present in sub-instance C. Note that, the more such constraints are added to the original ILP, the smaller is the search space of the resulting ILP model and, as a consequence, the easier it is for CPLEX to derive an optimal solution to the corresponding sub-instance. Note also that the output S^{cplex} of function SolveSubinstance(C, t_{ILP}) is—due to the computation time limit—not necessarily an optimal solution to the sub-instance.

The self-adaptive aspect of Adapt-CMSA is to be found in the dynamic change of parameters α_{bsf}, n_a, l_{size} and γ. In the first place, we will describe the adaptation of parameter α_{bsf}. First of all, Adapt-CMSA requires a lower bound α^{LB} and an upper bound α^{UB} for the value of α_{bsf} as input. In addition, the step size α_{red} for the reduction of α_{bsf} must also be given as input. Adapt-CMSA starts by setting α_{bsf} to the highest possible value α^{UB}; see line 4.[3] In case the resulting ILP can be solved in a computation time t_{solve} which is below a proportion t_{prop} of the maximally possible computation time t_{ILP}, the value of α_{bsf} is reduced by α_{red}; see line 14. The rationale behind this step is the following one. In case the resulting ILP can easily be solved to optimality, the search space is too small, caused by a rather low number of free variables. In order to have more free variables in the ILP, the solutions constructed in ProbabilisticSolution-Construction(S^{bsf}, α_{bsf}, l_{size}, d_{rate}, h_{rate}) should be more different to S^{bsf}, which can be achieved by reducing the value of α_{bsf}.

The adaptation of parameters n_a, l_{size} and γ is done in a similar way and with a similar purpose. These three parameters are set to their initial values, that is, $n_a := n^{init}$, $l_{size} = l^{init}_{size}$ and $\gamma = 0.1$ in function Initialize(n_a, l_{size}, γ), which may be called at three different occasions: (1) at the start of the algorithm (line 5), (2) whenever solution S^{cplex} is strictly better than S^{bsf} (line 16), and (3) whenever solution S^{cplex} is strictly worse than S^{bsf} and, at the same time, $n_a > n^{init}$ (line 20). On the other side, in those cases in which

---

[3] Remember that solutions constructed with a high value of α_{bsf} will be rather similar to S^{bsf}.

S^{cplex} and S^{bsf} are of the same quality, the algorithm can afford to generate larger sub-instances and therefore, the values of the three parameters are incremented in function $\text{Increment}(n_a, l_{\text{size}}, \gamma)$. More specifically, n_a in incremented by 1, l_{size} is incremented by $l_{\text{size}}^{\text{inc}}$, and γ is incremented by 0.1.

4.3   Solution Construction

When function $\text{ProbabilisticSolutionConstruction}(S^{\text{bsf}}, \alpha_{\text{bsf}}, l_{\text{size}}, d_{\text{rate}}, h_{\text{rate}})$ is called, first, a decision about the solution construction mechanism is made. In particular, a random number $r_1 \in [0,1]$ is drawn uniformly at random. In case $r_1 \leq h_{\text{rate}}$, our version of the C&W savings algorithm [8] is executed. Otherwise, our insertion algorithm is applied. Second, a decision is made about the algorithm variant. Both in the case of the C&W savings algorithm and in the case of the insertion algorithm, we have implemented two algorithm variants. The first one allows for battery and time window infeasibilities by simply accepting infeasible solution construction steps, while the second variant exclusively generates feasible solutions. The decision about the algorithm variant is made by drawing a second random number $r_2 \in [0,1]$ uniformly at random. In case $r_2 \leq \text{inf}_{\text{rate}}$, the first (infeasibility accepting) algorithm variant is executed. Otherwise, the variant that generates feasible solutions is applied. In the following both construction algorithms and their variants are described in detail.

Probabilistic Clark & Wright (C&W) Savings Algorithm. Our probabilistic version of the C&W savings algorithm starts by creating a set of direct routes $R = \{(0 - i - (N+1)) \mid i \in V\}$. Subsequently, a savings list L that contains all possible pairs (i, j) of nodes (customers and charging stations) together with their respective savings value σ_{ij} is generated. Hereby, σ_{ij} is calculated as follows:

$$\sigma_{ij} := d_{0i} + d_{0j} - \lambda d_{ij} + \mu |d_{0i} - d_{0j}| \tag{19}$$

The so-called *route shape parameter* λ adjusts the selection priority based on the distance between nodes i and j [36], while μ is used to scale the asymmetry between nodes i and j [30]. Note that well-working values for these parameters are obtained by parameter tuning which is presented in Sect. 5. Note also that the savings list L will, at all times, only contain entries (i, j) such that (1) node i and node j belong to different routes, and (2) both i and j are directly connected to the depot in the route to which they belong. For executing the C&W savings algorithm, the following list of steps is iterated until the savings list L is empty.

1. First, based on the current savings values of the entries in L, a new value q_{ij} is calculated for each entry $(i, j) \in L$ as follows:

$$q_{ij} := \begin{cases} (\sigma_{ij} + 1) \cdot \alpha_{\text{bsf}} & \text{if } s_{ij}^{\text{bsf}} = 1 \\ (\sigma_{ij} + 1) \cdot (1 - \alpha_{\text{bsf}}) & \text{otherwise} \end{cases} \tag{20}$$

The savings list L is then sorted according to non-increasing values of q_{ij}. Finally, a reduced saving list L_r that contains the first (maximally) l_{size} elements of the whole savings list is created.

2. Next, an entry (i, j) is chosen from L_r as follows. First, a random number $r \in [0, 1]$ is chosen uniformly at random. If $r \le d_{\text{rate}}$, the first entry is chosen from L_r. Otherwise, en entry is chosen according to the following probabilities:

$$\mathbf{p}(ij) := \frac{q_{ij}}{\sum_{(i', j') \in L_r} q_{i'j'}} \quad \forall\, (i, j) \in L_r \tag{21}$$

Note that, the higher the value of parameter $\alpha_{\text{bsf}} \in [0, 1]$, the stronger is the bias towards choosing edges that appear in the best-so-far solution S^{bsf}.

3. Then, the two routes corresponding to nodes i and j are merged. The four possible cases for merging two routes are as follows:

$Case1$: T_1 :< 0-i-...-N+1 > T_2 :< 0-j-...-N+1 > rev(T_1) − T_2 T_m :< 0-...-i-j-...-N+1 >
$Case2$: T_1 :< 0-i-...-N+1 > T_2 :< 0-...-j-N+1 > rev(T_1) − rev(T_2) T_m :< 0-...-i-j-...-N+1 >
$Case3$: T_1 :< 0-...-i-N+1 > T_2 :< 0-j-...-N+1 > T_1 − T_2 T_m :< 0-...-i-j-...-N+1 >
$Case4$: T_1 :< 0-...-i-N+1 > T_2 :< 0-...-j-N+1 > T_1 − rev(T_2) T_m :< 0-...-i-j-...-N+1 >

Based on the way in which nodes i and j are directly connected to the depot, one or both of the routes must be reversed in order to be able to connect nodes i and j. In this context, note that the reversed version of a route T_1 is denoted by rev(T_1). If the merged route T_m is infeasible in terms of vehicle capacity, the merged route is eliminated and the respective pair of nodes are removed from the savings list. A new candidate is selected following the procedure in the previous step. The variant of our C&W savings algorithms that allows for unfeasible solutions now proceeds with the next step. The other variant, however, first checks for possible time window infeasiblities in T_m. If such an infeasibility is detected, T_m is eliminated, and the respective pair of nodes are removed from the savings list. A new candidate is selected following the procedure in the previous step. If the merged route is battery infeasible, a charging station is inserted into the infeasible route. The corresponding procedure determines the first customer in the route at which the vehicle arrives with negative battery level and inserts the charging station between this customer and the previous customer. For this purpose, the charging station which least increases the route distance is selected and inserted between the respective nodes. If this insertion is not feasible, then the previous arcs are attempted in the same manner. In those cases in which the route is still infeasible after charging station insertion, it is eliminated, and the respective pair of nodes are removed from the savings list. A new candidate is selected following the procedure described in the previous step. This procedure is repeated while the savings list is not empty. After merging, some of the charging stations that were previously added to the routes may become redundant. Those charging stations are removed from the merged route.

4. The savings list L must be updated as described above.

Finally, the obtained set of routes is transformed into a solution S in matrix form.

Probabilistic Insertion Algorithm. This algorithm constructs a solution by sequentially inserting each customer into the available routes until no unvisited customer remains. The first route is initialized by inserting either the customer with the highest distance to the depot or the customer with the earliest deadline. Then, a costs list formed by each unvisited customer and all possible insertion positions together with their respective cost values is generated. The insertion cost of customer i between nodes j and k is calculated using the following equation:

$$c(j,i,k) = d_{ji} + d_{ik} - d_{jk} \tag{22}$$

Then, a uniform random number $r \in [0,1]$ is generated. If $r \leq d_{\text{rate}}$, the customer with the minimum insertion cost is inserted into the respective insertion position. Otherwise, a random customer is selected from the first l_{size} elements of the cost list and inserted into the respective position if the vehicle capacity allows for this. The version of the insertion algorithm that only generates feasible solutions also checks for time window and battery infeasibilities. That is, insertions are only allows if they result in feasible tours regarding the time windows. Moreover, in case of battery infeasibility, a charging station is inserted into the route as explained above in the description of the C&W savings algorithm. If the insertion leads to infeasibility in terms of vehicle load capacity, a new tour is initialized with the respective customer. Finally, the obtained set of routes is transformed into a solution S in matrix form.

5 Experimental Evaluation

All experiments were performed on a cluster of machines with Intel® Xeon® 5670 CPUs with 12 cores of 2.933 GHz and a minimum of 32 GB RAM. CPLEX version 20.1 was used in one-threaded mode within Adapt-CMSA for solving the respective sub-instances.

Problem Instances. The EVRP-TW problem instances introduced by [33] were used to test the performance of the proposed algorithm. These problem instances include 36 small-sized instances with 5, 10, and 15 customers and 56 large-sized instances with 100 customers and 21 charging stations. Each group includes three main classes of instances based on the distribution of customer locations. Instance name prefixes "c", "r" and "rc" indicate that the locations of customers in the respective instance are clustered, randomly distributed or random-clustered, respectively. Moreover, each class contains two sub-classes, namely, type 1 and type 2 instances which separate instances in terms of the length of the time windows, vehicle load, and battery capacities. Since the original problem instances only come with delivery demands, we had to modify them by adding pickup demands. For this purpose, the delivery demand of each customer was separated into a delivery and a pickup demand by using the approach from [32]. Based on this approach, we first calculate a ratio $\rho_i = \min\{\frac{x_i}{y_i}, \frac{y_i}{x_i}\}$ for each customer $i \in V$. Hereby, (x_i, y_i) is the location of customer i. Then, multiplying this ratio with the original demand δ_i, we calculate the delivery demand of the respective customer, $q_i = \delta_i * \rho_i$. Finally, subtracting the delivery demand from the original demand, we obtain the pickup demand, $p_i = \delta_i - q_i$.

Parameter Tuning. In order to find well-working parameter values for Adapt-CMSA we utilized the scientific tuning software irace [26]. Instances r107, r205, rc101, rc104, rc105, and rc205 were used for the tuning process. Note that in the case of numerical parameters, the precision of irace was fixed to two positions behind the comma. irace was applied with a budget of 3.500 algorithm applications. The time limit for each

Table 1. Parameters, their domains, and the values as determined by irace.

Parameter	Domain	Value	Description
λ	$[1,2]$	1.77	Route shape parameter (C&W algorithm)
μ	$[0,1]$	0.56	Asymmetry scaling (C&W algorithm)
d_{rate}	$[0,1]$	0.17	Determinism rate for solution construction
l_{size}^{init}	$\{10,15,20,50,100,200\}$	100	Initial list size value
l_{size}^{inc}	$\{10,15,20,50,100,200\}$	20	List size increment
n^{init}	$\{1,2,3,4,5\}$	1	Initial nr. of constructed solutions
t_{ILP}	$\{5,7,10,15,20,25,30,35,40\}$	20	CPLEX time limit (seconds)
α^{LB}	$[0.6,0.99]$	0.62	Lower bound for α_{bsf}
α^{UB}	$[0.6,0.99]$	0.76	Upper bound for α_{bsf}
α_{red}	$[0.01,0.1]$	0.06	Step size reduction for α_{bsf}
t_{prop}	$[0.1,0.8]$	0.44	Control parameter for bias reduction
h_{rate}	$[0,1]$	0.2	Probability to choose a heuristic
$\inf_{rate}$	$[0,1]$	0.22	Probability to choose infeasible construction

problem instance was set to 900 CPU seconds. A summary of the parameters, their domains, and values selected for the final experiments are provided in Table 1.

5.1 Computational Results

Due to space reasons we limit the experimental evaluation to the 12 small problem instances with 15 customers (see Table 2) and the 17 large clustered problem instances (see Table 3). In the context of the small problem instances we compare the performance of Adapt-CMSA with the standalone application of CPLEX. As CPLEX is not applicable in a standalone manner to the large-size problem instances, we compare Adapt-CMSA with our probabilistic C&W savings algorithm (pC&W) and our probabilistic sequential insertion algorithm (pSI). The parameters of both algorithms were set in the same way as for their application within Adapt-CMSA. Moreover, the same computation time limit was used as for Adapt-CMSA, that is, both algorithms were repeatedly applied until a computation time limit of 150 CPU seconds (small problem instances), respectively 900 CPU seconds (large problem instances), was reached. Moreover, Adapt-CMSA, pC&W and pSI were applied 10 times to each problem instance. Finally, note that we determined the cost of each vehicle used in a solution as 1000—that is, $M = 1000$—for the calculation of the objective function values.

The structure of the result tables is as follows. Instance names are given in the first column. Columns 'm' provide the number of vehicles utilized by the respective solutions. In the case of Adapt-CMSA, pC&W and pSI these numbers refer to the best solution found within 10 independent runs. In the case of Adapt-CMSA, pC&W, and pSI, columns 'best' show the objective function values of the best solutions found in 10 runs, while additional columns with heading 'avg.' provide the average objective function values of the best solutions of each of the 10 runs. Next, columns with heading 'time' show the computation time (in seconds) of CPLEX and the average computation times of Adapt-CMSA to find the best solutions in each run. Note that the time limit for CPLEX was set to two hours. Finally, columns 'gap(%)' provide the gap (in percent)

Table 2. Computational results for small-sized instances with 15 customers.

Instances name	CPLEX				Adapt-CMSA			
	m	Best	Time	Gap (%)	m	Best	Avg.	Time
c103C15	3	3348.46	7183.45	7.3	3	3348.46	3362.06	91.46
c106C15	3	3275.13	1.28	0	3	3275.13	3275.13	43.15
c202C15	2	2383.62	62.26	0	2	2383.62	2383.73	53.13
c208C15	2	2300.55	4.62	0	2	2300.55	2300.55	8.80
r102C15	5	5412.78	7183.67	20.6	5	5412.78	5412.78	23.94
r105C15	4	4336.15	7.60	0	4	4336.15	4336.15	1.77
r202C15	2	2358.00	1723.55	0	2	2358.00	2373.41	118.00
r209C15	1	1313.24	4396.03	0	1	1313.24	1313.95	112.08
rc103C15	4	4397.67	349.61	0	4	4397.67	4398.04	76.43
rc108C15	3	3370.25	1170.76	0	3	3370.25	3373.31	69.70
rc202C15	2	2394.39	859.43	0	2	2394.39	2394.39	59.51
rc204C15	1	1403.38	7183.65	28.7	1	**1402.61**	1420.50	104.07
average	–	3024.47	2510.49		–	**3024.40**	3028.67	63.50

between the best found solutions and the best lower bounds found by CPLEX. Note that, in case the gap value is zero, CPLEX has found an optimal solution.

The following observations can be made. First, CPLEX was able to solve 9 small-sized problem instances to optimality and provided feasible solutions for the remaining 3 instances without being able to prove optimality within the computation time limit. For all small instances, Adapt-CMSA was able to find the optimal solutions provided by CPLEX. In the case of the rc201C10 instance, Adapt-CMSA was even able to improve the solution obtained by CPLEX. Moreover, Adapt-CMSA showed this performance using considerably less computation time than CPLEX. More specficially, while CPLEX found its best solutions on average in 2510 s, Adapt-CMSA was able to do so in just 63.5 s.

Concerning the large-sized instances, Adapt-CMSA significantly outperforms both pC&W and pSI, both in terms of best-performance and average performance. The results also show that the average computation time of Adapt-CMSA is higher than the one of pC&W and pSI. This is because pC&W and pSI are, at some point, not able to improve their best-found solutions anymore, while Adapt-CMSA is still able to do so. Moreover, we would also like to stress that Adapt-CMSA, in 15 out of 17 cases, found solutions using fewer vehicles than the solutions identified by pC&W and pSI. In summary, we can say that Adapt-CMSA shows a very satisfactory performance both in the context of small and large problem instances.

Table 3. Computational results for large-sized clustered instances.

Instances name	pC&W				pSI				Adapt-CMSA			
	m	Best	Avg.	Time	m	Best	Avg.	Time	m	Best	Avg.	Time
c101	21	22854.30	23028.50	406.70	13	14788.00	15574.68	513.05	12	**13044.40**	**13067.22**	720.99
c102	19	20764.20	21008.31	350.05	13	14664.70	15366.11	397.20	12	**13058.00**	**13549.74**	832.64
c103	16	17548.30	17622.53	483.74	12	13641.10	14363.90	491.02	11	**12312.20**	**13398.36**	822.92
c104	13	14388.30	14428.41	500.50	11	12404.70	13195.66	415.63	11	**12126.70**	**12254.73**	710.92
c105	19	20679.90	21608.72	588.59	13	14622.90	14928.32	346.88	11	**12067.20**	**12931.48**	732.89
c106	18	19797.20	20626.70	386.43	13	14713.90	14770.94	349.02	11	**12037.30**	**12934.44**	790.80
c107	18	19842.00	20594.06	459.03	12	13685.10	14631.25	339.14	11	**12059.20**	**12836.45**	699.06
c108	16	17589.20	18105.10	402.57	13	14617.00	14693.32	472.70	12	**13029.70**	**13055.02**	831.32
c109	14	15520.10	16415.00	454.81	12	13534.10	13628.65	353.78	11	**12155.10**	**13065.85**	850.50
c201	10	11333.60	12051.53	448.41	5	6081.90	6184.08	506.19	4	**4629.95**	**4630.33**	505.19
c202	8	9256.17	9509.34	490.70	5	6136.20	6232.74	526.69	4	**4659.58**	**5396.94**	837.39
c203	7	8188.19	8245.88	308.89	5	6247.22	6352.57	573.05	4	**5217.73**	**5906.27**	839.30
c204	5	6159.23	6200.12	416.22	4	5314.79	5354.18	556.74	4	**5087.33**	**5146.24**	787.12
c205	8	9209.37	9472.50	412.12	5	6161.46	6277.49	527.66	4	**4629.95**	**4641.96**	770.57
c206	6	7234.69	7989.99	460.46	5	6269.37	6305.64	405.68	4	**4633.12**	**5450.54**	847.01
c207	7	8062.15	8134.61	397.72	5	6278.60	6329.67	486.32	4	**4679.33**	**5752.92**	836.27
c208	6	7167.66	7674.06	341.68	5	6225.26	6282.06	458.79	4	**4656.78**	**5295.45**	854.43
average	–	13858.50	14277.37	429.92	–	10316.84	10615.96	454.09		**8828.45**	**9371.41**	780.55

6   Conclusion and Outlook

This study described the application of a self-adaptive version of a recent hybrid meta-heuristics, Adapt-CMSA, for the electric vehicle routing problem with time windows, simultaneous pickup and deliveries, and partial recharging. At each iteration, the algorithm first creates a sub-instance of the tackled problem by merging the best-so-far solution with a number of solutions probabilistially generated using two different solution construction mechanisms, a C&W savings heuristic and an insertion heuristic. The resulting sub-instance is then solved by the application of the ILP solver CPLEX. Adapt-CMSA makes use of a self-adaptive mechanism to adjust some of the parameter values so that a specific tuning of those parameters for different types of problem instances is not necessary. This mechanism handles the dynamic control of the size of the sub-instances and prevents that sub-instances are too large to be solved by CPLEX. Computational experiments were performed on 12 small-sized and 17 large-sized benchmark instances. The proposed approach was evaluated and compared to CPLEX on the small-sized problem instances and to probabilistic versions of the C&W savings heuristic and the insertion heuristic on large-sized problem instances. Numerical results indicated that Adapt-CMSA exhibits a superior performance on both small and large-sized problem instances.

In future work, we aim to develop alternative ways to represent solution components and form the sub-instance in a different way. Thus, the problem can be formulated as, i.e., a set covering-based model and solved efficiently. We believe that this approach may make a significant improvement in the performance of CMSA for VRPs.

Acknowledgements. This paper was supported by grants PID2019-104156GB-I00 and TED2021-129319B-I00 funded by MCIN/AEI/10.13039 /501100011033. Moreover, M.A. Akbay and C.B. Kalayci acknowledge support from the Technological Research Council of Turkey (TUBITAK) under grant number 119M236. The corresponding author was funded by the Ministry of National Education, Turkey (Scholarship program: YLYS-2019).

References

1. Akbay, M.A., Blum, C.: Application of CMSA to the minimum positive influence dominating set problem. In: Artificial Intelligence Research and Development, pp. 17–26. IOS Press (2021)
2. Akbay, M.A., López Serrano, A., Blum, C.: A self-adaptive variant of CMSA: application to the minimum positive influence dominating set problem. Int. J. Comput. Intell. Syst. **15**(1), 1–13 (2022). Springer
3. Arora, D., Maini, P., Pinacho-Davidson, P., Blum, C.: Route planning for cooperative air-ground robots with fuel constraints: an approach based on CMSA. In: Proceedings of GECCO 2019 - Genetic and Evolutionary Computation Conference, pp. 207–214. Association for Computing Machinery, New York (2019)
4. Asghari, M., Al-e Hashem, S.M.J.M.: Green vehicle routing problem: a state-of-the-art review. Int. J. Prod. Econ. **231**, 107899 (2021)
5. Blum, C., Pinacho Davidson, P., López-Ibáñez, M., Lozano, J.A.: Construct, merge, solve & adapt: a new general algorithm for combinatorial optimization. Comput. Oper. Res. **68**, 75–88 (2016)
6. Blum, C., Raidl, G.R.: Hybrid Metaheuristics - Powerful Tools for Optimization. Artificial Intelligence: Foundations, Theory, and Algorithms, Springer, Switzerland (2016). https://doi.org/10.1007/978-3-319-30883-8
7. Boschetti, M.A., Maniezzo, V., Roffilli, M., Bolufé Röhler, A.: Matheuristics: optimization, simulation and control. In: Blesa, M.J., Blum, C., Di Gaspero, L., Roli, A., Sampels, M., Schaerf, A. (eds.) HM 2009. LNCS, vol. 5818, pp. 171–177. Springer, Heidelberg (2009). https://doi.org/10.1007/978-3-642-04918-7_13
8. Clarke, G., Wright, J.W.: Scheduling of vehicles from a central depot to a number of delivery points. Oper. Res. **12**(4), 568–581 (1964)
9. Conrad, R.G., Figliozzi, M.A.: The recharging vehicle routing problem. In: Proceedings of the 2011 Industrial Engineering Research Conference, p. 8. IISE Norcross, GA (2011)
10. Dantzig, G.B., Ramser, J.H.: The truck dispatching problem. Manage. Sci. **6**(1), 80–91 (1959)
11. Dethloff, J.: Vehicle routing and reverse logistics: the vehicle routing probleam with simultaneous delivery and pick-up. OR-Spektr. **23**(1), 79–96 (2001)
12. Duman, E.N., Taş, D., Çatay, B.: Branch-and-price-and-cut methods for the electric vehicle routing problem with time windows. Int. J. Prod. Res. **60**(17), 5332–5353 (2022). Taylor & Francis
13. Dupin, N., Talbi, E.G.: Matheuristics to optimize refueling and maintenance planning of nuclear power plants. J. Heurist. **27**(1), 63–105 (2021)
14. Elshaer, R., Awad, H.: A taxonomic review of metaheuristic algorithms for solving the vehicle routing problem and its variants. Comput. Industr. Eng. **140**, 106242 (2020)
15. Erdoğan, S., Miller-Hooks, E.: A green vehicle routing problem. Transp. Res. Part E: Logist. Transp. Rev. **48**(1), 100–114 (2012)
16. Felipe, Á., Ortuño, M.T., Righini, G., Tirado, G.: A heuristic approach for the green vehicle routing problem with multiple technologies and partial recharges. Transp. Res. Part E: Logist. Transp. Rev. **71**, 111–128 (2014)

17. Ferrer, J., Chicano, F., Ortega-Toro, J.A.: CMSA algorithm for solving the prioritized pairwise test data generation problem in software product lines. J. Heurist. **27**(1), 229–249 (2021)
18. Fischetti, M., Fischetti, M.: Matheuristics, pp. 121–153. Springer, Heidelberg (2018)
19. Golden, B.L., Raghavan, S., Wasil, E.A.: The Vehicle Routing Problem: Latest Advances and New Challenges, vol. 43. Springer, New York (2008). https://doi.org/10.1007/978-0-387-77778-8
20. Huerta-Rojo, A., Montero, E., Rojas-Morales, N.: An ant-based approach to solve the electric vehicle routing problem with time windows and partial recharges. In: 2021 40th International Conference of the Chilean Computer Science Society (SCCC), pp. 1–8. IEEE (2021)
21. Keskin, M., Çatay, B.: Partial recharge strategies for the electric vehicle routing problem with time windows. Transp. Res. Part C: Emerg. Technol. **65**, 111–127 (2016)
22. Keskin, M., Çatay, B.: A matheuristic method for the electric vehicle routing problem with time windows and fast chargers. Comput. Oper. Res. **100**, 172–188 (2018)
23. Keskin, M., Çatay, B., Laporte, G.: A simulation-based heuristic for the electric vehicle routing problem with time windows and stochastic waiting times at recharging stations. Comput. Oper. Res. **125**, 105060 (2021)
24. Koç, Ç., Laporte, G., Tükenmez, İ: A review of vehicle routing with simultaneous pickup and delivery. Comput. Oper. Res. **122**, 104987 (2020)
25. Lewis, R., Thiruvady, D., Morgan, K.: Finding happiness: an analysis of the maximum happy vertices problem. Comput. Oper. Res. **103**, 265–276 (2019)
26. López-Ibáñez, M., et al.: The irace package: iterated racing for automatic algorithm configuration. Oper. Res. Perspect. **3**, 43–58 (2016)
27. Moghdani, R., Salimifard, K., Demir, E., Benyettou, A.: The green vehicle routing problem: a systematic literature review. J. Clean. Prod. **279**, 123691 (2021)
28. Montoya, A., Guéret, C., Mendoza, J.E., Villegas, J.G.: The electric vehicle routing problem with nonlinear charging function. Transp. Res. Part B: Methodol. **103**, 87–110 (2017)
29. Montoya-Torres, J.R., Franco, J.L., Isaza, S.N., Jiménez, H.F., Herazo-Padilla, N.: A literature review on the vehicle routing problem with multiple depots. Comput. Industr. Eng. **79**, 115–129 (2015)
30. Paessens, H.: The savings algorithm for the vehicle routing problem. Eur. J. Oper. Res. **34**(3), 336–344 (1988)
31. Sadati, M.E.H., Çatay, B.: A hybrid variable neighborhood search approach for the multi-depot green vehicle routing problem. Transp. Res. Part E: Logist. Transp. Rev. **149**, 102293 (2021)
32. Salhi, S., Nagy, G.: A cluster insertion heuristic for single and multiple depot vehicle routing problems with backhauling. J. Oper. Res. Soc. **50**(10), 1034–1042 (1999)
33. Schneider, M., Stenger, A., Goeke, D.: The electric vehicle-routing problem with time windows and recharging stations. Transp. Sci. **48**(4), 500–520 (2014)
34. Toth, P., Vigo, D.: Vehicle Routing: Problems, Methods, and Applications. SIAM (2014)
35. Wassan, N.A., Wassan, A.H., Nagy, G.: A reactive tabu search algorithm for the vehicle routing problem with simultaneous pickups and deliveries. J. Comb. Optim. **15**(4), 368–386 (2008)
36. Yellow, P.: A computational modification to the savings method of vehicle scheduling. J. Oper. Res. Soc. **21**(2), 281–283 (1970)

A BRKGA with Implicit Path-Relinking for the Vehicle Routing Problem with Occasional Drivers and Time Windows

Paola Festa[1], Francesca Guerriero[2], Mauricio G. C. Resende[3,4], and Edoardo Scalzo[2(✉)]

[1] Department of Mathematics and Applications, University of Napoli Federico II, 80138 Napoli, Italy
paola.festa@unina.it
[2] Department of Mechanical, Energy and Management Engineering, University of Calabria, 87036 Rende, CS, Italy
{francesca.guerriero,edoardo.scalzo}@unical.it
[3] Amazon.com, Inc., 333 108th Ave NE, Bellevue, WA 98004, USA
[4] University of Washington, 3900 E Stevens Way NE, Seattle, WA 98195, USA
mgcr@berkeley.edu

Abstract. This paper describes a biased random-key genetic algorithm (BRKGA) with implicit path-relinking for the Vehicle Routing Problem with Occasional Drivers (VRPOD). After a review of the relevant literature, the paper describes a proposed decoder and how BRKGA parameters are set. Experimental results show the efficacy of the proposed approach.

Keywords: Occasional drivers · Vehicle routing problem · Biased random-key genetic algorithm · Restart strategy

1 Introduction

The various benefits of online transactions and the increased use of the web are contributing to a growth in online shopping around the world. As a result, e-commerce is increasingly becoming a fundamental and necessary element of retail sales. Delivering products to customers at a low cost and with high speed, at a convenient time and place, is crucial to being competitive and at the same time it is becoming very challenging. In addition, urban development and great pressure on delivery efficiency, caused by the competitiveness of the market, have encouraged big retailers to explore innovative and unconventional last-mile delivery systems to offer fast delivery services (same-day delivery) with total cost savings.

Crowd-shipping is one of those innovative solutions. It is part of the *sharing economy*, in which unused or underused resources are exploited. This concept

L. Di Gaspero et al. (Eds.): MIC 2022, LNCS 13838, pp. 17–29, 2023.
https://doi.org/10.1007/978-3-031-26504-4_2

arises in various sectors of the modern economy, including that of retail distribution. The main concept of crowd-shipping is to delegate the deliveries of some
orders to ordinary people, named *occasional drivers* (ODs), who are willing to
make some detours from their own path and have some compensation in return.
By registering on a dedicated crowd-shipping online application, these people
can accept and carry out delivery requests and receive compensation. Therefore,
they are willing to share three logistic assets: transport capacity, smartphone and
free time. Through the app the e-retailers share with the ODs the information
necessary to make deliveries directly to a home or to any place indicated by customer. In addition to the ODs, customers can also log into the app and indicate
a time window of availability in which to receive the delivery. With these conditions, a crowd-shipping solution is a variant of the capacitated vehicle routing
problem with time windows.

The remainder of this paper is structured as follows. In Sect. 2, we describe
the state of the art of the problem treated, i.e. the Vehicle Routing Problem with
Occasional Drivers (VRPOD), and of the genetic solution approach proposed to
solve it, i.e. the Biased Random-Key Genetic Algorithm (BRKGA). In Sect. 3,
we describe a mathematical programming model of the problem, while in Sect. 4
we give details about the considered genetic algorithm. In Sect. 5 we report
the computational results and in Sect. 6 summarize our conclusions and outline
possible future work along this research line.

2   State of the Art

In the literature, there are many interesting papers that describe the benefits and
advantages of crowd-shipping. Archetti et al. (2016) were first to consider occasional drivers (ODs) in the context of the vehicle routing problem. To solve the
problem, the authors presented a hybrid Variable Neighborhood Search (VNS)
and Tabu Search. The authors emphasized the importance and challenges associated with defining appropriate compensation schemes. Macrina et al. (2017)
extended the work of Archetti et al. (2016) by introducing time windows for
customers and ODs and also considered a system with multiple deliveries for
ODs, referring to this variant as VRPODTWmd. In addition, the authors also
considered a system with a split delivery policy. They showed the benefits of
crowd-shipping in various scenarios.

To solve the VRPODTWmd, Macrina et al. (2020a) proposed a VNS approach and tested the effectiveness of their heuristic on numerous benchmark
instances of Solomon (1987). Macrina et al. (2020b) introduced transshipment
nodes in the VRPODTWmd. These nodes are intermediate points, closer to
the delivery area than the central depot, supplied by traditional drivers. The
authors showed the advantages of transshipment nodes formulating the problem
as a particular instance of a two-echelon VRP.

Di Puglia Pugliese et al. (2021) extended the previous work introducing a
new scenario in which the intermediate points are activated as occasional depots
and therefore the owners are compensated only in case of need. Furthermore,

the authors classified the ODs in two sets according to the activities performed and policy compensation. They proposed a mixed-integer programming model and showed the benefits of this scenario.

Other variants of the VRPOD also consider electric vehicles (see, e.g., Macrina and Guerriero (2018) and Macrina et al. (2020c)) or consider information not necessarily available in advance, and for this reason are called stochastic/online variants (see Archetti et al. (2021)).

The difficulty of optimally solving the VRP has led researchers, more and more, to use metaheuristic methods and to elaborate new variants. Genetic algorithm, based on the fundamentals of natural selection, is one of these metaheuristic methods. Among the numerous variants of genetic algorithm, we focused our attention on Biased Random-Key Genetic Algorithms (BRKGA) which has been used with high-quality performance in several optimization problems (see, e.g., Gonçalves et al. (2011) and Resende et al. (2011)) and also successfully used for the VRP (see, e.g., Andrade et al. (2013) and Lopes et al. (2016)). The process behind the BRKGA is first introduced by Bean (1994) as Random-Key Genetic Algorithms (RKGA). A representation of random keys encodes solutions using random numbers, usually belonging to the interval $(0, 1]$ and with representation, the solution process is independent of the problem to be solved. Gonçalves et al. (2011) introduced bias in RKGA and defined a formal and detailed framework for BRKGA. They describe numerous applications of BRKGA. Lucena et al. (2014) introduced two variants of the BRKGA, one of which is the multi-parent version (BRKGA-MP). The authors considered more than two parents to generate offspring for the next generation and showed by computational results that this variant outperforms the standard two-parent variant. Andrade et al. (2021) extend the work of Lucena et al. (2014) developing a BRKGA with implicit path relinking procedures (BRKGA-MP-IPR). They showed the benefits of this variant in some real-world problems.

In this paper, we developed a BRKGA with implicit path relinking to solve the VRPOD with time windows and capacity constraints. In particular, we exploited two important strategies to enforce the standard procedure of the genetic algorithm. The first one is the bidirectional permutation-based implicit path-relinking (IPR-Per); the second strategy used is a restart procedure: if during a certain point of the evolution the algorithm is stuck and apparently does not move towards a better solution, we try to move to another zone of the solution space using a new seed for the random-keys generator.

3 The VRPODTW Description

In this section we report the mathematical programming model of the VRPODTW proposed by Macrina et al. (2017). Let C be the set of customers, let D be the set of company drivers, s their origin and t their destination node. Let K be the set of available occasional drivers and V the set of their destination nodes. We define the node set as $N := C \cup V \cup \{s, t\}$. Each node pair (i, j) has a positive cost c_{ij} and a travel time t_{ij}, which satisfy the triangle inequality.

Each customer has a request d_i. Each company driver and occasional driver k has a maximum transport capacity Q and Q_k respectively. Each $i \in C \cup D \cup K$ has a time window $[e_i, l_i]$. Let x_{ij} be a binary variable equal to 1 if and only if a company driver traverses the arc (i, j), and let r_{ij}^k be a binary variable equal to 1 if and only if occasional driver k traverses the arc (i, j). Let y_i and w_i^k be the available capacities of company driver and occasional driver, after delivering to $i \in C$ respectively. Let s_i and f_i^k be the arrival times of a company driver and the occasional driver k at customer i respectively. The problem that aims at minimizing the total cost can be formulated as follows:

$$\min \sum_{i \in C \cup \{s\}} \sum_{j \in C \cup \{t\}} c_{ij} x_{ij} + \sum_{k \in K} \rho \left(\sum_{i \in C \cup \{s\}} \sum_{j \in C} c_{ij} r_{ij}^k - c_{s v_k} r_{sj}^k \right) \tag{1}$$

s.t.

$$\sum_{j \in C \cup \{t\}} x_{ij} - \sum_{j \in C \cup \{s\}} x_{ji} = 0 \qquad \forall i \in C \tag{2}$$

$$\sum_{j \in C} x_{sj} - \sum_{j \in C} x_{jt} = 0 \tag{3}$$

$$y_j \geq y_i + d_j x_{ij} - Q(1 - x_{ij}) \qquad \forall i \in C \cup \{s\}, j \in C \cup \{t\} \tag{4}$$

$$y_s \leq Q \tag{5}$$

$$s_j \geq s_i + t_{ij} x_{ij} - M(1 - x_{ij}) \qquad \forall i, j \in C \tag{6}$$

$$e_i \leq s_i \leq l_i \qquad \forall i \in C \tag{7}$$

$$\sum_{j \in C} x_{sj} \leq |D| \tag{8}$$

$$\sum_{j \in C \cup \{v_k\}} r_{ij}^k - \sum_{j \in C \cup \{s\}} r_{ji}^k = 0 \qquad \forall i \in C, k \in K \tag{9}$$

$$\sum_{j \in C \cup \{v_k\}} r_{sj}^k - \sum_{j \in C \cup \{s\}} r_{j v_k}^k = 0 \qquad \forall k \in K \tag{10}$$

$$\sum_{k \in K} \sum_{j \in C \cup \{v_k\}} r_{sj}^k \leq |K| \tag{11}$$

$$\sum_{j \in C} r_{sj}^k \leq 1 \qquad \forall k \in K \tag{12}$$

$$w_j^k \geq w_i^k + d_i r_{ij}^k - Q_k(1 - r_{ij}^k) \qquad \forall i \in C \cup \{s\}, j \in C \cup \{v_k\}, k \in K \tag{13}$$

$$w_s^k \leq Q_k \qquad \forall k \in K \tag{14}$$

$$f_i^k + t_{ij} r_{ij}^k - M(1 - r_{ij}^k) \leq f_j^k \qquad \forall i \in C, k \in K, j \in C \cup \{v_k\} \tag{15}$$

$$f_i^k \geq e_{v_k} + t_{si} \qquad \forall i \in C, k \in K \tag{16}$$

$$f_{v_k}^k \leq l_{v_k} \qquad \forall k \in K \tag{17}$$

$$e_i \leq f_i^k \leq l_i \qquad \forall i \in C \tag{18}$$

$$\sum_{j \in C \cup \{t\}} x_{ij} + \sum_{j \in C \cup \{v_k\}} \sum_{k \in K} r_{ij}^k = 1 \qquad \forall i \in C \tag{19}$$

$$x_{ij} \in \{0, 1\} \qquad \forall i, j \in N \tag{20}$$

$$r_{ij}^k \in \{0, 1\} \qquad \forall i, j \in N, k \in K \tag{21}$$

$$y_i \in [0, Q] \qquad \forall i \in C \cup \{s, t\} \tag{22}$$

$$w_i^k \in [0, Q_k] \qquad \forall i \in C \cup \{s, v_k\}, k \in K \tag{23}$$

$$f_i^k \geq 0 \qquad \forall i \in C \cup \{s, v_k\}, k \in K \tag{24}$$

The objective function minimizes the total cost which consists of the routing cost of the company drivers and compensation cost of ODs. We describe two distinct sets of constraints. Equations (2)–(8) represent constraints of classical capacitated vehicle routing problem with time windows, while (9)–(19) manage the presence of ODs in the delivery system. In particular, constraints (2), (3) ensure flow conservation. Equations (4)–(5) are the capacity constraints. Constraints (6), (7) manage the arrival time and time windows constraints, respectively, where M is a large number. The last constraints (8) of the first set establish a maximum limit of available company vehicles. In the second set of constraints, Eqs. (9) and (10) represent flow conservation of the ODs. Constraints (11) establish a maximum limit of available ODs, while (12) ensure that each OD leaves their origin node at most once. Equations (13) and (14) are capacity constraints. Finally, constraints (15)–(18) manage the time windows of all nodes. In particular, constraints (15) compute the arrival time at node j, while constraints (16)–(18) ensure that each customer is visited and each OD makes deliveries within their own time windows. Constraints (19) ensure that each customer is served exactly once. Equations (20)–(24) define the domains of the variables.

4   Solution Approach Using BRKGA

In this paper we used the multi-parent biased random-key genetic algorithm with implicit path-relinking (Andrade et al. 2021), a variant of BRKGA, to solve the VRPODTW. In addition to the meta-intensification strategy IPR, we consider an evolution restart strategy.

The framework considered in our variant is as follows. Since the structure of the problem is articulated in two sets, the set of customers C and the set of drivers D, then we consider each chromosome as a vector consisting of $n :=|C| + |D|$ random-keys. Following the process of the BRKGA the procedure starts by creating the first generation of the evolutionary process, i.e. a set of random-key vectors, which is also called the initial population.

For each chromosome of the current generation, our decoder builds the solution and computes its fitness, that is the total solution cost to carry out all deliveries. The decoder in a BRKGA is the only part of the procedure which depends on the particular problem being analyzed. Our decoder sorts the random-keys in non-decreasing order and starts a customer processing phase. This way, each chromosome is divided into two rearranged sections: a customer and a driver sub-chromosome. Following the new ordering of genes, customers are processed by verifying, in terms of time windows, travel time and vehicle capacity, the feasibility of inserting the detour to make the delivery to customer in the path of the first available driver in the driver sub-chromosome. The phase for each customer stops as soon as a driver compatible with the customer is found or if no driver is able to deliver to the customer and in this case the solution is rejected because it is infeasible.

To make the decoder more effective we considered another parameter: probability of delivery ($prDel$). If the time and capacity checks are not violated, the procedure includes the customer at the end of the paths of a driver with a certain probability that depends on *prDel*. After obtaining the fitness of all chromosomes the permutation-based IPR is performed to try to improve the best solution. It receives as input two random-key vectors and returns a pair formed by a vector representing a new solution and its fitness value. For more details of this IPR see Andrade et al. (2021). If we suppose we have a generation in which the stop and restart strategy criteria are not reached, then the next step is to create a new generation as described below and the process is repeated by decoding a new population.

The size of a single population is calculated as $p := \alpha \cdot n$, where $\alpha \geq 1$ is called population size factor; the elite population is defined as $p_e := p \cdot pct_e$, where $pct_e \in [0.1, 0.25]$ is the elite percentage parameter; finally, the size of the mutant population is $p_m := p \cdot pct_m$, where $pct_m \in [0.1, 0.3]$ is the mutant percentage. After each decoding the population of the current generation is divided into two parts according to fitness: the elite population p_e containing the chromosome with the best fitness and the non-elite population p_{ne} which contains the rest of the chromosomes.

The next generation is produced by three operators. The first operator is the *copy repeat*, i.e. all elite chromosomes of the previous generation are copied to the populations of the next generation. With the *mutant* operator new random chromosomes are introduced in each iteration. The remaining part of the population, $p(1 - pct_e - pct_m)$, is generated by the *multi-parent crossover*. For this crossover it is necessary to choose three parameters, the number of the total of parents (π_t) and elite parents (π_e) to be selected; the probability that each parent has of passing genes on to their child. The probability is calculated taking into account the bias of the parent, which is defined by a pre-determined, non-increasing weighting bias function (ϕ) over its rank r. This operator allows to draw genes from a combination of different chromosomes. More details on this type of crossover and on the bias function was described in Andrade et al. (2021).

If after h iterations, the best solution has not been improved, then the procedure performs the restart of evolution, that is all chromosomes are discarded, except the best one, and a new seed is chosen for the random-keys generator. Finally, the whole procedure stops if it reaches a time limit or a number of consecutive iterations without improvement (wi). Table 1 summarizes the parameters of the proposed BRKGA, grouped into three sets: Operator; Path-relinking; Others. The last set includes the restart parameter h, population size factor and two parameters that refer to the VRPODTW, i.e. the compensation factor for occasional drivers and the probability of delivery.

5   Computational Study

In this section, we summarize the results of our computational experiments. The genetic algorithm described in this paper was implemented in C++. The computational tests were conducted using a 2.6 GHz Intel Core i7-3720QM processor and 8 GB 1600 MHz DDR3 of RAM running macOs Catalina 10.15.7. We

Table 1. Summary of the parameters grouped in three sets. In the first and second sets there are the operator parameters and the path-relinking parameters, respectively. The set Others describes the remaining parameters.

Operator		Path-relinking	
pct_e	Percentage of elite chromosomes	sel	Individual selection
pct_m	Percentage of mutant chromosomes	md	Minimum distance among chromosomes
π_t	Numbers of parents in the crossover	typ	Path-relinking type
π_e	Numbers of elite parents in the crossover	pct_p	Percentage of path size
ϕ	Bias function		
Others			
h	Number of iterations without improvements until restart		
α	Population size factor		
ρ	Compensation factor		
$prDel$	Probability of delivery		

performed some tuning phases before the experiments using the *irace* package (see (López-Ibáñez et al. 2016)). To evaluate the effectiveness of the BRKGA applied to VRPODTW we compared the results obtained by the proposed approach to the results found by the exact mathematical model for several small-size instances and results obtained with a VNS approach described in Macrina et al. (2020a) for other large-size instances. The exact mathematical model was coded in Java and solved to optimality using CPLEX 12.10 while the VNS was coded in C++.

5.1   Instances and Parameter Setting

We tested the effectiveness of the BRKGA to solve the VRPODTW, performing numerous computational experiments. We considered the clustered-type, random-type and mixed-type instances (see the Table 4) generated in Macrina et al. (2020a) and we divided them into four classes based on size. The first class (*Test*) is composed of 36 instances with 5, 10 and 15 customers, each of the other three classes consists of 15 instances with 25, 50 and 100 customers respectively (*Small, Medium* and *Large*). In the instances, the number of occasional drivers is chosen in the set {3, 5, 10, 15, 30}, while the number of company drivers in the set {3, 5, 8, 10}.

For the VNS, the only parameter to be set is the one associated with the stop criterion, that is the maximum number of iterations. It was set according to the size of the instances, in particular it was set to $4 \cdot 10^4$ for the class *Small*, 10^4 for the *Medium* and $5 \cdot 10^3$ for the *Large*.

For the BRKGA, we carried out some tuning phases using *irace* software for the following parameters: α, *prDel*, *Operator* and *Path-relinking* parameters. For all phases we grouped the classes of instances into two sets, the first one (T&S) includes classes *Test* and *Small*, while the second one includes *Medium* and *Large* and parameter tuning was done using ten random instances of each of the two sets. We set the compensation factor of the occasional drivers ρ to 0.6 for all experiments and we set *wi* according to the size of the population of evolution.

Instead, for the setting of the restart parameter h we analyzed the run time distribution in terms of iterations considering $h \in A := \{0, 100, 300, 500\}$ and we chose the best values. For each class, we randomly chose three instances and we studied the run time distribution, summed over the instances, considering the target values obtained from a test phase and stopping the algorithm when a solution at least as good as the target is found. We performed 30 independent runs on each instance, see Resende and Ribeiro (2011)) for more details on analyzing run time distribution. With this parameter the overview of the tuning phase was concluded and a summary of the tuned parameters is shown in Table 2.

Table 2. Tuned algorithm parameters.

Parameters	Operator					Path-relinking				Others		
	pct_e	pct_m	π_t	π_e	ϕ	sel	md	typ	pct_p	α	$prDel$	h
T&S	0.16	0.02	4	2	$\frac{1}{r^2}$	rand.	perm.	0.20	0.70	7	20	100
Medium	0.22	0.20	7	2	$\frac{1}{r^2}$	rand.	perm.	0.25	0.96	3	273	300
Large	0.22	0.20	7	2	$\frac{1}{r^2}$	rand.	perm.	0.25	0.96	3	273	100

5.2   Numerical Results

In Tables 3, 4, 5 and 6 in the $Cost_B$ and $Time_B$ columns there are the values of objective function and run time obtained with the genetic algorithm. While in the $Cost_C$, $Time_C$, $Cost_V$ and $Time_V$ columns there are the results and run time obtained with CPLEX and VNS respectively. We would like to clarify that all times shown in the tables are measured in seconds and all results of the two heuristics are the averages obtained over 30 independent runs. In the $Gap_\%$ column there are the values of the average gap on cost. As regards the first class, the values of the last column refer to the optimal gaps, while for the other three classes they refer to the gaps of the genetic algorithm with respect to the VNS. In the AVG row we report the averages of the values in the column. To evaluate the effectiveness of the BRKGA variant proposed applied to VRPODTW we made a comparison of the solutions found in two phases of experiments. We performed a first phase of experiments only on the class *Test* solving the instances to the optimum with CPLEX and with our genetic algorithm proposed and Table 3 shows the average results of this first phase. We set wi to 100, 1500, and 2500 for instances with 5, 10, and 15 customers respectively. The BRKGA is effective, in fact it found the optimal solution in all instances with 5 and 10 customers and in 9 (75%) instances with 15 customers with an average gap value of 0.81%. As regards run time, it is competitive on instances with 15 customers. In the second phase of experiments we studied the *Small*, *Medium* and *Large* classes. We set wi to 2500, 1500, and 1000 for instances with 25, 50, and 100 customers respectively. We decided to compare the results of these classes obtained from our genetic algorithm with those of the VNS since CPLEX is not able to solve the instances within one hour. As for the

Table 3. Average results obtained for the class *Test*.

$|C|$	$Cost_C$	$Time_C$	$Cost_B$	$Time_B$	$Gap_\%$
5	136.45	0.06	136.45	0.92	0.00%
10	223.13	0.61	223.13	31.22	0.00%
15	280.92	114.59	283.61	122.01	0.81%

quality of the solutions, the first important fact to observe is that the BRKGA is the most effective, on average, in fact all the three average $Gap_\%$ of classes are less than zero (see Tables 4, 5 and 6). Looking at the gaps in detail, we observed a different trend among the classes. In the class *Small* the BRKGA found better solutions than VNS in 12 (80%) instances and the value of the overall gap is about −4%. It is more effective than VNS in all clustered-type and mixed-type instances. In the best case, the gap value is about −15%, while in the worst case there is a positive gap of 10.66%. Furthermore, the best solutions, on average, are obtained on clustered-type instances, while the worst solutions are obtained on random-type instances. So, in conclusion, in this class the genetic algorithm is most effective, but not in all types of instances, in fact it is not more effective than the other two classes in the random-type instances. In the class *Medium* the quality of the BRKGA solutions is slightly lower than the previous one, in fact it found better solutions in 10 (67%) instances and the gap value is about −3.5%. In this class there is the instance (C105C50) with the best solution quality among all classes, with a gap value of about −21%. Furthermore, as in the previous class there is a difference between the types of instances. When customers are clustered the BRKGA almost always finds better solutions than the VNS with a high solution quality. Followed by the set of mixed-type instances and finally randomly-type instances. Although the size of the instances belonging to the class *Large* is greater than that of the others, an improvement in the quality of the solution has been observed in the random-type instances and as regard mixed-type instances, the BRKGA is more effective in this class than in the class *Medium*. Furthermore, there is more balance of effectiveness between the various types of instances. Also in this class, as in the first one, the genetic algorithm found better solutions in 12 (80%) instances, while the value of the overall gap is about −2%. Instead, the best case is a mixed-type instance with a gap value of about −7.5%.

In the two heuristics a single iteration performs different operations and this affects the overall execution times. Moreover, since the stopping criteria of both algorithms involve iterations, we decided to use a time-to-target test. So, in order to investigate the efficiency of the two approaches we performed multiple time-to-target plots (mttt-plots) described in Reyes and Ribeiro (2018). This tool is useful for studying the convergence speed of algorithms to given values. For each class we have considered 9 (60%) randomly instances and we performed for each algorithm 60 independent runs, until a solution at least as good as the fixed target was found. The Fig. 1 shows the mttt-plots for BRKGA versus VNS built using $6 \cdot 10^4$ points and summing the run time of the instances. On average, in

Table 4. Results obtained for the class Small. The letter "C" in the test name stands for clustered-type instance; "R" for random-type and "RC" for mixed-type.

Test	$Cost_V$	$Time_V$	$Cost_B$	$Time_B$	$Gap_\%$
C101C25	291.85	202.62	258.16	415.40	−11.54%
C102C25	330.28	169.51	280.37	375.56	−15.11%
C103C25	291.34	185.19	262.09	433.05	−10.04%
C104C25	291.61	205.37	257.30	386.79	−11.77%
C105C25	333.43	140.16	288.57	476.53	−13.45%
R101C25	336.46	159.06	335.03	337.09	−0.43%
R102C25	301.41	212.10	333.55	305.81	10.66%
R103C25	313.04	211.60	338.33	331.75	8.08%
R104C25	308.32	224.90	334.88	370.98	8.61%
R105C25	327.37	203.60	317.72	420.67	−2.95%
RC101C25	492.55	135.10	459.10	442.15	−6.79%
RC102C25	513.84	141.70	473.19	451.32	−7.91%
RC103C25	461.30	221.80	449.21	433.89	−2.62%
RC104C25	456.62	143.60	445.77	432.43	−2.38%
RC105C25	551.46	160.10	537.15	369.77	−2.60%
AVG	373.39	181.09	358.03	398.88	−4.01%

Table 5. Results obtained for the class *Medium*.

Test	$Cost_V$	$Time_V$	$Cost_B$	$Time_B$	$Gap_\%$
C101C50	486.13	214.69	407.06	485.89	−16.27%
C102C50	431.06	239.58	418.70	455.05	−2.87%
C103C50	443.85	256.72	458.83	381.86	3.37%
C104C50	429.44	253.22	381.15	537.35	−11.24%
C105C50	480.54	217.42	378.71	500.48	−21.19%
R101C50	754.06	150.13	680.16	379.81	−9.80%
R102C50	657.77	281.44	629.02	398.74	−4.37%
R103C50	554.83	284.57	567.55	301.15	2.29%
R104C50	501.46	289.13	590.90	483.83	17.84%
R105C50	661.90	255.30	625.66	557.50	−5.48%
RC101C50	537.43	233.95	601.74	467.42	11.97%
RC102C50	521.87	263.06	584.02	476.36	11.91%
RC103C50	559.17	278.16	545.29	547.35	−2.48%
RC104C50	621.92	238.68	535.62	428.10	−13.88%
RC105C50	624.69	243.74	559.77	511.20	−10.39%
AVG	551.07	246.65	530.95	460.81	−3.37%

Table 6. Results obtained for the class *Large*.

Test	$Cost_V$	$Time_V$	$Cost_B$	$Time_B$	$Gap_\%$
C101C100	1342.44	582.47	1265.58	736.18	−5.73%
C102C100	1240.08	675.70	1244.17	522.76	0.33%
C103C100	1197.58	667.10	1162.51	864.12	−2.93%
C104C100	1153.11	445.13	1128.94	704.22	−2.10%
C105C100	1336.12	475.66	1246.46	495.77	−6.71%
R101C100	1365.08	388.50	1273.47	802.34	−6.71%
R102C100	1272.33	421.55	1231.16	685.67	−3.24%
R103C100	1159.27	468.39	1144.23	664.41	−1.30%
R104C100	934.45	718.17	1082.31	514.60	15.82%
R105C100	1276.13	608.64	1184.51	708.55	−7.18%
RC101C100	1335.31	531.61	1241.41	784.90	−7.03%
RC102C100	1375.12	642.24	1274.27	716.03	−7.33%
RC103C100	1114.25	737.39	1156.43	502.74	3.79%
RC104C100	1152.15	746.65	1127.42	900.00	−2.15%
RC105C100	1244.29	648.46	1236.53	883.79	−0.62%
AVG	1233.18	583.84	1199.96	699.07	−2.21%

about 250 s the probability of reaching the target value in a single instance for the VNS is about 0%, while in the same time the probability of the BRKGA reaches even 90%. Also, to be sure of finding targets for all instances, expect a total run time of about 10730 s for the VNS and 8900 s for the BRKGA. We may conclude that, on average, the time taken by BRKGA to find solutions at least as good as the targets is less than the time required by VNS.

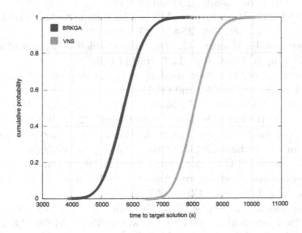

Fig. 1. Multiple time-to-target plots for BRKGA versus VNS for 9 instances for each class obtained by simulation with $6 \cdot 10^4$ points each.

6   Conclusions

In this paper we presented for the first time an application of the BRKGA-MP-IPR with a restart strategy on the VRPODTW. Since in the variant of VRP studied there are time windows for both customers and drivers and the BRKGA is based on random-keys to solve the problem, we built a decoder function to convert the chromosomes into driver paths starting from a check on the compatibility of time windows and travel times. To evaluate the performance of the proposed approach an extensive computational study was conducted considering instances of increasing size. A comparison with a VNS approach was also carried out using also the mttt-plots as a support tool. The results showed that the proposed BRKGA is more effective and efficient than the VNS in the three classes of instances considered. For future work we intend to develop a new variant of BRKGA, considering a variable mutant operator and a new decoder to improve the results of random-type of instances. Furthermore, we intend to implement a local search and extend the experimental tests and analyzes.

Acknowledgement. This work is supported by the Italian Ministry of University Education and Research (MIUR), project: "Innovative approaches for distribution logistics" - Code DOT1305451 - CUP H28D20000020006.

References

Andrade, C.E., Miyazawa, F.K., Resende, M.G.C.: Evolutionary algorithm for the k-interconnected multi-depot multi-traveling salesmen problem. In: Proceedings of the 15th Annual Conference on Genetic and Evolutionary Computation, GECCO 2013, New York, NY, USA, pp. 463–470 (2013)

Andrade, C.E., Toso, R.F., Gonçalves, J.F., Resende, M.G.C.: The multi-parent biased random-key genetic algorithm with implicit path-relinking and its real-world applications. Eur. J. Oper. Res. **289**(1), 17–30 (2021)

Archetti, C., Savelsbergh, M., Speranza, M.G.: The vehicle routing problem with occasional drivers. Eur. J. Oper. Res. **254**(2), 472–480 (2016)

Archetti, C., Guerriero, F., Macrina, G.: The online vehicle routing problem with occasional drivers. Comput. Oper. Res. **127**, 105144 (2021)

Bean, J.C.: Genetic algorithms and random keys for sequencing and optimization. INFORMS J. Comput. **6**(2), 154–160 (1994)

Di Puglia Pugliese, L., Guerriero, F., Macrina, G., Scalzo, E.: Crowd-shipping and occasional depots in the last mile delivery. In: Cerulli, R., Dell'Amico, M., Guerriero, F., Pacciarelli, D., Sforza, A. (eds.) Optimization and Decision Science. ASS, vol. 7, pp. 213–225. Springer, Cham (2021). https://doi.org/10.1007/978-3-030-86841-3_18

Gonçalves, J.F., Resende, M.G.C., Mendes, J.J.M.: A biased random-key genetic algorithm with forward-backward improvement for the resource constrained project scheduling problem. J. Heurist. **17**(5), 467–486 (2011)

Lopes, M.C., Andrade, C.E., Queiroz, T.A., Resende, M.G.C., Miyazawa, F.K.: Heuristics for a hub location-routing problem. Networks **68**(1), 54–90 (2016)

Lucena, M.L., Andrade, C.E., Resende, M.G.C., Miyazawa, F.K.: Some extensions of biased random-key genetic algorithms. In Proceedings of the Forty-Sixth Brazilian Symposium of Operational Research, pp. 2469–2480 (2014)

López-Ibáñez, M., Dubois-Lacoste, J., Cáceres, L.P., Birattari, M., Stützle, T.: The irace package: iterated racing for automatic algorithm configuration. Oper. Res. Perspect. **3**, 43–58 (2016)

Macrina, G., Guerriero, F.: The green vehicle routing problem with occasional drivers. In: Daniele, P., Scrimali, L. (eds.) New Trends in Emerging Complex Real Life Problems. ASS, vol. 1, pp. 357–366. Springer, Cham (2018). https://doi.org/10.1007/978-3-030-00473-6_38

Macrina, G., Di Puglia Pugliese, L., Guerriero, F., Laganà, D.: The vehicle routing problem with occasional drivers and time windows. In: Sforza, A., Sterle, C. (eds.) ODS 2017. SPMS, vol. 217, pp. 577–587. Springer, Cham (2017). https://doi.org/10.1007/978-3-319-67308-0_58

Macrina, G., Di Puglia Pugliese, L., Guerriero, F.: A variable neighborhood search for the vehicle routing problem with occasional drivers and time windows. In: Proceedings of the 9th International Conference on Operations Research and Enterprise Systems, vol. 1, pp. 270–277 (2020a)

Macrina, G., Di Puglia Pugliese, L., Guerriero, F., Laporte, G.: Crowd-shipping with time windows and transshipment nodes. Comput. Oper. Res. **113**, 104806 (2020b)

Macrina, G., Di Puglia Pugliese, L., Guerriero, F.: Crowd-shipping: a new efficient and eco-friendly delivery strategy. Proc. Manuf. **42**, 483–487 (2020c). International Conference on Industry 4.0 and Smart Manufacturing (ISM 2019)

Resende, M.G.C., Ribeiro, C.C.: Restart strategies for grasp with path-relinking heuristics. Optim. Lett. **5**, 467–478 (2011)

Resende, M.G.C., Toso, R.F., Gonçalves, J.F., Silva, R.M.: A biased random-key genetic algorithm for the Steiner triple covering problem. Optim. Lett. **6**, 605–619 (2011)

Reyes, A., Ribeiro, C.C.: Extending time-to-target plots to multiple instances. Int. Trans. Oper. Res. **25**, 1515–1536 (2018)

Solomon, M.M.: Algorithms for the vehicle routing and scheduling problems with time window constraints. Oper. Res. **35**, 254–265 (1987)

Metaheuristic Algorithms for UAV Trajectory Optimization in Mobile Networks

Valentina Cacchiani[1], Sara Ceschia[2](✉), Silvia Mignardi[1],
and Chiara Buratti[1]

[1] DEI, University of Bologna, Viale del Risorgimento 2, 40136 Bologna, Italy
{valentina.cacchiani,silvia.mignardi,c.buratti}@unibo.it
[2] DPIA, University of Udine, via delle Scienze 206, 33100 Udine, Italy
sara.ceschia@uniud.it

Abstract. We consider a mobile network in which traditional static terrestrial base stations are not capable of completely serving the existing user demand, due to the huge number of connected devices. In this setting, an equipped Unmanned Aerial Vehicle (UAV) can be employed to provide network connection where needed in a flexible way, thereby acting as an unmanned aerial base station. The goal is to determine the best UAV trajectory in order to serve as many users as possible. The UAV can move at different speeds and can serve users within its communication range, although the data rate depends on the positions of UAV and users. In addition, each user has a demand (e.g., the number of bits the user wants to download/upload from/to the network) and a time window during which requires the service. We propose a Biased Random-Key Genetic Algorithm (BRKGA) and a Simulated Annealing Algorithm (SAA), and compare them on realistic instances with more than 500 users in different settings.

Keywords: Unmanned aerial vehicles · Mobile networks · Genetic algorithm · Local search · Simulated annealing

1 Introduction

The continuous and rapid increase of the number of connected Internet of Things (IoT) devices, such as digital control systems, infrastructure sensors, monitoring systems, home appliances, voice controllers, as well as user devices such as smartphones and wearables, causes extremely high traffic demand and requires more and more flexibility in the mobile network to cope with demand peaks. In recent years, the use of Unmanned Aerial Vehicles (UAVs) as unmanned aerial base stations (i.e., mobile stations mounted on UAVs) has remarkably attracted interest due to their movement flexibility and low cost, allowing to serve traffic demand where and when needed. As reported in the recent survey [17], UAVs are utilized in many civil applications ranging from package delivery to infrastructure monitoring, precision agriculture, telecommunications, security and entertainment. In

L. Di Gaspero et al. (Eds.): MIC 2022, LNCS 13838, pp. 30–44, 2023.
https://doi.org/10.1007/978-3-031-26504-4_3

the field of telecommunications, drones (UAVs) can be used to establish connectivity with mobile devices that fall within their communication range, when acceptable levels of signal-to-noise ratio are guaranteed. According to the classification of [17], three main categories of applications that employ UAVs for mobile connections are identified: (i) UAVs assigned to stationary locations to act as intermediaries to connect mobile devices to base stations, (ii) UAVs flying along a trajectory, and (iii) drone-to-drone transmissions where mobile devices are connected one another via drones. Our study belongs to the second category, as we aim to optimize the UAV flying trajectory in order to maximize the number of served users. In the same category we find works that minimize the total energy used for the mobility of the UAVs to serve the IoT devices [14,23], evaluate the impact of the NarrowBand-IoT protocol on UAV-aided networks [13], maximize the minimum throughput over all ground users [22,24] or the average achievable rate for all users [16] or the number of served devices [12,19]. Most of the works from the literature propose heuristic approaches to solve this class of problems.

A recent tutorial on applications that employ UAVs for Wireless Networks is presented in [15]: it reports a comprehensive overview on how UAVs can be used as flying wireless base stations in different scenarios, such as providing coverage and capacity enhancement of beyond 5G networks, guaranteeing connectivity in natural disasters, allowing for wireless connectivity among a massive number of IoT devices. In [15], state-of-the-art methods and key research challenges are outlined, among which UAV trajectory optimization is listed. As specified in [15], several elements, such as air-to-ground channel model, channel variation due to the drone mobility, speed constraints, energy and battery consumption, user demands, have to be taken into account when determining the best UAV trajectory. For an overview of the most relevant air-to-ground channel models used in works on UAV placement and trajectory optimization, we refer the reader to the up-to-date survey [21].

In this work, we tackle the problem of UAV trajectory optimization in mobile networks, aiming at providing connectivity to the maximum number of IoT and user devices that cannot be served by static terrestrial base stations. Our contribution is twofold. First, we developed two metaheuristic algorithms to solve the problem: a Biased Random-Key Genetic Algorithm (BRKGA) and a Simulated Annealing Algorithm (SAA), that include many real-life features required in the studied application, such as the air-to-ground channel model and the corresponding data rate provided to the devices, the user demands and activation time windows, the different speeds of the UAV and its available battery. Second, we tested and compared the two algorithms on realistic instances with more than 500 users in different settings, showing the effectiveness of both methods and the superiority of SAA in terms of number of served users.

The remainder of this paper is organized as follows. In Sect. 2, the problem and its features are described. Section 3 presents the two metaheuristic algorithms that we propose as solution methods. Section 4 describes how the parameters used in BRKGA and SAA were tuned, and reports the results obtained by applying BRKGA and SAA on a set of realistic instances. Finally, the paper is concluded in Sect. 5.

2 Problem Description

We consider a set N of devices that cannot be connected through static terrestrial base stations. In the following, we simply call users all kinds of these devices. Each user $u \in N$ has a position (x_u, y_u) expressed by its coordinates, a demand d_u representing the number of bits the user wants to download/upload from/to the network, and an activation time window $[s_u, e_u]$ in which the user requires the connection service.

A UAV is present at a given position (x_{UAV}, y_{UAV}) (where charging takes place, also called depot), and has an available battery level TB expressed as its allowed flight time. The UAV must start from its depot and end its route again there. It can move in any direction and for any amount of time, thus allowing for infinitely many alternative routes. We consider the altitude of the UAV as fixed, hence it has not to be optimized, and we only have to consider 2D movements. In addition, the UAV can fly at different speeds: we assume a discrete set DS of speeds including the zero speed, which represents that the UAV is flying at its fixed altitude without moving in any direction (i.e., it is waiting). If the UAV cannot reach the depot even at maximum speed from an user u due to its activation time window and to the UAV battery level, then u does not need to be considered in the problem, because it would be impossible to serve it. Hence, these unreachable users are removed from N through a pre-processing step.

We have to determine a UAV route, described as a sequence of positions (x, y) of the area, and the corresponding sequence of speeds in DS, used by the UAV to move between these positions. The movement of the UAV from one position to another one is called *arc* and a UAV position is called *node* in the following. Note that (x, y) does not necessarily correspond to a position where a user is located: indeed, a user can be served even if the UAV does not reach its location, although the distance between UAV and user affects the connection. In particular, based on the respective UAV-user positions, we can compute the data rate that the UAV provides to the user.

The value of the data rate is determined by computing the link budget between the user and the UAV. This strongly depends on the link attenuation (hence, respective UAV-user positions) and on the channel model. We consider the air-to-ground channel model derived from [1] given by the 3GPP telecommunications standardization body for high frequency links and more recent technologies (see Sect. 4.2 for more details).

We assume that the data rate provided by the UAV when it travels along an arc connecting two consecutive positions is constant, and compute it according to the UAV final position of the arc. To compute the amount of demand that the UAV can satisfy for each user, the data rate is then multiplied by the minimum between the arc duration (that depends on the UAV speed) and a parameter T_{rate}. In order to satisfy a user demand, it is possible to partially serve it more than once. A user is considered to be served only if its demand is fully satisfied within its time window. The goal is to determine the UAV route and the corresponding speeds in order to fully serve as many users as possible before the UAV battery expires.

We call this problem the Drone Trajectory Optimization (DTO) problem. It is an NP-hard problem as it extends the Orienteering Problem with Time-Windows (OPTW, [9]). The OPTW, given a directed graph, in which each user has a non-negative score, and each arc has a non-negative travel time, calls for determining a route that visits a subset of the users, each at most once and within a given time-window, to maximize the total collected score, while satisfying a maximum route duration constraint. The DTO extends the OPTW by allowing more users to be served simultaneously (and based on the distance between the drone and the user), considering several drone speeds, permitting that a user is visited more than once and is served without exactly reaching its position.

3 Solution Methods

In this section, we describe the search methods developed in our study, namely BRKGA (Sect. 3.1) and SAA (Sect. 3.2).

3.1 Biased Random-Key Genetic Algorithm

The BRKGA [8] has been successfully applied to a wide range of combinatorial optimization problems that arise in real world situations [3]. In a BRKGA, each chromosome is represented as a vector of randomly generated real numbers in the interval $[0, 1]$. A problem dependent deterministic function, called *decoder*, associates to each chromosome a solution and its corresponding fitness value. Once the decoder has been defined, the BRKGA can be used as a *black-box* tool that evolves an initial population by applying mutation and crossover operators until a given number of generations is reached. More precisely, the initial population is composed by p chromosomes whose m alleles are independently generated at random with a uniform distribution. BRKGA implements an *elitist* strategy such that, at each generation, individuals are sorted according to their fitness values, and the best p_e ones are preserved for the next generation (with $p_e < p - p_e$). Conversely, diversification is ensured by two mechanisms: mutation and crossover. At each generation p_m mutant individuals are randomly generated and introduced into the population. The remainder of the population $p - p_e - p_m$ is completed through the process of mating: BRKGA randomly selects one parent from the elite list and the other from the non-elite list by employing a parametrized uniform crossover operator to generate offspring. In detail, each allele of the offspring is inherited from the elite element with probability ρ_e or from the non-elite element with probability $1 - \rho_e$. The population evolves for a fixed number of generations K.

BRKGA indirectly explores the solution space of the specific problem by searching the continuous m-dimensional unit hypercube, using the decoder to map chromosomes into solutions of the problem. As a consequence, we have to define the *problem-dependent* elements of the framework, i.e., the solution encoding and the chromosome decoder.

Solution Encoding. The BRKGA requires as input parameter the number m of alleles in each chromosome. In DTO, a solution is defined by a UAV route and the corresponding speeds to be used by the UAV along the arcs in the route. To encode a DTO solution in BRKGA we have to define a-priori the number of arcs in the UAV route. Thus, we assume that the UAV must fly maintaining the same direction (i.e., along its arc) for a fixed time step τ, and define the number of arcs ν of the route as the ratio between the maximum end time among all users' time windows and τ (recall that we removed from N users with an activation time that does not allow the UAV to reach the depot at the end of its route due to battery limitation). Each solution is encoded as a vector χ of random keys of length $m = 3\nu$, where the first 2ν keys are used to define the direction of the UAV movements in its route (since the UAV altitude is fixed, we only need to specify the 2D movement), and the last ν its speed along each arc of the route. In particular, each arc a is assigned two attributes, namely direction and speed, that are encoded in alleles $h = \nu k + a$, with $k = \{0, 1, 2\}$. The zero speed allows to represent the UAV waiting at a position (it is flying at its fixed altitude without moving in any direction). We observe that, compared to the general DTO problem setting, in BRKGA we limit the search space by fixing the number of arcs in a UAV route.

Chromosome Decoder. Each arc of the route represents a movement from the current UAV position to the next one. Each route starts and ends at the depot. The arc is identified by a direction expressed as an angle, and by a speed to be used along the arc. Since we fixed the time step, once the angle and the speed have been defined, the next position of the UAV can be deterministically computed. Therefore, when we have a chromosome χ with random values in interval $[0, 1]$, we define the UAV route and its speeds as follows. For each arc a, the cosine of the corresponding angle θ_a can be obtained as $\cos \theta_a = 2\chi[a] - 1$ and the sine as $\sin \theta_a = 2\chi[a + \nu] - 1$. Then the angle can be rebuilt using conveniently the inverse functions arc sine and arc cosine. The last ν keys of χ are used to select the speed $v_{\theta_a} \in DS$ for each arc. In detail, given $|DS|$ different speeds, the j-th speed value is selected for arc a if

$$\frac{j-1}{|DS|} < \chi[a + 2\nu] \le \frac{j}{|DS|} \qquad j \in \{1, \ldots, |DS|\}.$$

The initial position of the UAV is (x_{UAV}, y_{UAV}), while the coordinates of the following position can be obtained applying to its current position the following movement Δ decomposed in (x, y) components: $\Delta_x = v_{\theta_a} \tau \cos \theta_a$ and $\Delta_y = v_{\theta_a} \tau \sin \theta_a$. The last arc of the route forces the UAV to go back to the depot with the maximum speed. The fitness value of a chromosome corresponds to the number of served users computed according to the corresponding UAV route.

3.2   Simulated Annealing Algorithm

Our second solution method SAA is based on applying local search (LS) in a Simulated Annealing framework. In the following, we describe the fundamental elements of SAA, namely: (i) the search space and the initial solution strategy, (ii) the neighborhood relations, and (iii) the metaheuristic that guides the search.

Search Space and Initial Solution. The search space is composed by an integer-valued matrix and two integer-values arrays:

- the *direction* matrix stores, for each node of the route, two integer values that represent the 2D components of the move from node $i - 1$ to i;
- the *speed* vector stores the speed value from node $i - 1$ to i;
- the *waiting* time vector stores the possible waiting time for each node except for the final depot.

These structures are complemented by redundant data structures that help us in accelerating the computation of the objective function value and the local changes between neighborhood states provoked by the moves. In particular, we maintain an integer-valued *position* matrix that stores the (x, y) coordinates of the nodes representing the UAV route.

The main difference between the solution encoding of BRKGA and that of SAA is that, in the latter, the length of the route does not have to be fixed a-priori, while in the former the number of arcs is set to ν. Indeed, in SAA, the initial solution starts with only two nodes corresponding to the (initial and final) depot, then a constructive procedure adds new nodes to the route as long as the battery is not finished and there is still an active user to be served. The procedure adds one node i at the time by determining speed and direction from the current node $i - 1$. In particular, the speed value to fly from $i - 1$ to i is randomly selected in the DS set with uniform distribution. For the move direction, a user is randomly selected among those that are currently active (and not yet fully served), and the (x, y) components to move the UAV from its current position $i - 1$ towards that user are computed. All waiting times are initially set to zero. In addition, depending on the earliest user activation time window, the departure of the UAV from the depot can be delayed, if necessary.

In this way, the UAV route length is not fixed a-priori and is determined by construction. This allows an additional degree of freedom effectively exploited by SAA. During the neighborhood search, the cost function is composed by the number of served users and the distance to feasibility: indeed, we allow states that have a route duration longer than the UAV battery TB, but this violation is strongly penalized in the cost function with a suitable, fixed high weight.

Neighborhood Relations. We defined four atomic neighborhood relations:

- Change (C): The move $C\langle i, x, y, v\rangle$ changes the position of node i by replacing its current coordinates with (x, y). In addition the speed value to fly from node $i - 1$ to i is set to v.
  Preconditions: The node i is not the depot. The new position for node i is different from the old one or the new speed value is different from the old one.

- Add (A): The move $A\langle i, x, y, v \rangle$ adds a new i-th node to the route with (x, y) coordinates. The speed value to fly from node $i - 1$ to i is set to v.
  Preconditions: The old position of node i is different from the new one or the new speed value is different from the old one.
- Remove (R): The move $R\langle i \rangle$ removes node i from the route.
  Preconditions: The node i is not the depot.
- Wait (W): The move $W\langle i, t \rangle$ sets the waiting time for node i equal to t.Preconditions: The node i is not the final depot.

Neighborhood Selection. We implemented a composite neighborhood which is the union of the four atomic relations described above: $C \cup A \cup R \cup W$. To draw a random move from it, we first select one of the basic neighborhoods and then the specific attributes (e.g., position, speed) of the selected move. For the neighborhood selection we use fixed probabilities p_C, p_R and p_A, which represent the probability to draw a Change, Add and Remove move, respectively. Obviously, the probability of a Wait move is $1 - p_C - p_R - p_A$. Inside the single neighborhood, the values of the attributes are selected uniformly.

Simulated Annealing Strategy. The local search process is guided by Simulated Annealing (SA), which is an old-fashioned, but still very effective metaheuristic technique [11]. SA is the *state-of-the-art* method for various combinatorial optimization problems (see, e.g., [5]). For an up-to-date comprehensive presentation of different variants of SA and their current performances see [7].

The SA procedure starts from an initial solution generated as described above, and draws, at each iteration, a random move. As customary for SA, we introduce the notion of *temperature* T_0, such that a move is always accepted if it is improving or sideways (i.e., same cost), whereas worsening ones are accepted based on the time-decreasing exponential distribution $e^{-\Delta/T}$, where Δ is the difference of cost between the new and the current solution and T is the temperature.

The temperature is decreased after a fixed number of samples N_s is drawn according to the geometric cooling scheme, where α (with $0 < \alpha < 1$) is the *cooling rate*. In addition, in order to speed up the early stages of the search, we use the *cut-off* mechanism: to this aim we add a new parameter N_a, representing the maximum number of accepted moves at each temperature. The temperature is, thus, decreased as soon as one of the following two conditions occurs: (i) the number of sampled moves reaches N_s, (ii) the number of accepted moves reaches N_a. This allows us to save computational time in early stages, and exploit it later during the search.

To facilitate the tuning of the cut-off, we introduced the parameter $\rho = N_a/N_s$ (with $0 < \rho < 1$) which is the fraction of accepted moves with respect to the total number of moves sampled at each temperature.

The search is stopped when a total number of iterations $\mathcal{I}$ has been performed, which guarantees that the running time is equal for all configurations of the parameters. In order to keep the $\mathcal{I}$ fixed, we computed for each temperature T the number of sampled solutions N_s using the following formula:

$$N_s = \mathcal{I} \left/ \left(\frac{\log\left(T_f/T_0\right)}{\log \alpha} \right) \right.$$

where T_0 and T_f are the initial and final temperature, respectively.

4 Experimental Analysis

Our code is implemented in C++ and compiled using g++ v. 9.3.0. For BRKGA we used the brkgaAPI framework [18], while for SAA the EASYLOCAL++ framework [6]. The algorithm parameters were tuned, as described in Sect. 4.3, using the tool JSON2RUN [20] that implements the F-Race procedure [4]. All experiments ran on an Ubuntu Linux 20.04.2 LTS machine with 64 GB of RAM and 32 AMD Ryzen Threadripper PRO 3975WX (3.50 GHz) physical cores, hyper-threaded to 64 virtual cores. A single virtual core was dedicated to each experiment.

In Sect. 4.1 we describe the realistic case study (similar to the one used in [12]) from which we derived the tested instances, and in Sect. 4.2 we present the air-to-ground channel model. Section 4.3 reports the tuning method and the final parameter values. Finally, in Sect. 4.4, we present the comparison of BRKGA and SAA in different settings, i.e., different timeouts and different speeds.

4.1 Reference Scenario

We consider an area of size 1 km × 1 km in which 537 users are spread as represented in Fig. 1. These are the users that cannot be served by the available terrestrial base stations, which justifies the particular positions of the users. Each user has an associated demand (expressed in Mbit) represented as an integer number between 1 and 5. Each user time-window has a starting time between 0 and 1740 (in seconds), and duration 20 s. The UAV depot is placed in the middle of the area (black square). The UAV can use two speeds, i.e., 10 m/s and 20 m/s, and the zero speed. The available battery duration is equal to 1800 s. The value of T_{rate} is 5 s.

From this reference scenario, we derived 10 instances by varying the activation time of the users: in particular, s_u is generated with uniform distribution in $[0, 1740]$ $(u \in N)$.

4.2 Channel Model

As mentioned in Sect. 2, we consider the air-to-ground channel model derived from [1]. In particular, to properly formulate our problem, we assume that (i) links undergo non-line-of-sight propagation (i.e., the worst case, where communications obstacles between the UAV and the user may reduce the connection

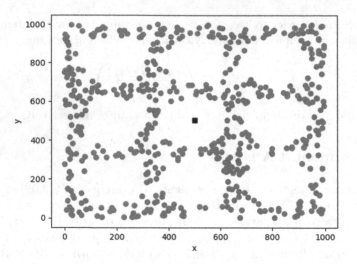

Fig. 1. Reference scenario.

quality), but (ii) no channel fluctuations affect the communication to avoid model randomicity. Then, the attenuation or path loss L becomes:

$$L = L^* + \eta$$

$$L^* = \begin{cases} 28 + 22 \log_{10}(d_{3D}) + 20 \log_{10}(f_c) & \text{if } 10 \leq d_{2D} \leq d_{BP} \\ 28 + 40 \log_{10}(d_{3D}) + 20 \log_{10}(f_c) & \\ \quad - 9 \log_{10}(d_{BP}^2 + (h - h_{ut})^2) & \text{otherwise} \end{cases}$$

where d_{3D} and d_{2D} are, respectively, the UAV-user distance in 3D and the distance projected on the 2D plane, $h = 100$ m and $h_{ut} = 1$ m are, respectively, the heights of UAV and user, $f_c = 27$ GHz is the carrier frequency, d_{BP} is the distance breakpoint given in [1], and $\eta = 20$ dB is an additional penetration loss. Finally, the data rate r, given the previously computed path loss L, and a bandwidth portion $B_{ch} = 1.44$ MHz, is:

$$r = B_{ch} \log_2(1 + 10^{\frac{\gamma(L)}{10}}),$$

where it holds $\gamma(L) = [P_{tx} + G_{tx} + G_{rx} - L] - P_{noise}$. In particular, $P_{tx} = 23$ dBm is the UAV transmit power, $G_{tx} = 16$ dB and $G_{rx} = 0$ dB are, respectively, the transmitter and receiver gains, and $P_{noise} = -116.4$ dBm is the noise power. Loss L and data rate r are then computed for each UAV-user link. Parameter γ corresponds to the Signal-to-Noise Ratio (SNR) that indicates the link quality by denoting a poor or strong connection. If the value of γ is below a connectivity threshold γ_{thr}, it means the communication cannot be established correctly and the user cannot be served. Therefore, higher values of γ_{thr} denote more strict link quality requirements, and consequently a smaller area of communication coverage for the UAV. On the contrary, if γ_{thr} is lower, the UAV coverage area gets larger. The minimum accepted SNR value γ_{thr}, that determines the extension of the UAV coverage radius, can take the following values: -10, -5, 0, 5 and 10 in dB units. This range of values was set in accordance with [2]. The described channel model is used by BRKGA and SAA to compute, during the search process, the value of the data rate and, hence, the number of served users.

4.3   Automatic Parameter Tuning

We defined the initial ranges for the selection of the best configuration for BRKGA following the recommendation of Gonçalves and Resende [8, Table 1]. For the BRKGA total number of generations is set to 10^3, while the total number of iterations for SA is $3.4 \cdot 10^6$, which corresponds to an average computational time of about $1000\,s$ seconds on our machine. For BRKGA, we fix the time step $\tau = T_{rate}$. For SA the ranges have been set according to preliminary experiments. In addition, experiments reveled that SA is not sensible to changes of the cooling rate, thus we fixed it to 0.99.

Table 1. Parameters and range of values for BRKGA and SA.

Description	Symbol	Range	Best value
BRKGA			
Size of population	p	100–550	532
Size of the elite population	p_e	$0.1p$–$0.25p$	$0.107p$
Size of the mutant population	p_m	$0.1p$–$0.3p$	$0.14p$
Elite allele inheritance probability	ρ_e	0.5–0.8	0.73
SAA			
Initial temperature	T_0	0.5–2	0.969
Final temperature	T_f	0.001–0.5	0.186
Accepted moves ratio	ρ	0.001–0.05	0.011
Probability of **Change** moves	p_C	0.5–0.7	0.69
Probability of **Add** moves	p_A	0.1–0.3	0.25
Probability of **Remove** moves	p_R	0.01–0.05	0.05

For each of two methods, we tested 50 different configurations extracted according to the *Hammersley point set* [10] on the testbed derived from the reference scenario. We compared the results using the F-Race procedure [4], which is based on the Friedman test to prove statistical significance, using the threshold of confidence set to 95% (p-value $= 0.05$). The best configurations are shown in Table 1. All of the following experiments were performed using these configurations for BRKGA and SAA.

4.4   Computational Results

We first compare BRKGA and SAA by setting the number of generations of BRKGA and the number of iterations of SAA so that they require comparable computing times. We consider in DS the zero speed and 20 m/s. Then, we report the results obtained when imposing several time limits and when allowing different speeds. Each algorithm is executed on each instance 10 times for each γ_{thr} value.

Comparative Results Between BRKGA and SAA. Table 2 reports the comparison between BRKGA and SAA with comparable computing times in seconds. In particular, for each instance and γ_{thr} value, we display the minimum, average and maximum number of served users obtained by BRKGA and SAA, respectively. As can be seen, SAA achieves better results than BRKGA both for the average and the maximum values, as expected since the search space of SAA is larger than that of BRKGA. In particular, we can see that SAA obtains significantly higher numbers of served users when $\gamma_{thr} \geq 5$. It is evident that smaller γ_{thr} values allow for serving more users, as expected. Finally, we can observe that the difference between the average and the maximum values are, for both algorithms, rather small, hence independent of the specific run.

Results with Different Timeouts. We investigate how shorter time limits impact on the results both for BRKGA and SAA. We set time limits from 10 s up to 640 s. In particular, very short time limits can be useful if the user positions and time windows are revealed in real-time. The average results on all instances for different γ_{thr} values are reported in Table 3. As expected, in most cases, longer time limits allow for finding better solutions for both algorithms, although reasonable numbers of users are served even within 10 secs of computing time. We can also observe that the improvement is more evident with shorter time limits (e.g., going from 10 to 20 s): indeed, the algorithms need some time for appropriately searching the solution space. Overall, SAA turns out to achieve the best performance also when short time limits are imposed.

Results with Different Speed Values. We evaluate whether allowing the UAV to use different speeds is useful to serve more users. In particular, we always consider the zero speed, and then include in DS speed 10 m/s, or both 10 and 20 m/s. The results obtained by SAA with these speeds are compared in Table 4 with those obtained by considering 20 m/s. As we can observe, in most cases it is better to only use 20 m/s for the tested instances. This can be due to the relatively low demand values and high time-window variability: indeed, usually the UAV does not require long time to satisfy a user demand, and quickly moves to serve another user. The rather short activation time of the users (20 s) requires the UAV to move to different zones of the area in order to satisfy the demands. However, we can also see that when both 10 and 20 m/s

Table 2. Computation results of BRKGA and SAA.

Inst.	γ_{thr}	BRKGA				SAA			
		Min	Avg	Max	Time [s]	Min	Avg	Max	Time [s]
Test1	−10	392	399.8	406	1002.3	200	511.1	526	931.2
	−5	365	372.5	384	981.0	484	506.2	514	943.0
	0	214	222.8	237	992.3	344	362.6	371	1010.9
	5	132	136.1	143	996.6	241	249.4	258	990.8
	10	71	77.7	83	998.3	175	184.5	193	943.6
Test2	−10	379	390.7	397	1000.1	508	516.8	521	981.5
	−5	362	367.5	373	980.3	497	504.5	510	964.3
	0	218	222.9	236	994.8	289	332.0	366	991.4
	5	130	135.1	145	995.7	248	260.4	270	1002.8
	10	66	73.5	79	997.5	169	186.7	200	958.5
Test3	−10	385	391.8	397	999.4	506	512.2	516	976.5
	−5	367	372.8	380	978.3	178	485.5	502	943.0
	0	218	224.2	233	994.1	349	359.5	366	1012.1
	5	133	142.6	158	995.6	246	257.9	268	973.9
	10	71	80.7	94	995.1	173	184.7	197	946.7
Test4	−10	386	395.3	408	1002.5	377	507.2	519	970.7
	−5	369	372.7	378	980.1	328	490.7	503	966.0
	0	214	224.0	234	997.4	294	358.2	369	1012.5
	5	136	137.8	141	997.8	242	257.3	263	973.4
	10	72	79.4	85	997.6	169	183.2	194	939.2
Test5	−10	382	390.5	399	997.1	510	514.1	518	995.8
	−5	362	366.9	376	977.0	496	500.7	505	977.9
	0	219	224.2	227	991.9	349	364.1	376	1023.5
	5	131	135.5	144	997.3	91	255.9	272	982.8
	10	70	75.7	80	990.8	172	185.0	197	943.2
Test6	−10	389	395.5	402	1004.8	470	504.2	512	971.5
	−5	363	372.4	382	985.2	461	493.6	500	954.6
	0	223	229.5	238	999.0	307	336.4	362	995.6
	5	128	135.7	143	997.6	243	257.4	269	981.6
	10	65	77.8	82	1002.1	177	188.1	199	927.9
Test7	−10	384	391.2	403	1042.1	497	510.1	515	982.4
	−5	360	366.4	373	1020.1	493	498.3	504	967.1
	0	217	226.8	237	1032.3	353	365.6	376	1018.8
	5	130	134.6	142	1034.6	254	264.0	272	993.9
	10	68	73.6	81	1033.2	174	189.4	199	941.2
Test8	−10	379	390.5	399	1002.2	509	514.7	520	984.4
	−5	356	367.7	384	977.6	496	501.5	508	976.2
	0	218	224.5	231	991.9	109	349.0	377	995.9
	5	127	135.8	147	994.8	251	264.5	276	1004.1
	10	72	77.8	83	995.6	170	185.6	196	952.6
Test9	−10	376	388.8	401	1003.4	508	513.3	517	985.0
	−5	361	367.4	372	980.1	495	500.7	506	970.6
	0	215	220.8	230	992.6	332	352.5	373	1020.5
	5	127	135.5	145	995.1	243	255.9	265	1002.2
	10	70	77.5	83	990.8	171	183.2	198	961.7
Test10	−10	387	394.7	401	1042.1	143	504.3	520	949.4
	−5	365	370.9	376	1019.9	493	501.3	507	964.4
	0	224	226.9	229	1036.7	245	343.0	366	1005.2
	5	128	136.0	144	1040.0	238	261.1	276	1002.5
	10	72	76.3	84	1041.2	176	186.9	197	947.3
Avg		233.0	240.2	248.2	1001.7	312.9	361.1	372.1	976.2

Table 3. Results obtained by BRKGA (B) and SAA (S) with different timeouts.

γ_{thr} secs	−10		−5		0		5		10	
	B	S	B	S	B	S	B	S	B	S
10	328.3	402.8	305.6	384.4	168.4	249.9	89.2	164.7	41.1	113.6
20	333.0	435.1	312.0	415.9	172.3	279.9	91.4	187.5	43.4	127.2
40	340.4	465.3	317.7	445.2	177.8	293.8	95.5	207.1	45.4	140.2
80	353.8	482.4	331.6	457.5	188.6	315.9	102.5	223.2	48.3	154.0
160	378.9	498.5	354.0	477.0	211.6	327.6	122.1	233.6	57.5	165.2
320	387.4	503.1	364.4	487.7	220.1	339.9	133.6	245.7	70.6	174.8
640	390.6	503.6	367.0	496.0	222.5	348.8	134.6	255.4	76.4	182.9

Table 4. Results obtained by SAA with different speed values in [m/s].

Inst.	γ_{thr}	10 [m/s] Avg	10–20 [m/s] Avg	20 [m/s] Avg	Inst.	γ_{thr}	10 [m/s] Avg	10–20 [m/s] Avg	20 [m/s] Avg
Test1	−10	449.1	512.1	511.1	Test6	−10	418.2	492.4	504.2
	−5	401.9	495.0	506.2		−5	372.3	453.3	493.6
	0	269.1	348.7	362.6		0	260.2	318.2	336.4
	5	178.7	237.7	249.4		5	180.5	249.4	257.4
	10	120.1	174.6	184.5		10	124.3	182.6	188.1
Test2	−10	395.3	512.9	516.8	Test7	−10	423.7	504.7	510.1
	−5	386.4	492.1	504.5		−5	430.0	490.3	498.3
	0	270.3	312.7	332.0		0	277.7	356.1	365.6
	5	184.8	247.6	260.4		5	183.6	252.0	264.0
	10	118.7	179.3	186.7		10	123.5	181.9	189.4
Test3	−10	449.6	506.3	512.2	Test8	−10	452.4	509.1	514.7
	−5	418.8	489.1	485.5		−5	429.7	492.0	501.5
	0	217.5	341.6	359.5		0	282.6	357.9	349.0
	5	182.4	249.7	257.9		5	175.3	251.5	264.5
	10	124.0	179.4	184.7		10	118.2	177.7	185.6
Test4	−10	456.4	506.1	507.2	Test9	−10	457.7	508.6	513.3
	−5	428.0	479.4	490.7		−5	431.1	492.9	500.7
	0	214.6	329.9	358.2		0	269.9	343.8	352.5
	5	181.9	249.4	257.3		5	177.0	245.0	255.9
	10	124.7	177.7	183.2		10	118.3	178.7	183.2
Test5	−10	450.0	508.3	514.1	Test10	−10	451.8	509.5	504.3
	−5	426.6	459.4	500.7		−5	429.1	490.8	501.3
	0	276.1	352.7	364.1		0	251.5	310.7	343.0
	5	185.5	253.6	255.9		5	171.6	249.1	261.1
	10	123.2	174.9	185.0		10	116.8	178.0	186.9

are possible, the results are similar to the best ones, and this feature can become relevant for other kinds of scenarios or when the UAV energy consumption is highly dependent on the its speed.

5   Conclusions

We studied a Drone Trajectory Optimization problem, in which a UAV is used as an unmanned aerial base station to provide connectivity to a set of users spread over an area. We proposed two metaheuristic algorithms to solve the problem: a Biased Random Key Genetic Algorithm (BRKGA) and a Simulated Annealing Algorithm (SAA). Both algorithms were tested on realistic instances with more than 500 users, and different time limits and UAV speeds. The SAA achieved the best performance, being able of serving about 400 users even with very short time limits when the connectivity threshold is low, and about 500 users with longer time limits.

Future research directions include (i) considering more UAVs to increase the covered demand, while imposing that they must fly with non-conflicting trajectories, (ii) dealing with an online problem in which the users' activation is unknown, (iii) handling the UAV energy consumption as a function of the speed.

Acknowledgements. Valentina Cacchiani acknowledges the support by the Air Force Office of Scientific Research under award number FA8655-20-1-7019.

References

1. Technical Specification Group Radio Access Network; Study on channel model for frequencies from 0.5 to 100 GHz (2019). 3GPP TR 38.901 version 16.1.0, Release 16
2. Technical Specification Group Radio Access Network; General aspects for Base Station (BS) Radio Frequency (RF) for NR (2020). 3GPP TS 38.817-02 version 15.9.0, Release 15
3. Andrade, C.E., Toso, R.F., Gonçalves, J.F., Resende, M.G.: The multi-parent biased random-key genetic algorithm with implicit path-relinking and its real-world applications. Eur. J. Oper. Res. **289**(1), 17–30 (2021)
4. Birattari, M., Yuan, Z., Balaprakash, P., Stützle, T.: F-Race and Iterated F-Race: an overview. In: Bartz-Beielstein, T., Chiarandini, M., Paquete, L., Preuss, M. (eds.) Experimental Methods for the Analysis of Optimization Algorithms, pp. 311–336. Springer, Heidelberg (2010). https://doi.org/10.1007/978-3-642-02538-9_13
5. Ceschia, S., Di Gaspero, L., Rosati, R.M., Schaerf, A.: Multi-neighborhood simulated annealing for the minimum interference frequency assignment problem. EURO J. Comput. Optim. 1–32 (2021). https://doi.org/10.1016/j.ejco.2021.100024
6. Di Gaspero, L., Schaerf, A.: EasyLocal++: an object-oriented framework for flexible design of local search algorithms. Softw. – Pract. Exp. **33**(8), 733–765 (2003)
7. Franzin, A., Stützle, T.: Revisiting simulated annealing: a component-based analysis. Comput. Oper. Res. **104**, 191 (2019)

8. Gonçalves, J.F., Resende, M.G.: Biased random-key genetic algorithms for combinatorial optimization. J. Heurist. **17**(5), 487–525 (2011)
9. Gunawan, A., Lau, H.C., Vansteenwegen, P.: Orienteering problem: a survey of recent variants, solution approaches and applications. Eur. J. Oper. Res. **255**(2), 315–332 (2016)
10. Hammersley, J.M., Handscomb, D.C.: Monte Carlo Methods. Chapman and Hall, London (1964)
11. Kirkpatrick, S., Gelatt, D., Vecchi, M.: Optimization by simulated annealing. Science **220**, 671–680 (1983)
12. Mignardi, S., Buratti, C., Cacchiani, V., Verdone, R.: Path optimization for unmanned aerial base stations with limited radio resources. In: 2018 IEEE 29th Annual International Symposium on Personal, Indoor and Mobile Radio Communications (PIMRC), pp. 328–332. IEEE (2018)
13. Mignardi, S., Mikhaylov, K., Cacchiani, V., Verdone, R., Buratti, C.: Unmanned aerial base stations for NB-IoT: trajectory design and performance analysis. In: 2020 IEEE 31st Annual International Symposium on Personal, Indoor and Mobile Radio Communications, pp. 1–6. IEEE (2020)
14. Mozaffari, M., Saad, W., Bennis, M., Debbah, M.: Mobile unmanned aerial vehicles (UAVs) for energy-efficient internet of things communications. IEEE Trans. Wireless Commun. **16**(11), 7574–7589 (2017)
15. Mozaffari, M., Saad, W., Bennis, M., Nam, Y.H., Debbah, M.: A tutorial on UAVs for wireless networks: applications, challenges, and open problems. IEEE Commun. Surv. Tutor. **21**(3), 2334–2360 (2019)
16. Na, Z., Wang, J., Liu, C., Guan, M., Gao, Z.: Join trajectory optimization and communication design for UAV-enabled OFDM networks. Ad Hoc Netw. **98**, 102031 (2020)
17. Otto, A., Agatz, N., Campbell, J., Golden, B., Pesch, E.: Optimization approaches for civil applications of Unmanned Aerial Vehicles (UAVs) or aerial drones: a survey. Networks **72**(4), 411–458 (2018)
18. Toso, R.F., Resende, M.G.: A C++ application programming interface for biased random-key genetic algorithms. Optim. Methods Softw. **30**(1), 81–93 (2015)
19. Tran, D.H., Nguyen, V.D., Chatzinotas, S., Vu, T.X., Ottersten, B.: UAV relay-assisted emergency communications in IoT networks: resource allocation and trajectory optimization. IEEE Trans. Wireless Commun. **21**(3), 1621–1637 (2021)
20. Urli, T.: Json2run: a tool for experiment design & analysis. CoRR abs/1305.1112 (2013)
21. Won, J., Kim, D.Y., Park, Y.I., Lee, J.W.: A survey on UAV placement and trajectory optimization in communication networks: from the perspective of air-to-ground channel models. ICT Express, 1–13 (2022, in press). https://doi.org/10.1016/j.icte.2022.01.015
22. Wu, Q., Zeng, Y., Zhang, R.: Joint trajectory and communication design for multi-UAV enabled wireless networks. IEEE Trans. Wireless Commun. **17**(3), 2109–2121 (2018)
23. Zeng, Y., Zhang, R.: Energy-efficient UAV communication with trajectory optimization. IEEE Trans. Wireless Commun. **16**(6), 3747–3760 (2017)
24. Zeng, Y., Zhang, R., Lim, T.J.: Throughput maximization for UAV-enabled mobile relaying systems. IEEE Trans. Commun. **64**(12), 4983–4996 (2016)

New Neighborhood Strategies
for the Bi-objective Vehicle Routing
Problem with Time Windows

Clément Legrand[1(✉)], Diego Cattaruzza[2], Laetitia Jourdan[1],
and Marie-Eléonore Kessaci[1]

[1] Univ. Lille, CNRS, Centrale Lille, UMR 9189 CRIStAL, 59000 Lille, France
{clement.legrand4.etu,laetitia.jourdan,
marie-eleonore.kessaci}@univ-lille.fr
[2] Univ. Lille, CNRS, Inria, Centrale Lille, UMR 9189 CRIStAL, 59000 Lille, France
diego.cattaruzza@centralelille.fr

Abstract. Local search (LS) algorithms are efficient metaheuristics to
solve vehicle routing problems (VRP). They are often used either individually or integrated into evolutionary algorithms. For example, the Multi-
Objective Evolutionary Algorithm based on Decomposition (MOEA/D)
can be enhanced with a local search replacing the mutation step based
on a single move operator traditionally. LS are based on an efficient
exploration of the neighborhoods of solutions. Many methods have been
developed over the years to improve the efficiency of LS. In particular, the
exploration strategy of the neighborhood and the pruning of irrelevant
neighborhoods are important concepts that are frequently considered
when designing a LS. In this paper, we focus on a bi-objective vehicle
routing problem with time windows (bVRPTW) where the total traveling cost and the total waiting time have to be minimized. We propose
two neighborhood strategies to improve an existing LS, efficient on the
single-objective VRPTW. First, we propose a new strategy to explore
the neighborhood of a solution. Second, we propose a new strategy for
pruning the solution neighborhood that takes into account the second
criterion of our bVRPTW namely the waiting time between customers.
Experiments on Solomon's instances show that using LS with our neighborhood strategies in the MOEA/D gives better performance. Moreover,
we can achieve some best-known solutions considering the traveling cost
minimization only.

Keywords: VRP · Multi-objective optimization · MOEA/D · Local
search

1 Introduction

Local search (LS) are known to be powerful algorithms used in evolutionary
algorithms to improve their performance [7]. Indeed, LS are able to intensify
the search by focusing on a specific region of the space. LS are based on neighborhood operators that link solution together and a neighborhood exploration

L. Di Gaspero et al. (Eds.): MIC 2022, LNCS 13838, pp. 45–60, 2023.
https://doi.org/10.1007/978-3-031-26504-4_4

strategy define how the neighbors are explored and when the exploration is stopped. Here, we are mainly interested in the Vehicle Routing Problem with Time Windows (VRPTW). It is a routing problem where time is considered as an important resource and where customers must be served within a fixed time interval. Some LS have been developed for this problem and consequently many neighborhoods are available. For our study, we consider the same neighborhood as defined in [16]. The operators are: relocate, swap and 2-opt*. These operators are commonly used in routing problems, since they are simple operators and they are able to produce a large neighborhood. However the LS steps are time-consuming, that is why different strategies exist to speed-up the search and reduce the time allocated to the neighborhood exploration. First LS can be applied following a probability, that is a parameter of the final algorithm. Indeed, not applying the LS may have a positive impact since it brings more diversity to the solutions. Second, the exploration of the neighborhood can be done entirely with strategy *best*, or partially with strategy *first*. For the strategy *best*, all neighbors are considered and the best one is selected. For the strategy *first*, the neighbor are evaluated one by one and the exploration is stopped as soon as an improving neighbor is found and selected. Since routing problems produce large neighborhood pruning techniques have been designed to avoid irrelevant moves. The most common one is probably the granular search [18]. It is based on the idea that two distant clients have a low chance to produce a relevant arc.

In this paper, we study a Bi-objective VRPTW (bVRPTW), that is a Multi-objective Combinatorial Optimization Problems (MoCOPs) [5]. Such problems are frequent in the industry where decision-makers are interested in optimizing several conflicting objectives at the same time. The objectives considered are the total traveling cost (a classical objective in routing problems), and the total waiting time incurred when drivers arrive before the opening of the time window. Although this objective has not received much attention in the literature [4,25], it is relevant when considering the transportation of people or medical goods. Indeed, when a patient has a medical appointment, we do not want that he waits too much. Note that, here we only consider the minimum possible waiting time incurred by time windows. Moreover, in real problems, there is more than one way to link two customers considering the traveled distance, and the traveling time. However in the Solomon's instances, that are used for our experiments, the traveling time between two customers corresponds to the distance between them, which is a strong hypothesis.

To solve this problem, we use MOEA/D, a Multi-Objective Evolutionary Algorithm based on Decomposition [24] where the mutation step is replaced by a local search. The contribution of the paper is to present neighborhood strategies that are better adapted to the bVRPTW. First, we present a new strategy to explore the neighborhood of bVRPTW solutions inspired from state of the art for permutation flowshop. Second, we propose a pruning technique that considers not only the distance between the clients, but also their respective time window.

The remaining of the paper is structured as follows. After a brief presentation of multi-objective problems, the bVRPTW studied is described in Sect. 2, as well as related works. Section 3 first focuses on the MOEA/D based framework used for this study, and then presents the different mechanisms proposed to improve the local search step. Section 4 describes the benchmark and how the algorithms were tuned. Then our experimental protocol is presented. Section 5 compares the results obtained for each combination of the mechanisms for the local search. Section 6 compares the results obtained with the best variant from Sect. 5, and the results obtained with state of the art algorithms for the VRPTW. Finally, Sect. 7 concludes and presents perspectives for this work.

2   Bi-objective Routing Problem with Time Windows

2.1   Multi-objective Optimization

In the following we formalize *Multi-objective Combinatorial Optimization Problems* (MoCOPs) [5]:

$$(MoCOP) = \begin{cases} Optimize\ F(x) = (f_1(x), f_2(x), \ldots, f_n(x)) \\ s.t.\ x \in \mathcal{D}, \end{cases} \tag{1}$$

where n is the number of objectives ($n \geq 2$), x is the vector of decision variables, $\mathcal{D}$ is the (discrete) set of feasible solutions and each objective function $f_i(x)$ has to be optimized (i.e. minimized or maximized). In multi-objective optimization the objective function F defines a so-called objective space denoted by $\mathcal{Z}$. For each solution $x \in \mathcal{D}$ there exists a point in $\mathcal{Z}$ defined by $F(x)$.

A *dominance* criterion is defined to compare solutions together: a solution x dominates a solution y, in a minimization context, if and only if for all $i \in [1 \ldots n]$, $f_i(x) \leq f_i(y)$ and there exists $j \in [1 \ldots n]$ such that $f_j(x) < f_j(y)$. A partial order is defined on the solutions by $x < y$ if and only if x dominates y.

Then a set of non dominated solutions is called a *Pareto front*. A feasible solution $x^* \in \mathcal{D}$ is called *Pareto optimal* if and only if there is no solution $x \in \mathcal{D}$ such that x dominates x^*. Resolving a MoCOP involves finding all the Pareto optimal solutions which form the *Pareto optimal set*. The *true Pareto front* of the problem is the image of the Pareto optimal set by the objective function.

Over the years, many metaheuristics based on local search techniques or using evolutionary algorithms [3] have been designed to solve multi-objective problems. Moreover, many tools [14] have been developed to assess and compare the performance of multi-objective algorithms. In this paper, we use the unary hypervolume (HV) [26], which is a metric defined relatively to a reference point $\mathcal{Z}_{ref}$. This indicator evaluates accuracy, diversity, and cardinality of the front, and it is the only indicator with this capability. Moreover, it can be used without knowing the true Pareto front of the problem. It reflects the volume covered by the members of a non dominated set of solutions. Thus, the larger the hypervolume, the better the set of solutions.

2.2   bVRPTW and Related Works

The bVRPTW [19] considered in this work is defined on a graph $G = (V, E)$, where $V = \{0, 1, \ldots, N\}$ is the set of vertices and $E = \{(i, j) \mid i, j \in V\}$ is the set of arcs. It is possible to travel from i to j, incurring in a travel cost c_{ij} and a travel time t_{ij}. Vertex 0 represents the depot where a fleet of K identical vehicles with limited capacity Q is based. Vertices $1, \ldots, N$ represent the customers to be served, each one having a demand q_i, a time window $[a_i, b_i]$ during which service must occur, and a service time s_i estimating the required time to perform the delivery. Vehicles may arrive before a_i. In that case, the driver has to wait until a_i to accomplish service incurring in a waiting time. Arriving later than b_i is not allowed. It is assumed that all inputs are nonnegative integers. We recall that a *route* r is an elementary cycle on G that contains the depot (that is vertex 0) and can be expressed as a sequence of vertexes $r = (v_0, v_1, \ldots, v_R, v_{R+1})$ where $v_0 = v_{R+1} = 0$ and vertexes $v_1, \ldots, v_R$ are all different. The cost c_r of a route r is then given as the sum of traveling costs on arcs used to visit subsequent vertexes, that is $\sum_{i=0}^{R} c_{v_i, v_{i+1}}$. A solution x can be represented as a set of (possibly empty) K routes, that is $x = \{r_1, \ldots, r_K\}$, and its cost is expressed as:

$$f_1(x) = \sum_{k=1}^{K} c_{r_k} \tag{2}$$

The waiting time W_i at a customer i is given as the maximum between 0 and difference between the opening of the TW a_i and the arrival time T_i at location i, that is $W_i = \max\{0, a_i - T_i\}$. Note that each route $r = (v_0, v_1, \ldots, v_R, v_{R+1})$ can be associated with a feasible (i.e., consistent with traveling times and TWs) arrival time vector $T_r = (T_{v_0}, T_{v_1}, \ldots, T_{v_R}, T_{v_{R+1}})$ and the total waiting time $W_r(T_r)$ on route r, with respect to T_r is given by $W_r(T_r) = \sum_{i=1}^{R} W_{v_i}$. Thus the total waiting time of a solution $x = \{r_1, \ldots, r_K\}$ on a graph G, given a time arrival vector for each route in the solution, i.e. $T_x = (T_{r_1}, \ldots, T_{r_K})$, is given by the following formula:

$$f_2(x, T_x) = \sum_{k=1}^{K} W_{r_k}(T_{r_k}) \tag{3}$$

The bVRPTW calls for the determination of at most K routes such that the traveling cost and waiting time are simultaneously minimized and the following conditions are satisfied: (a) each route starts and ends at the depot, (b) each customer is visited by exactly one route, (c) the sum of the demands of the customers in any route does not exceed Q, (d) time windows are respected. Note that a solution is represented as a permutation of the customers, and it is evaluated with the split algorithm detailed in [12].

The VRPTW, where only the traveling cost is minimized, received many interests in the literature. Nowadays, all Solomon's instances (of size 100) can be optimally solved using an exact algorithm [11], however the computational cost grows exponentially with the size of the instances (e.g. it takes 64105 s to solve the instance R208). In practice meta-heuristic algorithms can obtain a "good enough" solution in a short time and have the capacity to solve the large-scale complex problems, which is more suitable for applications. The NBD algorithm from Nagata et al. [10] is considered as a state of the art metaheuristic for the problem. Moreover, Schneider et al. [16] proposed different granular neighborhoods to improve an existing local search. Considering the multi-objective approaches the literature is more sparse. The second objective often minimized in the literature is the number of vehicles. Qi et al. [13] proposed a memetic algorithm based on MOEA/D to solve a bi-objective VRPTW. More recently, Moradi [9] integrated a learnable evolutionary model into a pareto evolutionary algorithm. The integration of learning mechanisms is known to be successful in both single-objective [1] and multi-objective algorithms [8]. In the following, we assume that the learning mechanism proposed is relevant for the studied problem, according to previous works [8].

3   Neighborhood Strategies

3.1   The Baseline MOEA/D

The MOEA/D [24] is a genetic algorithm that approximates the Pareto front by decomposing the multi-objective problem into several scalar objective subproblems, as illustrated in Fig. 1. MOEA/D is a simple algorithm that has already been studied a lot in the literature [23], making it a good candidate for our study. More precisely, the objective function of the i-th subproblem is defined with a weight vector $w^i = (w_1^i, w_2^i)$, such that $w_1^i + w_2^i = 1$, and is expressed as: $g_i = w_1^i \cdot f_1 + w_2^i \cdot f_2$, with f_1 and f_2 being the two objectives defined in Sect. 2.2. In the following we consider a uniform distribution on the weight vectors, and we assume that is enough to obtain diverse subproblems. Moreover an external archive stores nondominated solutions found during the search. These solutions are returned once the termination criteria is reached.

However, we do not use the basic MOEA/D, but a variant where learning is integrated. We will refer to this algorithm as $\mathcal{A}$. This algorithm contains four major mechanisms. Two of them belong to the genetic aspect (crossover and mutation), while the two others belong to the learning aspect (injection and extraction).

The crossover is a Partially Mapped Crossover (PMX) [21], that occurs with probability p_{cro}. It is performed between two solutions taken from close subproblems. Among the two solutions produced only one solution is randomly selected to undergo the injection step, which is a costly step.

The mutation, replaced here by the LS, is applied following a probability p_{mut}. Three neighborhood operators are applied: Relocate, Swap and 2-opt*, generating the same neighborhood as described in [16]. The operators are shuffled before applying them, so that they are not always applied in the same order. Two possible strategies are considered to explore the neighborhoods and will be described in Sect. 3.2. To perform an efficient exploration of the neighbors, we use sequences as defined in [22]. Once a local optimum has been reached for an operator, the next one is applied and so on, until all have been applied.

In order to present the extraction and injection steps, we have to briefly present the integrated learning mechanism. We refer to [8] for a complete description of the mechanism. The learning mechanism uses learning groups, noted $\mathcal{G}_i$. The learning group $\mathcal{G}_i$ is associated to the subproblem with weight vector w^i. Each group gathers knowledge that is relevant for its associated subproblem.

The learning groups are updated when the extraction step occurs. However, to ensure that knowledge is extracted from local optima only, the extraction can occur only when the local search has been applied during the iteration. In addition to that, the extraction occurs with probability p_{ext}. The extraction step is quite similar to the one performed in PILS [1]. Given one solution $x = \{r_1, \ldots, r_K\}$, patterns can be extracted from routes $r_1, \ldots, r_K$. These patterns are sequences of consecutive customers (not including the depot). The patterns have a size between 2 and $MaxSize$, which is a parameter of the algorithm.

Finally the injection step, following a probability p_{inj}, uses the knowledge stored in the groups to diversify the solutions. More precisely, any solution that undergoes the injection step will receive at most $N_{Injected}$ patterns from one learning group randomly chosen. A pattern is kept only if it improves the solution. Each pattern is selected as follows. First the size of the pattern is randomly chosen, and then it is selected among the $N_{Frequent}$ most frequent patterns of the same size in the corresponding group. Figure 2 illustrates how the injection is performed. First the pattern is formed by deleting adjacent vertices, and then the pieces of route created are put together to form the best possible solution.

Algorithm 1 presents the framework of $\mathcal{A}$. Initially the external archive is empty as well as the learning groups. The initial population is randomly generated, and undergo the LS (still with its own probability). Then, until the termination criteria is reached, subproblems are solved one at a time. The crossover is the first operator applied, followed by the injection and the LS. The extraction is performed only if the LS occurred. Then neighboring subproblems have their solutions updated if necessary, as well as the archive.

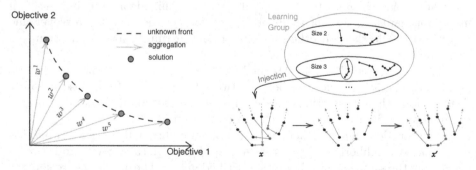

Fig. 1. A bi-objective problem decomposed into five scalar problems with MOEA/D.

Fig. 2. Injection of a frequent pattern of size 3 in a VRPTW solution.

Algorithm 1: The $\mathcal{A}$ framework.

Input: M weight vectors $w^1, \ldots, w^M$ and the size m of each neighborhood.
Output: The external archive A
```
/* Initialisation                                                     */
```
1 $A \leftarrow \emptyset$
2 $P \leftarrow$ random initial population (x^i for the i-th subproblem)
3 **for** $i \in \{1, \ldots, M\}$ **do**
4 $\quad \mathcal{N}(i) \leftarrow$ indexes of the m closest weight vectors to w^i
5 $\quad x^i \leftarrow LS(x^i)$
6 $\quad Obj^i \leftarrow F(x^i)$
7 $\quad \mathcal{G}_i \leftarrow \emptyset$

```
/* Core of the algorithm                                              */
```
8 **while** *not stopping criteria satisfied* **do**
9 $\quad$ **for** $i \in \{1, \ldots, M\}$ **do**
10 $\quad\quad (k, l) \leftarrow$ select randomly two indexes from $\mathcal{N}(i)$
11 $\quad\quad x_c \leftarrow Crossover(x^k, x^l)$
12 $\quad\quad x_{inj} \leftarrow Injection(s, x_c)$
13 $\quad\quad x' \leftarrow LS(x_{inj})$
14 $\quad\quad$ **if** *LS applied* **then**
15 $\quad\quad\quad \mathcal{K} \leftarrow Extraction(x')$
16 $\quad\quad\quad \mathcal{G}_1, \ldots, \mathcal{G}_M \leftarrow$ update with $\mathcal{K}$
17 $\quad\quad$ **for** $j \in \mathcal{N}(i)$ **do**
18 $\quad\quad\quad$ **if** $g_j(x') \leq g_j(x^j)$ **then**
19 $\quad\quad\quad\quad x^j \leftarrow x'$
20 $\quad\quad\quad\quad Obj^j \leftarrow F(x')$
21 $\quad\quad A \leftarrow Update(A, x')$

22 **return** A

3.2 Strategy of Exploration

In this section we give more details about the two exploration strategies considered in the local search. In routing problems, the most commonly used neighborhood exploration strategy is the classical *best* strategy, where the best move found by the operator is applied. That is why, we consider this strategy as the reference. Although this exploration allows a fast convergence towards a local optimum it requires an entire exploration of the neighborhoods before applying a single move, that is time consuming.

Here we propose a *first-best* strategy, which is inspired from [15]. This method is commonly used to solve flowshop problems. Algorithm 2 gives the pseudo-code of the *first-best* procedure. The procedure requires a neighborhood operator (e.g. Swap, Relocate or 2-opt*), and the solution x which undergoes the LS. For the given operator we try to insert each customer to its best location, considering the possible moves allowed by the operator. These moves are given through the procedure *generate_moves* (l.7 of Algorithm 2). We repeat the process until no more improving moves are found for any customer.

The two strategies considered, *best* and *first-best*, lead to two variants of the algorithm $\mathcal{A}$, that are respectively $\mathcal{A}^{best}$ and $\mathcal{A}^{first-best}$.

Algorithm 2: The $First - Best$ procedure.

Input: A solution x and a neighborhood operator $\mathcal{N}$
Output: A local optimum

1 $improve \leftarrow True$
2 **while** $improve$ **do**
3      $improve \leftarrow False$
4      $indexes \leftarrow shuffle([1 \ldots N])$
5   **for** $customer \in indexes$ **do**
6          $x' \leftarrow$ remove $customer$ from x
7          $moves \leftarrow generate\_moves(customer, \mathcal{N})$
8          $x' \leftarrow$ best solution obtained by applying a move from moves
9   **if** $g(x') < g(x)$ **then**
10             $x \leftarrow x'$
11             $improve \leftarrow True$

12 **return** x

3.3 Granularity and Pruning of Neighborhoods

In routing problems, many moves of a neighborhood operator can be a priori classified as irrelevant, and thus should not be considered during the neighborhood exploration. Most of the time these moves consider customers that are "far" distant. Having a method that restricts the neighborhood to relevant moves is interesting to spare time and resources during the LS. However, such a method requires a way to quantify the closeness between customers. In [18], the closeness between two customers is evaluated according to the distance between them. If it is enough for single-objective problems, it might not be adapted for multi-objective problems. In particular for our bi-objective VRPTW, close customers can incur a big waiting time if they are visited in the same route. Once a metric between customers is defined, a natural way to prune the neighborhood is to consider moves including the δ nearest customers for the metric defined.

For our study we compare two different metrics. The first metric, called d_1, is the classical metric used in single-objective routing problems: the distance between two customers is simply the euclidean distance between them. The second metric, d_2, is an aggregation of both objectives. More precisely, each subproblem generated in MOEA/D, with weight vector $w = (w_1, w_2)$, has its own metric that is defined as: $d_2^w(u, v) = w_1 \cdot distance(u, v) + w_2 \cdot WT(u, v)$. The value $WT(u, v)$ is the waiting time incurred by going to v from u. If $[a_u, b_u]$ (resp. $[a_v, b_v]$) is the time window of customer u (resp. v), s_u the service time of customer u and t_{uv} the traveling time from u to v, then WT is expressed as follows: $WT(u,v) = max(0, a_v - (a_u + s_u + t_{uv}))$.

The strategies presented in this section lead to four variants of $\mathcal{A}$ following the exploration strategy and the distance metrics used by the neighborhood operators: $\mathcal{A}_{d_1}^{best}$, $\mathcal{A}_{d_1}^{first-best}$, $\mathcal{A}_{d_2}^{best}$, and $\mathcal{A}_{d_2}^{first-best}$.

4   Experimental Setup

4.1   The Solomon's Benchmark

We use the Solomon's instances [17] to evaluate the performance of the four variants presented in Sect. 3. The set contains 56 instances divided into three categories according to the type of generation used, either R (random), C (clustered) or RC (random-clustered). The generation R randomly places customers in the grid, while the generation C tends to create clusters of customers. The generation RC mixes both generations. Each category is itself divided into two classes, either $1XX$ or $2XX$, according to the width of time windows. Instances of class $1XX$ have wider time windows than instances of class $2XX$, meaning that instances $2XX$ are more constrained. All 56 instances exist in three sizes: 25, 50 and 100. However, instances of size 25 and 50 are restrictions of instances of size 100. For our experiments instances of size 25 are discarded, since they are too small. Although this set was originally created to evaluate single-objective algorithms, it is used in the literature to evaluate the performance of multi-objective algorithms [6,9,13].

4.2   Setup and Tuning

We recall that the four variants compared are: $\mathcal{A}_{d_1}^{best}$, $\mathcal{A}_{d_1}^{first-best}$, $\mathcal{A}_{d_2}^{best}$ and $\mathcal{A}_{d_2}^{first-best}$. Note that the algorithm $\mathcal{A}_{d_1}^{best}$ will be our referent algorithm during the experiments, since it uses state of the art mechanisms.

Each algorithm is tuned to find a good setting of the parameters. To perform the tuning, we generated 96 new instances of sizes 50 and 100, by using the method described by Uchoa et al. [20] to mimic the Solomon's instances.

Each variant uses 10 parameters: M, the number of subproblems considered and m the size of the neighborhood of each subproblem. The four probabilities associated to each mechanism: p_{cro}, p_{inj}, p_{mut}, p_{ext}. The granularity parameter δ used to prune the neighborhood during LS. The maximal size $MaxSize$ of the patterns extracted, and the number $N_{Injected}$ of patterns injected, chosen among the $N_{Frequent}$ most frequent patterns. The parameters obtained after tuning are reported in Table 1.

The experiments are performed on two computers "Intel(R) Xeon(R) CPU E5-2687W v4 @ 3.00 GHz", with 24 cores each, in parallel (with slurm). The variants have been implemented using the jMetalPy framework [2].

Table 1. The configurations returned by irace for each variant and for both sizes of instance.

Parameters	$\mathcal{A}_{d_1}^{best}$		$\mathcal{A}_{d_1}^{first-best}$		$\mathcal{A}_{d_2}^{best}$		$\mathcal{A}_{d_2}^{first-best}$	
	50	100	50	100	50	100	50	100
M	13	68	31	50	42	15	29	15
m	4	26	8	15	6	4	11	4
δ	21	51	25	75	16	19	36	31
p_{cro}	0.94	0.30	0.88	0.86	0.93	0.35	0.94	0.67
p_{mut}	0.06	0.05	0.42	0.55	0.05	0.06	0.11	0.21
p_{ext}	0.50	0.96	0.48	0.60	0.55	0.90	0.86	0.83
$MaxSize$	2	3	5	5	2	4	2	5
p_{inj}	0.70	0.88	0.83	0.93	0.89	0.59	0.86	0.70
$N_{Frequent}$	73	165	74	135	52	175	66	115
$N_{Injected}$	33	80	17	74	10	63	18	31

4.3   Experimental Protocol

In our experimentation, we investigate the efficiency of the mechanisms proposed, and their impact on the quality of the solutions returned.

To that aim, all the variants use a same termination criteria, being the maximum running time allowed. It is fixed to $N \times 6$ s, where N is the size of the instance. Each variant is executed 30 times on each instance of the Solomon's benchmark (56 instances of size 50, and 56 instances of size 100). For each algorithm, the k-th run of an instance is executed with the same seed being $10 \times (k - 1)$. To compare the results obtained, we use the hypervolume metric, since we do not know the true Pareto fronts of the instances. Note that, for the experiments we use the same values to normalize the objectives of the solutions returned by all variants. These values are simply the best and worst values for each objective, obtained among all the executions.

To complete the results obtained, the gap between the best-known and the best solution found by each algorithm is given, as well as the average gap over the 30 runs. The optimal solutions are available on CVRPlib.

Finally we compare our best variant, considering the results obtained, to state of the art single-objective algorithms: the TS_{tw} from Schneider et al. [16] and NBD from Nagata et al. [10], but also to competitive multi-objective algorithms: the M-MOEAD from Qi et al. [13] and the MODLEM from Moradi [9].

5   Analysis of Neighborhood Strategies

The Table 2 regroups the average hypervolume obtained on all classes of instance for all the variants. One can see that two variants stand out from the others: $\mathcal{A}_{d_1}^{first-best}$ and $\mathcal{A}_{d_2}^{first-best}$. Meaning that the exploration strategy has a positive impact on the performance of the algorithm, and thus it is better than the strategy $best$. However the variant $\mathcal{A}_{d_1}^{first-best}$ returns slightly higher hypervolumes than $\mathcal{A}_{d_2}^{first-best}$ on most instances, and clearly outperforms $\mathcal{A}_{d_2}^{first-best}$ on few instances (e.g. RC1 of size 50 and C1 of size 100). Indeed, the subproblems which mainly focus on the waiting time will "forget" the distance between customers. That can worsen the hypervolume since we would like to obtain the minimal cost with the minimal possible waiting time in our Pareto front. Knowing that, the d_2 metric can be improved.

Now we analyze the gaps between the best solutions returned for the total cost objective and the best-knowns. The gaps obtained on instances of size 100 are reported in Table 3 (R instances), Table 4 (RC instances) and Table 5 (C instances). For each variant, the first column is the gap between the best-known and the best solution returned, while the second column is the average gap considering the solutions returned on all 30 runs. One can notice that the variants $\mathcal{A}_{d_1}^{first-best}$ and $\mathcal{A}_{d_2}^{first-best}$ still outperform the two other variants. However, this is the variant $\mathcal{A}_{d_2}^{first-best}$ which returns the best results on most instances.

Table 2. Average hypervolume obtained with the four variants on all classes of instance of both sizes.

Class	Size	$\mathcal{A}_{d_1}^{best}$	$\mathcal{A}_{d_1}^{first-best}$	$\mathcal{A}_{d_2}^{best}$	$\mathcal{A}_{d_2}^{first-best}$
R1	50	0.716	0.768	0.729	**0.783**
R2	50	0.666	**0.747**	0.690	0.743
R1	100	0.773	**0.842**	0.720	0.833
R2	100	0.626	**0.760**	0.615	0.747
RC1	50	0.604	**0.760**	0.611	0.689
RC2	50	0.637	0.682	0.647	**0.692**
RC1	100	0.682	**0.758**	0.631	0.739
RC2	100	0.658	0.766	0.662	**0.769**
C1	50	0.519	**0.574**	0.523	0.550
C2	50	0.404	0.408	**0.414**	0.403
C1	100	0.881	**0.945**	0.846	0.882
C2	100	0.899	**0.967**	0.858	0.954

Table 3. Gaps (%) obtained for the total cost objective, relatively to the best-known on instances of class R. For each algorithm, the first column gives the gap with the best solution found. The second column contains the average gap over the 30 runs.

Instance	Size	Reference	$\mathcal{A}_{d_1}^{best}$		$\mathcal{A}_{d_1}^{first-best}$		$\mathcal{A}_{d_2}^{best}$		$\mathcal{A}_{d_2}^{first-best}$	
R101	100	1637.7	0.2	0.9	0.1	0.2	0.1	0.8	0.1	0.2
R102	100	1466.6	0.7	2.0	0.5	1.0	0.4	2.0	0.0	1.0
R103	100	1208.7	2.2	4.6	1.5	3.6	2.3	4.7	1.9	3.4
R104	100	971.5	4.7	8.4	4.8	7.5	5.3	9.2	3.6	7.2
R105	100	1355.3	0.6	2.2	0.4	1.4	0.4	2.4	0.4	1.4
R106	100	1234.6	0.9	4.1	2.3	3.6	2.6	4.2	1.3	3.4
R107	100	1064.6	1.9	6.2	2.5	6.0	4.2	7.2	3.4	5.3
R108	100	932.1	4.1	8.3	5.1	7.2	5.0	9.3	4.8	7.4
R109	100	1146.9	2.7	5.7	1.6	3.1	2.1	6.2	0.9	3.6
R110	100	1068.0	4.7	7.5	3.7	5.7	6.0	9.2	3.9	6.3
R111	100	1048.7	3.7	7.3	3.1	5.6	5.1	7.5	2.5	5.8
R112	100	948.6	4.0	8.6	4.0	7.7	4.6	10.6	3.5	8.1
Mean gap			2.5	5.5	2.5	**4.4**	3.2	6.1	**2.2**	4.4
R201	100	1143.2	4.2	9.5	2.2	5.1	3.3	6.9	2.9	4.9
R202	100	1029.6	6.2	9.6	3.8	6.5	1.5	8.6	3.1	6.0
R203	100	870.8	6.0	11.4	3.4	6.3	3.6	9.0	1.7	6.4
R204	100	731.3	3.3	9.6	3.0	5.9	1.9	9.0	3.4	5.6
R205	100	949.8	3.4	7.1	0.9	4.9	3.5	7.2	0.8	4.0
R206	100	875.9	3.4	6.8	2.6	5.3	2.7	6.9	2.3	6.2
R207	100	794.0	5.0	8.9	3.7	6.0	2.7	9.2	1.8	5.9
R208	100	701.0	4.8	8.4	3.0	6.2	3.7	8.4	2.3	6.2
R209	100	854.8	2.8	7.0	2.2	4.9	3.8	6.8	1.7	4.6
R210	100	900.5	6.1	9.6	4.1	6.6	4.2	8.4	3.0	6.7
R211	100	746.7	5.5	8.6	2.4	5.2	5.4	9.9	2.5	5.9
Mean gap			4.6	8.8	2.8	**5.7**	3.3	8.2	**2.3**	5.7

6   Comparison with State of the Art Algorithms

Considering the results obtained in the former section, we decide to compare the variant $\mathcal{A}_{d_2}^{first-best}$ to the other state of the art algorithms. Table 6 compares the average value of the best traveling cost obtained by different algorithms on each class of instance of size 100. We recall that, there are two single-objective algorithms: TS_{tw} [16] and NBD [10], and two multi-objective algorithms: M-MOEA/D [13] and MODLEM [9]. Note that MODLEM integrates a learning

Table 4. Gaps (%) obtained for the total cost objective, relatively to the best-known on instances of class RC. For each algorithm, the first column gives the gap with the best solution found. The second column contains the average gap over the 30 runs.

Instance	Size	Reference	$\mathcal{A}_{d_1}^{best}$	$\mathcal{A}_{d_1}^{first-best}$		$\mathcal{A}_{d_2}^{best}$		$\mathcal{A}_{d_2}^{first-best}$	
RC101	100	1619.8	2.4	4.0	1.5	3.2	1.3	4.2	1.6	3.2
RC102	100	1457.4	2.5	4.8	2.4	3.5	2.1	5.2	2.3	4.2
RC103	100	1258.0	7.2	10.3	7.6	9.5	7.5	11.0	7.3	9.2
RC104	100	1132.3	2.7	8.9	4.9	8.9	6.6	10.5	5.5	9.1
RC105	100	1513.7	4.6	7.0	3.4	5.5	3.3	6.7	2.4	5.3
RC106	100	1372.7	5.4	8.2	3.2	5.6	4.2	7.9	3.7	6.2
RC107	100	1207.8	6.9	11.0	5.6	10.0	8.0	12.1	4.8	10.4
RC108	100	1114.2	5.9	10.3	4.3	10.0	4.8	12.8	5.5	10.7
Mean gap			4.7	8.1	**4.1**	**7.0**	4.7	8.8	**4.1**	7.3
RC201	100	1261.8	2.0	7.5	2.0	4.2	2.3	6.8	1.2	3.9
RC202	100	1092.3	2.4	8.7	2.0	4.0	1.6	7.7	1.4	3.8
RC203	100	923.7	5.5	11.4	2.5	5.7	4.8	9.3	2.1	5.5
RC204	100	783.5	4.1	8.3	1.3	5.0	2.1	7.0	1.5	4.3
RC205	100	1154.0	4.9	10.7	2.3	5.7	2.6	7.6	0.5	4.6
RC206	100	1051.1	3.8	8.0	2.5	5.3	2.9	7.0	2.0	4.7
RC207	100	962.9	1.3	6.9	1.0	5.2	3.5	6.7	2.9	5.0
RC208	100	776.1	4.5	7.8	2.5	5.3	5.0	8.4	0.8	5.3
Mean gap			3.6	8.7	2.0	5.0	3.1	7.6	**1.5**	**4.6**

mechanism, that is a learnable evolution model based on decision trees. Moreover the algorithms that solve the VRPTW in a single-objective context, first minimize the number of vehicles and then the traveled distance. To be fair, we add in brackets the average number of vehicles contained in the solutions returned by our algorithm.

Since our algorithm did not focus on the number of vehicles, it seems normal that the average number of vehicles used in the solutions returned is much higher than the one found by other algorithms. However, our algorithm is able to reach competitive results on C instances.

Table 5. Gaps (%) obtained for the total cost objective, relatively to the best-known on instances of class C. For each algorithm, the first column gives the gap with the best solution found. The second column contains the average gap over the 30 runs.

Instance	Size	Reference	$\mathcal{A}_{d_1}^{best}$		$\mathcal{A}_{d_1}^{first-best}$		$\mathcal{A}_{d_2}^{best}$		$\mathcal{A}_{d_2}^{first-best}$	
C101	100	827.3	0.0	0.8	0.0	0.0	0.0	0.0	0.0	0.0
C102	100	827.3	0.0	1.7	0.0	0.2	0.0	0.6	0.0	1.0
C103	100	826.3	0.0	7.3	0.1	5.1	0.0	12.0	0.0	9.3
C104	100	822.9	0.1	10.5	1.6	12.4	0.9	16.7	0.4	12.1
C105	100	827.3	0.0	2.3	0.0	0.0	0.0	0.9	0.0	1.0
C106	100	827.3	0.0	1.4	0.0	0.0	0.0	1.0	0.0	1.4
C107	100	827.3	0.0	2.2	0.0	0.2	0.0	3.0	0.0	2.4
C108	100	827.3	0.0	1.7	0.0	0.6	0.0	5.5	0.0	2.7
C109	100	827.3	0.0	2.7	0.0	1.5	0.0	5.8	0.0	4.4
Mean gap			0.0	3.4	0.2	**2.2**	0.1	5.1	**0.0**	3.8
C201	100	589.1	0.0	0.0	0.0	0.0	0.0	0.0	0.0	0.0
C202	100	589.1	0.0	0.3	0.0	0.0	0.0	0.3	0.0	0.0
C203	100	588.7	0.0	1.4	0.0	0.8	0.0	1.9	0.0	1.4
C204	100	588.1	0.6	5.2	0.0	2.1	1.1	7.2	0.0	2.5
C205	100	586.4	0.0	0.7	0.0	0.1	0.0	0.4	0.0	0.1
C206	100	586.0	0.0	0.2	0.0	0.0	0.0	0.5	0.0	0.0
C207	100	585.8	0.0	0.1	0.0	0.0	0.0	0.2	0.0	0.0
C208	100	585.8	0.0	0.2	0.0	0.0	0.0	0.5	0.0	0.0
Mean gap			0.1	1.0	**0.0**	0.4	0.1	1.4	**0.0**	0.5

Table 6. Comparison of the average of the best traveling cost obtained on instances of size 100 between four state of the art algorithms and our algorithm $\mathcal{A}_{d_2}^{first-best}$. The corresponding average number of vehicles used is given in brackets.

Class	NBD [10]	TS_{tw} [16]	M-MOEA/D [13]	MODLEM [9]	$\mathcal{A}_{d_2}^{first-best}$
R1	1210.34 (11.9)	1220.83 (11.9)	1216.73 (12.4)	1210.40 (11.9)	1196.22 (13.8)
R2	951.03 (2.7)	959.86 (2.7)	924.18 (3.1)	916.95 (4.6)	892.85 (5.0)
RC1	1384.16 (11.5)	1392.54 (11.5)	1390.35 (11.9)	1384.17 (11.5)	1387.11 (13.8)
RC2	1119.24 (3.3)	1140.13 (3.3)	1119.93 (3.4)	1074.67 (4.0)	1015.76 (5.8)
C1	828.38 (10.0)	828.38 (10.0)	828.38 (10.0)	828.38 (10.0)	**827.02** (10.0)
C2	589.86 (3.0)	589.86 (3.0)	589.86 (3.0)	589.86 (3.0)	**587.38** (3.0)

7   Conclusion

LS are commonly used in evolutionary algorithms to improve the performance. In this paper we considered a LS from [16], adapted to the VRPTW, that uses a *best* strategy for exploration. That strategy has been compared to a *first-best* strategy inspired from ones used for other combinatorial problems like flowshops.

Through our experiments, conducted on the Solomon's instances, we showed that the adapted *first-best* strategy performs better than the *best* strategy, on the bVRPTW. We also investigated a new method for pruning the solution neighborhood taking into account the second criterion of our bVRPTW, being the waiting times. Our pruning method is able to reach similar results than the original one, but with smaller neighborhoods. The experimental results show also the benefit of our pruning method to reach better solutions when considering the first criterion only. The performance compared to state-of-the-art algorithms for both single- and bi-objective VRPTW show the interest of our new neighborhood strategies. Future works will investigate the neighborhood exploration strategy for other variants of routing problems. Moreover, we will analyze the impact of the weights of our pruning method on the Pareto front.

References

1. Arnold, F., Santana, Í., Sörensen, K., Vidal, T.: PILS: exploring high-order neighborhoods by pattern mining and injection. Pattern Recognit. **116**, 107957 (2021)
2. Benitez-Hidalgo, A., Nebro, A.J., Garcia-Nieto, J., Oregi, I., Del Ser, J.: jMetalPy: a python framework for multi-objective optimization with metaheuristics. Swarm Evol. Comput. **51**, 100598 (2019)
3. Blot, A., Kessaci, M.É., Jourdan, L.: Survey and unification of local search techniques in metaheuristics for multi-objective combinatorial optimisation. J. Heurist. **24**(6), 853–877 (2018). https://doi.org/10.1007/s10732-018-9381-1
4. Castro-Gutierrez, J., Landa-Silva, D., Pérez, J.M.: Nature of real-world multi-objective vehicle routing with evolutionary algorithms. In: 2011 IEEE International Conference on Systems, Man, and Cybernetics. IEEE (2011)
5. Coello, C.A.C., Dhaenens, C., Jourdan, L.: Multi-objective combinatorial optimization: problematic and context. In: Coello Coello, C.A., Dhaenens, C., Jourdan, L. (eds.) Advances in Multi-Objective Nature Inspired Computing. Studies in Computational Intelligence, vol. 272, pp. 1–21. Springer, Heidelberg (2010). https://doi.org/10.1007/978-3-642-11218-8_1
6. Ghoseiri, K., Ghannadpour, S.F.: Multi-objective vehicle routing problem with time windows using goal programming and genetic algorithm. Appl. Soft Comput. **10**(4), 1096–1107 (2010)
7. Knowles, J.D.: Local-search and hybrid evolutionary algorithms for Pareto optimization. Ph.D. thesis, University of Reading Reading (2002)
8. Legrand, C., Cattaruzza, D., Jourdan, L., Kessaci, M.-E.: Enhancing MOEA/D with learning: application to routing problems with time windows. In: Proceedings of the GECCO Companion (2022)
9. Moradi, B.: The new optimization algorithm for the vehicle routing problem with time windows using multi-objective discrete learnable evolution model. Soft. Comput. **24**(9), 6741–6769 (2020)
10. Nagata, Y., Bräysy, O., Dullaert, W.: A penalty-based edge assembly memetic algorithm for the vehicle routing problem with time windows. Comput. Oper. Res. **37**(4), 724–737 (2010)
11. Pecin, D., Contardo, C., Desaulniers, G., Uchoa, E.: New enhancements for the exact solution of the vehicle routing problem with time windows. INFORMS J. Comput. **29**(3), 489–502 (2017)

12. Prins, C.: A simple and effective evolutionary algorithm for the vehicle routing problem. Comput. Oper. Res. **31**(12), 1985–2002 (2004)
13. Qi, Y., Hou, Z., Li, H., Huang, J., Li, X.: A decomposition based memetic algorithm for multi-objective vehicle routing problem with time windows. Comput. Oper. Res. **62**, 61–77 (2015)
14. Riquelme, N., Von Lücken, C., Baran, B.: Performance metrics in multi-objective optimization. In: 2015 Latin American computing conference (CLEI), pp. 1–11. IEEE (2015)
15. Ruiz, R., Stützle, T.: A simple and effective iterated greedy algorithm for the permutation flowshop scheduling problem. Eur. J. Oper. Res. **177**(3), 2033–2049 (2007)
16. Schneider, M., Schwahn, F., Vigo, D.: Designing granular solution methods for routing problems with time windows. Eur. J. Oper. Res. **263**(2), 493–509 (2017)
17. Solomon, M.M.: Algorithms for the vehicle routing and scheduling problems with time window constraints. Oper. Res. **35**(2), 254–265 (1987)
18. Toth, P., Vigo, D.: The granular tabu search and its application to the vehicle-routing problem. INFORMS J. Comput. **15**(4), 333–346 (2003)
19. Toth, P., Vigo, D.: Vehicle Routing: Problems, Methods, and Applications. SIAM (2014)
20. Uchoa, E., Pecin, D., Pessoa, A., Poggi, M., Vidal, T., Subramanian, A.: New benchmark instances for the capacitated vehicle routing problem. Eur. J. Oper. Res. **257**(3), 845–858 (2017)
21. Varun Kumar, S., Panneerselvam, R.: A study of crossover operators for genetic algorithms to solve VRP and its variants and new sinusoidal motion crossover operator. Int. J. Comput. Intell. Res. **13**(7), 1717–1733 (2017)
22. Vidal, T., Crainic, T.G., Gendreau, M., Prins, C.: A hybrid genetic algorithm with adaptive diversity management for a large class of vehicle routing problems with time-windows. Comput. Oper. Res. **40**, 1 (2013)
23. Xu, Q., Xu, Z., Ma, T.: A survey of multiobjective evolutionary algorithms based on decomposition: variants, challenges and future directions. IEEE Access **8**, 41588–41614 (2020)
24. Zhang, Q., Li, H.: MOEA/D: a multiobjective evolutionary algorithm based on decomposition. IEEE Trans. Evol. Comput. **11**, 6 (2007)
25. Zhou, Y., Wang, J.: A local search-based multiobjective optimization algorithm for multiobjective vehicle routing problem with time windows. IEEE Syst. J. **9**(3), 1100–1113 (2014)
26. Zitzler, E., Thiele, L., Laumanns, M., Fonseca, C.M., Da Fonseca, V.G.: Performance assessment of multiobjective optimizers: an analysis and review. IEEE Trans. Evol. Comput. **7**(2), 117–132 (2003)

Tabu Search with Multiple Decision Levels for Solving Heterogeneous Fleet Pollution Routing Problem

Bryan F. Salcedo-Moncada[1,2], Daniel Morillo-Torres[2(✉)],
and Gustavo Gatica[3]

[1] Universidad Pontificia Bolivariana, Palmira, Colombia
bryan.salcedo@upb.edu.co
[2] Pontificia Universidad Javeriana Cali, Cali, Colombia
{bryansalcedo,daniel.morillo}@javerianacali.edu.co
[3] Universidad Andres Bello, Santiago, Chile
ggatica@unab.cl

Abstract. Organizations, in order to gain a competitive advantage, must improve their logistics performance along with the planning and distribution of their goods. Thus, they face significant challenges in managing their orders to be delivered on time. However, transportation is responsible for 79% of the CO_2 emissions of the total polluting gases in the atmosphere. Therefore, there is a growing interest to investigate methods to optimize logistics and to consider environmental aspects. However, the literature only considers realistic system characteristics such as: different vehicles and speeds, time windows and route inclination. For this reason, the focus is on the solution of an extension with a heterogeneous fleet and discrete speeds of the Vehicle Routing Pollution Problem (PRP), whose objective is the reduction of greenhouse gases (GHG). Based on the MEET model, the main polluting gases with the greatest impact on health are measured: carbon dioxide (CO_2), nitrogen dioxide (NO_X) and carbon monoxide (CO). For its solution, a Tabu Search metaheuristic is proposed with different decision levels: node sequence, assigned speeds and vehicles used, from different neighborhood structures. Finally, the balance between exploration and exploitation is achieved by incorporating favorable attributes to the created solutions. The proposed metaheuristic achieves efficient results both in total logistic cost and in emissions released to the environment.

Keywords: Pollution Routing Problem · Heterogeneous fleet · Tabu search

1 Introduction

Transportation is one of the main ways of generating damaging impacts on the environment from greenhouse gases (GHG). These gases, particularly carbon dioxide (CO_2), generate high volumes of pollution, affecting human health and negatively impacting the ozone layer [5]. The United States Environmental Protection Agency (EPA), through its inventory of greenhouse gas emissions and sinks in the USA, places transportation as the sector with the highest energy consumption in 2020. Because it releases 1 565 million metric tons (MMT) of CO_2 into the atmosphere, which represents 79% of the total GHG released [32].

The damage caused by the effects of transportation on the environment calls for the evolution of green supply chain management, logistics and green transportation. It is receiving increasing attention from companies, academia and governments [14]. Consequently, the literature starts from the central problems of logistics, such as the Vehicle Routing Problem (VRP), which usually focuses on the minimization of distances or route costs, to incorporate environmental components. Thus, the Pollution Routing Problem (PRP) is defined, which is the environmental evolution of the VRP [14]. By reducing the distance, it may be thought that fuel consumption is also being minimized and, likewise, the emission of polluting gases. However, this is not really the case, as there are several contributing factors, such as the type of vehicle, the relationship between its capacity and the load transported, the slopes of the roads, the speed of the vehicle during the trip, among others [24].

There is concern about the environmental consequences of the greenhouse effect in the world, and deterrent measures must be implemented [2]. The challenge for Colombia is great, there are several fronts that can lead to climate change mitigation and help a sustainable environment; among them is the improvement of the transportation sector to reduce GHG emissions. According to the Colombian Ministry of Mines and Energy, in 2017, transportation is the sector with the highest energy consumption in Colombia, its consumption is 39.8% with respect to the total of sectors. It represents 11 812 kt/year (kilotons of oil equivalent).

As a result, between 1990–2012, the country's emissions increased by 15%, with the industry, mining and energy, and transportation sectors showing the greatest increase according to [7,15]. On the other hand, Colombia is interested not only in improving its logistics processes, but also in implementing strategies to reduce pollution. According to Colombia's National Planning Department, at least half of the companies develop green logistics actions, with emissions reduction being the fourth most important [6].

The literature presents high-precision models that involve some of the problems mentioned above. There are few works that incorporate several realistic characteristics simultaneously. For example, discretized speed between routes (low, medium and high speed), time windows for receiving customer orders, the gradient between routes (the slope of the tracks) and the load transported between nodes. For this reason, this research addresses a heterogeneous extension of the PRP that includes all the aforementioned characteristics. For its solution, a Tabu Search metaheuristic is proposed, with the intention of reducing the costs associated with the use of different vehicles, the payment to drivers and the costs related not only to the emission of carbon dioxide (CO_2), but also Carbon Monoxide (CO) and nitrogen dioxide (NO_X). These are the most harmful pollutants emitted mainly by vehicle exhaust gases [21].

The rest of this document is organized as follows. Section 2 describes the problem addressed. Section 3 contains the most relevant background information, followed by Sect. 4 where the proposed metaheuristic is detailed. Section 5 presents the computational results and finally the conclusions in Sect. 6.

2   Problem Description

One of the central problems in logistics is the VRP. In general, this consists of finding routes (Hamiltonian cycles) for a subset of nodes for each available vehicle with a defined capacity. Thus, all customers are visited and returned to the starting point (known as depot). The total distance traveled is minimized. Due to its complexity, the VRP has demonstrated as part of the NP-hard [18] set. One of the extensions of this problem is the one that considers not only the distance as an objective to be minimized, but also considers the reduction of elements that contribute to pollution. Generally, these elements are greenhouse gases. This problem is the known as PRP, a detailed description of which will be given in this section.

Formally, the PRP addressed in this research can be defined by a vehicle circulation network represented by the graph $G = (N, A)$, with a set $N = \{0, \cdots, n, \cdots, n'\}$ of vertices and with a set $A = \{(i, j) \in N^2, i \neq j\}$ of arcs. N includes both nodes (customers) and the depots ($n = 0$). Each arc $(i, j) \in A$ has an associated travel distance D_{ij} [km], and a track gradient, called road gradient G_{ij} [%]. A value of 0% indicates a flat road and a value of 10% indicates a maximum slope. The sign of the parameter indicates whether the slope is ascending or descending. In addition, the slope is considered symmetric, i.e., if the $G_{ab} = \%$, then $G_{ba} = -\%$.

The use of a heterogeneous fleet $K = \{1, 2, 3\}$ with different capacities Q_k [kg] is considered, not only to make the problem more realistic but also because the use of a heterogeneous fleet can help in the overall reduction of emissions [4]. It will also be taken into account that each customer has a time constraint for his visit with a non-negative demand Dem_j [kg] for a single product. Thus, each customer j has a time window [$Tmin_j, Tmax_j$] in seconds, where $Tmin_j$ and $Tmax_j$ represent the earliest and latest delivery time. Customers must be served exclusively in this time window, where waiting time is allowed. In addition, each customer has an associated service time Ts_j [s].

Each of these vehicles, on each arc $(i, j) \in A$ can choose the speed that it will travel on that arc. The speed $Vel_v, v \in V = \{1, 2, 3\}$ is discretized in three levels: slow, medium and fast. According to the regulations in Colombia these speeds are: 30, 55 and 80 [km/h]. [22]. Finally, each vehicle will generate CO_2, NO_X and CO pollutant emissions, depending on the chosen speed, the gradient of the road and the current load on that stretch.

The objective of the problem is to find the route for each type of vehicle, with its respective speed allocation in each arc and the sequence of customers (nodes) to visit to satisfy the demand and time windows. Thus, reducing the costs of the use of the different vehicles, the payment of the drivers and the costs of signal emissions. As an example, Fig. 1 shows the best solution for a set of 10 nodes according to the above characteristics; the data correspond to instance PB-UK10_01-B. The instances are described in Sect. 5.

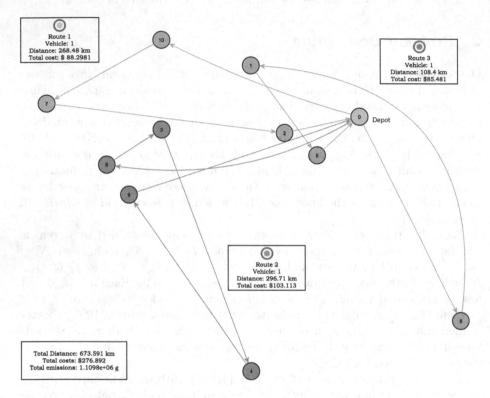

Fig. 1. A optimal solution for PB-UK10_01-B instance.

3   Literature Review

In recent years, research in the area of logistics has focused on global environmental concerns [34]. Consequently, research associated with vehicle routing, such as VRP, which seeks only the minimization of total transportation costs, evolves through an environmental approach, these problems are known as Green-Vehicle Routing Problem (GVRP).

There are three categories of GVRP: PRP, Green-VRP (GVRP), and reverse logistics VRP. All categories coincide in the minimization of economic and environmental costs [19]. To distinguish between the first two, the PRP focuses on minimizing greenhouse gas emissions and the GVRP on minimizing the total distance of alternative fuel vehicles instead of conventional vehicles, usually with charging stations.

Naderipour and Alinaghian in 2016 seek to solve the Open Time Dependent Vehicle Routing Problem (OTDVRP) using an improved Particle Swarm Optimization Algorithm (PSO) algorithm. In this model, despite not considering a heterogeneous fleet, the authors propose an emission measurement model based on the Methodologies for Estimating Emissions of Air Pollutants from Transport (MEET). [13]. Where vehicle speed, load and road gradient are the main factors affecting pollutant emissions [25].

For their part, Ren et al. in 2020 consider that GHG production damages the environment, threatens human life and health. Therefore, he studies the Ecological VRP with mixed vehicles also based on the MEET model to simultaneously calculate five types of pollutants: CO, CO_2, COV, NO_X and PM. However, it does not consider the slope of the roads, as they are relatively flat in the application area of their research [30]. Liu et al. in 2020, in contrast, focuses solely on the MEET model to calculate carbon emissions and fuel consumption of vehicles [20].

Thus, they propose a model that addresses the Time Dependent Vehicle Routing Problem With Time Windows (TDVRPTW). It considers time-dependent vehicle speeds, vehicle capacity and customer time constraints. It seeks to minimize driver costs, fuel consumption and carbon emissions using Ant Colony Optimization (ACO). Moryadee et al. in 2019 adds to the approach (TDVRP, Time Dependent Vehicle Routing Problem) in the city of Bangkok and propose, for CO_2 emission reduction, a comprehensive emissions model based on [5], where GHG emissions are directly converted from fuel consumption [24].

For these reasons, various strategies for the solution can be approached and classified according to the characteristics of PRP. In principle as exact methods that guarantee the optimality of the problem such as Integer-Mixed Linear Programming [8,12,14]. Secondly, with the use of heuristics, as methods of limited exploration of the search space for the generation of solutions with reasonable computational times [3,29]. Finally, through metaheuristics, which perform a robust search in the solution space escaping local optima [20,25].

Several authors share heuristic and metaheuristic algorithms for different extensions of the PRP, who sacrifice accuracy for computational efficiency to achieve results. TS and VNS stand out for solution in real cases [26]. A summary of the main investigations is presented in Table 1, where Ho: homogeneous fleet, He: heterogeneous fleet, Sing: single depot, Mult: multiple depot, V: velocity, C: capacity, TD: time dependence, and TW: time window.

Based on this review, despite the growth of research in recent years, there are few who address the PRP with a heterogeneous fleet (HPRP) extension with discrete velocities, time windows, the gradient of the tracks and with a reservoir. For this reason, this work has as main contributions: first, characterizing the PRP with the characteristics described above, and second the development of a novel Tabu search with multiple decision levels for its solution.

Table 1. Main research related to PRP.

Paper	Fleet		Depot		V	C	TD	TW	Solution approach
	Ho.	He.	Sing.	Mult.					
[5]	✓		✓		✓	✓		✓	MILP - Branch-and-Cut
[8]	✓		✓			✓			MILP - CW - CBCA
[26]	✓		✓			✓		✓	TS
[25]	✓		✓		✓	✓	✓		PSO - IPSO
[10]	✓		✓		✓	✓		✓	MINLP - Disjunctive convex programming
[14]		✓	✓		✓	✓			MILP - GA
[28]	✓		✓			✓			MCDA - TVa-PSOGMO
[24]	✓		✓		✓	✓	✓	✓	GA - TS
[33]	✓		✓		✓	✓		✓	MINLP - NSGA-II
[23]		✓	✓			✓		✓	MILP - VNS - TS
[20]	✓		✓		✓	✓	✓	✓	ACO
[30]		✓	✓		✓	✓		✓	VNS
[12]	✓		✓		✓	✓		✓	MILP

4   Proposed Solution Methodology

This section describes the methodology proposed to solve the HPRP, the objective is to reduce the costs of the use of the different vehicles, the cost of the driver and the environmental costs incurred in the emission of CO, CO_2 and NO_X, and thus reduce their emission. Initially, the method for measuring and quantifying emissions is defined, based on the MEET adapted from [25]. Then, the variant that includes a heterogeneous fleet with three different types of vehicles is detailed, based on [14], and discretized velocity and time windows, taking the study of [12]. Subsequently, the generation of the initial solution, the coding and decoding used, and the design of the Tabu Search are described.

4.1   Measurement of Emissions

The MEET pollutant measurement model is used [25]. Where vehicle speed, load and road slope are the main factors affecting pollutant emissions. This model is used to measure carbon dioxide (CO_2), carbon monoxide (CO) and nitrogen dioxide (NO_X). The model is selected for its adaptability in on-road measurements and for being suitable for vehicle types of less than 7.5 tons [20]. The Eq. (1) represents the total emissions (E) of CO_2, NO_X and CO [g] generated on a route. Where e_i is the emission factor type i [g/km] and D is the distance traveled [km] by vehicles on a route.

$$E = \sum_{i=1}^{3} e_i \cdot D \tag{1}$$

Equations (2), (3) and (4) express the relationship between the emission factors of CO_2, NO_X and CO, respectively. Where GC_i represents the road gradient correction coefficient (Eqs. (5), (6) and (7)) and LC_i represents the vehicle load coefficient correction between nodes (Eqs. (8), (9) and (10)). On the other hand, v indicates the vehicle speed $[km/h]$, γ the gradient of the road $[\%]$, finally $x \in [0,1]$ the ratio between the vehicle load and its capacity between nodes.

$$e_1 = \left(110 + 0.000375v^3 + \frac{8702}{v}\right) \cdot GC_1 \cdot LC_1 \tag{2}$$

$$e_2 = \left(0.508 + 3.87^{-6}v^3 + \frac{92.5}{v} - \frac{77.3}{v^2}\right) \cdot GC_2 \cdot LC_2 \tag{3}$$

$$e_3 = \left(1.5 - 0.0595v + 0.001119v^2 - 6.16^{-6}v^3 + \frac{58.8}{v}\right) \cdot GC_3 \cdot LC_3 \tag{4}$$

$$GC_1 = e^{\left((0.0059v^2 - 0.0775v + 11.936)\gamma\right)} \tag{5}$$

$$GC_2 = e^{\left((0.0062v^2 - 0.0427v + 11.301)\gamma\right)} \tag{6}$$

$$GC_3 = e^{\left((0.001v^2 - 0.0442v + 6.1207)\gamma\right)} \tag{7}$$

$$LC1 = (0.27)x + 1 + 0.0614\gamma x - 0.0011\gamma^3 x - 0.00235vx - \left(\frac{1.33}{v}\right)x \tag{8}$$

$$LC2 = (0.26)x + 1 + 0.0672\gamma x - 0.00117\gamma^3 x - 1.90^{-5}v^2 x - \left(\frac{1.6}{v}\right)x \tag{9}$$

$$LC3 = (0.09)x + 1 + 0.037\gamma x - 5.29^{-4}\gamma^3 x - 1.52^{-7}v^3 x \tag{10}$$

Parameters used are based on [13] y [30]. The monetary cost of the three pollutants is calculated based on [5], where one gram of emission is equivalent to 0.000027 dollars. The rate is used since Colombia does not have a procedure for this calculation [12].

4.2 Heterogeneous Fleet

This paper extends the heterogeneous fleet variant with limited vehicles, taking into account the fixed costs $[CF_k]$ in dollars for using k vehicles according to their Q_k capacity. The payment to drivers is considered, according to a standard $[Pt]$ payment of 3 $[\$UDS/hour]$ [25]. If a vehicle leaves the depot at time t_1 and returns to the depot at time t_2 the worker's pay will be equal to $Pt \cdot (t_2 - t_1)$. The fleet characteristics are detailed in Table 2.

Table 2. Heterogeneous fleet available.

Vehicle k	Empty weight [kg]	Capacity Q_k [kg]	Fixed cost CF_k [$]	Variable cost Pt $/h
1	4 100	2 218	60	3
2	5 200	2 986	81	3
3	7 500	4 782	129	3

4.3 Initial Solution

The initial routing solution is based on the idea of the Nearest Neighbor Algorithm (NNA) [9], used extensively in other research to construct the initial solution of the problem [16, 27, 31]. In this case, the NNA adapts the Least Pollution Neighbor (LPN) taking into account the distance and gradient variables. The less distance traveled and the less gradient the road has, the less pollution the vehicle will generate. The inputs to perform the VMC are the matrix of distances and gradients, travel time, service time and customer time windows.

Initially, it takes into account the number of available vehicles and performs a random assignment of their departure from the depot. Then the least polluting neighbor is assigned with respect to the previous node with a random speed, once a new assignment cannot be made due to capacity limitation or non-compliance with time sales, another vehicle is taken randomly. To find the nearest neighbor, the expression (11) must be calculated which normalizes the respective distance and gradient. Where D_{ij} represents the distance between nodes i and j, D_{max} is the maximum distance between all the nodes, γ_{ij} is the road gradient between nodes i and j, and γ_{max} is the maximum gradient. A weighting of 50% was used.

$$VMC_{ij} = \frac{D_{ij}}{D_{max}} \cdot 0.5 + \frac{\gamma_{ij}}{\gamma_{max}} \cdot 0.5 \qquad (11)$$

4.4 Coding and Decoding

The coding consists of three vectors: one for the routing priority between clients $vCus$, the second refers to the speed $vSpeed$ used in each arc of $vCus$, and the third represents the type of vehicles available $vVehi$. It is important to emphasize that $vVehi$ should not correspond to the size of $vCus$, but to the number of available vehicles. In Fig. 2, the three-vector encoding scheme is shown.

The decoding starts with the selection of a vehicle of the type that stores $vVehi$, then starts building the route with the corresponding $vSpeed$ and $vCus$ sequence, until the vehicle capacity is covered or else the impossibility to visit more customers due to time windows. Upon completion of a route, the distances and emissions of each component are calculated for the route, according to the methodology detailed in Sect. 4.1. Then the next $vVehi$ vehicle is chosen and the process is repeated. At the end, the total distances and emissions of the solution, and the logistic cost (cost of using the vehicles, drivers' pay and environmental cost) are calculated. If there are not enough vehicle types available, they are chosen randomly.

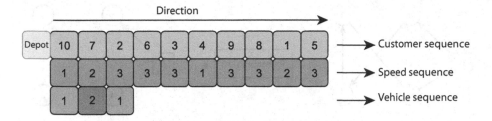

Fig. 2. Solution coding.

4.5 Tabu Search Proposed

Among the metaheuristics proposed for the PRP, Tabu Search (TS) has been shown to be very effective, because of its relationship between solution quality and computational time [26]. As such, Tabu Search is expected to be effective for the present PRP variant. Tabu Search proposed by Glover in 1989, basically performs a local search by generating neighborhood structures under a greedy criterion, the main feature is the use of short and long term memory to avoid converging only to local optima [11].

The main differential of the developed TS is the use of three decision levels, built to find the best possible combinations. Thus, three Tabu Lists (TL) that interact with the search simultaneously are also defined. One list for each neighborhood structure ($kmax = 3$). The first level considers the sequence of nodes to be visited by the vehicles, related to $vCus$; for this level it is proposed, with the intention of reducing the computational effort, to store the attributes of the movements performed as the nodes involved and their initial positions in the routes before moving to the next solution. The second level, considers the assigned speeds, level related to $vSpeed$, and the third level, takes care of the search among the assigned vehicles, related to $vVehi$. For the last two, the final sequence is stored for each solution. To measure its performance, fitness is defined as the sum of the costs of using the different types of vehicles, the costs of the drivers and the emissions released to the environment.

The TS explores in each iteration the neighboring solutions of a current solution S since they share a meaningful S structure. The neighbors are obtained by applying the different types of moves defined in the three decision levels N_k. Then, the solution representing the best neighbor for each decision level is found. The best solution will be chosen as the new current solution. The order of the neighborhoods does not matter, as all must be performed and compared to each other a total of iterations. Neighborhood structures adapted to the coding are described below:

- $N_{k=1}$: the first neighborhood structure, corresponding to the first decision level, contemplates the sequence of nodes to be visited by the vehicles. Each neighbor is found by means of the Exchange [1] movement. The basic idea is to randomly select two nodes from the sequence vector and exchange the position between them. An existing path can be modified or created. An example of this operator is shown in figure Fig. 3.

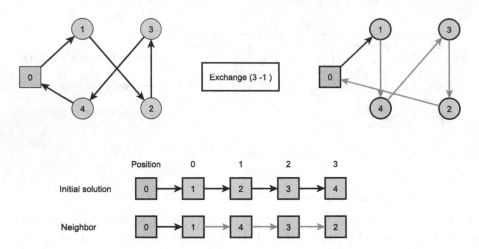

Fig. 3. Example of the exchange operator on the sequence of nodes.

- $N_{k=2}$: This second neighborhood structure is intended to alter the speeds on the routes. This is achieved by generating a perturbation movement in the assigned speeds, a 30% of the total number of customers in a current S solution is considered randomly. The perturbation consists of changing the speed of the affected node to a different one. Three speeds are available and, always, the random change happens on the two available speeds. The 30% random change is performed with the intention of generating new neighbors, but they must contain characteristics of the corresponding current solution.
- $N_{k=3}$: the third structure corresponds to the level related to the type of vehicle used. The operator uses a perturbation movement on the vehicles randomly, over a percentage equal to the previous structure to generate new neighbors. When generating the change of vehicles, it takes into account those available vehicles that have not been previously used in the solution and also, it considers the case of not having available vehicles of a specific type.

To prevent the search from returning to previously visited solutions and to direct the search to regions of the solution space not yet explored, the attributes of the perturbations generated from the three decision structures are declared as tabu and remain in the corresponding LT. Their permanence, called tabu tenure, is five occurrences unless the aspiration criterion is met. That is, if a tabu move, in any of the structures, improves the best solution found up to that point, the forbidden move must be accepted. Tabu Tenure (list length) is selected due to the successful work of several studies, among them [26].

Thus, a more efficient intensification search is generated, which succeeds in exploring areas of the solution space through a cycle of small periods. Algorithm 1, depicted the proposed pseudocode of the metaheuristic.

Algorithm 1. Proposed Tabu search pseudocode.

Input: Initial solution ($S0$)
Output: Best solution (S_{best})
1: **procedure** TS
2:     Initiation of LTs, $S_{best} = S0$
3:     **while** $i \leq Itera_{max}$ **do**
4:         **for** $k = 0, k \leq 3, k++$ **do**
5:             $Neighborhood_k = N_k$ (decision level k)
6:             $SN_k^* = $ best of $Neighborhood_k$
7:             Update Taboo lists
8:             $SA = \max_{k \in K} SN_k^*$
9:         **end for**
10:        **if** $SA >= S_{best}$ **then**
11:            $S_{best} = SA$
12:        **end if**
13:    **end while**
14:    **return** S_{best}
15: **end procedure**

Finally, as a stopping rule based on computational experimentation runtimes, an iteration counter is set and the limit $Itera_{max}$ is defined at 1 000 repetitions.

5   Results

To evaluate the proposed Tabu Search for solving the HPRP and discrete velocities, we adapt the instances proposed by [12] derived in turn from the instances of [17] (available at https://github.com/dmorill/HPRP_instnaces). In total there are 360 original instances, divided into 18 sets of 20 instances each. Because of the complexity of the model, this paper addresses 100 instances composed of the 10, 20, 50, 75, and 100 customer sets (each with 20 instances).

The proposed algorithm has been implemented in C++ language under Windows 10 pro. The computational tests are run on a computer with Intel Core i5 (2.30 GHz) processor with 8 GB of RAM. The results are summarized in Table 3, where each row represents a set of 20 instances with a defined number of clients and type of time windows. Each reported value represents the average found among the 20 instances in each set.

Table 3 shows that the proposed TS manages to obtain the total costs of the routes, composed of the costs of using the different types of vehicles, the payment to the drivers, and the total emissions with a minimum of 7 s for the smallest instances and 42 min in those instances of 100 nodes. While the time is acceptable, the performance of the operators in considering all possible neighbors makes the scaling of the algorithm not so attractive. To the best of our knowledge, no other methods applied to this specific problem, as presented in Table 1, have been found in the literature. Therefore, this would be the first approach that solves it according to the characteristics described in Sect. 2.

Table 3. Computational results of the TS in the selected set of instances.

Number of customers	Rute cost [$]	Total distance [km]	Emission factors [g/km]			Total emission [t]	Used vehicles			Execution time [s]
			CO_2	NO_x	CO		1	2	3	
10	298,3	715,1	5 179,5	48,4	23,9	1.1	4	2	1	7,4
20	552,9	1 315,6	10 249,3	96,3	47,2	2.4	4	3	0	36,66
50	1 276,1	2 656,6	22 607,7	211,1	111,7	4.5	7	6	2	373,7
75	2 009,9	4 228,5	31 217,5	288,5	163,3	6.6	10	9	3	1 126,7
100	2 682,3	5 444,6	42 458,7	391,9	221,2	8.7	12	12	7	2545,1

Also evident is the ratio of kilograms of GHG type per kilometer traveled. The CO_2 stands out, reaching 98.6% of the total gases measured. On the other hand, it can be seen that the use of 1 type vehicles is much higher in relation to the other types. The cause may be that they do not need to use all their capacity, given the restriction of the time windows, and must return to the depot.

On the other hand, the percentage of improvement with respect to the solution found by the S_{best} TS and the initial S_0 solution was quantified. For all sets of instances, an improvement from 85% to 152% was found. As for the total emissions [t], it is reduced on average for all instances by 249%. At the same time, it is possible to indicate that the number of vehicles used decreases significantly compared to the initial solution, as the number of customers increases. This is because the larger the customer sequence list, the greater the possible combinations that achieve the routing while decreasing the number of vehicles.

Likewise, it was observed among all the iterations performed, which ones found in higher proportion the best neighbor for each instance size. It was found that, as the number of clients increases, the neighborhood structure N_1 (Exchange movement) presents most of the best solutions found. For sizes of 100 customers, it achieves 98.9% of the best solutions. On the other hand, the neighborhood structure N_2 (Speed disturbance), manages to obtain better solutions only for cases of 10 and 20 nodes, with percentages of 23% and 2% respectively. As for the perturbation on vehicles (N_3), it presents at best 20% of the best solutions in small instances (10). The above behaviour is comparable with the execution times of each of the structures as N_1 takes much longer as the number of clients grows.

Finally, to compare the solution of the problem with a different approach, the reduction of distance instead of pollutants emitted is targeted. From this, it can be determined how the distance is reduced, on average for all instances, by 16%. However, CO_2, NO_X and CO suffer a considerable increase. Total emissions [t] increase up to four times more because of using vehicles with higher load and capacity, which manage to include a larger number of customers in their route and thus return to the depot using a shorter distance. This results in higher emissions of pollutants released.

6   Conclusions

In recent years, the VRP has evolved in many ways, thus requiring the inclusion of more and more realistic models. In this way, it contributes to respond to the need for actions that positively impact the preservation and sustainability of the environment in the distribution of goods in realistic environments. Thus, a novel approach to PRP is proposed, where the heterogeneous fleet, discrete speeds and time windows are considered in order to reduce the costs associated with the use of vehicles and the pollutant emissions released into the atmosphere; the latter calculated from the MEET methodology in the gases CO_2, NO_X and CO.

In terms of the resolution methodology, a Tabu Search metaheuristic is proposed that performs exploration over multiple neighborhood structures. These are used to improve both routing and assignment of vehicles and their speeds. The selection of neighborhoods is dynamically adjusted by several Tabu lists throughout the search to maintain the balance between exploration and exploitation, thus, incorporating good attributes to the created solutions. For validation, the set of instances referenced in the literature was used, in all cases the efficiency of the algorithm is appreciated in reference to pollution, logistic costs and vehicle usage. The results contribute to provide a solution that requires fewer vehicles to efficiently perform the same amount of work. This is important in urban logistics systems, as it contributes to reducing the presence of vehicles on city streets and, therefore, their negative impact on congestion and the environment.

In short, the improvement of the performance of the transportation logistics operation with an environmental approach needs further research and development. Therefore, it is proposed as future work to improve the algorithm by incorporating other neighborhood structures in the proposed decision levels to achieve more efficient results or to incorporate different metaheuristic layers creating hybrid algorithms, which have shown high performance in other routing problems. Also, to continue contributing to the organizations in the path of competitiveness and environmental responsibility.

References

1. Aarts, E., Lenstra, J. (eds.): Local Search in Combinatorial Optimization. Wiley-Interscience Series in Discrete Mathematics and Optimization. Wiley-Interscience (1997)
2. Abdi, A., Abdi, A., Akbarpour, N., Amiri, A.S., Hajiaghaei-Keshteli, M.: Innovative approaches to design and address green supply chain network with simultaneous pick-up and split delivery. J. Clean. Prod. **250**, 119437 (2020). https://doi.org/10.1016/j.jclepro.2019.119437
3. Arboleda Zúñiga, J., Gaviria-Gómez, J.A., Álvarez-Romero, J.A.: Propuesta de ruteo de vehículos con flota heterogénea y ventanas de tiempo (HFVRPTW) aplicada a una comercializadora pyme de la ciudad de Cali. Revista de Investigación **11**(1), 39–55 (2018). https://doi.org/10.29097/2011-639x.178
4. Behnke, M., Kirschstein, T.: The impact of path selection on GHG emissions in city logistics. Transp. Res. Part E: Logist. Transp. Rev. **106**, 320–336 (2017). https://doi.org/10.1016/j.tre.2017.08.011

5. Bektaş, T., Laporte, G.: The pollution-routing problem. Transp. Res. Part B: Methodol. **45**(8), 1232–1250 (2011). https://doi.org/10.1016/j.trb.2011.02.004
6. El Departamento Nacional de Planeación (DNP), Colombia: Encuesta Nacional Logística. Technical report (2020)
7. Ministerio de minas y energía, C.: Informe de gestión 2016–2017. Technical report (2017)
8. Erdogan, S., Miller-Hooks, E.: A green vehicle routing problem. Transp. Res. Part E: Logist. Transp. Rev. **48**(1), 100–114 (2012). https://doi.org/10.1016/j.tre.2011.08.001
9. Flood, M.M.: The traveling-salesman problem. Oper. Res. **4**(1), 61–75 (1956). https://EconPapers.repec.org/RePEc:inm:oropre:v:4:y:1956:i:1:p:61-75
10. Fukasawa, R., He, Q., Song, Y.: A disjunctive convex programming approach to the pollution-routing problem. Transp. Res. Part B: Methodol. **94**, 61–79 (2016). https://doi.org/10.1016/j.trb.2016.09.006
11. Glover, F.: Tabu search-part I. ORSA J. Comput. **1**(3), 190–206 (1989). https://doi.org/10.1287/ijoc.1.3.190
12. Gutiérrez-Padilla, M.V., Morillo-Torres, D., Gatica, G.: A novel mathematical model for a discrete speed pollution routing problem with time windows in a Colombian context. IFAC-PapersOnLine **54**(1), 229–235 (2021). https://doi.org/10.1016/j.ifacol.2021.08.027
13. Hitckman: Project Report Se/491/98 Methodology for Calculating Transport Emissions (1999)
14. Hsueh, C.F.: A vehicle routing problem with consideration of green transportation. J. Manag. Sustain. **7**(4), 89 (2017). https://doi.org/10.5539/jms.v7n4p89
15. Instituto de Hidrología, Meteorología y Estudios Ambientales (IDEAM) de Colombia: Tercera comunicación nacional de Colombia - A la convención marco de las Naciones Unidas sobre el cambio climático. Technical report (2017)
16. Karakostas, P., Sifaleras, A., Georgiadis, M.C.: Adaptive variable neighborhood search solution methods for the fleet size and mix pollution location-inventory-routing problem. Expert Syst. Appl. **153** (2020). https://doi.org/10.1016/j.eswa.2020.113444
17. Kramer, R., Subramanian, A., Vidal, T., Cabral, L.D.A.F.: A matheuristic approach for the pollution-routing problem. Eur. J. Oper. Res. **243**(2), 523–539 (2015). https://doi.org/10.1016/j.ejor.2014.12.009
18. Lenstra, J.K., Kan, A.H.G.R.: Complexity of vehicle routing and scheduling problems. Networks **11**(2), 221–227 (1981). https://doi.org/10.1002/net.3230110211
19. Lin, C., Choy, K., Ho, G., Chung, S., Lam, H.: Survey of green vehicle routing problem: past and future trends. Expert Syst. Appl. **41**(4, Part 1), 1118–1138 (2014). https://doi.org/10.1016/j.eswa.2013.07.107. https://www.sciencedirect.com/science/article/pii/S095741741300609X
20. Liu, C., Kou, G., Zhou, X., Peng, Y., Sheng, H., Alsaadi, F.E.: Time-dependent vehicle routing problem with time windows of city logistics with a congestion avoidance approach. Knowl.-Based Syst. **188** (2021) (2020). https://doi.org/10.1016/j.knosys.2019.06.021
21. Matsumoto, R., Umezawa, N., Karaushi, M., Yonemochi, S.I., Sakamoto, K.: Comparison of ammonium deposition flux at roadside and at an agricultural area for long-term monitoring: emission of ammonia from vehicles. Water Air Soil Pollut. **173**(1–4), 355–371 (2006). https://doi.org/10.1007/s11270-006-9088-z
22. Ministero de Transporte de Colombia: Ley 1239 (2008)

23. Molina, J.C., Salmeron, J.L., Eguia, I., Racero, J.: The heterogeneous vehicle routing problem with time windows and a limited number of resources. Eng. Appl. Artif. Intell. **94**, 103745 (2020). https://doi.org/10.1016/j.engappai.2020.103745
24. Moryadee, C., Aunyawong, W., Shaharudin, M.R.: Congestion and pollution, vehicle routing problem of a logistics provider in Thailand. Open Transp. J. **13**(1), 203–212 (2019). https://doi.org/10.2174/1874447801913010203
25. Naderipour, M., Alinaghian, M.: Measurement, evaluation and minimization of CO_2, NOx, and CO emissions in the open time dependent vehicle routing problem. Measur.: J. Int. Measur. Confederation **90**(x), 443–452 (2016). https://doi.org/10.1016/j.measurement.2016.04.043
26. Nguyen, P.K., Crainic, T.G., Toulouse, M.: A tabu search for time-dependent multizone multi-trip vehicle routing problem with time windows. Eur. J. Oper. Res. **231**(1), 43–56 (2013). https://doi.org/10.1016/j.ejor.2013.05.026
27. Normasari, N.M.E., Yu, V.F., Bachtiyar, C., Sukoyo: A simulated annealing heuristic for the capacitated green vehicle routing problem. Math. Probl. Eng. **2019** (2019). https://doi.org/10.1155/2019/2358258
28. Poonthalir, G., Nadarajan, R.: A fuel efficient green vehicle routing problem with varying speed constraint (F-GVRP). Expert Syst. Appl. **100**, 131–144 (2018). https://doi.org/10.1016/j.eswa.2018.01.052
29. Puenayán, D.E., Londoño, J.C., Escobar, J.W., Linfati, R.: Un algoritmo basado en búsqueda tabú granular para la solución de un problema de ruteo de vehículos considerando flota heterogénea. Revista Ingenierías Universidad de Medellín **13**(25), 81–98 (2014). https://doi.org/10.22395/rium.v13n25a6
30. Ren, X., Huang, H., Feng, S., Liang, G.: An improved variable neighborhood search for bi-objective mixed-energy fleet vehicle routing problem. J. Clean. Prod. **275**, 124155 (2020). https://doi.org/10.1016/j.jclepro.2020.124155
31. Toth, P., Vigo, D.: Vehicle Routing Probelms, Methods, and Applications (2014)
32. United States Environmental Protection Agency (EPA): Inventory of U.S. Greenhpuse Gas Emissions and Sinks: 1990–2020. Technical report (2022)
33. Xu, Z., Elomri, A., Pokharel, S., Mutlu, F.: A model for capacitated green vehicle routing problem with the time-varying vehicle speed and soft time windows. Comput. Ind. Eng. **137**, 106011 (2019). https://doi.org/10.1016/j.cie.2019.106011
34. Zhang, W., Gajpal, Y., Appadoo, S.S., Wei, Q.: Multi-depot green vehicle routing problem to minimize carbon emissions. Sustainability (Switzerland) **12**(8), 1–19 (2020). https://doi.org/10.3390/SU12083500

A Learning Metaheuristic Algorithm
for a Scheduling Application

Nazgol Niroumandrad[✉], Nadia Lahrichi, and Andrea Lodi

CIRRELT and Department of Mathematics and Industrial Engineering,
Polytechnique Montréal, CP6079 Succursale Centre-ville, Montréal,
QC H3C 3A7, Canada
nazgol.niroumandrad@polymtl.ca

Abstract. Tabu Search is among one of the metaheuristic algorithms
that are widely recognized as efficient approaches to solve many combinatorial problems. Studies to improve the performance of metaheuristics
have increasingly relied on the use of various methods, either combining
different metaheuristics or originating outside of the metaheuristic field.
This paper presents a learning algorithm to improve the performance
of tabu search by reducing its search space and the evaluation effort.
We study its performance using classification methods in an attempt to
select moves through the search space more intelligently. The experimental results demonstrate the benefit of using a learning mechanism under
deterministic environment and with uncertainty conditions.

Keywords: Learning tabu search · Combinatorial problems · Logistic
regression

1 Introduction

Metaheuristics are popular and demanded optimization methods for many hard
optimization problems because they are flexible and can be adapted to complex
applications. These algorithms were first introduced by [1] and refer to approximate algorithms that obtain near optimal solutions for combinatorial or integer
programming problems.

Metaheuristics generate a lot of dynamic data during the iterative search process. However, they do not use explicit knowledge discovered during the search.
The main idea of this paper revolves around using the data generated during the
search process to show that metaheuristics, and more specifically tabu search,
can behave more intelligently and efficiently. Using advanced machine learning
(ML) models can be helpful to extract valuable knowledge that will guide and
enhance the search performance to move smarter through the search space. Having a better exploration of the search space and exploring solutions that might
not be reachable without these learning algorithms leads to optimizing computation time in complex problems.

L. Di Gaspero et al. (Eds.): MIC 2022, LNCS 13838, pp. 76–87, 2023.
https://doi.org/10.1007/978-3-031-26504-4_6

Since tabu search (TS) has been successful in solving many hard optimization problems, we chose to study the effects of learning methods on the performance of a tabu search algorithm. We are proposing a novel learning tabu search algorithm using logistic regression in the context of a physician scheduling problem. To the best of our knowledge, there is no comprehensive study on integrating ML techniques into TS to explore the search space. Most studies on learning metaheuristics evolve in the context of clustering or intensification and diversification strategies.

Although there are a number of studies on improving the performance of TS, the literature lacks a comprehensive study on how ML techniques can be integrated into TS to enhance its search procedure. This paper provides an intensive study on the use of ML techniques in the design of TS, which does not rely on a simple diversification/intensification strategy to improve the search procedure. We believe this paper is beneficial for both academic and industry experts engaged in solving hard combinatorial problems. Our proposed method is used to study a scheduling problem, and our results are compared with those of [2]. Since the original tabu search presented in [2] proved to be very efficient and optimized under deterministic conditions, this paper focuses on improving, by learning, the performance of the TS with uncertainty conditions.

The rest of the paper is organized as follows. In Sect. 2, we review the related studies. In Sect. 3, we explain the problem along with the proposed method in detail. We describe the instance generation and present the results in Sect. 4. A discussion on the results and different testing strategies is presented in Sect. 5 and Sect. 6 provides concluding remarks.

2   Related Literature

The study of learning within metaheuristics has not been given the attention it deserves. However, benefiting from machine learning (ML) techniques to solve combinatorial problems received a lot of attention recently. More precisely, [3] and [4] demonstrated that employing ML during the search process can improve the performance of heuristic algorithms. [5] described algorithms that improve the search performance by learning an evaluation function that predicts the outcome of a local search algorithm from features of states visited during the search. Other studies on learning evaluation functions are presented in [6,7] and [8]. Along the same subject, [9] is another example of a recent study on learning metaheuristics. This study proposed a learning variable neighborhood search (LVNS) that identifies quality features simultaneously. This information is then used to guide the search towards promising areas of the solution space. The LVNS learning mechanism relies on a set of trails, where the algorithm measures the quality of the solutions.

ML was also employed to prune the search space of large-scale optimization problems by developing pre-processing techniques [10,11]. Some other studies used ML-based methods to directly predict a high-quality solution [12,13]. Moreover, [14] and [15] provided a review of studies where metaheuristic algorithms benefited from ML and the potential future work.

Building upon these previous studies, we propose a learning tabu search algorithm enhanced with a logistic regression method to guide the search through the solution space of hard combinatorial problems. We observe that the emphasis on adaptive memory within tabu search represents the nature of its learning mechanism. However, most previous works focused on clustering [1] or intensification/diversification [16] strategies. In metaheuristics, intensification and diversification strategies play important roles in the quality of solutions. In a recent study, [17] presented a relaxation-adaptive memory programming algorithm on a resource-constrained scheduling problem. In that approach, primal-dual relationships help to effectively explore the interplay between intensification, diversification, and learning (IDL). The algorithm is designed to integrate the current most effective Lagrangian-based heuristic with a simple tabu search. The authors aimed to present a study on the IDL relationship when dual information is added to the search. In [18], the authors presented a learning tabu search algorithm for a truck-allocation problem in which they considered a trail system (in the ant colony optimization, a trail system is inspired from the pheromone trails of ants to mark a path) for the combination of important characteristics. In this study, a diversification mechanism is introduced to help visit new solution space regions. The authors proposed diversifying the search by performing "good" moves that were not often performed in the previous cycles.

There are also numerous studies found in the literature on improving the performance of tabu search. In particular, [19] emphasized selecting particular attributes of solutions and determining conditions that help to find the prohibited moves, in order to produce high-quality solutions. Following that work, [20] and [21] employed a balance among more commonly used attributes. The presented computational experiments showed that considering these attributes can significantly outperform all other methods. These outcomes underpin researchers' ongoing strategy of identifying attributes that lead to more effective methods. However, to the best of our knowledge, our study in favor of learning the characteristics of the search space during the tabu search algorithm's search procedure is a novel contribution to the literature. We aim to use a classification model to fill this gap. Our contribution focuses on how ML helps to learn the best neighborhoods to build or change a solution during the search process.

3   Problem Statement and Proposed Learning Algorithm

In a nutshell, tabu search [22] is an iterative procedure that starts from an initial solution x_0 (possibly infeasible). From each current solution x ($x = x_0$ at the beginning of the procedure), it moves to a neighbor solution x'. The neighborhood is defined by all solutions that can be reached from $x \in X$, where X is the solution space, after applying a specific move. Let us denote this neighborhood by $N(x)$. The next solution x' is the best non-tabu solution in the neighborhood $N(x)$ (an exception is made if a move is tabu but it improves upon the current best solution x^*, i.e., through the so-called aspiration criteria).

The function $f(x)$ is defined to evaluate each solution. To prevent cycling, a tabu list (T_{list}) containing attributes of recently visited solutions (or attributes of moves) is maintained. The associated solutions cannot be revisited for a specific number of iterations. Various strategies can be applied to search the neighborhood solutions. The moves and strategies to search the solution space are the main ingredients of tabu search methods. Hence, the adequate combination and sequence have a great impact on the quality of the results. Tabu search is used in multiple applications and is adapted to handle uncertainty. This is commonly done by considering several scenarios, typically generated using historical data or a probability distribution. Instead of moving to the best solution in the neighborhood (deterministic environment), the best solution in average (with uncertainty conditions) is preferred.

A naïve approach to choose a solution x' is to evaluate and analyze all possibilities, which would be computationally expensive. Instead, we can characterize the neighborhood using experiments. In particular, the guiding principle of our investigation is that using a learning method can help us find good solutions faster. ML techniques allow us to extract knowledge from good solutions and use it to generate even better solutions. This knowledge can be in the form of a set of rules or patterns [23]. Table 1 shows how applying a learning method during the search process to reduce the space in a metaheuristic algorithm has computational impact. This table presents an example where we have $|P|$ number of physicians, $|M|$ number of patients and $|K|$ number of time blocks in a physician scheduling problem. However, this can be generalized to other problems.

Table 1. Examples of computational impact of reduction of search space

Size of the problem	Complexity of the problem without learning	Predicting element	Complexity of the problem with learning
$|P| \times |M| \times |K|$	$O(n^3)$	$p \in P$	$O(n^2)$
		$p \in P \ \& \ m \in M$	$O(n)$
		$p \in P \ \& \ m \in M \ \& \ k \in K$	$O(1)$

We studied the impact of deploying a learning procedure to answer this situation. We chose logistic regression for this purpose, as it is one of the most important models that can be applied to analyze categorical data. The learning procedure is employed in two phases of training and application.

- Training phase: this phase is divided in stages T_1 and T_2. In T_1, the original TS collects data related to the structure of the search space. Once enough data is collected, training begins with the logistic regression model for a specific number of iterations (T_2). Training starts at iteration I_{st} and ends at I_{end};
- Application phase: in this phase, an action will be taken based on the prediction of the trained model and after validating the elements of a tabu list (T_{list}).

The learning methods are very sensitive to the input information. Thus, extracting the features that have the greatest impact on the solution space is the first and most important step. These features represent important characteristics of the solutions space and moves. Table 2 presents these features in different categories.

Table 2. Input/output features for the training model

	Input		Output
Category	Features	Format	Label
Cost improvement	Δ_x	$\mathbb{R}$	Observed Output
	$\Delta_x > 0$	$\mathbb{B}$	
Tabu	$x_i \in T_{list}$	$\mathbb{B}$	
	$|t_i| = T_{list}(x)$	$\mathbb{Z}$	
	$Freq_{-ID}$	$\mathbb{Z}$	
Solution	x	$M_{\mathbb{Z}}$	
	D^*	$M_{\mathbb{B}}$	
	D'	$M_{\mathbb{B}}$	
Move	Attractiveness	$\mathbb{Q}$	
	Trail of the moves	$M_{\mathbb{Q}}$	

Let x be the current solution, x^* the best known solution, and x' the next solution. Each solution is represented as a matrix of integers. In the category of cost improvement, we are first considering the improvement in cost, which is presented as $\Delta_x = f(x') - f(x)$. The first feature is the value of Δ_x and the second reflects if the solution is improved ($\Delta_x > 0$ in the context of maximizing). In the tabu category, we have a binary variable that determines if the accepted move belongs to the T_{list} and whether it has already been visited along with the tabu value for the move, meaning when it will be free and can be considered again. The frequency of the move is also considered, which indicates how many times the move has been visited so far. The next category represents the characteristics of solution. First, the solution x itself. A binary matrix (D^*) denotes the difference between the obtained solution and the last best known solution. A value 1 indicates if the corresponding entry is equal, 0 otherwise. Also, a binary matrix (D') denotes the difference between the obtained solution and the previous solution, with the same meaning for 1's and 0's. The final category is related to the move characteristics, i.e., the attractiveness of the move and a trail matrix of the moves. These features were inspired by the recent work of [9]. Here, a trail system influences the decisions made by the ants in the Ant Algorithms, and the notion of move attractiveness shows that moves with high attractiveness values have a higher chance to be performed.

As tabu search can be used in situations with uncertainty conditions, the set of input features under these conditions is modified by considering the average of $f(x)$ over all scenarios at each iteration. In other words, we need to find the move that is the best in average over all the scenarios. Thus, our objective is to find the solution $x' \in N_v(x)$ that minimizes the average solution over all scenarios $\left(\frac{\sum_w f_w(x'_w)}{|W|} \right)$.

The learning tabu search (L-TS) model differs from the original TS mostly in the search space. We seek to reduce the number of evaluations in the search process. Thus, we predict the subset of promising moves and evaluate only those neighbors instead of evaluating every possible neighbor of $N(x)$.

The L-TS algorithm starts with the original TS to collect the necessary data for I_{st}^D iterations. We train the learning method for a specific number of iterations (stage T_2), evaluate the result of the learning algorithm, and update the set of input features to encourage the method to search more promising regions. In the application phase, we use the last trained model at iteration $I_{end}^D - 1$ to identify (by prediction) promising moves to build $N'(x) \subset N(x)$ (note that superscript D stands for deterministic and S will be used for the case with uncertainty conditions). The total procedure ends when the stopping criterion ($Stop_{max}$) is reached. Our criterion is 1h of CPU time. Details of the parameter initialization step are documented in Sect. 4.

4   Experiments

In this section, we show the advantage of using a learning algorithm during the TS procedure for both deterministic environment and with uncertainty conditions.

We compare the results with a benchmark previously published in [2] for both deterministic environment and an environment with uncertainty in which the performance was compared with CPLEX. We refer to this previous work as "original TS" in the remainder of the paper. In [2], the authors studied a tabu search algorithm in a physician scheduling problem. The goal is to find a weekly cyclic schedule for physicians in a radiotherapy department and to assign the arriving patients to the best possible available specialist for their cancer type. In most radiotherapy centers, the physician schedule is task based. Each day is divided into one or multiple periods, and each period is dedicated to a single task. The goal is to minimize the duration of the pre-treatment phase for patients. This is defined as the time from the patient's arrival day to the day the final task is finished before the treatment starts. Taking inspiration from this work, we use 21 generated pseudo-real instances where we vary the number of new patients arriving each week and the number of available physicians. The number of physicians varies from 6 to 10 and the number of patients from 7 to 60, ranging from small instances to real-world applications. Each instance is labeled $pr - (\#$ of physicians, # of patients). In the deterministic case, we select one scenario to obtain a typical schedule, and in the situation with uncertainty, we consider a subset of W for different scenarios. We refer the readers to [2] for more details on instance generation.

We report and analyze results in deterministic and uncertain environments. All results were compared and evaluated with respect to the original TS method and a random approach (in which $N'(x) \subset N(x)$ was randomly chosen) to test the performance of the L-TS method. Comparing these three approaches helps us to see the performance of each and confirm the advantage of choosing TS over the random approach and of choosing L-TS over TS. It shows that we are learning during the process, and the results are not achieved by chance.

First, we wish to validate our L-TS algorithm and determine the value of the parameters, i.e., the size of the T_{list}, the number of iterations in each neighborhood, and the iteration to start the training phase and application phase. The values tested are all related to the size of the instances (i.e., number of patients, number of physicians, and number of time blocks). For the deterministic case, we use $I_1^D = 1$, $I_2^D = 2|J| + |I| + |5n|$, $I_3^D = 1$, $I_{st}^D = 3\sqrt{|J| \times |I| \times |D|}$, $I_{end}^D = 2|I_{st}^D|$, $Stop_{max} = 1h$, and we set $\theta^D = 2|J| + |I| + |D|$.

To evaluate the solution obtained from the learning tabu search algorithm under uncertainty conditions, we proceed as follows:

– Generate a set A of scenarios (up to 50 different scenarios);
– For each instance (i.e., pr-(6,7) to pr-(10,60)), run the algorithm using 10 or 50 scenarios from set A (using one scenario is equivalent to the deterministic case).

To evaluate the performance of the learning algorithm with uncertainty conditions, the values of the parameters I_1^S, I_3^S and θ^S are the same, except that I_2^S is now equal to $|J|$, the number of patients, $I_{st}^S = \sqrt{|J| \times |I| \times |D|}$, and $Stop_{max} = 5h$.

Table 3. Comparing the performance of different methods

	Tests	Deterministic case				Uncertainty case							
		GAP - Best (%)		GAP - Avg (%)		GAP - Best (%)				GAP - Avg (%)			
						10 Scenarios		50 Scenarios		10 Scenarios		50 Scenarios	
		Random	L-TS	Random	L-TS	Random	L-TS	Random	L-TS	Random	L-TS	Random	L-TS
Small	pr-(6,7)	0.5	−1.0	2.3	−0.7	1.3	−0.3	1.0	−0.7	1.1	0.2	0.9	−1.1
	pr-(6,9)	1.0	0.8	1.7	−1.4	0.8	0.0	1.5	0.3	0.9	−1.2	0.5	−2.2
	pr-(6,11)	2.4	0.3	2.1	−0.2	2.2	−0.3	1.1	−1.3	1.0	−1.1	2.6	−2.6
	pr-(6,12)	2.9	0.5	3.1	0.3	0.8	−1.4	0.8	−1.6	0.3	−1.3	1.6	1.3
	pr-(8,7)	2.7	−0.7	2.4	−0.1	1.3	−0.5	−0.4	−1.1	0.4	−1.0	2.1	−1.3
	pr-(8,9)	2.2	−0.4	2.5	−0.2	1.7	−0.4	1.4	−0.4	1.6	−0.6	0.7	−1.4
	pr-(10,7)	1.5	0.1	2.3	0.1	1.0	−0.7	1.0	0.1	1.3	−0.4	0.7	−0.9
Medium	pr-(6,20)	1.2	0.9	3.6	1.3	0.7	−1.3	2.7	0.6	1.6	−1.4	0.5	0.6
	pr-(8,11)	3.0	0.6	2.9	0.4	1.0	−0.6	0.9	−0.6	1.3	−1.3	0.3	−1.6
	pr-(8,12)	1.5	0.0	2.3	0.0	1.2	−0.8	0.0	−1.5	0.6	−0.8	0.0	−2.0
	pr-(8,20)	3.3	0.8	3.7	−0.1	2.4	−0.7	1.8	−0.4	−1.3	−1.4	1.7	−2.1
	pr-(10,9)	2.3	0.3	2.1	0.1	1.5	−0.1	1.0	−0.4	1.5	−0.4	0.3	−1.6
	pr-(10,11)	1.7	−0.3	2.2	0.2	0.4	−1.0	0.9	0.3	1.5	−1.0	−0.4	−1.9
	pr-(10,12)	1.5	0.1	3.0	0.4	1.3	−0.3	0.4	−0.4	1.2	−0.3	0.1	−2.2
Large	pr-(6,40)	2.6	−4.0	4.5	−0.4	5.2	4.6	−1.3	−3.5	2.2	1.9	4.3	−0.8
	pr-(6,60)	9.4	−0.7	5.8	−3.6	−10.7	−25.0	3.8	−1.5	16.3	−44.7	0.1	−3.2
	pr-(8,40)	2.3	−0.5	4.4	1.7	1.3	−0.6	0.8	0.0	2.8	−0.4	0.5	−2.4
	pr-(8,60)	8.2	1.3	8.5	3.3	0.0	−3.2	1.7	−0.3	3.3	−4.8	0.1	−3.5
	pr-(10,20)	3.9	0.2	4.1	0.4	1.1	−0.7	1.5	0.0	1.2	−1.0	0.2	−4.6
	pr-(10,40)	3.4	0.5	4.4	0.1	0.2	−2.1	0.9	−0.4	0.4	−1.0	0.1	−1.5
	pr-(10,60)	1.5	0.4	6.2	2.2	2.0	0.3	0.9	0.4	1.5	4.3	2.8	−1.9
	Average:	2.81	−0.03	3.53	0.18	0.79	−1.67	1.07	−0.59	1.93	−2.75	0.94	−1.75

Table 3 compares the cost (same definition as in [2]) values of different methods and the gap columns show improvements with respect to the original TS. In this table, "GAP - Best" compares the best solution obtained from ten different runs of each method and represents the improvements from the best solution obtained from the original TS. Conversely, "GAP - Avg" represents the average values from ten different runs. A negative value in the GAP columns indicates that the learning tabu search has improved the solution on average. It can be observed that logistic regression succeeded in slightly improving the cost, by 0.03% on average. We see more improvement in large instances where the algorithm has more flexibility.

It can be observed that the learning method improved tabu search in uncertainty case globally. We can see the most improvement in cases using 10 scenarios, with -1.67% on average for L-TS.

The value of the cost function alone cannot represent the advantage of using each approach. Hence, we measured the primal integral value for all methods to compare the progress of the primal bound's convergence towards the best-known solution over the entire solving time. Figure 1 illustrates this measure for all instances.

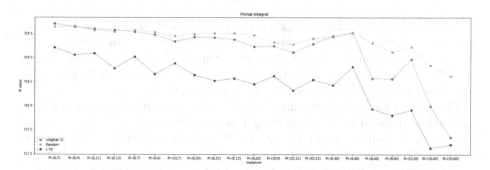

Fig. 1. Comparing the convergence speed for different methods in the deterministic case

The idea of the primal integral [24] is that the smaller the primal integral value is, the better the expected quality of the solution will be if we stop the solver at an arbitrary point in time. It can be observed from Fig. 1 that the logistic regression method has better primal integral values than the other methods for all instances. Again, this figure shows that the learning method, presented by the green line, performs better than the original TS and Random algorithms. Figure 2 compares the number of evaluations at each iteration, the total number of evaluations, and the total number of iterations until we reach the stopping criterion for different methods. It clearly shows a decrease in the number of calculations in the L-TS. We can also observe that the random method performs same number of evaluations but requires more iterations (compared with L-TS) to find the solution, due to its poor performance. This behavior was expected from the random approach since it is evaluating $N'(x)$, a subset of $N(x)$, randomly.

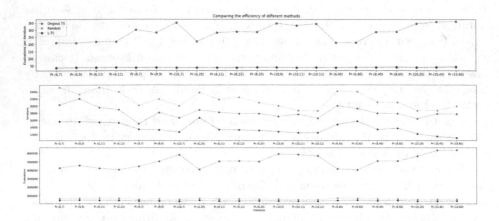

Fig. 2. Comparing computing performance for all methods in the deterministic case

All of these illustrations, including the improvement in the objective value presented in Table 3, show the advantage of employing the learning TS idea. More precisely, given the similar computing time and number of iterations to reach the best solution, as well as the improvement in the cost value, the logistic regression method demonstrates superior performance.

The performance of the learning algorithm in an uncertain environment was also evaluated based on its computing time. Figure 3 compares the time per iteration by method for 10 and 50 scenarios. It is clear that increasing the number of scenarios increases the computing time.

We observe that with 10 scenarios, the average gap improved by 1.67% with the logistic regression model. Also, with 50 scenarios, we have 0.59% improvements in average gap for L-TS model.

5   Discussion

The proposed L-TS algorithm show improvements in both deterministic and uncertain environments. The performance of this algorithm was evaluated with extensive number of experiments and this paper presents part of these experiments to demonstrate the advantage of employing a learning mechanism within TS.

This study was initially started by applying more sophisticated methods, such as classification, decision trees and neural networks, to predict the promising neighbors. However, the methods were time consuming and might not be applicable to large combinatorial and real case problems. Sometimes, a simpler

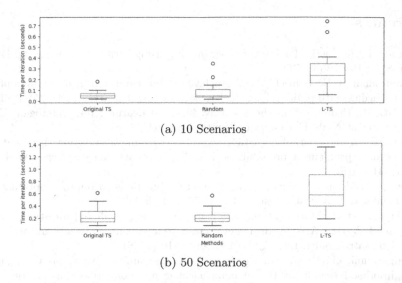

(a) 10 Scenarios

(b) 50 Scenarios

Fig. 3. Comparing computing time per iteration for all methods with uncertainty

method is able to achieve same results with less computational efforts. The performance of these methods was also evaluated through different configurations; in most cases, the logistic regression model outperforms the other ones.

Additionally, in the case with uncertainty conditions, we performed tests where we varied the number of scenarios from 10 to 50. We observed no significant impact with 30 scenarios. We also performed a sensitivity analysis to set the parameters, e.g. $Stop_{max} = 2.5, 5, 7.5$ and $10\,\mathrm{h}$.

6 Conclusion

In this paper, we proposed a learning tabu search method and studied its performance on a physician scheduling problem. The performance of the proposed algorithm was evaluated using the benchmark instances.

We evaluated the new method in both deterministic and uncertain environments. We showed that our method is very efficient compared with the original tabu search and a random method, especially in the case with uncertainty conditions where there is more space to learn. Over 21 instances, the average gap improved by 1.67% with logistic regression in the case with 10 scenarios. The learning method obtained best solutions faster than the original TS and random methods over the computing time. Although we studied the application of this method in a scheduling problem, tabu search has already been used to solve pretty much all optimization problems. Thus, the learning tabu search algorithm can be adapted to other applications. Future work could employ learning tabu search to solve other optimization problems by generalizing several ingredients for which we gave special attention to our specific application.

References

1. Glover, F.: Heuristics for integer programming using surrogate constraints. Decis. Sci. **8**(1), 156–166 (1977)
2. Niroumandrad, N., Lahrichi, N.: A stochastic tabu search algorithm to align physician schedule with patient flow. Health Care Manag. Sci. **21**(2), 244–258 (2018)
3. Battiti, R., Brunato, M.: The lion way. Machine Learning plus Intelligent Optimization. LIONlab, University of Trento, Italy, 94 (2014)
4. Hafiz, F., Abdennour, A.: Particle swarm algorithm variants for the quadratic assignment problems-a probabilistic learning approach. Expert Syst. Appl. **44**, 413–431 (2016)
5. Boyan, J., Moore, A.W.: Learning evaluation functions to improve optimization by local search. J. Mach. Learn. Res. **1**, 77–112 (2000)
6. Baluja, S., et al.: Statistical machine learning for large-scale optimization (2000)
7. Boyan, J., Moore, A.W.: Learning evaluation functions for global optimization and Boolean satisfiability. In: AAAI/IAAI, pp. 3–10 (1998)
8. Bongiovanni, C., Kaspi, M., Cordeau, J.-F., Geroliminis, N.: A predictive large neighborhood search for the dynamic electric autonomous dial-a-ride problem. Technical report (2020)
9. Thevenin, S., Zufferey, N.: Learning variable neighborhood search for a scheduling problem with time windows and rejections. Discret. Appl. Math. **261**, 344–353 (2019)
10. Sun, Y., Li, X., Ernst, A.: Using statistical measures and machine learning for graph reduction to solve maximum weight clique problems. IEEE Trans. Pattern Anal. Mach. Intell. **43**, 1746–1760 (2019)
11. Lauri, J., Dutta, S.: Fine-grained search space classification for hard enumeration variants of subset problems. In: Proceedings of the AAAI Conference on Artificial Intelligence, vol. 33, pp. 2314–2321 (2019)
12. Abbasi, B., Babaei, T., Hosseinifard, Z., Smith-Miles, K., Dehghani, M.: Predicting solutions of large-scale optimization problems via machine learning: a case study in blood supply chain management. Comput. Oper. Res. **119**, 104941 (2020)
13. Fischetti, M., Fraccaro, M.: Machine learning meets mathematical optimization to predict the optimal production of offshore wind parks. Comput. Oper. Res. **106**, 289–297 (2019)
14. Karimi-Mamaghan, M., Mohammadi, M., Meyer, P., Karimi-Mamaghan, A.M., Talbi, E.-G.: Machine learning at the service of meta-heuristics for solving combinatorial optimization problems: a state-of-the-art. Eur. J. Oper. Res. **296**, 393–422 (2021)
15. Talbi, E.-G.: Machine learning into metaheuristics: a survey and taxonomy of data-driven metaheuristics (2020)
16. Glover, F., Hao, J.-K.: Diversification-based learning in computing and optimization. J. Heuristics **25**(4–5), 521–537 (2019)
17. Christopher, R., Riley, L., Rego, C.: Intensification, diversification, and learning via relaxation adaptive memory programming: a case study on resource constrained project scheduling. J. Heuristics **25**(4–5), 793–807 (2019)
18. Schindl, D., Zufferey, N.: A learning tabu search for a truck allocation problem with linear and nonlinear cost components. Nav. Res. Logist. (NRL) **62**(1), 32–45 (2015)
19. Glover, F., Laguna, M.: Tabu Search. Wiley, Hoboken (1993)

20. Wang, Y., Qinghua, W., Glover, F.: Effective metaheuristic algorithms for the minimum differential dispersion problem. Eur. J. Oper. Res. **258**(3), 829–843 (2017)
21. Wu, Q., Wang, Y., Glover, F.: Advanced algorithms for bipartite Boolean quadratic programs guided by tabu search, strategic oscillation and path relinking (2017)
22. Glover, F.: Tabu search: a tutorial. Interfaces **20**(4), 74–94 (1990)
23. Arnold, F., Santana, Í., Sörensen, K., Vidal, T.: PILS: exploring high-order neighborhoods by pattern mining and injection. Pattern Recogn. **116**, 107957 (2021)
24. Berthold, T.: Measuring the impact of primal heuristics. Oper. Res. Lett. **41**(6), 611–614 (2013)

MineReduce-Based Metaheuristic for the Minimum Latency Problem

Marcelo Rodrigues de Holanda Maia[1,2]([✉]) [iD], Ítalo Santana[3] [iD],
Isabel Rosseti[1] [iD], Uéverton dos Santos Souza[1] [iD], and Alexandre Plastino[1] [iD]

[1] Instituto de Computação, Universidade Federal Fluminense, Niterói, RJ, Brazil
{mmaia,rosseti,ueverton,plastino}@ic.uff.br
[2] Instituto Brasileiro de Geografia e Estatística, Rio de Janeiro, RJ, Brazil
marcelo.h.maia@ibge.gov.br
[3] Departamento de Informática, Pontifícia Universidade Católica do Rio de Janeiro,
Rio de Janeiro, RJ, Brazil
isantana@inf.puc-rio.br

Abstract. The minimum latency problem is a variant of the well-known travelling salesperson problem where the objective is to minimize the sum of arrival times at vertices. Recently, a proposal that incorporates data mining into a state-of-the-art metaheuristic by injecting patterns from high-quality solutions has consistently led to improved results in terms of solution quality and running time for this problem. This paper extends that proposal by leveraging data mining to contract portions of the problem frequently found in high-quality solutions. Our proposal aims at mitigating the burden of searching for improving solutions by periodically solving a reduced version of the original problem. Computational experiments conducted on a well-diversified set of instances demonstrate that our proposal improved solution quality without increasing computational time, introducing 11 new best solutions to the literature.

Keywords: Metaheuristics · Data mining · Size reduction · MLP

1 Introduction

The minimum latency problem (MLP) is a variant of the well-known travelling salesperson problem where the objective is to minimize the sum of arrival times at vertices in a Hamiltonian cycle. It can model several real-world applications like distribution logistics, machine scheduling and disaster relief [3,5,7].

This work was supported by Conselho Nacional de Desenvolvimento Científico e Tecnológico (CNPq, Brazil) [grants 310444/2018-7, 310624/2018-5, 309832/2020-9], Fundação Carlos Chagas Filho de Amparo à Pesquisa do Estado do Rio de Janeiro (FAPERJ, Brazil) [grant E-26/201.344/2021], Coordenação de Aperfeiçoamento de Pessoal de Nível Superior (CAPES, Brazil) [grant 88887.646206/2021-00], and Instituto Brasileiro de Geografia e Estatística (IBGE, Brazil).

Recently, a hybrid metaheuristic named MDM-GILS-RVND [18] has appeared as a high-performance algorithm for the MLP. This hybrid algorithm was a result of combining the *Multi Data Mining* (MDM) approach [14] and GILS-RVND [19], a state-of-the-art hybrid metaheuristic for the MLP that combines components of Greedy Randomized Adaptive Search Procedures (GRASP) [6], Iterated Local Search (ILS) [10], and Random Variable Neighborhood Descent (RVND) [20]. The combination with MDM, which relies on data mining to extract patterns from high-quality solutions followed by their insertion into initial solutions, made GILS-RVND significantly improve its state-of-the-art solution quality and computational time results.

In this paper, we propose another improvement through the application of the MineReduce approach, which has achieved promising results for variants of vehicle routing [12] and vertex cover [11] problems.

In the MineReduce approach, the patterns mined from an elite set of solutions are used to perform problem size reduction. A problem instance is reduced to a smaller-size version by contracting or deleting elements that appear in a mined pattern – as they are assumed to be part of a solution for the original instance. Then, the reduced instance is solved, and the solution found is expanded to become a solution for the original instance.

The results of our computational experiments, reported in this paper, show that the proposed MineReduce-based metaheuristic for the MLP, named MR-GILS-RVND, overcomes MDM-GILS-RVND, achieving higher solution quality without increasing CPU running time, particularly for larger instances. It found new best solutions for 11 out of 56 benchmark instances.

The remainder of this paper is organized as follows. Section 2 defines the problem and lists relevant methods from the literature to solve it. Section 3 describes the MDM-GILS-RVND metaheuristic for the MLP proposed in [18]. The MineReduce-based metaheuristic for the MLP proposed in this work is introduced in Sect. 4. Section 5 reports our experimental results. Finally, conclusions and directions for future work are presented in Sect. 6.

2   The Minimum Latency Problem

The MLP, described as follows, is a variant of the well-known travelling salesperson problem (TSP) and NP-hard as well [3]. Let $G = (V, A)$ be a directed graph, where $V = \{0, 1, ..., n\}$ is a set composed of $n+1$ vertices and $A = \{(i, j) : i, j \in V, i \neq j\}$ is the set of arcs. Vertex 0 is the depot from where the salesperson departs, whereas the set $V' = V \backslash \{0\}$ consists of the remaining vertices representing the n customers. For each arc $(i, j) \in A$, there is an associated travel time t_{ij}. The aim is to find a Hamiltonian cycle $(i_0, i_1, ..., i_{n+1})$ in G, where $i_0 = i_{n+1} = 0$ (i.e., the cycle starts and ends at the depot), that minimizes the sum of arrival times, given by $\sum_{k=1}^{n+1} l(i_k)$, where $l(i_k) = \sum_{m=0}^{k-1} t_{i_m i_{m+1}}$ represents the latency of vertex i_k (i.e., the total travel time to reach i_k).

We present a toy example in Fig. 1 to illustrate an MLP solution and the computation of its cost. That is, we show a sequence of customer visits that forms a Hamiltonian cycle S (Fig. 1a) and its associated cost (Fig. 1b), or $f(S)$, which is 164. One can note that MLP is more challenging than TSP since minor changes in the ordering of customers in S can impact drastically $f(S)$ due to the sum of all cumulative costs of each customer, while a TSP solution cost is obtained by a simple sum of all traversed arcs.

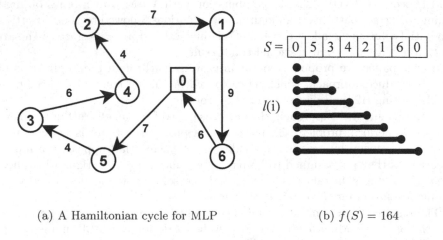

(a) A Hamiltonian cycle for MLP   (b) $f(S) = 164$

Fig. 1. Example of an MLP solution

In the literature, several exact algorithms were proposed for the MLP [1,2,4, 7,16]. Thanks to these methods, existing instances with up to 200 customers can be solved optimally at the cost of significant computational time. In contrast, heuristics and metaheuristics are alternative methods that consistently find high-quality solutions in controllable running time. The best-performing ones are usually able to efficiently solve instances with up to 1000 customers [5,7,13,17–19]. In particular, GILS-RVND [19] has achieved state-of-the-art results for MLP and was further improved with data mining as MDM-GILS-RVND [18].

3   The MDM-GILS-RVND Metaheuristic

The MDM-GILS-RVND metaheuristic [18] is an algorithm resulting from the incorporation of data mining into GILS-RVND [19], a state-of-the-art hybrid metaheuristic for the MLP that combines components of GRASP, ILS and RVND. This hybrid algorithm applies an approach known as Multi Data Mining (MDM), which uses frequent patterns extracted from good solutions by a data mining process to guide the search, initially proposed for a hybrid version of the GRASP metaheuristic (MDM-GRASP) [14].

In MDM, an *elite set* E keeps the d best solutions found during the execution of the metaheuristic, and a data mining method is periodically executed to extract a set of patterns P from E. These mined patterns are then used in the solution initialization process. The data mining method is based on the FPmax* algorithm [8], which mines maximal frequent itemsets. An itemset is considered frequent if it achieves a given minimum support value, i.e., if it is present in at least a given minimum number of the elite set solutions. Hence, mined patterns are composed of items that frequently appear together in the sub-optimal solutions of the elite set. Intuitively, it is assumed that these items should likely be part of the best solutions to the problem and, thus, they can favour the overall searching process when included in initial solutions. In MDM-GILS-RVND, the set of items representing each solution used for mining refers to the set of all arcs in the solution. Therefore, the mined patterns are sets of frequent paths between customers in the elite solutions.

Usually, in the MDM approach, the data mining method is invoked whenever E is considered *stable* (unchanged for a number of consecutive multi-start iterations). However, the stabilization criterion was not used in MDM-GILS-RVND because the number of multi-start iterations performed by the GILS-RVND metaheuristic is too small (only ten). Instead, it invokes the data mining method once half of the multi-start iterations are completed and, afterwards, whenever the elite set has been updated in the previous multi-start iteration.

The high-level structure of MDM-GILS-RVND is shown in Algorithm 1, where $f(s)$ denotes the cost of a solution s.

Algorithm 1. MDM-GILS-RVND

1: $f(s^*) \leftarrow \infty$
2: **for** $i \leftarrow 1, \ldots, I_{Max}/2$ **do**
3:      $s \leftarrow$ ConstructiveProc()
4:      $s' \leftarrow$ ILS(s)
5:      **if** $f(s') < f(s^*)$ **then**
6:          $s^* \leftarrow s'$
7: **for** $i \leftarrow 1, \ldots, I_{Max}/2$ **do**
8:      **if** $i = 1$ or E was updated in iteration $i - 1$ **then**
9:          $P \leftarrow$ Mine($E, sup_{min}, MaxP$)
10:     $s \leftarrow$ HybridConstructiveProc($p \in P$)
11:     $s' \leftarrow$ ILS(s)
12:     **if** $f(s') < f(s^*)$ **then**
13:         $s^* \leftarrow s'$
14: **return** s^*

In the first half of the multi-start iterations (lines 2–6), the algorithm's structure is the same as GILS-RVND, where each iteration builds an initial solution s using a GRASP-like constructive process, which depends on a parameter α that controls the balance between greediness and randomness (line 3). Then, it runs an ILS component to obtain a locally optimum solution s' (line 4) and updates the best solution s^* in case of improvement (line 6). In this first half of iterations, the elite set E is updated whenever a new eligible solution is found within the ILS component. A new solution is inserted into E if it is different from all solutions in E and cheaper than the worst of them. In the second half (lines 7–13), the data mining method is invoked at the first iteration and whenever E has been updated in the previous iteration, returning a set P containing the largest $MaxP$ patterns with a relative minimum support value sup_{min} (line 9). An initial solution s is built by a hybrid constructive process based on one of the mined patterns selected from P (line 10), which starts by inserting all pattern elements in the partial solution and completes it using the original constructive method strategies. The remaining steps are the same as in the first half iterations, including the insertion of new solutions into E whenever all requirements are met. Once all multi-start iterations are finished, the algorithm returns the best solution found (line 14).

One key aspect of MDM-GILS-RVND is a move evaluation procedure inherited from GILS-RVND. It consists of a framework that uses preprocessed data structures to compute costs of neighbor solutions in constant amortized time operations [9,22]. In practice, three data structures are used to store the partial costs of each subsequence of vertices of a local minimum solution, where the cost of every neighbor solution is reached by computing their partial costs on a "by concatenation" fashion. We describe these data structures and how concatenation is performed as follows:

- The *duration* $T(\sigma)$ of a sequence σ, which is the total travel time to perform the visits in the sequence.
- The *cost* $C(\sigma)$ to perform a sequence σ, when starting at time 0.
- The *delay* $W(\sigma)$ associated with a sequence σ, which is the number of customers visited in the sequence.

Let $T(i)$, $C(i)$ and $W(i)$ denote the values of the re-optimization data structures corresponding to a subsequence with only a single vertex i. In this case: $T(i) = 0$ and $C(i) = 0$ since there is no travel time; and $W(i) = 1$ if i is a customer, otherwise $W(i) = 0$. These values can be computed on larger subsequences by induction on the concatenation operator $\oplus$ as follows. Let $\sigma = (\sigma_u, \ldots, \sigma_v)$ and $\sigma' = (\sigma'_w, \ldots, \sigma'_x)$ be two subsequences. The subsequence $\sigma \oplus \sigma' = (\sigma_u, \ldots, \sigma_v, \sigma'_w, \ldots, \sigma'_x)$ is characterized by the following values:

- $T(\sigma \oplus \sigma') = T(\sigma) + t_{\sigma_v \sigma'_w} + T(\sigma')$
- $C(\sigma \oplus \sigma') = C(\sigma) + max(W(\sigma'), 1)(T(\sigma) + t_{\sigma_v \sigma'_w}) + C(\sigma')$
- $W(\sigma \oplus \sigma') = W(\sigma) + W(\sigma')$.

4   MineReduce-Based Metaheuristic for the MLP

4.1   The MineReduce Approach

The MineReduce approach builds upon the ideas introduced by previous approaches for incorporating data mining into metaheuristics, like the MDM approach [14]. Since the mined patterns are assumed to likely be part of the best solutions to a problem instance, they are well-suited for reducing its size.

For example, the items in a pattern (temporarily fixed in the solution) can be deleted from the instance or merged in a condensed representation.

MineReduce's first steps are to *build an elite set* of solutions and to *mine* patterns from this set. These steps should be carried out like in the MDM approach, i.e., the best solutions found are stored in the elite set until the data mining method is invoked. The subsequent steps compose a problem size reduction (PSR) process intended to replace a multi-start metaheuristic's initial solution generation method. The *Reduce* step uses a pattern p to transform a problem instance I into a reduced-size instance I'. The *Optimize* step is accomplished through the application of the metaheuristic's original optimization procedures to I'. The *Expand* step transforms the solution to I' into a solution to I, which concludes the MineReduce-based generation of an initial solution.

MineReduce has been successfully applied in metaheuristics for problems such as vehicle routing and vertex cover variants, with considerable improvements in solution quality and computational time, especially compared with MDM-based metaheuristics [11, 12].

According to Talbi's taxonomy for metaheuristics that incorporate machine learning (ML) in their design [21], MineReduce-based methods are primarily classified as *problem-level ML-supported metaheuristics* since this approach uses data mining for hierarchical problem decomposition (defining and solving smaller subproblems). In addition, they can also be classified as *low-level ML-supported metaheuristics* given that data mining is used in a process that drives the initialization of solutions. Finally, regarding the *learning time* criteria, they are classified as *online* ML-supported metaheuristics since they gather knowledge during the search while solving the problem.

4.2   MineReduce-Based GILS-RVND

The reduction process adopted for this problem is similar to that adopted for a vehicle routing problem [12]. In this case, a mined pattern is a set of subsequences of customers. These subsequences can be contracted by replacing all vertices in a subsequence with a single vertex.

Let $G = (V, A)$ be a directed graph associated with an MLP instance and p a pattern consisting of a set of subsequences of customer vertices in that instance. Let $G^* = (V^*, A^*)$ be a directed graph associated with the corresponding reduced instance based on p. Such a reduced version can be obtained as follows. Initially, G^* is defined as a copy of G. For each subsequence $\sigma = (i_1, i_2, ..., i_{|\sigma|}) \in p$ selected to be contracted, each of the customers in σ is removed from G^* – that is, the vertex corresponding to the customer is removed from V^* and the arcs that connect that vertex to the others are removed from A^*. Then, a customer vertex i_σ corresponding to the subsequence is added to V^* and arcs connecting i_σ to the other vertices in V^* are added to A^*. The travel time from each vertex $i^* \in V^*$ to i_σ is given by $t_{i^* i_\sigma} = t_{i^* i_1}$, that is, the travel time from i^* to i_1 (the first customer in σ). The travel time from i_σ to each vertex $i^* \in V^*$ is given by $t_{i_\sigma i^*} = t_{i_{|\sigma|} i^*}$, that is, the travel time from $i_{|\sigma|}$ (the last customer in σ) to i^*.

The values in the "by concatenation" framework structures for a subsequence with only the single vertex i_σ are defined as $T(i_\sigma) = T(\sigma)$, $C(i_\sigma) = C(\sigma)$ and $W(i_\sigma) = W(\sigma)$. Note that the need to adapt the "by concatenation" framework structures to work seamlessly with reduced instances made this application of MineReduce challenging even though the approach had previously been applied to another routing problem.

The structure of the MineReduce-based version of GILS-RVND, called MR-GILS-RVND, is depicted in Algorithm 2. The difference to Algorithm 1 is the use of a MineReduce-based constructive process instead of the hybrid constructive process from MDM-GILS-RVND in the last β iterations (line 13).

Algorithm 2. MR-GILS-RVND

1: $f(s^*) \leftarrow \infty$
2: **for** $i \leftarrow 1, \ldots, I_{Max}/2$ **do**
3:     $s \leftarrow$ ConstructiveProc()
4:     $s' \leftarrow$ ILS(s)
5:     **if** $f(s') < f(s^*)$ **then**
6:         $s^* \leftarrow s'$
7: **for** $i \leftarrow 1, \ldots, I_{Max}/2$ **do**
8:     **if** $i = 1$ or E was updated in iteration $i - 1$ **then**
9:         $P \leftarrow$ Mine($E, sup_{min}, MaxP$)
10:     **if** $i \leq (I_{Max}/2) - \beta$ **then**
11:         $s \leftarrow$ HybridConstructiveProc($p \in P$)
12:     **else**
13:         $s \leftarrow$ MR-ConstructiveProc($p \in P$)
14:     $s' \leftarrow$ ILS(s)
15:     **if** $f(s') < f(s^*)$ **then**
16:         $s^* \leftarrow s'$
17: **return** s^*

The MineReduce-based constructive process, presented in Algorithm 3, is a PSR process based on a pattern, as defined by the MineReduce approach.

Algorithm 3. MR-ConstructiveProc(p)

1: ReduceInstance(p, γ)
2: $s \leftarrow$ ConstructiveProc()
3: $s' \leftarrow$ ILS(s)
4: $s_0 \leftarrow$ ExpandSolution(s')
5: **return** s_0

In this implementation, we sort all subsequences in a pattern in decreasing size (number of traversed arcs) order. Then, we contract subsequences from the largest to the smallest until we have used a portion γ of all pattern's arcs. The adoption of this strategy was motivated by preliminary tests showing that the mined patterns contained too many arcs, producing reduced instances that were too small. Hence, after expanding solutions found for the reduced instances, a considerable effort was still necessary for the local search on the original instance. Using only a portion of the arcs in a pattern adds control to the reduction factor. Finally, we chose to favour larger subsequences because they are less likely to occur than small subsequences given the same minimum support. Therefore, they represent more robust and relevant portions of the patterns.

In Algorithm 3, the instance is reduced based on the provided pattern p (line 1). Then, a solution for the reduced instance is obtained by applying the original constructive and ILS methods from GILS-RVND (lines 2–3). Finally, the solution found is expanded, producing a solution for the original instance (line 4), which is returned (line 5).

Figure 2 illustrates the application of MineReduce's PSR process to an MLP instance. Let S_1 and S_2 be two solutions composing an elite set (Fig. 2a and Fig. 2b, respectively) and $sup_{min} = 100\%$. The mined pattern is depicted in dashed lines in Fig. 2c. The largest subsequences in these patterns – $(8, 6, 1)$ and $(10, 4, 5)$ – are contracted into vertices a and b, respectively, resulting in the reduced instance I' shown in Fig. 2d. Then, a solution for I' (Fig. 2e) is obtained using the original construction and search methods from GILS-RVND and expanded to become a solution for the original instance (Fig. 2f).

5   Computational Results

We have assessed the performance of our proposed method, MR-GILS-RVND, by running computational experiments comparing it to the original state-of-the-art MDM-GILS-RVND metaheuristic [18]. We have built MR-GILS-RVND upon the original MDM-GILS-RVND source code. Both were implemented in C++ and compiled with g++ 4.4.7. The experiments were run in a single thread on an Intel® Core™ i7-5500U 2.40 GHz CPU.

In these experiments, we used a set composed of 56 benchmark instances with 120 to 1379 customers from TSPLIB [15], which was also used to compare MDM-GILS-RVND and GILS-RVND in [18]. We ran both algorithms on ten tests using different random seeds for each instance.

The configuration of the MDM-GILS-RVND parameters adopted in [18] was used for both methods in our experiments: $I_{Max} = 10$ (the number of multi-start iterations); $I_{ILS} = min(100, n)$ (the number of ILS iterations); $R = \{0.00, 0.01, \ldots, 0.25\}$ (the possible values for α, a value that controls the greediness level of the constructive process, randomly chosen for each multi-start iteration); $sup_{min} = 70\%$ (the relative minimum support of the mined patterns); $d = 10$ (the capacity of the elite set); and $MaxP = 5$ (the maximum number of patterns returned by the data mining process).

The parameters introduced in MR-GILS-RVND – β (the number of multi-start iterations applying the MineReduce-based constructive process) and γ, the portion of all arcs in a pattern that are used for contraction – had their values

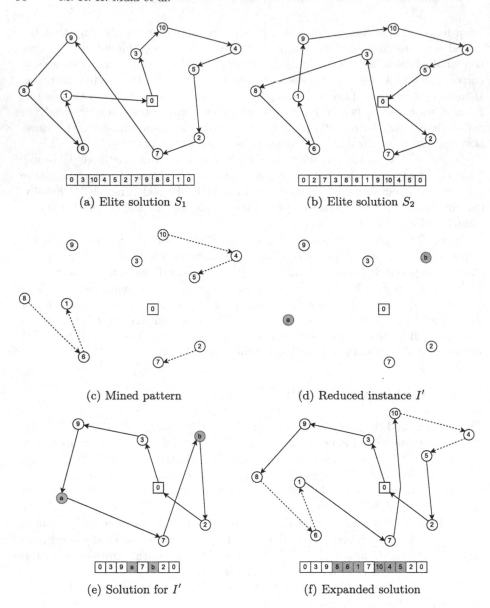

(a) Elite solution S_1

(b) Elite solution S_2

(c) Mined pattern

(d) Reduced instance I'

(e) Solution for I'

(f) Expanded solution

Fig. 2. MineReduce's PSR applied to an MLP instance

tuned based on the best trade-off between solution quality and computational time found in tests run on a sample with 12 out of the 56 benchmark instances (with 262 to 1291 customers). We refer to these 12 instances as the *training set*. The following values were considered for tuning the parameters: $\{1, 2, \ldots, 5\}$ for β, and $\{60\%, 2/3, 70\%, 80\%\}$ for γ. The best values found were $\beta = 4$ and $\gamma = 2/3$. Hence, we used these values in the experiments on the remaining

44 instances – the *validation set*. MR-GILS-RVND, with this best parameter configuration, found new best solutions for 6 out of the 12 instances in the training set, which are reported in Table 1.

Table 1. New best solutions found by MR-GILS-RVND for training instances

Instance	Solution	Instance	Solution	Instance	Solution
gil262	285,043	rd400	2,762,336	gr431	21,154,740
si535	12,246,046	gr666	63,454,259	vm1084	94,608,098

Table 2 summarizes the results obtained in the experiments using the instances in the validation set. MDM-GILS-RVND and MR-GILS-RVND are compared regarding the wins in best cost, average cost and average CPU running time, the number of new best solutions found, and the summed number of best known solutions (BKS) and new best solutions found.

Table 2. Results summary (all validation instances)

	MDM-GILS-RVND	MR-GILS-RVND
Wins best cost	4	6
Wins avg. cost	12	18
Wins avg. time	25	19
New best	–	5
Nb BKS + new best	37	39

The comparison on all 44 benchmark instances shows that MR-GILS-RVND overcomes MDM-GILS-RVND regarding solution quality, obtaining better solutions for most instances. Furthermore, MR-GILS-RVND found new best solutions for five instances in this set.

On the other hand, MDM-GILS-RVND obtained more wins in average time. This can be explained by the fact that all 15 instances with $n \leq 195$ are in the validation set. These small instances are easier than the others, and they all have been solved optimally by exact methods. The original MDM-GILS-RVND finds their optimal solutions in a few seconds. Thus, the slight computational overhead introduced by applying the size reduction process in MR-GILS-RVND is not compensated by a convergence speedup as usual since the solutions cannot be further improved.

Therefore, a separate comparison is presented in Table 3 considering only the instances with $n > 195$ (the 29 largest instances). For these larger instances, both methods are technically tied regarding average computational time, whereas the superiority of MR-GILS-RVND regarding solution quality is further evidenced, with about twice the number of wins of MDM-GILS-RVND in average cost. Hence, these results show that the MineReduce approach, applied in MR-GILS-RVND, improved solution quality without increasing computational time over MDM-GILS-RVND.

Table 3. Results summary ($n > 195$)

	MDM-GILS-RVND	MR-GILS-RVND
Wins best cost	4	6
Wins avg. cost	9	17
Wins avg. time	15	14
New best	–	5
Nb BKS + new best	22	25

Table 4 presents the detailed solution cost comparison results for the validation set. For each instance, we report the BKS cost from the literature and the best and average costs obtained by each method over the ten runs. Winning values in the comparison are in bold, solution costs matching the BKS are presented in italics, and new best solution costs are underlined.

Table 5 presents the detailed running time comparison results for the validation set. We report the average CPU running time in seconds obtained by each method over the ten runs for each instance. Again, winning values in the comparison are presented in bold. As it can be noticed, the overhead in terms of running time generated by the application of the MineReduce approach is almost negligible. The most expensive process in this approach is the mining of maximal frequent itemsets, which is NP-hard [23]. However, this step is already present in MDM-GILS-RVND. Besides, its combinatorial explosion is kept under control since the input is a dataset with only ten transactions (the solutions in the elite set). The instance reduction step is performed in $O(n^2)$ since its most costly operation is the reconstruction of the travel time matrix, and the solution expansion in $O(n)$ since it simply replaces the contracted vertices with the corresponding subsequences. These steps are performed only once for each of the last β iterations. Hence, their computational cost becomes irrelevant when compared to the search process. Finally, the problem decomposition provided by the reduction process makes the search more efficient in harder instances, compensating for the slight overhead.

Table 4. Solution cost comparison

Instance	BKS	MDM-GILS-RVND		MR-GILS-RVND	
		Best cost	Avg. cost	Best cost	Avg. cost
gr120	363,454[a,b]	363,454	363,454.0	363,454	363,700.3
pr124	3,154,346[a,b]	3,154,346	3,154,346.0	3,154,346	3,154,346.0
bier127	4,545,005[a,b]	4,545,005	4,545,691.9	4,545,005	**4,545,005.0**
ch130	349,874[a,b]	349,874	349,903.5	349,874	349,903.5
pr136	6,199,268[a,b]	6,199,268	**6,200,041.6**	6,199,268	6,200,360.9
gr137	4,061,498[a,b]	4,061,498	4,061,498.0	4,061,498	4,061,498.0
pr144	3,846,137[a,b]	3,846,137	3,846,137.0	3,846,137	3,846,137.0
ch150	444,424[a,b]	444,424	444,424.0	444,424	444,424.0
kroA150	1,825,769[a,c]	1,825,769	1,825,769.0	1,825,769	1,825,769.0
kroB150	1,786,546[a,c]	1,786,546	1,786,546.0	1,786,546	1,786,546.0
pr152	5,064,566[a,b]	5,064,566	5,064,566.0	5,064,566	5,064,566.0
u159	2,972,030[a,c]	2,972,030	2,972,204.2	2,972,030	2,972,204.2
si175	1,808,532[a,c]	1,808,532	1,808,532.0	1,808,532	1,808,532.0
brg180	174,750[a,c]	174,750	174,750.0	174,750	174,750.0
rat195	218,632[a,c]	218,632	**218,736.6**	218,632	218,771.3
d198	1,186,049[c]	1,186,049	1,186,273.3	1,186,049	1,186,049.0
kroA200	2,672,437[c]	2,672,437	2,672,444.2	2,672,437	2,672,437.0
kroB200	2,669,515[c]	2,669,515	**2,675,993.6**	2,669,515	2,676,444.9
gr202	2,909,247[d]	2,909,247	**2,913,368.4**	2,909,247	2,913,957.8
ts225	13,240,046[d]	13,240,046	13,240,046.0	13,240,046	13,240,533.0
tsp225	402,783[d]	402,783	402,933.3	402,783	**402,925.4**
pr226	7,196,869[d]	7,196,869	7,196,869.0	7,196,869	7,196,869.0
gr229	10,725,914[d]	10,725,914	10,731,249.9	10,725,914	**10,729,841.4**
pr264	5,471,615[d]	5,471,615	5,471,615.0	5,471,615	5,471,615.0
a280	346,989[d]	346,989	347,106.9	346,989	**347,098.9**
pr299	6,556,628[d]	6,556,628	**6,559,030.8**	6,556,628	6,559,653.7
lin318	5,619,810[d]	5,619,810	5,630,590.5	5,619,810	5,630,590.5
fl417	1,874,242[d]	1,874,242	1,874,242.0	1,874,242	1,874,246.0
pr439	17,829,541[d]	17,829,541	17,868,632.7	17,829,541	**17,866,993.1**
pcb442	10,301,705[d]	10,301,705	10,321,465.7	10,301,705	**10,321,299.8**
d493	6,677,458[d]	6,677,458	6,687,268.2	6,680,576	**6,685,445.4**
ali535	31,860,679[d]	31,860,679	31,910,477.9	31,860,679	**31,907,551.1**
pa561	658,870[d]	660,590	661,790.6	**660,127**	**661,757.9**
p654	7,827,273[d]	7,827,273	**7,827,867.8**	7,827,273	7,827,953.4
d657	14,112,540[d]	14,112,540	14,195,797.6	14,112,540	**14,194,627.6**
u724	13,504,408[d]	13,504,408	**13,537,514.7**	**13,491,599**	13,543,353.9
dsj1000	7,642,715,113[d]	7,642,715,113	7,664,531,851.2	**7,642,418,952**	**7,662,310,139.5**
dsj1000ceil	7,646,395,679[d]	7,646,395,679	7,676,973,751.4	**7,632,965,540**	**7,674,650,570.0**
si1032	46,896,355[d]	46,896,355	**46,896,783.6**	46,896,355	46,897,212.2
u1060	102,508,056[d]	102,539,819	102,759,493.6	**102,436,120**	102,681,878.9
pcb1173	30,890,385[d]	30,890,385	30,957,008.7	30,891,243	**30,945,301.5**
rl1304	144,592,447[d]	144,592,447	145,398,549.2	**144,585,587**	145,320,131.5
rl1323	155,697,857[d]	**155,719,283**	156,273,365.5	155,762,567	**156,229,029.1**
nrw1379	35,291,795[d]	35,291,795	**35,456,093.0**	35,329,106	35,461,487.4
Wins		4	12	6	18
Wins ($n > 195$)		4	9	6	17

[a] Optimality proven.
[b] From [16].
[c] From [4].
[d] From [18].

Table 5. Average running time comparison

Instance	MDM-GILS-RVND	MR-GILS-RVND
gr120	**10.73**	10.88
pr124	**6.83**	7.14
bier127	**10.31**	10.65
ch130	**11.99**	12.90
pr136	19.30	**17.29**
gr137	**9.25**	9.79
pr144	**11.31**	11.43
ch150	**14.19**	14.69
kroA150	20.77	**20.42**
kroB150	19.41	**18.68**
pr152	**13.13**	14.09
u159	**16.59**	16.70
si175	**19.04**	20.98
brg180	23.84	**23.29**
rat195	45.90	**45.39**
d198	42.79	**38.52**
kroA200	45.14	**43.15**
kroB200	49.30	**44.68**
gr202	39.28	**38.36**
ts225	**35.15**	47.88
tsp225	**56.59**	56.65
pr226	**37.91**	39.55
gr229	**53.37**	53.63
pr264	**51.31**	52.99
a280	117.64	**107.57**
pr299	**104.23**	106.88
lin318	**129.40**	133.96
fl417	**469.94**	499.70
pr439	405.55	**399.02**
pcb442	604.91	**601.88**
d493	1,034.42	**883.64**
ali535	**1,345.69**	1,361.16
pa561	**1,649.07**	1,728.66
p654	1,825.05	**1,742.79**
d657	2,571.72	**2,508.68**
u724	**4,209.10**	4,430.35
dsj1000	**17,630.28**	17,792.34
dsj1000ceil	17,185.99	**17,033.65**
si1032	2,794.02	**2,661.32**
u1060	**13,336.35**	14,128.16
pcb1173	19,192.65	**19,024.96**
rl1304	**17,636.57**	18,013.06
rl1323	22,081.21	**21,322.80**
nrw1379	**45,325.74**	48,402.52
Wins	**25**	19
Wins ($n > 195$)	**15**	14

6   Conclusion

This paper proposes a hybrid metaheuristic for the MLP based on the MineReduce approach, which uses patterns extracted from an elite set of solutions using data mining to reduce the size of problem instances.

The proposed method, named MR-GILS-RVND, was built through the application of the MineReduce approach on MDM-GILS-RVND [18], a state-of-the-art algorithm that applies another approach for incorporating data mining into metaheuristics, which consists in inserting mined patterns in initial solutions.

We conducted computational experiments with 56 benchmark instances from TSPLIB to compare MDM-GILS-RVND and MR-GILS-RVND. The reported results evidence that our proposed MR-GILS-RVND overcomes MDM-GILS-RVND, achieving better solutions for most instances without increasing computational time. Furthermore, the results show a more evident superiority of the MineReduce-based method in more challenging instances (with $n > 195$).

These results reinforce the potential of the MineReduce approach for improving the performance of metaheuristics within a framework already applied to other combinatorial optimization problems. We shall extend this work with additional experiments run on a larger number of different problem instances to draw more general conclusions. Also, in future work, further investigation on the current application can be made to other challenging problem variants (e.g., time windows) that may require specialized design in the "by concatenation" framework. Finally, we expect that the contributions made in this work lead to a better comprehension of challenges involving problem size reduction for hard combinatorial optimization problems.

References

1. Abeledo, H., Fukasawa, R., Pessoa, A., Uchoa, E.: The time dependent traveling salesman problem: polyhedra and algorithm. Math. Program. Comput. **5**, 27–55 (2013). https://doi.org/10.1007/s12532-012-0047-y
2. Angel-Bello, F., Alvarez, A., García, I.: Two improved formulations for the minimum latency problem. Appl. Math. Model. **37**(4), 2257–2266 (2013). https://doi.org/10.1016/j.apm.2012.05.026
3. Blum, A., Chalasani, P., Coppersmith, D., Pulleyblank, B., Raghavan, P., Sudan, M.: The minimum latency problem. In: Proceedings of the Twenty-Sixth Annual ACM Symposium on Theory of Computing, STOC 1994, pp. 163–171. Association for Computing Machinery, New York (1994). https://doi.org/10.1145/195058.195125
4. Bulhões, T., Sadykov, R., Uchoa, E.: A branch-and-price algorithm for the minimum latency problem. Comput. Oper. Res. **93**, 66–78 (2018). https://doi.org/10.1016/j.cor.2018.01.016
5. Campbell, A.M., Vandenbussche, D., Hermann, W.: Routing for relief efforts. Transp. Sci. **42**(2), 127–145 (2008). https://doi.org/10.1287/trsc.1070.0209
6. Feo, T.A., Resende, M.G.C.: Greedy randomized adaptive search procedures. J. Global Optim. **6**(2), 109–133 (1995). https://doi.org/10.1007/BF01096763

7. Fischetti, M., Laporte, G., Martello, S.: The delivery man problem and cumulative matroids. Oper. Res. **41**(6), 1055–1064 (1993). https://doi.org/10.1287/opre.41.6. 1055

8. Grahne, G., Zhu, J.: Efficiently using prefix-trees in mining frequent itemsets. In: Goethals, B., Zaki, M.J. (eds.) Proceedings of the IEEE ICDM Workshop on Frequent Itemset Mining Implementations (2003)

9. Kindervater, G.A.P., Savelsbergh, M.W.P.: Vehicle routing: handling edge exchanges, pp. 337–360. Princeton University Press (2018). https://doi.org/10. 1515/9780691187563-013

10. Lourenço, H.R., Martin, O.C., Stützle, T.: Iterated Local Search, pp. 320–353. Springer, Boston (2003). https://doi.org/10.1007/0-306-48056-5_11

11. Maia, M.R.H., Plastino, A., Souza, U.S.: MineReduce for the minimum weight vertex cover problem. In: Proceedings of the International Conference on Optimization and Learning (OLA 2020), pp. 11–22 (2020)

12. Maia, M.R.H., Plastino, A., Penna, P.H.V.: MineReduce: an approach based on data mining for problem size reduction. Comput. Oper. Res. **122**, 104995 (2020). https://doi.org/10.1016/j.cor.2020.104995

13. Mladenović, N., Urošević, D., Goos, P., Hanafi, S.: Variable neighborhood search for the travelling deliveryman problem. 4OR **11**, 57–73 (2013). https://doi.org/10. 1007/s10288-012-0212-1

14. Plastino, A., Barbalho, H., Santos, L.F.M., Fuchshuber, R., Martins, S.L.: Adaptive and multi-mining versions of the DM-GRASP hybrid metaheuristic. J. Heurist. **20**, 1899–1911 (2014). https://doi.org/10.1007/s10732-013-9231-0

15. Reinelt, G.: TSPLIB-a traveling salesman problem library. ORSA J. Comput. **3**(4), 376–384 (1991). https://doi.org/10.1287/ijoc.3.4.376

16. Roberti, R., Mingozzi, A.: Dynamic ng-path relaxation for the delivery man problem. Transp. Sci. **48**(3), 413–424 (2014). https://doi.org/10.1287/trsc.2013.0474

17. Salehipour, A., Sörensen, K., Goos, P., Bräysy, O.: Efficient GRASP+VND and GRASP+VNS metaheuristics for the traveling repairman problem. 4OR **9**(2), 189–209 (2011). https://doi.org/10.1007/s10288-011-0153-0

18. Santana, I., Plastino, A., Rosseti, I.: Improving a state-of-the-art heuristic for the minimum latency problem with data mining. Int. Trans. Oper. Res. **29**(2), 959–986 (2022). https://doi.org/10.1111/itor.12774

19. Silva, M.M., Subramanian, A., Vidal, T., Ochi, L.S.: A simple and effective metaheuristic for the minimum latency problem. Eur. J. Oper. Res. **221**(3), 513–520 (2012). https://doi.org/10.1016/j.ejor.2012.03.044

20. Subramanian, A., Drummond, L., Bentes, C., Ochi, L., Farias, R.: A parallel heuristic for the vehicle routing problem with simultaneous pickup and delivery. Comput. Oper. Res. **37**(11), 1899–1911 (2010). https://doi.org/10.1016/j.cor.2009.10.011

21. Talbi, E.G.: Machine learning into metaheuristics: a survey and taxonomy. ACM Comput. Surv. **54**(6), 1–32 (2021). https://doi.org/10.1145/3459664

22. Vidal, T., Crainic, T.G., Gendreau, M., Prins, C.: A unifying view on timing problems and algorithms. Technical report, CIRRELT (2011)

23. Yang, G.: The complexity of mining maximal frequent itemsets and maximal frequent patterns. In: Proceedings of the Tenth ACM SIGKDD International Conference on Knowledge Discovery and Data Mining, KDD 2004, pp. 344–353. Association for Computing Machinery, New York (2004). https://doi.org/10.1145/ 1014052.1014091

Optimizing Multi-variable Time Series Forecasting Using Metaheuristics

Francesco Zito[✉], Vincenzo Cutello, and Mario Pavone

Department of Mathematics and Computer Science, University of Catania,
v.le Andrea Doria 6, 95125 Catania, Italy
francesco.zito@phd.unict.it, cutello@unict.it, mpavone@dmi.unict.it

Abstract. Multi-variable time series forecasting is one of several applications of machine learning. Creating an artificial environment capable of replicating real-world behavior is useful for understanding the intrinsic relationship between variables. However, selecting a predictor that ensures good performance for variables of different natures is not always a simple process. An algorithmic approach based on metaheuristics could be a good alternative to find the best predictive model for variables. Each predictor is optimized for forecasting a particular variable in a multi-agent artificial environment to improve the overall performance. The resulting environment is compared with other solutions that use only the same type of predictor for each variable. Finally, we can assert that using a multi-agent environment can improve the performance, accuracy, and generalization of our model.

Keywords: Metaheuristic · Optimization · Machine learning · Multi-variable time series prediction · Neural network · Artificial forecasting

1 Introduction

Metaheuristics are nowadays widely used methodologies in combinatorial optimization to solve hard and complex problems [6,12,13,17,18]. Their ability to find good approximate solutions in reasonable times makes them an excellent alternative to traditional techniques [4,9]. The wide range of metaheuristic methods published in literature has allowed researchers to revise classical problems by adapting metaheuristic methods as needed [21]. However, in recent years, the use of the metaheuristics, and their applications, has changed. Indeed, with the advent and expansion of artificial intelligence technologies in various application domains, several research works have been conducted and published that integrate metaheuristics with these new techniques to provide more accurate solutions. For instance, many research proposals that integrate metaheuristics and machine learning have been made. On the one hand, machine learning models are incorporated into metaheuristics to analyze and extract useful information from the many data generated by metaheuristics with the aim of guiding and

L. Di Gaspero et al. (Eds.): MIC 2022, LNCS 13838, pp. 103–117, 2023.
https://doi.org/10.1007/978-3-031-26504-4_8

enhancing their performance [3,11]. It is known, indeed, that metaheuristics do not take properly advantage of the many knowledge discovered during the search process, therefore the use of machine learning into the metaheuristics allows to exploit all information collected during the search process to guide the algorithm toward optimal solutions faster [20] On the other hand, the metaheuristics incorporated into machine learning are useful to this last [1] to design more efficient and reliable Deep Learning techniques and architectures [19], as well as to adjust hyperparameters and improve then prediction accuracy [5].

In this research work, a metaheuristic is presented to automatically set the machine learning configurations with the aim to increase the prediction accuracy. To achieve this, an optimization technique is applied to a multi-variable time series forecasting problem to validate the proposed algorithm. Considering that the variables in real-life evolve based on unknown functions, the ideal predictor must be able to accurately replicate real-world behavior using only real-world observations as training data [8]. The predictor forecasts a time series of values for each variable. This type of problem is often used to understand the relationship between variables and how they relate and interact to each other [15]. An example is the genetic inference problem where a gene regulatory network is constructed based on the observations of the real environment that reveals the relationships between genes [10,22].

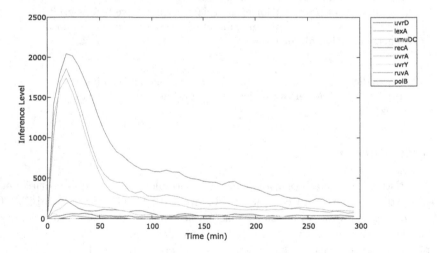

Fig. 1. The target time series. It is obtained from the average of all SOS DNA Repair experiments.

For validating the goodness and efficiency of the proposed approach a gene expression dataset was considered, which is obtained by observing the change in concentration of genes in a specific time window, as reported in [7]. Specifically, the SOS DNA repair reported in [14] was used, which consists of four experiments, each of which was recorded with fifty observations of eight genes, every six minutes. Considering that SOS DNA Repair dataset contains four experiments with significantly different values, due to measurement noise, the first two experiments were considered as training set, whilst the third one as validation set. The outcomes presented in this research paper were evaluated considering the average of all four experiments. Figure 1 displays the target time series of all SOS DNA Repair experiments. From the several experiments performed and from the analysis of the outcomes obtained, which were evaluated using the Cosine Similarity and Pearson Correlation, appears how the proposed approach significantly improves the prediction quality that in most cases coincides with the target one.

2   Method, Model and Optimization

In this section, a method for generating an artificial environment with the highest similarity index is presented. The description of the method is divided into three parts. The first part describes our model; in the second part, a method for creating an environment with the appropriate configuration is explained; and finally in the last part, an optimization algorithm is presented to determine the best configuration of the environment to improve performance.

2.1   Model

Developing a methodology to predict the behavior of a real environment, based solely on observations, is one of the most interesting challenges in computer science. Basically, a real environment consists of several entities, also called variables, that change over time. Figure 2 represents the environment schema under consideration.

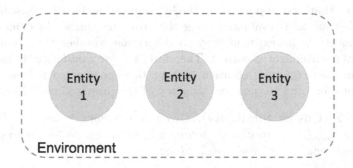

Fig. 2. Environment architecture

Definition 1 (State of the environment). *Let E be an environment with k variables, the state of the environment at time t, denoted by $s^{(t)}$, is defined as* a *k-vector in which the generic element* $s_i^{(t)}$ *is the value of the i-th variable at time t:*

$$s^{(t)} = \left(s_1^{(t)}, \ldots, s_k^{(t)}\right). \tag{1}$$

Let therefore $s^{(t)} = \left[s_1^{(t)}, s_2^{(t)}, \ldots, s_k^{(t)}\right]$ be the current state of the environment; the subsequent state of environment is then computed as follows:

$$s^{(t+1)} = f_E\left(s^{(t)}\right), \tag{2}$$

where f_E is the state transaction function and defines the behavior of the environment E. The state of the environment evolves throughout time. Considering that each variable in the proposed model is predicted by a single predictor, also called *agent*, the Eq. 2 can be decomposed into k components, one for each variable, each of which is responsible for predicting the value of a single variable, given the state of the environment at the previous time:

$$f_E = (f_{E1} \ldots, f_{Ek}) \tag{3}$$

with f_{Ei} the agent function that returns the i-th component of the state at time $t+1$, given the environment's state at time t. According to Eq. 3, it follows that Eq. 2 can be rewritten as follows:

$$s^{(t+1)} = \left(f_{E1}\left(s^{(t)}\right), \ldots, f_{Ek}\left(s^{(t)}\right)\right) \tag{4}$$

Each agent in the environment can be considered as a *black-box function* with its own architecture. They are in fact autonomous in terms of architecture, even though they influence each other through the state of the environment. Basically, the problem of environment prediction is to create an artificial environment $\hat{E}$ to replicate the behavior of the real environment. From a mathematical perspective, a function $f_{\hat{E}}$ that best approximates the hypothesis function f_E must be found. The resulting model is evaluated using similarity measures. As demonstrated in [2], similarity is an excellent way to determine whether the prediction of the artificial environment is correct. Therefore, *Cosine Similarity* (Definition 2) and *Pearson Correlation* (Definition 3) have been considered to calculate the environment score and to show the accuracy of the proposed model [16].

Definition 2 (Cosine Similarity). *Given two vectors, x and $\hat{x}$, the cosine similarity, $S_{cosine}(x, \hat{x})$, measures the angle between two vectors represented in a multidimensional space, and is defined as:*

$$S_{cosine}(x, \hat{x}) = \frac{\sum_{j=1}^{n} x_j \hat{x}_j}{\sqrt{\sum_{j=1}^{n} x_j^2}\sqrt{\sum_{j=1}^{n} \hat{x}_j^2}}, \tag{5}$$

where x_j and $\hat{x}_j$ are components of vector x and $\hat{x}$ respectively, while n is the size of vectors. The smaller the angle between the vectors, the higher the cosine similarity scores.

Definition 3 (Pearson Correlation). *Given two vectors, x and $\hat{x}$, the Pearson correlation, $S_{pearson}(x, \hat{x})$, measures the linear relationship between the two vectors and is computed as:*

$$S_{pearson}(x, \hat{x}) = \frac{\sum_{j=1}^{n} (x_j - \mu_x)(\hat{x}_j - \mu_{\hat{x}})}{\sqrt{\sum_{j=1}^{n} (x_j - \mu_x)^2} \sqrt{\sum_{j=1}^{n} (\hat{x}_j - \mu_{\hat{x}})^2}} \tag{6}$$

where x_j and $\hat{x}_j$ are components of the vector x and $\hat{x}$, respectively; μ_x and $\mu_{\hat{x}}$ are the average of the vectors x and $\hat{x}$; and n is the size of the vectors.

Supposing that $\hat{X} \in \mathbb{R}^{k \times n}$ represents the time series data predicted from an artificial environment $\hat{E}$ and $X \in \mathbb{R}^{k \times n}$ the time series data from the real environment E, the *artificial environment score* is defined as follows:

$$R\left(E, \hat{E}\right) = \sum_{i=1}^{k} \left| S(X_i, \hat{X}_i) \right|, \tag{7}$$

where k is the number of variables; $\hat{X}_i$ is a n-vector representing the time course of the i-th variable estimated by the artificial environment; X_i is a n-vector indicating the time course of the i-th variable in the real world; and $S(X_i, \hat{X}_i)$ is one of the similarity metrics stated above. As can be seen from Eq. 7, the score of the environment ranges between 0 an k. The quality of the environment will be expressed in percentage terms, as specified below, to facilitate its interpretation.

Definition 4 (Similarity Index). *Let $\hat{E}$ be an artificial environment that approximate the real environment E. The similarity index of $\hat{E}$ is a percentage value that expresses how similar the artificial and real environments are, and it is calculated as follows:*

$$R_\%(E, \hat{E}) = \frac{R\left(E, \hat{E}\right)}{k} 100. \tag{8}$$

2.2 Configuration

As described above, an environment is composed of a given number of agents, each of which able to predict the value of a variable over time. To achieve this, a possible collection of agents capable of accurately simulating the behavior of these variables must be identified. Agents with different architectures can coexist in the same environment. Each agent is formed according to the configuration selected when the environment was created. Different types of predictors have been used to explore different possible configurations of the artificial environment. In Table 1 are reported the available configurations for each predictor, and specifically:

- **Predictor Name:** that indicates the type of predictor used to predict the variable. A predictor can be a neural network or another predictor model, such as a linear regression model;

Table 1. Agent configurations for each type of predictor

Predictor name	Machine learning task	Agent prediction schema	Training parameters
Fully Connected Neural Network (FCNN)	Classification (L)/Regression	Simple/Delta Prediction	Epochs, Mini Batch Size, Learning Rate
Recurrent Neural Network (RNN-LSTM)	Classification (L)/Regression	Simple/Delta Prediction	Epochs, Mini Batch Size, Learning Rate
Convolutional Neural Network (CNN)	Regression	Simple/Delta Prediction	Epochs, Mini Batch Size, Learning Rate
Simple Linear Regression (SLR)	Regression	Simple/Delta Prediction	

- **Machine Learning Tasks:** as it is well known, there are two types of machine learning tasks, such as regression and classification. In the regression task, the predictor forecasts a real value; otherwise, it predicts a class. Conversely, in the classification task, the training data must be prepared before training to divide the dataset into classes. The number of classes that must be used is specified by the parameter L;
- **Agent Prediction Schema:** in this proposed model two different types of prediction schemes are provided. In the first scheme, an agent predicts a value directly, which is called *Simple Prediction*, while in the second one, an agent predicts the offset between the subsequent and current values, and it is called *Delta Prediction*;
- **Training Parameters:** to train a model certain training parameters must be specified. For instance, to train a neural network it must specify the number of epochs, the minimum stack size, the learning rate, and so on.

Figure 3 depicts an example of an environment with three distinct agents, each with their own configuration and the responsibility of estimating the corresponding variable.

2.3   Optimization

In this section, the proposed metaheuristic is presented with the main goal of determining the optimal configuration of the environment from one or more initial configurations. In particular, a local-search algorithm (LS) was developed to improve the initial solution and create an artificial environment with the highest similarity index. At each iteration of the algorithm, new feasible solutions are discovered and only the highest scoring solutions are promoted to the next iteration. The solutions discovered in one iteration are used to create feasible solutions for the next iteration through the following operators, each of these performed

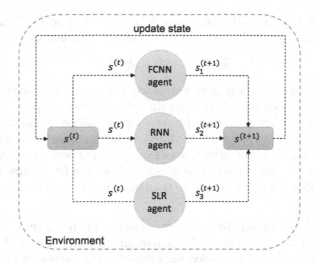

Fig. 3. State update process. The state vector is updated using the outputs of all agents. Each agent predicts the value of the corresponding variable based on the state of the environment at the previous time.

with a given probability: *mutation, generation, propagation,* and *combination.* Only a limited number of solutions pass to the next iteration. The parameters of the algorithm are listed in Table 2. For simplicity, the current interaction is denoted by i and the set of feasible solutions explored in the i-th iteration by U_i.

Table 2. Parameters considered for the proposed metaheuristic

Name	Description	Value
ρ	The probability that a feasible solution will be mutated	0.5
α	The probability that a new feasible solution will be generated at random during the current iteration	0.4
γ	The probability that a candidate will be propagated at the new iteration	0.8
β	The probability that two or more feasible solution will be merged each other	0.7
ν	Maximum number of interactions that can be performed	4
η	The number of feasible solutions that can be passed to the next iteration	10

Mutation: the mutation operator is applied to a feasible solution subset $\hat{U}_{i-1} \subseteq U_{i-1}$ in order to generate new solutions in U_i. A feasible solution in $\hat{U}_{i-1}$ can be mutated with a probability of ρ. The mutation affects only agents who earned a low rating during the evaluation (Sect. 2.1). According to Eq. 7, the rating of the agent A_h with $h = 1, 2, \ldots, k$ can be calculated as follows:

$$R(A_h) = |S(\hat{x}_{hj}, x_{hj})|. \tag{9}$$

Knowing that an agent is defined by some specific properties, as shown in Table 1, the mutation process randomly changes the value of these properties. Because of the mutation, a new agent is created to replace the old one in the artificial environment. When an environment is mutated, only the agents that have changed need to be trained. This operation allows exploration of new environment configurations that are identical to the parent one except for a few agents. In addition, the mutation operator can create more than one mutated solution from the same solution in U_i.

Creation: with the probability α, a new artificial environment can be created at each iteration and added to U_i. The environment configuration thus obtained is created simply by randomly selecting values for the agents' properties.

Combination: during iteration, two or more solutions in U_{i-1} can be merged together with probability β to create an artificial environment containing the best agents for each variable. Equation 9 is used to calculate the score of all agents that refer to the same variable but are in different environment configurations. For each variable, the best agent with the highest score is selected and assigned to a new environment. The resulting environment is then inserted into U_i.

Propagation: in propagation, the best solutions from the current iteration $i-1$ are promoted to the following iteration i. A solution contained in U_{i-1} is inserted into U_i with a probability γ. To prevent U_i from growing to infinity, all solutions with a low score are discarded at the end of each iteration.

3   Results

In this section, we present the results obtained by applying the method described in Sect. 2 to the dataset, shown in Fig. 1. As mentioned earlier, a collection of initial solutions must be defined for the optimization algorithm, and therefore four environment configurations were used as initial solutions and are listed in the Table 3. As shown in the table, it is assumed that all agents in an artificial environment initially have the same architecture. Using the optimization algorithms presented in Sect. 2.3, it was possible to improve the initial configurations and identify the appropriate artificial environment with the highest similarity index.

    Figure 4 shows how the optimization algorithm explores the search space iteration by iteration, starting from the initial configurations described in Table 3 (nodes 1 to 4) and ending with the best environment (node 90) whose final configuration is reported in Table 4. Each node in the search graph represents a potential environment configuration and thus a possible solution for the algorithm. Each node contains the following attribute: (1) similarity index (or fitness); (2) environment configuration; (3) identifier; (4) parent identifiers; and

Table 3. Initial environment configurations

Configuration name	Predictor name	Machine learning task	Agent prediction schema
Type 1	CNN	Regression	Simple Prediction
Type 2	FCNN	Classification (100)	Simple Prediction
Type 3	RNN-LSTM	Classification (100)	Simple Prediction
Type 4	SLR	Regression	Simple Prediction

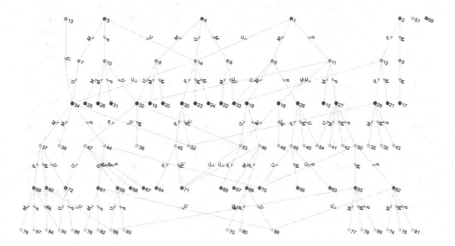

Fig. 4. Solutions discovered during the search process. Red nodes represent the initial solutions. The color of a node represents the number of iterations through which the algorithm explores that node. All nodes with the same color were visited in the same iteration. The label on the edge of the graph indicates the type of operation by which the node was created: Mutation (M), Propagation (P), Combination (U). The red mark indicates the best solution. (Color figure online)

Table 4. Configuration of the best artificial environment founded by the optimization algorithmic

Variable	Predictor name	Machine learning task	Agent prediction schema
uvrD	FCNN	Classification (100)	Simple Prediction
lexA	CNN	Regression	Simple Prediction
umuDC	CNN	Regression	Simple Prediction
recA	FCNN	Classification (150)	Delta Prediction
uvrA	CNN	Regression	Simple Prediction
uvrY	RNN	Regression	Delta Prediction
ruvA	FCNN	Regression	Delta Prediction
polB	FCNN	Classification (100)	Simple Prediction

(5) number of iterations. The metaheuristic operators, such as mutation, propagation, or combination, are used to move from one node in the search space to another one. Nodes can be visited during each iteration in two ways: using

Table 5. Performance comparison of the best artificial environment with the initial environments using both cosine similarity (5a) and person correlation (5b).

Variable	Type 1 (CNN)	Type 2 (FCNN)	Type 3 (RNN)	Type 4 (SLR)	Optimized
(a) COSINE SIMILARITY					
uvrD	0.906836	**0.955288**	0.907892	0.839389	0.945584
lexA	0.979805	**0.994548**	0.848259	0.633996	0.984051
umuDC	0.882852	0.874044	0.785110	0.710735	**0.991632**
recA	0.959850	0.986910	0.868888	0.729974	**0.991369**
uvrA	0.990824	**0.995476**	0.939292	0.616042	0.983343
uvrY	0.723409	0.832653	0.891370	0.806082	**0.946544**
ruvA	0.671901	0.712845	0.625257	0.680062	**0.846692**
polB	0.813158	0.938896	0.903303	0.794426	**0.956507**
Similarity Index	86.61%	91.13%	84.62%	72.63%	**95.57%**
(b) PEARSON CORRELATION					
Variable	Type 1 (CNN)	Type 2 (FCNN)	Type 3 (RNN)	Type 4 (SLR)	Optimized
uvrD	0.840166	**0.891629**	0.596472	0.025073	0.831584
lexA	0.982959	**0.991858**	0.865443	0.027845	0.976797
umuDC	0.777615	0.686630	0.349825	0.148003	**0.980669**
recA	0.962543	0.969059	0.785229	0.134906	**0.987008**
uvrA	0.988535	0.993400	0.960460	0.249609	**0.994350**
uvrY	0.641409	0.569276	0.524529	0.038941	**0.891394**
ruvA	0.555691	0.628533	0.223569	0.066942	**0.725781**
polB	0.547234	0.786681	0.647060	0.068000	**0.879271**
Similarity Index	78.70%	81.46%	61.91%	9.49%	**90.84%**

the metaheuristic operations described above, or randomly to expand the search space and explore thus more solutions. As can be seen analyzing the Fig. 4, the best solution is node 90 (red marked), which is created by combining the parent nodes, which are mutations or combinations of their parent nodes. Reading this graph, the bottom to the top, it is possible to see that node 90 is a permutation of the agents in the initial solution with some mutations. Furthermore, inspecting Table 4 is possible to note that the *Delta Prediction* scheme was preferred over the *Simple Prediction* one for a subset of variables.

In Table 5 is reported the similarity score calculated for each variable given the configurations listed in Table 3 and the best configuration reported in Table 4. Note that the Eq. 9 was used to calculate the similarity score for each variable. From the inspection of this table, it is possible to see that the proposed optimization process increases the similarity index of the environment by about five percentage points. In fact, a similarity index of about 95% (or 90% when using Pearson correlation) is obtained. However, it is worth to also point out that if an environment where all agents are FCNN is considered, a similarity index of 91% is obtained, which is higher than all other initial configurations. Then, from the overall investigation of both Tables 4 and 5 is possible to assert as outcome of this research paper that using multiple agent architectures allows the creation of artificial environments capable of accurately replicating real-world behavior. However, it is also possible to note the score of most agents in the best configuration is higher than their score in the initial solutions, except for uvrD, lexA,

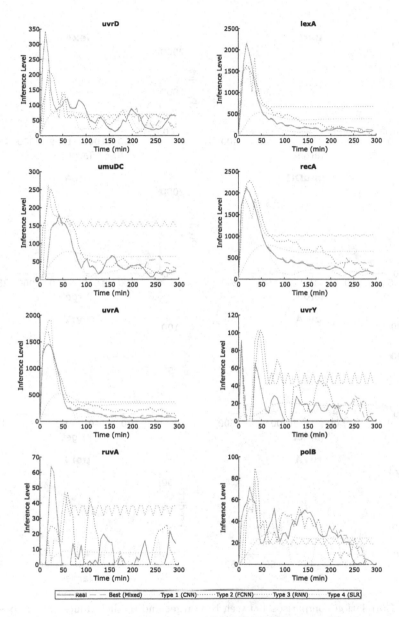

Fig. 5. The prediction results of 8 genes in SOS DNA Repair. The blue line represents the real gene expression data, while the other lines represent the forecasting results from the artificial environments. The orange line denotes the best environment identified by the optimization algorithm. (Color figure online)

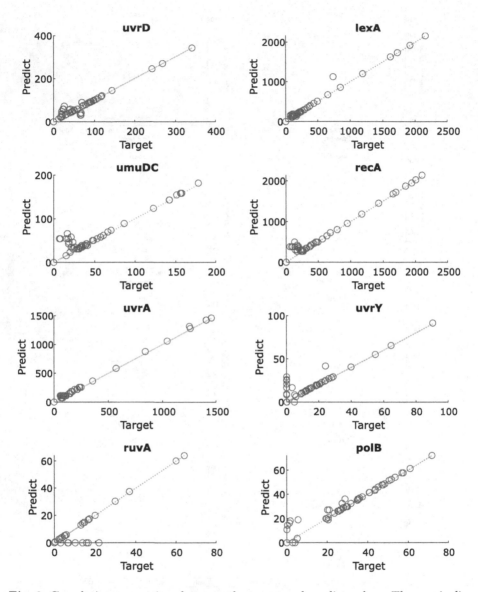

Fig. 6. Correlation comparison between the target and predict values. The x-axis displays the predicted values, while the y-axis displays the actual values. The red diagonal line represents the ideal values where the cosine similarity is equal to 1. (Color figure online)

and uvrA. Indeed, the score of the agents in the best solution is slightly lower for these agents. One reason for explaining this behavior is that the agent score is computed considering the shared state of the environment, as seen in Eq. 4. As a result, an agent with poor performance but high correlation with other variables may impact on all other agents.

Figures 5 and 6 compare the performance of the best environment with the target value. In particular, Fig. 5 shows the prediction results of the 8 genes in SOS DNA Repair, where appears evident how the curve of the optimized approach is almost always coincident with the target one in all variables, whilst in Fig. 6 is displayed the correlation comparison between the target and predict values. In these plots, the red diagonal line represents the ideal values where the cosine similarity is equal to 1. Also, these plots confirm the good prediction quality of the proposed approach since the predicted values almost always fall on the diagonal line of the ideal values.

4 Conclusions and Future Works

The integration of machine learning and metaheuristics is a new area of computer science that needs to be further explored by researchers. In this paper, a meta-heuristic local search was developed for determining the optimal configuration of the artificial environment considered starting from one or more initial configurations. For assessing the goodness and reliability of the presented method, the *Cosine Similarity* and *Pearson Correlation* considered, from which is also possible to estimate the accuracy of the method. From the inspection of all outcomes presented, appears how the proposed optimization method significantly increases the similarity indices, and consequently, allows to create artificial environments able to accurately replicate real-world behavior. This proposed optimization method shows, and primarily proves, how metaheuristics and machine learning can be used together to improve the overall performance on this type of problem.

As future research works, the proposed metaheuristics will be applied to design a neural network fully automatically, such as, for instance, selecting its basic architecture; appropriate activation functions; layer parameters (such as the output size in fully connected layers or the probability of dropping out input elements in the dropout layers); and so on.

References

1. Akay, B., Karaboga, D., Akay, R.: A comprehensive survey on optimizing deep learning models by metaheuristics. Artif. Intell. Rev. **55**(2), 829–894 (2022). https://doi.org/10.1007/s10462-021-09992-0
2. Almu, A., Bello, Z.: An experimental study on the accuracy and efficiency of some similarity measures for collaborative filtering recommender systems (2021). https://oer.udusok.edu.ng/xmlui/handle/123456789/948

3. Bengio, Y., Lodi, A., Prouvost, A.: Machine learning for combinatorial optimization: a methodological tour d'Horizon. Eur. J. Oper. Res. **290**(2), 405–421 (2021). https://doi.org/10.1016/j.ejor.2020.07.063. https://www.sciencedirect.com/science/article/pii/S0377221720306895

4. Bianchi, L., Dorigo, M., Gambardella, L.M., Gutjahr, W.J.: A survey on metaheuristics for stochastic combinatorial optimization. Nat. Comput. **8**(2), 239–287 (2009). https://doi.org/10.1007/s11047-008-9098-4

5. Bibaeva, V.: Using metaheuristics for hyper-parameter optimization of convolutional neural networks. In: 2018 IEEE 28th International Workshop on Machine Learning for Signal Processing (MLSP), pp. 1–6 (2018). https://doi.org/10.1109/MLSP.2018.8516989

6. Cutello, V., Fargetta, G., Pavone, M., Scollo, R.A.: Optimization algorithms for detection of social interactions. Algorithms **13**(6) (2020). https://doi.org/10.3390/a13060139. https://www.mdpi.com/1999-4893/13/6/139

7. Gebert, J., Radde, N., Weber, G.W.: Modeling gene regulatory networks with piecewise linear differential equations. Eur. J. Oper. Res. **181**(3), 1148–1165 (2007). https://doi.org/10.1016/j.ejor.2005.11.044. https://www.sciencedirect.com/science/article/pii/S0377221706001512

8. Ghahremaninahr, J., Nozari, H., Sadeghi, M.E.: Artificial intelligence and machine learning for real-world problems (a survey). Int. J. Innov. Eng. **1**(3), 38–47 (2021). https://ijie.ir/index.php/ijie/article/view/27

9. Greco, S., Pavone, M.F., Talbi, E.-G., Vigo, D. (eds.): MESS 2018. AISC, vol. 1332, p. XI, 57. Springer, Cham (2021). https://doi.org/10.1007/978-3-030-68520-1

10. Hecker, M., Lambeck, S., Toepfer, S., van Someren, E., Guthke, R.: Gene regulatory network inference: data integration in dynamic models-a review. Biosystems **96**(1), 86–103 (2009). https://doi.org/10.1016/j.biosystems.2008.12.004. https://www.sciencedirect.com/science/article/pii/S0303264708002608

11. Karimi-Mamaghan, M., Mohammadi, M., Meyer, P., Karimi-Mamaghan, A.M., Talbi, E.G.: Machine learning at the service of meta-heuristics for solving combinatorial optimization problems: a state-of-the-art. Eur. J. Oper. Res. **296**(2), 393–422 (2022). https://doi.org/10.1016/j.ejor.2021.04.032. https://www.sciencedirect.com/science/article/pii/S0377221721003623

12. Plebe, A., Cutello, V., Pavone, M.: Optimizing costs and quality of interior lighting by genetic algorithm. In: Sabourin, C., Merelo, J.J., Madani, K., Warwick, K. (eds.) IJCCI 2017. SCI, vol. 829, pp. 19–39. Springer, Cham (2019). https://doi.org/10.1007/978-3-030-16469-0_2

13. Plebe, A., Pavone, M.: Multi-objective genetic algorithm for interior lighting design. In: Nicosia, G., Pardalos, P., Giuffrida, G., Umeton, R. (eds.) MOD 2017. LNCS, vol. 10710, pp. 222–233. Springer, Cham (2018). https://doi.org/10.1007/978-3-319-72926-8_19

14. Raza, K., Alam, M.: Recurrent neural network based hybrid model for reconstructing gene regulatory network. Comput. Biol. Chem. **64**, 322–334 (2016). https://doi.org/10.1016/j.compbiolchem.2016.08.002. https://www.sciencedirect.com/science/article/pii/S1476927116300147

15. Sarker, I.H.: Machine learning: algorithms, real-world applications and research directions. SN Comput. Sci. **2**(3), 160 (2021). https://doi.org/10.1007/s42979-021-00592-x

16. Schober, P., Boer, C., Schwarte, L.A.: Correlation coefficients: appropriate use and interpretation. Anesth. Analg. **126**(5) (2018). https://journals.lww.com/anesthesia-analgesia/Fulltext/2018/05000/Correlation_Coefficients_Appropriate_Use_and.50.aspx

17. Talbi, E.G.: Metaheuristics: From Design to Implementation. Wiley, Hoboken (2008)
18. Talbi, E.G.: Combining metaheuristics with mathematical programming, constraint programming and machine learning. Ann. Oper. Res. **240**(1), 171–215 (2016). https://doi.org/10.1007/s10479-015-2034-y
19. Talbi, E.G.: Automated design of deep neural networks: a survey and unified taxonomy. ACM Comput. Surv. (CSUR) **54**(2), 1–37 (2021)
20. Talbi, E.G.: Machine learning into metaheuristics: a survey and taxonomy. ACM Comput. Surv. **54**(6) (2021). https://doi.org/10.1145/3459664
21. Wong, W., Ming, C.I.: A review on metaheuristic algorithms: recent trends, benchmarking and applications. In: 2019 7th International Conference on Smart Computing Communications (ICSCC), pp. 1–5 (2019). https://doi.org/10.1109/ICSCC.2019.8843624
22. Zito, F., Cutello, V., Pavone, M.: A novel reverse engineering approach for gene regulatory networks. In: Cherifi, H., Mantegna, R.N., Rocha, L.M., Cherifi, C., Miccichè, S. (eds.) COMPLEX NETWORKS 2022. SCI, vol. 1077, pp. 310–321. Springer, Cham (2023). https://doi.org/10.1007/978-3-031-21127-0_26

Unsupervised Machine Learning for the Quadratic Assignment Problem

Thé Van Luong[1] and Éric D. Taillard[2(✉)]

[1] University of Lausanne, Service de la recherche, Bâtiment Amphipôle, CH-1015 Lausanne, Switzerland
The-Van.Luong@unil.ch
[2] University of Applied Sciences and Arts of Western Switzerland, HEIG-VD, Department of Information and Communication Technologies, Route de Cheseaux 1, CH-1401 Yverdon-les-Bains, Switzerland
eric.taillard@heig-vd.ch

Abstract. An unsupervised machine learning method based on association rule is studied for the Quadratic Assignment Problem. Parallel extraction of itemsets and local search algorithms are proposed. The extraction of frequent itemsets in the context of local search is shown to produce good results for a few problem instances. Negative results of the proposed learning mechanism are reported for other instances. This result contrasts with other hard optimization problems for which efficient learning processes are known in the context of local search.

Keywords: Machine learning · Big data · Metaheuristics · Quadratic assignment

1 Introduction

In the past few years, big data has captured the attention of analysts and researchers since there is a strong demand to analyze large data collected from monitoring systems to understand behaviors and identify hidden trends. Science, business, industry, government and society have already undergone a change with the influence of big data. In [27], the authors are exposing opportunities and challenges that represent big data analytics.

On the one hand, with the increase of computational power, machine learning has emerged as the leading research field in artificial intelligence for dealing with big data and more generally with data science [13]. Machine learning techniques have given rise to huge societal impacts in a wide range of applications such as computer vision, natural language understanding and health.

On the other hand, metaheuristics such as genetic algorithms or local search are iterative methods in operations research that have been successfully applied to solve hard combinatorial optimization problems in the past. One of their main goals is to support decision-making processes in complex scenarios and provide near-optimal solutions to industrial problems.

The hybridization of metaheuristics with machine learning techniques is a promising research field for the operations research community [4]. The major

L. Di Gaspero et al. (Eds.): MIC 2022, LNCS 13838, pp. 118–132, 2023.
https://doi.org/10.1007/978-3-031-26504-4_9

interest in using machine learning techniques is to extract useful knowledge from the history of the search in order to improve the efficiency and the effectiveness of a metaheuristic [7].

This paper focuses on the association rule learning, which is an unsupervised machine learning method for discovering interesting relations between variables in very large databases [3]. Agrawal et al. [1] proposed frequent itemset mining for discovering similarities between products in a large-scale transaction data for supermarket chain stores. Initially designed for data mining, finding association rules is now widely generalized in many fields including web research, intrusion detection and bioinformatics.

We propose to incorporate the extraction of frequent itemsets in the context of local search metaheuristics. A similar work comes from Ribeiro et Al. in [19] to improve a GRASP metaheuristic where the learning process consists of extracting different patterns (i.e. subsets of frequent itemsets) from an elite set of 10 solutions and takes few seconds to provide a new generation.

The motivation of our work goes further, and its application is more appropriate to a big data context with gigabytes of data. The goal is to investigate if one can learn anything from the execution of thousands of local search algorithms to generate new sets of improved solutions. Hence, we propose reproducible strategies based on the extraction of millions frequent itemsets, i.e. extending the training phase to last one day and considering thousands of solutions performed in parallel across many generations.

The quadratic assignment problem (QAP) is considered in this study. This problem is hard to solve, even for instances of moderate size (less than 100 elements). This contrasts, for instance, with the travelling salesman problem (TSP) for which fairly large instances can be solved optimally. For the TSP, the set of edges composed by the union of a few locally optimal solutions of moderate quality may contain a very large proportion of the edges of the best solution known [24,25]. A goal of this paper is to evaluate if learning with locally optimal solutions is as successful for the QAP as it is for the TSP.

The objective values of solutions obtained by machine learning techniques for hard optimization problems are generally far from the values that can be obtained by direct heuristic algorithms. For the QAP, the reader is referred to [26] for a comparison of different methods based on neural graph machine network.

The remaining of this paper is organized as follows. Section 2 describes some technical background to understand the traditional local search algorithm, the quadratic assignment problem used for the experiments and frequent itemsets in associative rule learning. Section 3 introduces the extraction of frequent itemsets and its parallelization for local search algorithms. The experimental results are reported in Sect. 4. Finally, Sect. 5 concludes and proposes future research avenues.

2   Technical Background

2.1   The Quadratic Assignment Problem

To put in practice the different learning mechanisms proposed in this paper, the popular quadratic assignment problem (QAP) [12] has been investigated.

The QAP [5] arises in many applications such as facility location or data analysis. Let $A = (a_{ij})$ and $B = (b_{ij})$ be $n \times n$ matrices of positive integers. In the context of local search, the most convenient solution representation is by a permutation: The objective of the QAP is to find a permutation π of the set $\{1, 2, \ldots, n\}$ that minimizes the function:

$$z(\pi) = \sum_{i=1}^{n} \sum_{j=1}^{n} a_{ij} b_{\pi(i)\pi(j)}$$

The evaluation function has a $O(n^2)$ time complexity where n is the instance size. A neighborhood based on exchanging 2 elements ($\frac{n \times (n-1)}{2}$ neighbors) has been considered. Hence, for each iteration of a local search, $\frac{(n-2) \times (n-3)}{2}$ neighbors can be evaluated in $O(1)$ and $2n - 3$ can be evaluated in $O(n)$. The condition for an efficient neighborhood evaluation is to store all Δ differences of values between neighbor solutions at each iteration of the local search. This requires a memory space that increases quadratically with the size of the examples. Evaluating all the Δ for the first time takes an effort in $O(n^3)$ but an effort only in $O(n^2)$ for each of the next local search iteration [21].

A complete review of the most successful algorithms to solve the QAP is proposed in [15].

2.2   Frequent Itemsets in Associative Rule Learning

In associate rule learning, the existence of very large databases requires determining groups of items that frequently appear together in transactions, called *itemsets* [2]. From any itemset, one can determine an association rule that predicts how frequently an itemset is likely to occur in a transaction.

For example, a retail organization provides thousands of products and services [1]. The number of possible combinations of these products and services is potentially huge. The enumeration of all possible combinations is impractical, and methods are needed to concentrate efforts on those itemsets that are recognized as important to an organization. The most used measure of an itemset is its *support*, which is calculated as the percentage of all transactions that contain the itemset. Itemsets that meet a minimum support threshold are referred to as frequent itemsets.

An itemset which contains k items is a k-itemset. So, it can be said that an itemset is frequent if the corresponding support count is greater than a minimum support count.

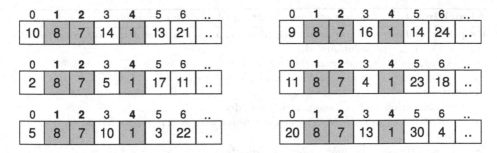

Fig. 1. Extraction of one frequent itemset of size 3. In all solutions, elements 8, 7 and 1 appear at positions 1, 2 and 4.

3   Frequent Itemsets for Local Search Algorithms

The motivation of this research work is to investigate if one can learn anything from the solutions found in local search algorithms. One observation is that some elements from local optima may be found at the exact same positions of the global optimum, meaning that elements that frequently appear at particular positions may also be discovered in good solutions. Unlike other works on machine learning for combinatorial optimization, we do not consider the value of the objective function in the learning mechanism. We only want to see if it is possible to learn something from the structure of locally optimal solutions.

One tool to achieve this is to extract all the frequent itemsets from a set of solutions. In the context of combinatorial optimization, each itemset can be represented by pairs of one element associated with one position. Figure 1 illustrates an extraction for a 3-itemset.

Once all frequent itemsets are known, a new generation of solutions can be constructed from these itemsets.

3.1   Extraction and Combination of Frequent Itemsets

The global process used in this paper can be divided into two phases: the extraction of frequent itemsets and their combination to generate new solutions. Algorithm 1 gives an insight of how this global process works.

The initial set of solutions is obtained from the execution of multi-start local search algorithms (lines 1 to 4). For each local search, the initial solution is randomly generated and the selection of a better neighbor is done according to the best improvement strategy (steepest descent).

In the main loop, the first phase consists in extracting all frequent itemsets from the current set of solutions (line 6) with Apriori algorithm [2]. Since the worst-case time complexity of Apriori algorithm is exponential according to the number of items, min_sup and $itemsets\_limit$ are user-defined parameters to control the number of candidate itemsets to retain in practice. The second phase is a procedure that combines these itemsets to construct new solutions that can be improved afterwards by the same local search algorithm (lines 7 to 10). The process is repeated for a given number of generations.

Algorithm 1. Extraction and combination of frequent itemsets

Require: *instance_data, nb_solutions, nb_generations, min_sup* and *itemsets_limit*
1: **for** $i \leftarrow 1, \ldots, nb\_solutions$ **do**
2:     $solutions[i] \leftarrow random\_initialization()$
3:     $solutions[i] \leftarrow local\_search(instance\_data, solutions[i])$
4: **end for**
5: **for** $generation \leftarrow 1, \ldots, nb\_generations$ **do**
6:     $all\_itemsets \leftarrow extract\_itemsets(solutions, min\_sup, itemsets\_limit)$
7:     **for** $i \leftarrow 1, \ldots, nb\_solutions$ **do**
8:         $solutions[i] \leftarrow combine\_itemsets(all\_itemsets)$
9:         $solutions[i] \leftarrow local\_search(instance\_data, solutions[i])$
10:     **end for**
11: **end for**
Ensure: *solutions*

3.2   Apriori Algorithm for Extracting Itemsets from a Set of Solutions

The Apriori algorithm is used in this paper to extract all frequent itemsets from a set of solutions. It was originally designed to operate on databases containing transactions [2]. Basically, Apriori performs a bottom-up approach where frequent subsets are extended one item at a time (groups of candidates) and tested with the data. The algorithm finishes when no further successful extensions can be discovered.

Even if it is not the fastest method to directly extract k-itemsets in comparison with other approaches [11,18], its application seems the most appropriate since all frequent itemsets of any size are required here. More important, Apriori does not make any assumption of the size of the dataset and it perfectly fits in the context of big data algorithms.

Algorithm 2. Apriori algorithm for the extraction of frequent itemsets

Require: *solutions, min_sup* and *itemsets_limit*
1: $k \leftarrow 1$
2: $C_k \leftarrow generate\_itemsets(solutions, \emptyset, \emptyset)$
3: $L_k \leftarrow filter\_itemsets(C_k, min\_sup, \emptyset)$
4: $all\_itemsets \leftarrow L_k$
5: **while** $L_k \neq \emptyset$ **do**
6:     $C_{k+1} \leftarrow generate\_itemsets(solutions, L_k, L_1)$
7:     $L_{k+1} \leftarrow filter\_itemsets(C_{k+1}, min\_sup, itemsets\_limit)$
8:     $all\_itemsets \leftarrow all\_itemsets \cup L_{k+1}$
9:     $k \leftarrow k + 1$
10: **end while**
Ensure: *all_itemsets*

Algorithm 2 describes the major steps of Apriori used in the *extract_itemsets* procedure of Algorithm 1. The first step consists in generating the list of all candidate itemsets of size 1 (lines 1 and 2). In the case of combinatorial optimization, a 1-itemset is exactly a pair of one element associated with one position. The candidate list is then pruned according to the *minimum support* (i.e. minimum number of times that an itemset must appear in all solutions) defined by the user (line 3). From the resulting filtered list of 1-itemsets, all candidate itemsets of size 2 are investigated (line 6) where a 2-itemset represents two pairs of one element associated with one position. A list of $k + 1$-itemsets is produced by extending the filtered list of k-itemsets with the 1-itemsets in all possible ways. The process is repeated until a candidate list cannot be constructed.

At each generation, all extracted k-itemsets are conserved in a list (lines 4 and 8) that will be later used to construct new solutions in the *combine_itemsets* procedure of Algorithm 1.

Limiting the number of retained itemsets (e.g. keeping one million itemsets that are among the most frequent ones) is necessary to reduce the computational and space complexities when generating new candidates for further generations.

3.3   Combining Itemsets for Creating a New Set of Solutions

The goal of the combine phase is to create a new set of solutions from all the frequent itemsets extracted during the previous generation.

Each solution is constructed by exploring all frequent itemsets. In this paper, two main strategies are taken into account regarding how itemsets are explored:

1. Random exploration of all frequent itemsets (REFI). In this strategy, every retained itemset has the same probability to be applied during the construction of a new solution.
2. Exploration based on sorted frequent itemsets (ESFI). All itemsets are sorted according to their support in decreasing order. The probability of applying an itemset to a solution (i.e. fixing elements at different positions) depends on the itemset support. For instance, a 2-itemset (e.g. element 5 at position 10 and element 1 at position 7) that appears in 2% of all previous solutions has also a probability of 2% to be in a new solution.

If the current solution cannot be completely constructed from the exploration of all itemsets, all unassigned elements will be randomly added at unassigned positions.

4   Performance Evaluation

The amount of computing required for machine learning is extremely high. Traditionally, neural networks use GPUs with thousands of processors. In the context of this study, where the type of calculation is very different (local search), we have observed in other works [16] that the use of GPUs offers only moderate speed-ups. This is why we parallelized the calculations on the CPU. By using dynamic scheduling to perform independent local searches and by storing intermediate files that serve as buffers, we were able to limit computation time and RAM requirements. Indeed, the candidate itemsets for one generation representing up to a dozen of gigabytes of data, they are written in a file and a buffer storing 10,000 candidate itemsets is reused accordingly. This allowed us to achieve satisfactory speedups. However, the computation time to get a generation is close to one day. The processing of a problem instance can therefore be counted in weeks.

The computational results presented in this section have been obtained on a PC running on Linux and equipped with an AMD Ryzen Threadripper 1950X 3.4Ghz (16 cores/32 threads). The algorithms introduced in Sect. 2 have been implemented in C++ using the OpenMP Library for the parallelization. This parallelization approach results in very good speed-ups (from 12× to 15× according to the number of candidate itemsets).

4.1 QAP Instances

The QAPLIB repository [6] contains 136 instances and has been enriched by hundreds other ones freely available on http://mistic.heig-vd.ch/taillard/problemes. dir/qap.dir/qap.html. Since it was not practically possible to conduct our numerical experiments for all instances, only 12 QAP instances have been carefully selected. All selected instances are widely studied in the literature [15].

The selected instances cover a large panel of the flows/distances matrices structures that can be found in the literature. Their size (n between 45 and 64) is large enough so that an exact method cannot solve most of them on modern computers. When possible, we selected 2 or 3 examples of the same type but of slightly different size. This was done in order to verify a certain consistency in the results presented. Much larger numerical experiments would have been necessary to confirm the results in an unambiguous way. On the one hand, the volume of calculation required would be prohibitive and on the other hand, the results obtained on the selected examples show a very high volatility in the quality of the solutions. As we will see later, it seems therefore that our unsupervised learning approach is not usable for all types of problems.

The first 3 instances are from Skorin-Kapov [20] (sko49, sko56 and sko64). No optimal solution has been proven yet for these instances. The distances are Manhattan on a rectangular grid, and the flows are pseudo-random numbers. These instances are similar to Nugent et al. [17] ones, but larger. Due to symmetries in the distance matrix, multiples of 4 or 8 optimal solutions exists.

Then, 3 asymmetrical instances from Li and Pardalos (lipa50a, lipa60a and lipa50b) were selected. These instances were generated so that the optimal solutions are known [14].

Then, 2 symmetrical instances with flows and distances randomly, uniformly generated have been selected (tai50a and tai60a) [21]. These instances are similar to Roucairol's ones, but larger.

Then, 2 asymmetrical instances non-uniformly generated (tai50b and tai60b) comes from [22]. An instance for generating grey patterns (tai64c) proposed in the same article has also been selected. This instance is not specially hard, but has a very large number of optimal solutions, spread all over the solutions' space.

Finally, a symmetrical and structured instance (tai45e01) proposed in [9] was selected. This instance was generated in such a way that a number of local search based methods have difficulties to find a moderately good solution.

4.2   Parameters for the Experiments

The algorithms of this paper rely on extracting most frequent itemsets from all solutions then combining them to create a new set of solutions.

In Algorithm 1 the number of generations is set to 8 and 10,000 local searches are executed per generation.

Regarding the combining phase, the first set of experiments are based on the random exploration of frequent itemsets (REFI) whereas the second one is on the exploration on sorted frequent itemsets (ESFI). A multi-start with 90,000 local search algorithms from random solutions is also considered. Even if the execution time differs, it is used as an indicator of comparison where no learning process is implemented. Disregarding the time needed for selecting the itemsets and building starting solutions, all the methods are indeed performing 90,000 local searches.

The default minimum support for the extraction of itemsets is set to 0.1% (i.e. keep itemsets which appear in 10 out of 10,000 solutions). The itemsets limit is set to one million for each k-itemset. These parameters have been tuned in such a way that each generation does not exceed one day of calculation.

4.3   Quality of Solutions

For optimization problems, the main criteria to be evaluated is the quality of the solutions. The last is measured relative to the value of the best solution known to date (bvk), which is optimal for a few instances (lipa, tai64c and tai45e01) or believed to be optimal for the other ones. The distribution of the quality of the solutions is visualized with the proportion of runs having reached a solution below a given percentage above best known.

The quality of the solutions for the instances are graphically illustrated in Fig. 2. All the solutions compared to the bvk are represented for the 90,000 solutions found by the multi-start algorithm (dash-dotted line) and the 8 generations of REFI (plain line) and ESFI (dotted line) learning methods.

For the instance sko49, the distribution reveals that most solutions produced by REFI and ESFI algorithms are, respectively, about 0.5% and 1% above the bvk whereas multi-start produces solutions with a normal spread around 3% above the bvk. A similar observation can be made for the instance sko56. The phenomenon is more pronounced for the instance sko64 where the REFI algorithm was able to produce solutions very close to the bvk.

The benefits of the learning phase are also prominent for the lipa50a instance, where the multi-start from random solutions was unable to find the optimum.

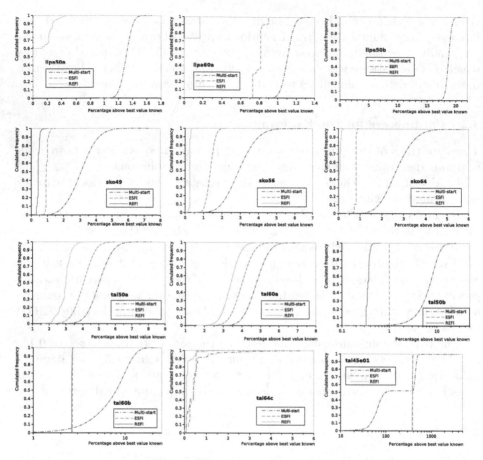

Fig. 2. Multi-start, ESFI and REFI distribution of solutions quality for lipa50a, lipa60a, lipa50b (asymmetric with known optimal solutions), sko49, sko56, sko64 (Manhattan distances on a square grid), tai50a, tai60a (uniformly generated), tai50b, tai60b (asymmetric and randomly generated), tai64c, tai45e01 (structured).

A similar behavior stands for the lipa60a instance, except that REFI is the only algorithm able to reach the known optimum.

Regarding the lipa50b instance, the difference of quality is very important since most REFI and ESFI solutions are optimal whereas multi-start solutions are between 15 and 20% above the bvk. For the 11 selected instances, this is the only one for which learning with itemsets is highly successful.

For the tai50a instance, there is a moderate trend indicating that most of the REFI and ESFI runs are able to learn something. Unsurprisingly, the learning is less pronounced for randomly, uniformly generated instances. A similar observation can be made for the tai60a instance.

Regarding the instance tai50b, most ESFI and REFI solutions are between 0.5 and 1% above the bvk. The multi-start algorithm produces solutions that are spread 7.3% above the bvk with a standard deviation of 3.3%.

A similar observation can be made for the instance tai60b. The multi-start algorithm solutions are spread 8% above the bvk with a standard deviation of 3.4%. For this type of instances, learning with itemsets is possible, but not as successful as it is for lipa..b instances.

Regarding the structured instance tai45e01, the ESFI and REFI algorithms are completely unable to learn something interesting. These algorithms are just focusing on solutions that are very far from the optimal one. The learning techniques based on the frequent itemsets seem to be inefficient for dealing with such structured instances. The population of solutions just converges too early. We were rather surprised by this result, since various metaheuristics combining a local search with a learning mechanism are perfectly able to reach the optimum, for instance GRASP with path relinking, late acceptance local search, genetic hybrids or ant systems [9].

For tai64c, most solutions being below 1% above the bvk, it is not clear that a learning algorithm outperforms a simple multi-start. It might be explained by the fact that the instance has multiple global optima, and it is easy to solve it optimally [8].

4.4 Additional Information for the Positions of Solutions

Another criterion to assess is the similarity of the solutions produced by the algorithms with a target solution with bvk. The similarity can be measured by the number of elements in that are at the same position. These results are reported in Table 1. The third column provides the mean and the standard deviation of the number of positions identical to the target solution. The next two columns are the percentage of solutions under 5% above the bvk including those sharing at least 10% of common positions with the target. The next column provides the percentage of different solutions. This proportion gives an indication of the population diversity. Finally, the number of itemsets revealing all the patterns discovered during the exploration phase is reported.

Table 1 shows for the sko49 instance that the number of positions identical to the target is almost non-existent for all algorithms (between 1 and 2 on the average). It is an easy instance since more than 98% of solutions are under 5% above the bvk, including the simple multi-start from random solutions. As shown in Fig. 2, the learning mechanism helps to improve the last percentages above the bvk.

A similar observation can be made for sko56. The main difference is that among all the solutions that are under 5% above the bvk, there is a significant percentage of solutions (61.88% for REFI and 24.61% for ESFI) that share more than 10% of common positions with the target. The same remark occurs for the instance sko64 but the diversity of REFI solutions is pretty low (3.04%).

Regarding lipa50a and lipa60a instances (asymmetric with known optimal solutions), the number of shared positions of REFI and ESFI with the target is

Table 1. Additional results for the positions of produced solutions for sko49, sko56, sko64, lipa50a, lipa60a, lipa50b tai50a, tai60a, tai50b, tai60b, tai45e01 and tai64c instances: number of positions that are identical (mean and standard deviation) to a target solution, percentage of solutions under 5% above the best value known (bvk) and, for those that share at least 10% of common positions with the target, percentage of different solutions and total number of itemsets discovered.

Instance	Algorithm	# positions identical to the target	% solutions under 5% above the bvk		% different solutions	# itemsets
			all	pos > 10%		
sko49	REFI	$1.3_{1.5}$	99.80%	3.85%	51.35%	66.8×10^6
	ESFI	$1.1_{1.4}$	99.74%	2.69%	69.78%	58.9×10^6
	multi-start	$1.8_{1.7}$	98.32%	7.20%	100.00%	–
sko56	REFI	$7.6_{4.9}$	99.92%	61.88%	46.13%	83.1×10^6
	ESFI	$4.1_{2.2}$	99.93%	24.61%	83.80%	58.9×10^6
	multi-start	$2.0_{2.0}$	99.27%	6.01%	100.00%	–
sko64	REFI	$7.4_{7.2}$	100.00%	49.74%	3.04%	136.4×10^6
	ESFI	$4.6_{2.3}$	99.99%	17.16%	67.41%	84.3×10^6
	multi-start	$2.2_{2.0}$	99.94%	3.55%	100.00%	–
lipa50a	REFI	$30.3_{21.1}$	100.00%	71.57%	45.46%	79.9×10^6
	ESFI	$22.4_{16.9}$	100.00%	70.06%	59.93%	101.4×10^6
	multi-start	$1.3_{1.5}$	100.00%	1.87%	100.00%	–
lipa60a	REFI	$32.7_{27.0}$	100.00%	66.31%	50.06%	176.7×10^6
	ESFI	$10.1_{8.2}$	100.00%	53.88%	77.09%	63.8×10^6
	multi-start	$1.2_{1.4}$	100.00%	0.61%	100.00%	–
lipa50b	REFI	$50.0_{0.0}$	100.00%	100.00%	0.00%	37.6×10^6
	ESFI	$49.1_{6.2}$	98.03%	98.03%	1.97%	23.0×10^6
	multi-start	$1.9_{3.7}$	0.41%	0.41%	99.59%	–
tai50a	REFI	$1.9_{1.4}$	79.93%	0.71%	88.39%	20.5×10^6
	ESFI	$1.6_{1.3}$	76.29%	0.64%	100.00%	40.2×10^6
	multi-start	$1.1_{1.2}$	48.76%	0.25%	100.00%	–
tai60a	REFI	$1.8_{1.4}$	88.77%	0.49%	99.99%	33.0×10^6
	ESFI	$1.6_{1.3}$	85.50%	0.26%	100.00%	24.2×10^6
	multi-start	$1.2_{1.2}$	66.41%	0.09%	100.00%	–
tai50b	REFI	$22.9_{11.4}$	87.58%	78.93%	12.86%	106.0×10^6
	ESFI	$1.2_{1.7}$	86.11%	1.56%	38.59%	108.2×10^6
	multi-start	$2.1_{2.6}$	26.70%	4.28%	100.00%	–
tai60b	REFI	$1.8_{3.5}$	97.67%	3.70%	24.40%	123.4×10^6
	ESFI	$2.5_{2.0}$	94.27%	1.88%	25.53%	226.9×10^6
	multi-start	$3.1_{3.1}$	21.91%	6.89%	100.00%	–
tai45e01	REFI	$0.9_{3.6}$	0.02%	0.02%	10.68%	33.2×10^6
	ESFI	$0.5_{3.1}$	0.03%	0.03%	9.48%	109.4×10^6
	multi-start	$3.3_{5.0}$	0.01%	0.01%	96.76%	–
tai64c	REFI	$30.0_{15.1}$	99.98%	55.01%	100.00%	71.6×10^6
	ESFI	$22.8_{17.3}$	99.98%	33.73%	94.41%	45.1×10^6
	multi-start	$10.4_{13.7}$	99.98%	6.14%	100.00%	–

prominent (around 10 and 30). But it is not a difficult instance since 100% of solutions are under 5% above the bvk. The learning phase is also determinant for improving the last percentages above the bvk.

The lipa50b case (high values for matrix entries) is interesting since only 0.41% of multi-start solutions are under 5% above the bvk. The benefits of learning mechanisms are meaningful for this instance since most REFI and ESFI solutions converge to the target.

For tai50a and tai60a instances, Table 1 shows that the number of positions identical to the target is also almost non-existent. Indeed, the produced solutions that share 10% of common positions with the target and that are under 5% above the bvk is less than 1%. The number of itemsets (patterns) discovered for both tai50a and tai60a is lower than the other instances.

Things are quite different for the tai50b (asymmetric and randomly generated) where the percentage of different solutions is rather low (less than 40%), meaning that many solutions converge to the same local optima. On the one hand, the solutions produced by REFI share an important number of common positions with the target (22.9 in average). On the other hand, ESFI has very little in common with the target. In both cases, the number of discovered itemsets is rather high (more than 100 millions) and 85% solutions are under 5% above the bvk. It is really significant in comparison with a multi-start where only 26.7% solutions are within the same quality.

A similar observation can be made for the tai60b instance. Even if the number of positions shared with the target is pretty low (less than 2.5), more than 94% of produced solutions by a learning algorithm are under 5% above the bvk, whereas a simple multi-start has only 21.91% under this level. Interestingly this instance has generated the highest number of different patterns.

Regarding the instance tai45e01, the diversity of the population of solutions is also significantly low. It represents less than 11% of different solutions even if the number of itemsets is significant. The number of positions identical to the target is even lower than a multi-start with 90, 000 random solutions. The number of solutions under 5% above the bvk is close to 0%. It seems that the learning mechanisms studied in this article are not really efficient for such a structured instance. The itemsets produced are just focusing on bad quality local optima, very far from the global optimum. It can be noticed that both REFI and SEFI got the optimal solution at the first generation, but this solution was lost for the remaining generations.

The structured tai64c is easy to solve since 12, 715 different global optima were found during the different runs. Since global optima are spread all over the solutions' space, it is not clear whether something can be learned with itemsets or not. Anyway, since 99.98% solutions are under 5% above the bvk for all algorithms, the benefits of a learning process are not really meaningful in the context of optimization.

5 Conclusions

The main interest in combining the unsupervised association rule learning with metaheuristics is to discover useful knowledge about the history of the search in order to enhance the produced solutions.

In this paper, we proposed to incorporate the extraction of frequent itemsets for parallel local search algorithms in a big data context. The global process can be iterated through two phases: the extraction of millions frequent itemsets and their combination for generating new solutions.

For QAP, learning mechanisms using association rules produce very contrasting results depending on the problem examples. For some, we observe a very good convergence towards the best known solutions. For other examples, the technique can focus on a restricted space of very bad solutions. The ESFI method, which biases the selection of itemsets based on their frequency of occurrence, is not as good as the REFI method, which selects them in a uniform manner. In the context of metaheuristics, this paper shows that, for QAP, it is very important to direct the search by taking into account a fitness function and to use it to filter the solution patterns.

This contrasts with works on other optimization problems like the travelling salesman. Indeed, for this problem, a few dozen of a very fast randomized local search is able to extract most of the components of target solutions. Since learning techniques can be very efficient for this optimization problem, it would be interesting to study its behavior for other problems where a permutation is sought, such as the flowshop scheduling problem.

In contrast with metaheuristics dedicated to a specific optimization problem, the advantage of these learning techniques is that they are rather simple to design and do not require a priori knowledge of the problem at hand. The drawback is that they take a full day on a single machine to train one generation of solutions. In comparison, the dedicated robust taboo search [21], genetic hybrid [22] or the fast ant systems [23] will find better solutions in just a few minutes. The results obtained in this study can be seen as an a posteriori justification of certain strategies used by these methods. Indeed, the pheromone traces of the artificial ant colonies are statistics collected on 1-itemsets (frequency of occurrence of an item at a position, weighted by the fitness of the solution). The taboo list prohibits returning to solutions with certain 2-itemsets. The diversification mechanism of Ro-TS forces the presence of 1-itemsets whose frequency of occurrence is low. Finally, the crossover operator of a genetic algorithm copies the largest itemset present in 2 solutions and randomly selects 1-itemsets present in these solutions.

Let us mention that, in the context of big data, one day of calculation on a single machine is still reasonable regarding usual machine and deep learning trainings that may take a couple of weeks on a cluster of GPU-based machines [10].

A research avenue could be a finer tuning of parameters (i.e. minimum support, itemsets limit and number of solutions) to see how they can influence the search process and to control the duration of the execution according to the scenario. For example, a low minimum support allows limiting the training phase to couple minutes, while a higher number of solutions will make it last a week. Another perspective could be to investigate how machine learning can enhance state-of-the-art metaheuristics for the QAP. It would also be necessary to study supervised learning mechanisms, in particular by taking into account the fitness of the solutions.

The general conclusion of this paper is that the quality of the solution produced with these learning techniques is not competitive compared to state-of-the-art metaheuristics and that there is still a long way till general learning techniques will surpass more direct optimization techniques for the QAP.

References

1. Agrawal, R., Imieliński, T., Swami, A.: Mining association rules between sets of items in large databases. In: Proceedings of the 1993 ACM SIGMOD International Conference on Management of Data, pp. 207–216 (1993)
2. Agrawal, R., Srikant, R.: Fast algorithms for mining association rules. In: International Conference on Very Large Databases (VLDB), pp. 487–499 (1994)
3. Alpaydin, E.: Introduction to Machine Learning. MIT Press, Cambridge (2020)
4. Birattari, M., Kacprzyk, J.: Tuning Metaheuristics: A Machine Learning Perspective, vol. 197. Springer, Heidelberg (2009)
5. Burkard, R.E., Çela, E., Rote, G., Woeginger, G.J.: The quadratic assignment problem with a monotone anti-Monge and a symmetric Toeplitz matrix: easy and hard cases. Math. Program. **82**, 125–158 (1998)
6. Burkard, R.E., Karisch, S.E., Rendl, F.: QAPLIB-a quadratic assignment problem library. J. Glob. Optim. **10**(4), 391–403 (1997). https://coral.ise.lehigh.edu/data-sets/qaplib/
7. Calvet, L., de Armas, J., Masip, D., Juan, A.A.: Learnheuristics: hybridizing metaheuristics with machine learning for optimization with dynamic inputs. Open Math. **15**(1), 261–280 (2017)
8. Drezner, Z.: Finding a cluster of points and the grey pattern quadratic assignment problem. OR Spectr. **28**(3), 417–436 (2006). https://doi.org/10.1007/s00291-005-0010-7
9. Drezner, Z., Hahn, P.M., Taillard, É.D.: Recent advances for the quadratic assignment problem with special emphasis on instances that are difficult for meta-heuristic methods. Ann. OR **139**(1), 65–94 (2005). https://doi.org/10.1007/s10479-005-3444-z
10. Erhan, D., Courville, A., Bengio, Y., Vincent, P.: Why does unsupervised pre-training help deep learning? In: Proceedings of the Thirteenth International Conference on Artificial Intelligence and Statistics, pp. 201–208 (2010)
11. Han, J., Pei, J., Yin, Y.: Mining frequent patterns without candidate generation. In: Proceedings of the 2000 ACM SIGMOD International Conference on Management of Data, SIGMOD 2000, pp. 1–12. Association for Computing Machinery, New York (2000). https://doi.org/10.1145/342009.335372
12. Koopmans, T.C., Beckmann, M.: Assignment problems and the location of economic activities. Econometrica **25**(1), 53–76 (1957). https://www.jstor.org/stable/1907742
13. L'heureux, A., Grolinger, K., Elyamany, H.F., Capretz, M.A.: Machine learning with big data: challenges and approaches. IEEE Access **5**, 7776–7797 (2017)

14. Li, Y., Pardalos, P.M.: Generating quadratic assignment test problems with known optimal permutations. Comput. Optim. Appl. **1**(2), 163–184 (1992)
15. Loiola, E.M., de Abreu, N.M.M., Boaventura-Netto, P.O., Hahn, P., Querido, T.: A survey for the quadratic assignment problem. Eur. J. Oper. Res. **176**(2), 657–690 (2007)
16. Luong, T.V., Melab, N., Taillard, É.D., Talbi, E.G.: Parallelization strategies for hybrid metaheuristics using a single GPU and multi-core resources. In: 12th International Conference on Parallel Problem Solving From Nature (PPSN) Proceedings (2012). http://mistic.heig-vd.ch/taillard/articles.dir/LuongMTT2012.pdf. Only preprint technical report available
17. Nugent, C.E., Vollmann, T.E., Ruml, J.: An experimental comparison of techniques for the assignment of facilities to locations. Oper. Res. **16**(1), 150–173 (1968). https://doi.org/10.1287/opre.16.1.150
18. Park, J.S., Chen, M.S., Yu, P.S.: An effective hash-based algorithm for mining association rules. ACM SIGMOD Rec. **24**(2), 175–186 (1995)
19. Ribeiro, M., Plastino, A., Martins, S.: Hybridization of grasp metaheuristic with data mining techniques. J. Math. Model. Algorithms **5**, 23–41 (2006)
20. Skorin-Kapov, J.: Tabu search applied to the quadratic assignment problem. ORSA J. Comput. **2**(1), 33–45 (1990)
21. Taillard, É.D.: Robust taboo search for the quadratic assignment problem. Parallel Comput. **17**(4–5), 443–455 (1991)
22. Taillard, É.D.: Comparison of iterative searches for the quadratic assignment problem. Location Sci. **3**(2), 87–105 (1995). https://www.sciencedirect.com/science/article/pii/0966834995000086
23. Taillard, É.D.: Fant: Fast ant system. Technical report. IDSIA-46-98, Istituto Dalle Molle Di Studi sull'Intelligenza Artificiale (1998)
24. Taillard, É.D.: A linearithmic heuristic for the travelling salesman problem. EURO J. Oper. Res. **297**(2), 442–450 (2022). https://doi.org/10.1016/j.ejor.2021.05.034. Available online June 2021
25. Taillard, É.D., Heslgaun, K.: POPMUSIC for the travelling salesman problem. EURO J. Oper. Res. **272**(2), 420–429 (2019). https://doi.org/10.1016/j.ejor.2018.06.039
26. Wang, R., Yan, J., Yang, X.: Neural graph matching network: learning Lawler's quadratic assignment problem with extension to hypergraph and multiple-graph matching (2020)
27. Zhou, Z.H., Chawla, N.V., Jin, Y., Williams, G.J.: Big data opportunities and challenges: Discussions from data analytics perspectives [discussion forum]. IEEE Comput. Intell. Mag. **9**(4), 62–74 (2014)

On Optimizing the Structure of Neural Networks Through a Compact Codification of Their Architecture

Marcos Lupión[1](✉), N. C. Cruz[2] iD, B. Paechter[3] iD, and P. M. Ortigosa[1] iD

[1] Department of Informatics, University of Almería, ceiA3 Excellence Agri-food Campus, Almería, Spain
{marcoslupion,ortigosa}@ual.es
[2] Department of Computer Architecture and Technology-CITIC, University of Granada, Granada, Spain
ncalvocruz@ugr.es
[3] School of Computing, Edinburgh Napier University, Edinburgh, Scotland, UK
B.Paechter@napier.ac.uk

Abstract. Neural networks stand out in Artificial Intelligence for their capacity of being applied to multiple challenging tasks such as image classification. However, designing a neural network to address a particular problem is also a demanding task that requires expertise and time-consuming trial-and-error stages. The design of methods to automate the designing of neural networks define a research field that generally relies on different optimization algorithms, such as population meta-heuristics. This work studies utilizing Teaching-Learning-based Optimization (TLBO), which had not been used before for this purpose up to the authors' knowledge. It is widespread and does not have specific parameters. Besides, it would be compatible with deep neural network design, i.e., architectures with many layers, due to its conception as a large-scale optimizer. A new encoding scheme has been proposed to make this continuous optimizer compatible with neural network design. This method, which is of general application, i.e., not linked to TLBO, can represent different network architectures with a plain vector of real values. A compatible objective function that links the optimizer and the representation of solutions has also been developed. The performance of this framework has been studied by addressing the design of an image classification neural network based on the CIFAR-10 dataset. The achieved result outperforms the initial solutions designed by humans after letting them evolve.

Keywords: Artificial intelligence · Neural network architecture optimization · Meta-heuristics · Teaching-Learning-based optimization

1 Introduction

Deep neural networks simulate the behavior of the human brain to perform different tasks, such as image classification, object recognition, language pro-

L. Di Gaspero et al. (Eds.): MIC 2022, LNCS 13838, pp. 133–142, 2023.
https://doi.org/10.1007/978-3-031-26504-4_10

cessing, and anomaly detection [13]. There exist multiple kinds of fundamental neural networks (e.g., recurrent, convolutional, and fully connected), and they are known to stand out in different tasks. However, designing a neural network to address a particular problem is not trivial. It is not only necessary to select the type of network but also its configuration or architecture, which requires expertise and takes multiple attempts.

There have been proposed neural networks developed throughout the years specially designed for particular problems. For instance, Yolo [12] aims at object detection, ResNet [5] and VGG [15] have been conceived for image processing, while Pix2Pix [6] focuses on image domain conversion. Regardless, pre-arranged designs do not always fit well with the working dataset. There is usually room for improvement in the architecture of neural networks tailored for real applications. It might even be a necessity to achieve competitive results. Therefore, finding the most appropriate neural network architecture remains an open problem. It results in an active research field known as Neural Architecture Search (NAS) [8], in which Google was a pioneer in designing algorithms for this purpose [20].

In this context, population-based algorithms are widely used to address the underlying optimization problem [8,11]. Since these methods are meta-heuristics, the computational effort will be significant, and the results might be sub-optimal. Nevertheless, their exploration capabilities significantly outperform human designers, whose search process will be tedious and potentially biased. Besides, the gain increases with the problem size, as the options to try rapidly become overwhelming for human beings.

Population-based algorithms work with sets of candidate solutions and treat them as individuals that evolve by interacting with each other. Each solution represents a feasible network architecture and has an associated value representing its quality or performance (fitness). Hence, it is necessary to define how to encode all the relevant properties of network architectures. The representation of solutions and the particular optimizer ultimately determine how the evolution of individuals is simulated. More specifically, the encoding of every layer [14] and block (set of layers) [2] is critical. So far, most strategies propose fixing the network size [9,18], limiting the attainable configurations. Nevertheless, some proposals also let the size of candidate networks vary during the search [16], which allows exploring more options automatically. Regarding the population-based methods considered, it is possible to mention Particle-Swarm Optimization (PSO) [16] and Ant-Colony Optimization (ACO) [1], which find designs comparable to those defined by human experts.

This work proposes using Teaching-Learning-based Optimization (TLBO), a population-based meta-heuristic, for neural network architecture optimization. This algorithm was designed for large-scale problems, i.e., with many variables. In contrast, the performance of methods such as PSO and ACO decreases with the number of variables [17]. Furthermore, it is virtually parameter-less since the optimizer only expects the population size and cycles to run. Hence, TLBO is more suitable than other population-based strategies to design deep neural networks. Coupled with TLBO, this work presents a compact representation of candidate network architectures. It encodes any possible configuration as a real vector with two components per layer. The first one represents the class of layer, which can be of any type, while the second one is mapped to the virtual space of parameters. A convolutional neural network for image classification has been designed to assess the effectiveness of these proposals. As another novelty, batch normalization and dropout layers, which attenuate network over-fitting, are considered. Hence, to summarize, this work tries TLBO for a NAS problem for the first time up to our knowledge, presents a new neural network architecture encoding scheme, and assesses them in a widespread benchmark problem.

The rest of the paper is structured as follows: Sect. 2 describes the optimization algorithm, the proposed solution encoding method, and the fundamentals of an appropriate objective function linking both. Section 3 presents the problem addressed in this work to assess the effectiveness of the proposals. Section 4 shows the experimentation carried out. Finally, Sect. 5 draws conclusions and states future work.

2   Proposed Methodology

This section outlines the optimizer suggested for neural network architecture optimization, the solution encoding proposed, and the objective function connecting the optimization algorithm with the solution encoding, and the main aspects of their assessment.

2.1   Teaching-Learning-Based Optimization (TLBO)

Teaching-Learning-based Optimization is a population-based meta-heuristic proposed in [10] for large-scale and derivative-free continuous optimization. This method is widely used due to its simplicity of implementation, configuration, and high performance [4]. It simulates a class of students that improve their skills through academic interaction. In practical terms, each solution is a candidate solution to the problem at hand. As introduced, this method lacks specific parameters and only expects the population size and number of cycles to run. Based on this information, TLBO creates and evaluates as many solutions as allowed. They will generally be random feasible points in the search space, yet some promising solutions can be injected to orientate the seek. Once it has an initial population, TLBO executes its main loop in as many cycles as requested. It consists of two consecutive stages, the teacher and the learner.

The teacher stage simulates how students learn from their professor. The best candidate solution is taken as the teacher, T, and the other solutions, i.e., points in the search space, are shifted towards it. More specifically, after identifying T, TLBO computes a vector M containing the mean for each component in the population. Then, for each solution S, the optimizer computes a modified version, S'. It applies Eq. (1), which is expressed in terms of each component or problem dimension, i. T_F is a random integer that can be either 1 or 2, and r_i is a random real number in the range $[0, 1]$. Both, r_i and T_F are fixed until the next iteration. At the end of this stage, all the modified solutions that outperform their original versions replace them.

$$S_i' = S_i + r_i \left(T_i - T_F M_i\right) \tag{1}$$

The learner stage models how students learn from each other. For this purpose, TBLO pairs every candidate solution, S, with another partner, P_S, to create a modified version of it, S', according to Eq. (2). It is also expressed in vectorial terms, so r_i is a real random number in the range $[0, 1]$ for the component i. Again, these stage-specific random factors are fixed until being recomputed at the next iteration. As can be seen, this stage attempts a local shift for every S in the (improving) direction that it defines with its pair. At the end of this stage, the altered solutions that outperform their original versions replace them.

$$S_i' = \begin{cases} S_i + r_i \left(S_i - P_{S_i}\right) & \text{if } S \text{ better than } P_S \\ S_i + r_i \left(P_{S_i} - S_i\right) & \text{otherwise} \end{cases} \tag{2}$$

After having executed all the iterations allowed, TLBO returns the best solution in its population.

2.2 Solution Encoding

TLBO is a method for continuous optimization, which implies that its candidate solutions must be vectors in $\mathbb{R}^N$, being N the number of dimensions of the search space. Accordingly, as explained in the next section, the objective or cost function that measures the fitness of solutions must be a function $f : \mathbb{R}^N \to \mathbb{R}$.

Every candidate solution must represent a particular network architecture to be ranked. For this purpose, and considering the previous requirements, they are encoded as real vectors. One can think of a network architecture definition as a sequence of layers. Each layer is a set of artificial neurons of a particular type and configuration. Hence, there are two main pieces of data to encode: the type of layer and its internal configuration. The proposed encoding represents this information using a single real value per layer, i. The integer part, I_i, defines the type, and the decimal part, D_i, is mapped to its corresponding space of possible configurations. The user must define the number of layers, and it ultimately defines the dimension of the optimization problem for TLBO, i.e., N.

Assigning any I_i to a particular type of layer is straightforward. For instance, if there are 3 types of layers, the encoded value will correspond to the first, second or third type depending of if the real value is in the range $[0, 1)$, $[1, 2)$, or $[2, 3)$, respectively. As can be seen, the type is ultimately defined by the integer part of the layer since the decimal part will always be between 0 and 0.9999.... The types available must be defined by the user, depending on the application. However, it is advisable to reserve a type for disabled layers. This way makes the search more flexible, and the optimizer might find simpler architectures, i.e., with fewer layers. Hence, the number of layers, N, is an upper bound in reality. This strategy overcomes the fixed conception of the number of dimensions of most optimization algorithms, like TLBO. A similar approach is followed in [3], where the authors look for the optimal configuration of a set of solar trackers. They assign the value 0 to those to be deactivated so that the number of dimensions is fixed from the optimization perspective.

The interpretation of the decimal part linked to any I_i, D_i is more sophisticated than the previous one. It encodes the configuration of the type of layer defined by I_i, but the number of parameters (and ranges) significantly vary from one type of layer to another. The proposed strategy for mapping the space of parameters to a single decimal part is based on how N-dimensional arrays can be represented in a single-dimensional vector in Computer Science. In that situation, the vector has a certain number of positions (length) resulting from multiplying the size of each dimension. The referred length can be ultimately scaled in the range $[0, 1]$. Undoing the conversion (re-scaling) requires rounding values. Thus, near decimal values will have the same interpretation or decoding. Nevertheless, like PSO and ACO, TLBO is a derivative-free optimizer, so it is not affected by occasional and reduced plateaus in the search space.

For example, let us think of a specific type of layer that expects two different integer parameters. The first one is the activation function and has 4 possible values. The second one is the number of neurons used, taking 500 different values. In this situation, the space of configurations can be seen as a 4×500 matrix (interpretation matrix). The resulting vector would have 2000 positions, from 0 to 1999, as usually considered in Computer Science. Figure 1 contains a graphical representation of this example and the interpretation of 0.85 as a sample decimal part. Since the integer part would have selected the type of layer, it is known that there are up to 2000 virtual positions in this case. Thus, decoding starts by re-scaling the decimal part from the real range $[0, 1.0) \in \mathbb{R}$ to $[0, 1999] \in \mathbb{Z}$, which is a virtual vector containing a component per possible value. Finally, the latter is known to come from a two-dimensional vector for this layer, so the one-dimensional index is mapped to a matrix of 4 rows (possible activation function) and 500 columns (number of neurons). The row results from dividing the one-dimensional index by the number of columns, and the column is equal to the modulus of that division.

2.3 Cost Function

As previously mentioned, the cost or objective function must be of the form $f : \mathbb{R}^N \to \mathbb{R}$ and evaluate the quality of a given solution. Before defining this function, it is first necessary to determine if feasible and infeasible neural network architectures exist. If every architecture is feasible, the objective function can be the average quality of the resulting network with a validation dataset after being trained with a training one.

If not every architecture is feasible, feasible architectures can be evaluated as described above. However, it is also necessary to define a way to assign them a value that tags them as infeasible (i.e., not desired), yet it must also distinguish between those that are better than the others. This way of treating infeasibility

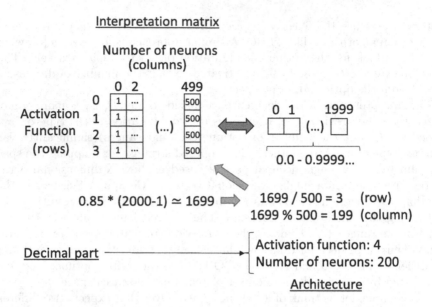

Fig. 1. Example of interpretation of decimal part as a neural network architecture.

is the recommended approach to handle constraints with population-based meta-heuristics [19].

3   Application Example

As introduced, the proposed architecture optimization strategy has been applied to an image classification problem addressed with convolutional neural networks. The CIFAR-10 dataset [7] has been used. It contains 60 000 images, 50 000 for training, and 10 000 for validation. In this context, 6 kinds of layers have been considered. Table 1 shows them and their associate parameters. The maximum number of layers allowed, i.e., N in terms of problem dimensionality, is 10, and each dimension is defined in the range $[0, 6)$.

For this problem, not all combinations of layers are possible. For instance, after 3 convolutional layers, the size of the resulting images is 5×5. In this situation, trying to apply another convolutional layer with filter size 7 would not be feasible since the remaining dimensions would not be greater than 0. More specifically, the following constraints are considered:

- The first layer must be convolutional.
- A convolutional layer cannot follow a fully-connected layer.
- As a feed-forward network, any layer must be compatible with its preceding one.

As previously explained, these constraints will be handled by gradually penalizing the fitness of the associated individuals. Besides, some of the initial solu-

Table 1. Types of layers and parameters for the application example.

ID	Layer type	Configurations	
		Name	Parameter ranges
0	Disabled	–	No parameters
1	Convolutional	Number of filters	[1–512]
		Filter X coordinate	[1–8]
		Filter Y coordinate	[1–8]
		Strides X coordinate	[1–4]
		Strides Y coordinate	[1–4]
		Activation Function	[ReLu, Tanh, Sigmoid, LeakyReLu]
2	Pooling	Type of Pooling	[MaxPooling, AveragePooling]
		Strides X coordinate	[2–4]
		Strides Y coordinate	[2–4]
3	Dropout	% of Hidden units	[1–90]
4	Batch Normalization	–	No parameters
5	Fully Connected	Number of neurons	[1–500]
		Activation Function	[ReLu, Tanh, Sigmoid, LeakyReLu]

tions created by TLBO are explicitly modified to be feasible. Otherwise, the optimizer would need significantly larger populations and many more iterations.

Regarding the objective function, feasible candidate architectures to evaluate are trained using backpropagation and the Adam optimizer with a batch size of 64 and a learning rate of 0.0004. The stopping criterion is not fixed to a given number of iterations or epochs. Instead, training stops when the neural network is not able to reduce its error with the validation dataset after two consecutive epochs. This strategy is implicitly self-adapted to the quality of each candidate's solution. It is also well aligned with the current trends in training neural networks, which try to avoid overfitting. Therefore, the value of feasible solutions, i.e., the standard objective function to minimize, can be expressed as follows:

$$cost = CrossEntropy + 10^{-6} params \qquad (3)$$

where $Cross\_Entropy$ is the error of the studied (and trained) network with the validation dataset, and $params$ is the number of parameters to train in the network. By adding this appendix to the plain error, simpler architectures are prioritized since they are harder to train and more computationally demanding at deployment. The factor of 10^6 wights the relevance of this aspect, and it can be adapted depending on the range of values and the relevance of simplicity. In several cases such as the development of deep learning models to be embedded in Internet of Things devices, the size of the model is a key factor, due to the limitations of memory and processor. Therefore, lightweight architectures have to be produced.

Concerning infeasible architectures, their value is 999 minus the number of layers not violating any constraints.

4 Experimentation and Results

The experiments have been run in the cluster of the University of Almería. In this cluster, a node with 2 NVIDIA TESLA V10 GPUs was used. It features 2 x AMD EPYC 7302 16-Core Processor with 512 GB of DDR4 (3200 MHz MHz) RAM. CUDA 11.0.2 and TensorFlow 2.4.1 have been used under CentOS 8.2 (OpenHPC 2) as the operating system.

After preliminary experimentation, TLBO has been configured to work with a population of 200 individuals and execute 400 cycles. This configuration resulted in 160 000 function evaluations approximately, which took 80 h. The best architecture found is shown in Table 2.

Table 2. Best design found at network architecture optimization.

Layer type	Configuration
conv	Filters = 61, FilterX = 5, FilterY = 5 StridesX = 1, StridesY = 1, Activation = Tanh
Pooling	Type = Avg, StridesX = 4, StridesY = 2
Pooling	Type = Avg, StridesX = 2, StridesY = 4
Dropout	Hidden = 55
BatchNorm	–
Fully Connected	Neurons = 61, Activation = ReLu
Fully Connected	Neurons = 10, Activation = Softmax

The resulting neural network architecture has 7 layers and 96 897 parameters. The value of the associate individual in terms of the objective function is 1.134. Its accuracy rate with images of the evaluation dataset is 64%. One of the initial solutions solution manually created for the problem at hand and loaded into the optimizer was a LeNet-CNN with 9 layers and 63 386 parameters, and it achieved an accuracy rate of 55% with the validation dataset. Another one was a VGG net with 9 layers and 124 330 parameters featuring an accuracy rate of 61% in this same context. Therefore, the proposed method has evolved its initial population of candidate architectures until outperforming them in terms of performance (and with less complexity, if compared to the VGG).

It is also necessary to highlight the impact of the cost function. If the number of parameters was ignored, the resulting neural networks could achieve higher accuracy, but the number of parameters would be significantly larger. For example, the following architecture was obtained in a preliminary experiment that did not penalize the number of parameters: Conv - Conv - BatchNorm - Conv - Dropout - BatchNorm - FullyConnected - FullyConnected. Its accuracy was 71% but at the expense of consisting of 4 365 253 parameters. However, there exist networks featuring 95% of accuracy in this dataset with that number of parameters.

5   Conclusions

This work has proposed and studied the use of TLBO, a widespread population-based optimizer, to optimize the architecture of neural networks for particular applications. Since the algorithm is for continuous optimization, a new encoding mechanism for network architectures has been developed. The method is based on how N-dimensional arrays can be represented in a single-dimensional vector in Computer Science. It can represent any architecture with a single vector of real values and a component per layer. The number of layers and their associated parameters can be extended as needed. Thus, the encoding mechanism is of general interest and can be coupled with any other continuous optimizer and used for any problem. The price to pay is the occasional noise or instability at decoding: Some different encoding might ultimately result in the same architecture and value. Nevertheless, small plateaus in the search space do not affect population-based meta-heuristics. Besides, the possibility of benefiting from the vast set of meta-heuristic optimizers for continuous problems compensates for this drawback.

The proposal has been studied by optimizing a convolutional neural network for image classification based on the CIFAR-10 dataset. There have been found neural network architectures that outperform expert-based configurations for this dataset. These results support the proposal and let us think about the possibility of replacing tedious architecture design stages by human experts with automated optimizers.

For future work, the objective function will be further studied. Namely, different stopping criteria will be compared. Similarly, reducing the training dataset to speedup evaluations will be considered. Additionally, the possibility of directly working with blocks of layers will be analyzed.

Acknowledgements. This research has been funded by the R+D+i project RTI2018-095993-B-I00, financed by MCIN/AEI/10.13039/501100011033/ and ERDF "A way to make Europe"; by the Junta de Andalucá with reference P18-RT-1193; by the University of Almería with reference UAL18-TIC-A020-B and by the Department of Informatics of the University of Almería. M. Lupión is supported by FPU program of the Spanish Ministry of Education (FPU19/02756). N.C. Cruz is supported by the Ministry of Economic Transformation, Industry, Knowledge and Universities from the Andalusian government.

References

1. Byla, E., Pang, W.: DeepSwarm: optimising convolutional neural networks using swarm intelligence. In: Ju, Z., Yang, L., Yang, C., Gegov, A., Zhou, D. (eds.) UKCI 2019. AISC, vol. 1043, pp. 119–130. Springer, Cham (2020). https://doi.org/10.1007/978-3-030-29933-0_10
2. Chen, Z., Zhou, Y., Huang, Z.: Auto-creation of effective neural network architecture by evolutionary algorithm and resnet for image classification. In: 2019 IEEE International Conference on Systems, Man and Cybernetics (SMC), pp. 3895–3900 (2019)

3. Cruz, N.C., Álvarez, J.D., Redondo, J.L., Berenguel, M., Ortigosa, P.M.: A two-layered solution for automatic heliostat aiming. Eng. Appl. Artif. Intell. **72**, 253–266 (2018)
4. Cruz, N.C., Marín, M., Redondo, J.L., Ortigosa, E.M., Ortigosa, P.M.: A comparative study of stochastic optimizers for fitting neuron models. Application to the cerebellar granule cell. Informatica **32**(3), 477–498 (2021)
5. He, K., Zhang, X., Ren, S., Sun, J.: Deep residual learning for image recognition. In: Proceedings of the IEEE Conference on Computer Vision and Pattern Recognition, pp. 770–778 (2016)
6. Isola, P., Zhu, J.Ya., Zhou, T., Efros, A.A.: Image-to-image translation with conditional adversarial networks. In: Proceedings of the IEEE Conference on Computer Vision and Pattern Recognition, pp. 1125–1134 (2017)
7. Krizhevsky, A., Hinton, G., et al.: Learning multiple layers of features from tiny images (2009)
8. Liu, Y., Sun, Y., Xue, B., Zhang, M., Yen, G., Tan, K.: A survey on evolutionary neural architecture search. IEEE Trans. Neural Netw. Learn. Syst. **PP**, 1–21 (2021)
9. Lu, Z., et al.: NSGA-Net: neural architecture search using multi-objective genetic algorithm. In: Proceedings of the Genetic and Evolutionary Computation Conference, pp. 419–427 (2019)
10. Rao, R.V., Savsani, V.J., Vakharia, D.P.: Teaching-learning-based optimization: an optimization method for continuous non-linear large scale problems. Inf. Sci. **183**(1), 1–15 (2012)
11. Real, E., et al.: Large-scale evolution of image classifiers. In: International Conference on Machine Learning, pp. 2902–2911. PMLR (2017)
12. Redmon, J., Divvala, S., Girshick, R., Farhadi, A.: You only look once: unified, real-time object detection. In: Proceedings of the IEEE Conference on Computer Vision and Pattern Recognition, pp. 779–788 (2016)
13. Sharma, N., Sharma, R., Jindal, N.: Machine learning and deep learning applications - a vision. Glob. Transit. Proc. **2**(1), 24–28 (2021)
14. Shu, H., Wang, Y.: Automatically searching for u-net image translator architecture (2020)
15. Simonyan, K., Zisserman, A.: Very deep convolutional networks for large-scale image recognition (2014)
16. Wang, B., Sun, Y., Xue, B., Zhang, M.: Evolving deep convolutional neural networks by variable-length particle swarm optimization for image classification. In: 2018 IEEE Congress on Evolutionary Computation (CEC), pp. 1–8. IEEE (2018)
17. Yang, Z., Li, K., Guo, Y., Ma, H., Zheng, M.: Compact real-valued teaching-learning based optimization with the applications to neural network training. Knowl.-Based Syst. **159**, 51–62 (2018)
18. Ye, F.: Particle swarm optimization-based automatic parameter selection for deep neural networks and its applications in large-scale and high-dimensional data. PLoS ONE **12**(12), e0188746 (2017)
19. Yeniay, Ö.: Penalty function methods for constrained optimization with genetic algorithms. Math. Comput. Appl. **10**(1), 45–56 (2005)
20. Zoph, B., Le, Q.V.: Neural architecture search with reinforcement learning (2016)

Neural Architecture Search Using Differential Evolution in MAML Framework for Few-Shot Classification Problems

Ayla Gülcü[1]([⊠])(iD) and Zeki Kuş[2](iD)

[1] Department of Software Engineering, Bahçeşehir University, Istanbul, Turkey
`ayla.gulcu@eng.bau.edu.tr`
[2] Department of Computer Engineering, Fatih Sultan Mehmet Vaqif University, Istanbul, Turkey
`zkus@fsm.edu.tr`

Abstract. Model-Agnostic Meta-Learning (MAML) algorithm is an optimization based meta-learning algorithm which aims to find a good initial state of the neural network that can then be adapted to any novel task using a few optimization steps. In this study, we take MAML with a simple four-block convolution architecture as our baseline, and try to improve its few-shot classification performance by using an architecture generated automatically through the neural architecture search process. We use differential evolution algorithm as the search strategy for searching over cells within a predefined search space. We have performed our experiments using two well-known few-shot classification datasets, mini-ImageNet and FC100 dataset. For each of those datasets, the performance of the original MAML is compared to the performance of our MAML-NAS model under both 1-shot 5-way and 5-shot 5-way settings. The results reveal that MAML-NAS results in better or at least comparable accuracy values for both of the datasets in all settings. More importantly, this performance is achieved by much simpler architectures, that is architectures requiring less floating-point operations.

Keywords: Meta-learning · Neural architecture search · Differential evolution

1 Introduction

Convolutional Neural Networks (CNNs) are known to achieve excellent results for a wide variety of computer vision tasks provided that *(i)* proper architecture search and hyper-parameter tuning is performed, and *(ii)* there is abundant data.

Performing architecture engineering and hyper-parameter tuning manually can be a solution for the first problem; however, this process can become

This work was supported by the Scientific and Technological Research Council of Turkey (TÜBİTAK) under grant number 121E240.

computationally infeasible as the architectures get deeper and more complex. This also requires involvement of an expert from the domain because, deciding on the architectural structure and the hyper-parameters to tune, and defining value domains for each of those elements require meticulous design and experiment processes. *Neural Architecture Search (NAS)* which is a subfield of *AutoML* aims to automate the tedious architecture design process. It is often confused with *Hyper-parameter Optimization (HPO)* which is another subfield of AutoML. HPO basically takes an architecture as given and tries to optimize its parameters like the learning rate, activation function and the batch size. In contrast, NAS focuses on optimizing architecture-related parameters like the number of layers, number of units and also connection types among those units. In NAS studies, in order to reduce the computational burden, the search space is defined over small building blocks, called *cells*, instead of the whole architecture. These cells are then stacked on top of each other to form the final architecture. The computation procedure for an architecture of a single cell is represented as a directed acyclic graph in which nodes represent the connections and the edges represent the operations between the nodes. The search space is then defined by the maximum number of nodes and edges and the types of operations in a single cell. This freedom of connections among different units may result in better performing architectures than following the traditional convolutional blocks patterns in CNNs.

The performance of CNNs drop significantly under limited data regime because, they require a huge amount of labeled data in order to be able to generalize well. However, in some circumstances finding labeled data can be a problem by itself. The type of problems for which there are only a few training examples available is referred to as *Few-Shot Learning* (FSL) problems [8,11,12], and these problems have led the development of new specialized methods such as meta-learning. Machine learning approaches that aim to make learning more generalizable with the help of meta-knowledge obtained from previous tasks, so that new tasks can be learned very quickly, are known as *learning to learn* or *meta-learning* [1,20]. Lu et al. [15] suggest that FSL research can be roughly divided into two periods, the first one being the period of methods without deep learning (from 2000 to 2015) and the second one being the period of methods with deep learning (from 2015 to now). In [5], meta-learning architectures for FSL are categorized into three main classes: memory-based methods, metric-based methods and optimization-based methods. In memory-based methods, a meta-learner is trained with memory to learn novel concept whereas in metric-based methods a meta-learner learns a deep representation with a metric in feature space. Model-Agnostic Meta-Learning (MAML) algorithm [9] that we use in this paper is an optimization-based algorithm which aims to find a good initial state of the neural network that can then be adapted to any novel task using a few optimization steps.

MAML [9] uses a four-block CNN architecture as its backbone. The architecture consists of four convolution blocks each comprising a 64-filter 3×3 convolution, a batch normalization layer, a ReLU nonlinearity and a 2×2 max-pooling layer. We take MAML with simple four-block architecture as our baseline method, and try to improve its classification performance by using an architecture generated automatically through Neural Architecture Search (NAS) process as its backbone.

We use Differential Evolution (DE) algorithm which is an Evolutionary Algorithm (EA) as the search strategy in NAS. The use of DE in NAS for optimizing deep neural networks architecture is not new; however, using this method as a search strategy in NAS for optimizing the architecture in a meta-learning algorithm is new. Since the training and evaluation of meta-learning algorithms is different, we believe this study provide useful insights to the researchers in this area. It is also important to note that a fair performance comparison with respect to the accuracy values can only be performed between the two models if the complexity of them which is best measured in terms of *floating-point operations per second (FLOPs)* are similar. Although the number of FLOPs is not taken as a constraint or as a second objective during NAS; the results state that the complexity of the architectures generated by NAS is simpler than that of the four-block CNN architecture. Moreover, these architectures result in the same or even better accuracy values in some settings.

2 Neural Architecture Search in MAML

In this section, we first provide a brief discussion on NAS methods, then explain the MAML framework on which NAS will be performed. Finally, we discuss DE which defines the NAS search strategy in this study.

2.1 Neural Architecture Search

Due to the increasing interest for the topic, several NAS methods have been introduced in the literature. These methods differ from each other in terms of the adopted search space, selected search strategy and performance prediction strategy [7]. The search strategy aims to find a good architecture within a the predefined search space subject to a computational budget. The prediction strategy returns the predicted performance of a new architecture for the search strategy (for a comprehensive survey on NAS please see [22]).

Earliest studies in the field use Reinforcement Learning (RL) based approaches. Although the work of Zoph and Le [28] achieved excellent results using RL in the image classification domain, their work is impossible to replicate due to its computational requirements. Searching for an architecture for CIFAR-10 dataset [10] required 800 GPUs for 28 days. Zoph et al. [29] propose searching over neural building blocks instead of whole architectures with the hope of reducing required computational resources. They propose a new search space, *NASNet*, consisting of small building blocks, called *cells*. Following NasNet, Liu et al. propose another cell-based search space called *DARTS*. Following these, several NAS benchmarks like NAS-Bench-101 [27], NAS-Bench-201 [6] and NAS-Bench-301 [23] have been introduced with the aim of enabling a fair comparison among different NAS algorithms and also reducing the computational burden of the process.

Evolutionary Algorithms (EA)-based methods are among the successful methods for NAS. Real et al. [21] used Regularized Evolution (RE), and showed for the first time that the architectures generated by EAs surpass hand-engineered architectures on the *NASNet* search space. They achieved comparable results with much fewer parameters when compared to the work in [29]. It is also shown [27] that RE yields in excellent performance for NAS-Bench-101. It is later shown in [3] that another EA-based method, Differential Evolution (DE), yields in better solutions than RE for several benchmarks. It is later shown in [23] that DE also had remarkable search performance. Interested readers may refer to [14] for a survey on Evolutionary NAS methods.

There is also another line of work focusing on predictor-based NAS algorithms which aim to estimate the performance of a previously unseen architecture to a high accuracy. As there is no benchmark available for meta-learning in *k-shot* setting and therefore we will not be searching on a standard search space, we will try to estimate the performance of an architecture by running it for a small number of iterations.

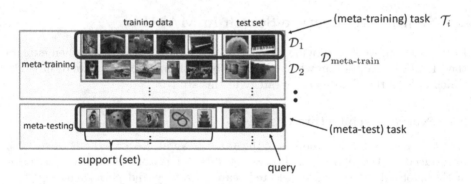

Fig. 1. Representation of a meta-dataset with a meta-training and a meta-testing datasets consisting of several small datasets, D_i, each of which is also called a *task*. Source: [20]

2.2 MAML Framework

Model-Agnostic Meta-Learning (MAML) algorithm is a model-agnostic algorithm in the sense that it can be applied to any model trained with gradient descent [9], and can be used to solve any problem like classification, regression and reinforcement learning problems.

A generic notion of a learning task can be described as follows [20]: In typical machine learning setting, a dataset D is usually splitted as D_{train} and D_{test}. In meta-learning, there is a meta-dataset D consisting of a training meta-dataset, $\mathcal{D}_{train}$, and a test meta-dataset $\mathcal{D}_{test}$ which is used to evaluate the generalization performance of the meta-model. Each meta-dataset consists of multiple datasets, $\mathcal{D}_i$, each of which is known as a *task* which is shown in Fig. 1. Each task is further splitted into a training and a test set known as a *support* and a *query* set, respectively. During MAML training, the meta-model with parameters θ is adapted with a number of gradient descent updates on the support set of a given task $\mathcal{T}_i$, and the model's parameters become θ'_i. This model is tested on the query set of the same task, and the loss is recorded. For each task, adaptation starts with the same initial model parameters θ and those parameters are only updated during the meta-optimization stage which starts after all tasks in a given batch are complete. Meta-model parameters are updated using the loss computed using the temporary model parameters. In meta-testing phase, the ability of the meta-learner to learn new tasks of novel classes is evaluated by training it on only a few images in the support set and testing on the query set. Usually, this process is repeated for several tasks, and the average test performance with 95% confidence interval is reported. In this default version of MAML, meta-update requires computing a gradient through a gradient. The authors also proposed a variant called *first-order MAML*, in which the second gradient computation is ignored. Although some gradient information is lost, it is also shown in the study that the first-order MAML works nearly as well as MAML and thus will be used in this study. There are also several variants of MAML algorithm in the literature. Nichol and Schulman [16] propose an algorithm called *Reptile* which is a variant of *first-order MAML*. Raghu et al. [19] introduce *ANIL* (Almost No Inner Loop) algorithm which is a simplification of MAML.

MAML [9] basically use a four-block CNN architecture given in [26] as its backbone. The architecture consists of four blocks each comprising a 64-filter 3×3 convolution, a batch normalization layer, a ReLU nonlinearity and a 2×2 max-pooling layer. The last layer is then fed into a softmax layer. This simple architecture is also used in another meta-learning algorithms, namely, Prototypical networks [24]. More complex backbones like Conv-6, Resnet-10, Resnet-18 and Resnet-34 are also used in other studies as the backbone; however, it is clearly shown in [4] that the performance gap among different meta-learning methods drastically reduces as the backbone gets deeper. As the performance difference of the algorithms can best be observed when a simple backbone model is used, a good architecture should help the meta-learning algorithm to achieve high accuracy without increasing the complexity.

2.3   Differential Evolution

Differential Evolution (DE) algorithm [18] is an evolutionary algorithm which consists of a population of solutions that evolve via several operators like selection, crossover and mutation. Each individual in a DE population represents a candidate solution which is usually encoded as a D-dimensional real-valued vector. For example, an individual $i \in \{1, 2, .., NP\}$ at generation 0 is represented by $\mathbf{x}_{i,0} = (x^1_{i,0}, .., x^D_{i,0})$, where NP is the size of the population which remains constant from iteration to iteration.

In general, the initial population consists of randomly initialized individuals. For each dimension in a given individual, a random value within the predefined range of that dimension is selected. A new generation is generated from the current population with the help of crossover, mutation and selection operations. At a generation g, each individual $x_{i,g}$ is a *target* vector, and a *mutant* vector, $\mathbf{v}_{i,g}$ is created for each of those target vectors via the selected mutation method. There are many mutation strategies in the literature, and the most frequently-used ones are listed below:

$$DE/rand/1 : \mathbf{v}_{i,g} = \mathbf{x}_{r1,g} + F(\mathbf{x}_{r2,g} - \mathbf{x}_{r3,g}) \tag{1}$$

$$DE/best/1 : \mathbf{v}_{i,g} = \mathbf{x}_{best,g} + F(\mathbf{x}_{r1,g} - \mathbf{x}_{r2,g}) \tag{2}$$

$$DE/current\text{-}to\text{-}best/1 : \mathbf{v}_{i,g} = \mathbf{x}_{i,g} + F(\mathbf{x}_{best,g} - \mathbf{x}_{i,g}) + F(\mathbf{x}_{r1,g} - \mathbf{x}_{r2,g}) \tag{3}$$

$$DE/rand/2 : \mathbf{v}_{i,g} = \mathbf{x}_{r1,g} + F(\mathbf{x}_{r2,g} - \mathbf{x}_{r3,g}) + F(\mathbf{x}_{r4,g} - \mathbf{x}_{r5,g}) \tag{4}$$

$$DE/best/2 : \mathbf{v}_{i,g} = \mathbf{x}_{best,g} + F(\mathbf{x}_{r1,g} - x_{r2,g}) + F(\mathbf{x}_{r3,g} - \mathbf{x}_{r4,g}) \tag{5}$$

In the shorthand notation in Eqs. 1–5, the first term after "DE" specifies the strategy to select the target vector, and the next term specifies the number of vector differences contributing to the differential. F is a positive control parameter used to scale the difference vector, and ri is a random integer selected from $[1, NP]$. In a given mutation method, all these random integers should be unique.

At a generation g, each target vector is subjected to a crossover operation with its corresponding mutant vector to generate a *trial* vector, $u_{i,g}$. The binomial crossover operation which is the most widely used one is given in Eq. 6, where CR is the crossover rate parameter $\in [0, 1)$. For each dimension, a random number $rand \in [0, 1]$ is compared to CR, and if it is less than CR, trial inherits that dimension value from the mutant; otherwise it inherits from the target. It is also ensured that at least one dimension is inherited from the mutant using n_j which is also a random number $\in \{1, ..D\}$.

$$u_{i,g}^j = \begin{cases} v_{i,g}^j, & \text{if } rand \leq CR \text{ or } j = n_j \\ x_{i,g}^j, & \text{otherwise} \end{cases}, \text{for } j = \{1, ..D\} \tag{6}$$

After the mutation and crossover operations, the resultant trial vectors should be checked for feasibility. A boundary check mechanism should be applied in order to restore feasibility of each trial vector. The encoding scheme and the other DE-related parameters will be given in detail in Sect. 3.4.

3   Experimental Setting

3.1   Datasets

MiniImageNet Dataset [26] is a common benchmark for few-shot learning algorithms. It contains 100 classes randomly sampled from the very large-scale ImageNet dataset. Each class contains 600 images of size 84×84. There are two different kinds of splits for this dataset and we use the splits from Ravi et al. [20] in this work. According to this split, the dataset is divided into 3 parts as train, validation, and test sets each of which contains 64, 16 and 20 classes, respectively.

Fewshot-CIFAR100 Dataset which is also known as FC100 was introduced by Oreshkin et al. [17] based on CIFAR100 for few-shot learning. The original CIFAR100 dataset consists of 32×32 color images belonging to 100 different classes, and each class contains 600 images. The 100 classes are further grouped into 20 superclasses in order to minimize the information overlap. The train split contains 60 classes belonging to 12 superclasses, the validation and test splits contain 20 classes belonging to 5 superclasses each. This dataset presents a more challenging few-shot learning problem than miniImageNet due to the reduced image size in addition to having less information overlap between train and test datasets.

3.2   Training and Evaluation

MAML requires several hyper-parameters like the number of inner and outer iterations, batch size and the learning rate to be tuned. As miniImageNet dataset has already been used by Finn et al. [9] in their MAML paper, we adopt the same hyper-parameters while evaluating the performance of MAML with a given architecture. The following hyper-parameters are used for training and testing the MAML algorithm:

- meta batch size = 4 for 1-shot, 2 for 5-shot
- meta learning rate and optimizer = 0.001 and Adam
- fast learning rate $\alpha = 0.01$
- train adaptation steps = 5
- test adaptation steps = 10
- number of train iterations = 60000

MAML algorithm with a given model is trained for 60000 iterations (meta-training phase) using Adam optimizer with a meta learning rate of 0.001. At each iteration, the model is trained for 5 gradient steps (train adaptation steps) with step size α of 0.01 (fast learning rate). The number of tasks is selected as 4 and 2 for the 1-shot 5-way and 5-shot 5-way setting, respectively. During the test time, the number of test adaptation steps is selected as 10; and the test results on each query set which contains 5 examples per class is recorded. 600 test episodes each of which contains a batch of tasks are created. The average accuracy with 95% confidence interval over these test episodes are reported. In our evaluation, we have used the same number of examples per class for both the support and the query set. Except this, all the hyper-parameters are the same as in [9].

In [9], MAML hyper-parameter values were tuned considering the CNN4 backbone. We use the same values for our MAML-NAS algorithm whose backbone is generated by NAS with the hope of performing a fair comparison. Moreover, we believe that the effect of using a backbone generated by NAS in MAML can be best evaluated if the other MAML hyper-parameters are kept constant even if this is a disadvantage for us.

During the execution of DE, the performance of a newly generated architecture needs to be evaluated so that it can be compared to another architecture.

The simplest approach is training the new model for 60000 iterations, testing it on 600 episodes, and then taking the average accuracy as the objective function value of that model. Although being reliable, a complete training of a model is computationally infeasible because hundreds of models need to be evaluated during a single DE run. We followed the most basic function approximation approach of training for smaller iterations. This pessimistic estimation does not effect the evolution of the good models, since our aim is be able to perform a fair comparison between different models rather than to obtain their actual accuracy values during the execution of DE. Thus, instead of 60000 iterations, we used 1000 iterations during meta-training. The average accuracy obtained over 16 tasks selected from the validation set is recorded as the objective function value of that model. We also used 5 test adaptation steps instead of 10 during test time. For FC100 Dataset, we used the same MAML hyper-parameters as in miniImageNet. After DE is complete, we have selected top 3 models with the best estimation performance and subjected each one them for a complete training. Then, the test performance of the best model is reported.

3.3   Search Space for MAML

Instead of the four-block architecture stated in Sect. 2.2, we use the architecture generated automatically using NAS for MAML. Following Zoph et al. [29] and Real et al. [21], we propose searching over neural building blocks called *cells* instead of the whole architectures.

The computation procedure for an architecture of a single cell is represented as a directed acyclic graph in which edges represent the connections and the nodes represent the operations between the nodes. In our study, we have adopted a new cell search space which is created considering the search spaces of both the DARTS [13] and NAS-Bench-101 [27]. The number of nodes are limited to seven including one input and one output nodes. These nodes can be connected with a maximum of nine edges with the following nine operations:

- Two separable convolution operations: 3×3 sep_conv, 5×5 sep_conv
- Two dilated separable convolution operations: 3×3 dil_conv, 5×5 dil_conv
- Four convolution operations: 1×1 conv_2d, 3×3 conv_2d, 5×5 conv_2d, 7×7 conv_2d
- Skip connection operation: skip_connect

In NAS-Bench-101 [27] benchmark, each cell is stacked three times which is followed by a downsampling layer. This pattern is then repeated three times in order to generate the final architecture. In order to keep the complexity of the resultant architecture at a reasonable level, we allowed only one cell in a stack which is followed by downsampling, and this pattern is repeated three times as shown in Fig. 2 (left). On the right, the inner representation of a cell which consists of all possible connections between the nodes is illustrated.

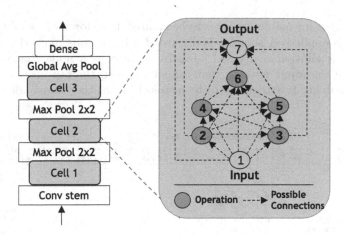

Fig. 2. Solution representation

3.4   Architecture Encoding and DE Parameters

In our study, we adopt a continuous search space as in [3] with the hope of maintaining diversity. Each solution x_i, where $i \in \{1, 2, .., NP\}$, is represented by a 26-dimensional real-valued vector that takes on continuous values in [0, 1] as illustrated in Fig. 3. As the mutation strategy, *rand1* mutation shown in Eq. 1 which was also used in both [3,23] is employed. After the mutation, *projection to boundary* method is applied on each dimension in order to bring the dimension value value into [0, 1]. We used the binomial crossover given in Eq. 6.

In Fig. 3, the first 21 dimensions represent the possible connections between each pair of nodes. As there are 7 nodes, there can be at most 21 edges in a given architecture. An upper triangular square matrix representation for the connected nodes is also given in the same figure. Each continuous value in these dimensions should be mapped to a discrete value with 1 denoting a connection and 0 denoting a no-connection, but these discretized copies of the individuals are only created for evaluation purposes. For a given i^{th} individual $\mathbf{x}_i = (x_i^1, .., x_i^D)$, discretized copy $x\_copy_i^j = 1$ if $x_i^j \geq 0.5$, and $x\_copy_i^j = 0$, otherwise, for each $j \in \{1, ..D\}$.

The operations on the intermediary nodes starting from node 2 up to node 6 in the order are represented by the remaining 5 dimensions. 9 equal-sized bins are created within [0, 1] in order to map these continuous values into 9 discrete values, for 9 different operations. For $j \in \{1, ..D\}$, if $x\_copy_i^j$ is $\in [0, 0.11)$ then skip_connect; else if it is $\in [0.11, 0.22)$ then 1×1 conv_2d; else if it is $\in [0.22, 0.33)$ then 3×3 conv_2d; else if it is $\in [0.33, 0.44)$ then 5×5 conv_2d; else if it is $\in [0.44, 0.55)$ then 7×7 conv_2d; else if it is $\in [0.55, 0.66)$ then 3×3 sep_conv; else if it is $\in [0.66, 0.77)$ then 5×5 sep_conv; else if it is $\in [0.77, 0.88)$ then 3×3 dil_conv; otherwise $\in [0.88, 1)$ then 5×5 dil_conv is applied.

Although there are 21 possible edges in a given architecture, edge constraint ensures that the total number of edges does not exceed 9. Our encoding scheme does not prevent infeasible solutions to be created; therefore, a feasibility control mechanism is applied. A solution is discarded if it contains more than 9 edges. In addition, a solution is deemed infeasible if there is a node that cannot be reached from the input node or that cannot reach to the output node. In this case, the feasibility is attempted to be restored by deleting that node.

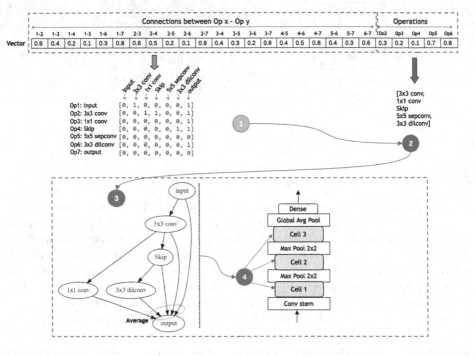

Fig. 3. DE encoding

In a given DE generation g, each target individual $\mathbf{x}_{i,g}$ competes with its corresponding trial vector $\mathbf{u}_{i,g}$ for being included in the next generation. If the objective value of $\mathbf{x}_{i,g}$ is better than the objective value of $\mathbf{u}_{i,g}$, then $\mathbf{x}_{i,g}$ is selected as a parent for the next generation; otherwise, trial vector replaces the target vector. The number of generations can be used as the stopping criterion, but we prefer to use the total number of feasible solutions generated as the stopping criterion due to our computational budget. We set the population size, NP, to 20, and DE stops as soon as 200 feasible solutions are generated. DE also requires F and CR parameters to be defined. According to Storn et al. [25], DE is much more sensitive to the choice of F than it is to the choice of CR, and the suggested value ranges for F are $[0.5, 1]$. In our experiments, we use a value of 0.5 for both the F and CR.

4   Results and Discussion

We have performed our experiments using two well-known few-shot classification datasets, miniImageNet and FC100 datasets. Taking the MAML implementation in learn2learn (l2l) PyTorch meta-learning library [2] as our basis, we have implemented our MAML-NAS using DE algorithm in Python programming language. All the experiments have been carried out using a single 24 GB RTX 3090 GPU.

Table 1. Results on miniImageNet and FC100 datasets

		Accuracy*		FLOPs	
		MAML-NAS	MAML**	MAML-NAS	MAML
miniImageNet	1-shot	51.32% ± 0.01	45.23% ± 1.84	320 M	488.86 M
	5-shot	63.42% ± 0.01	61.52% ± 0.98	378 M	488.86 M
FC100	1-shot	35.21% ± 0.01	38.30% ± 1.88	9.96 M	71.22 M
	5-shot	46.60% ± 0.01	46.29% ± 0.96	46.58 M	71.22 M

* Average accuracy over 600 test episodes with 95% confidence interval
** Results obtained by re-running MAML in l2l

For each few-shot setting, 1-shot and 5-shot, we have first executed MAML using l2l implementation with the original hyper-parameters stated in Sect. 3.2. Although these hyper-parameters were tuned in [9] for MAML, and thus may not be the best fit for our MAML-NAS, we decided not to modify them in order to best reveal the effect of the backbone modification.

DE is executed for each dataset and for each few-shot setting separately. During a DE execution, each candidate backbone architecture represented by a DE individual is used as the backbone in MAML and this model, MAML-NAS, is trained and evaluated exactly the same way as MAML. Although the hyper-parameters and the task settings are the same, we trained each MAML-NAS using only a small number of iterations because of the computational burden of a complete training process. After the DE execution, top three MAML-NAS models are selected and each of them is subjected to a complete training. These models are then tested on 600 test episodes, and the average accuracy along with the 95% confidence interval for each dataset under each few-shot setting is reported.

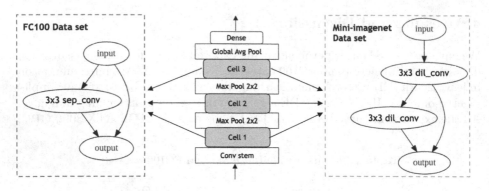

Fig. 4. The best MAML-NAS models generated for different data sets by DE.

An illustration of the architecture of the best MAML-NAS models for each of the datasets is given in Fig. 2. It can be seen in the figure that the resultant architectures are quite simple. Even these cells are repeated three times by stacking them vertically, the complexity of the whole architecture with respect to the FLOPs is still way smaller than the complexity of the backbone used in MAML as stated in Table 1. For the miniImageNet dataset, there is significant performance improvement in MAML-NAS over MAML. The improvement in 5-shot setting is less, though. For the FC100 dataset, a similar effect on the accuracy is present only in 5-shot setting. There is a slight accuracy improvement over MAML, but this slight improvement has been achieved by a much simpler MAML-NAS model. In 1-shot setting, MAML-NAS performs worse than MAML in terms of accuracy; however, the number of FLOPs in MAML-NAS model is less than one seventh of the FLOPs in MAML.

Table 2. Results on FC100 under 1-shot setting with different training iterations

	2000	5000	8000
Batch size 4	38.04 ± 0.01 (64 M)	37.80 ± 0.01 (456 M)	37.31 ± 0.010 (403 M)

FC100 dataset is more difficult than miniImageNet dataset, and especially 1-shot setting makes the problem even harder. For 1-shot setting, the best model found in DE does not yield in the best accuracy after a complete training. We suspect that evaluating the performance of the models during DE based on a small number of training iterations may not provide a good estimate for the actual performance of those models. When we increase the number of training iterations from 1000 to 2000, we were able to find a MAML-NAS model that achieves the same accuracy as MAML model, again with fewer parameters. However, increasing training iterations even more did not yield in models with better accuracy values as shown in Table 2.

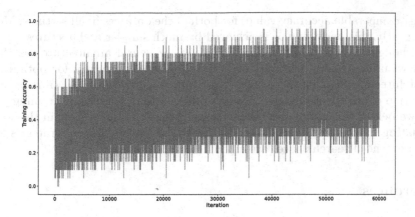

Fig. 5. Training steps

We believe that this is due to the task definitions used in training meta-learning models. As illustrated in Fig. 5, meta-learning algorithms learn slowly, because their aim is not to specialize in a given task; rather to find a good initial state so that they can be adapted to any novel task using a few optimization steps. The search performance of DE algorithm may be further evaluated by applying different mutation strategies and using different population sizes. In addition, we believe it would help to improve performance using hyperparameters especially tuned for MAML-NAS. It should also be noted that these results are obtained by the MAML-NAS models generated within a limited search space due to our limited computational resources. Extending the search space might help to increase the probability of obtaining better models.

5   Conclusion

In this study, we try to improve the few-shot classification performance of the one of the most well-known optimization based meta-learning algorithms, namely, Model-Agnostic Meta-Learning (MAML), with the help of the neural architecture search. The original four-block convolution architecture is replaced with the architectures automatically generated through the neural architecture search process with the aim of improved performance. Differential evolution algorithm is adopted as the search strategy for searching over cells within a predefined search space. We have performed our experiments using two well-known few-shot classification datasets, miniImageNet and FC100 dataset. For each of those datasets, the performance of the original MAML is compared to the performance of our MAML-NAS model under both 1-shot 5-way and 5-shot 5-way settings. It is also important to note that a fair performance comparison with respect to the accuracy values can only be performed between the two models if the complexity of them which is best measured in terms of floating-point operations per second are similar. The results reveal that MAML-NAS results in better or at least comparable accuracy values for both of the datasets in all settings. More importantly, this performance is achieved by much simpler architectures.

Although the use of differential evolution for neural architecture search is not new; using a meta-heuristic method as the search strategy for optimizing the architecture in a meta-learning algorithm is new. Since the training and evaluation of meta-learning algorithms is different than traditional training process, we believe this study provide useful insights to the researchers in this area. Evaluating the search performance of DE with different mutation strategies and within extended search spaces are among our future studies.

References

1. Andrychowicz, M., et al.: Learning to learn by gradient descent by gradient descent. Adv. Neural Inf. Process. Syst. **29** (2016)
2. Arnold, S.M.R., Mahajan, P., Datta, D., Bunner, I., Zarkias, K.S.: learn2learn: a library for meta-learning research. arXiv preprint arXiv:2008.12284 (2020)
3. Awad, N., Mallik, N., Hutter, F.: Differential evolution for neural architecture search. arXiv preprint arXiv:2012.06400 (2020)
4. Chen, W.Y., Liu, Y.C., Kira, Z., Wang, Y.C.F., Huang, J.B.: A closer look at few-shot classification. arXiv preprint arXiv:1904.04232 (2019)
5. Chen, Y., Wang, X., Liu, Z., Xu, H., Darrell, T.: A new meta-baseline for few-shot learning (2020)
6. Dong, X., Yang, Y.: NAS-Bench-201: extending the scope of reproducible neural architecture search. arXiv preprint arXiv:2001.00326 (2020)
7. Elsken, T., Metzen, J.H., Hutter, F.: Neural architecture search: a survey. J. Mach. Learn. Res. **20**(1), 1997–2017 (2019)
8. Fei-Fei, L., Fergus, R., Perona, P.: One-shot learning of object categories. IEEE Trans. Pattern Anal. Mach. Intell. **28**(4), 594–611 (2006)
9. Finn, C., Abbeel, P., Levine, S.: Model-agnostic meta-learning for fast adaptation of deep networks. In: International Conference on Machine Learning, pp. 1126–1135. PMLR (2017)
10. Krizhevsky, A., Hinton, G., et al.: Learning multiple layers of features from tiny images (2009)
11. Lake, B., Salakhutdinov, R., Gross, J., Tenenbaum, J.: One shot learning of simple visual concepts. In: Proceedings of the Annual Meeting of the Cognitive Science Society, vol. 33 (2011)
12. Lake, B.M., Salakhutdinov, R., Tenenbaum, J.B.: Human-level concept learning through probabilistic program induction. Science **350**(6266), 1332–1338 (2015)
13. Liu, H., Simonyan, K., Yang, Y.: DARTS: differentiable architecture search. arXiv preprint arXiv:1806.09055 (2018)
14. Liu, Y., Sun, Y., Xue, B., Zhang, M., Yen, G.G., Tan, K.C.: A survey on evolutionary neural architecture search. IEEE Trans. Neural Netw. Learn. Syst. (2021)
15. Lu, J., Gong, P., Ye, J., Zhang, C.: Learning from very few samples: a survey. arXiv preprint arXiv:2009.02653 (2020)
16. Nichol, A., Achiam, J., Schulman, J.: On first-order meta-learning algorithms. arXiv preprint arXiv:1803.02999 (2018)
17. Oreshkin, B., Rodríguez López, P., Lacoste, A.: TADAM: task dependent adaptive metric for improved few-shot learning. Adv. Neural Inf. Process. Syst. **31** (2018)

18. Price, K.V., Storn, R.M., Lampinen, J.A.: Differential Evolution. NCS, Springer, Heidelberg (2005). https://doi.org/10.1007/3-540-31306-0
19. Raghu, A., Raghu, M., Bengio, S., Vinyals, O.: Rapid learning or feature reuse? Towards understanding the effectiveness of MAML. arXiv preprint arXiv:1909.09157 (2019)
20. Ravi, S., Larochelle, H.: Optimization as a model for few-shot learning. In: 5th International Conference on Learning Representations (2017)
21. Real, E., Aggarwal, A., Huang, Y., Le, Q.V.: Regularized evolution for image classifier architecture search. In: Proceedings of the AAAI Conference on Artificial Intelligence, vol. 33, pp. 4780–4789 (2019)
22. Ren, P., et al.: A comprehensive survey of neural architecture search: challenges and solutions. ACM Comput. Surv. **54**(4), 1–34 (2021)
23. Siems, J., Zimmer, L., Zela, A., Lukasik, J., Keuper, M., Hutter, F.: NAS-Bench-301 and the case for surrogate benchmarks for neural architecture search. arXiv preprint arXiv:2008.09777 (2020)
24. Snell, J., Swersky, K., Zemel, R.: Prototypical networks for few-shot learning. Adv. Neural Inf. Process. Syst. **30** (2017)
25. Storn, R., Price, K.: Differential evolution – a simple and efficient heuristic for global optimization over continuous spaces. J. Global Optim. **11**(4), 341–359 (1997)
26. Vinyals, O., Blundell, C., Lillicrap, T., Wierstra, D., et al.: Matching networks for one shot learning. Adv. Neural Inf. Process. Syst. **29** (2016)
27. Ying, C., Klein, A., Christiansen, E., Real, E., Murphy, K., Hutter, F.: NAS-Bench-101: towards reproducible neural architecture search. In: International Conference on Machine Learning, pp. 7105–7114. PMLR (2019)
28. Zoph, B., Le, Q.V.: Neural architecture search with reinforcement learning. arXiv preprint arXiv:1611.01578 (2016)
29. Zoph, B., Vasudevan, V., Shlens, J., Le, Q.V.: Learning transferable architectures for scalable image recognition. In: Proceedings of the IEEE Conference on Computer Vision and Pattern Recognition, pp. 8697–8710 (2018)

Neural Architecture Search Using Metaheuristics for Automated Cell Segmentation

Zeki Kuş[✉], Musa Aydın, Berna Kiraz, and Burhanettin Can

Department of Computer Engineering, Fatih Sultan Mehmet Vakif University,
Istanbul, Turkey
{zkus,maydin,bkiraz,bcan}@fsm.edu.tr

Abstract. Deep neural networks give successful results for segmentation of medical images. The need for optimizing many hyper-parameters presents itself as a significant limitation hampering the effectiveness of deep neural network based segmentation task. Manual selection of these hyper-parameters is not feasible as the search space increases. At the same time, these generated networks are problem-specific. Recently, studies that perform segmentation of medical images using Neural Architecture Search (NAS) have been proposed. However, these studies significantly limit the possible network structures and search space. In this study, we proposed a structure called UNAS-Net that brings together the advantages of successful NAS studies and is more flexible in terms of the networks that can be created. The UNAS-Net structure has been optimized using metaheuristics including Differential Evolution (DE) and Local Search (LS), and the generated networks have been tested on Optofil and Cell Nuclei data sets. When the results are examined, it is seen that the networks produced by the heuristic methods improve the performance of the U-Net structure in terms of both segmentation performance and computational complexity. As a result, the proposed structure can be used when the automatic generation of neural networks that provide fast inference as well as successful segmentation performance is desired.

Keywords: Neural architecture search · Cell segmentation · Metaheuristics · Deep learning

1 Introduction

Cell is the basic structural, functional, and biological unit in all living organisms. Imaging cells and collecting information from them are important for various scientific fields [5] such as image cytometry, flow cytometry, cell sorters and time-lapse cytometers. More precise imaging of cells and subcellular parts has become possible because of recent improvements in high-resolution fluorescence microscopy [32]. Increasing computational capability of computers and developments in the field of deep learning have accelerated the studies in medical

L. Di Gaspero et al. (Eds.): MIC 2022, LNCS 13838, pp. 158–171, 2023.
https://doi.org/10.1007/978-3-031-26504-4_12

imaging [8]. Automatic analysis of cellular images provides cell counting and segmentation with higher accuracy, efficiency and reproducible information. On the other hand, manual segmentation performed by human experts is more time-consuming and is prone to more error due to subjective interpretations.

In previous studies, different image-processing techniques have been proposed for segmentation of cells in 2D images [8,18]. For example, watershed-based image segmentation [28,33] and level set methods [10,20] have been used to separate the overlapping cells. Besides, active contour models and Snake's algorithm [34] are among the traditional image processing techniques used in image segmentation.

In addition to traditional image processing techniques, machine learning approaches are used for pixel-based classification and segmentation of cells in bright field and phase contrast images [31,39]. Recently, deep learning networks have been used in different problems such as segmentation and classification of biomedical images [13,14,19]. In the study proposed by Song et al. [30], a convolutional neural network-based deep learning approach has been proposed for the segmentation of cell images. However, the clustered cells could not be separated clearly in this study. Delgado-Ortet et al. [7] propose another method based on the convolutional neural network (CNN) which is used for segmentation of red blood cell images. They use a deep neural network and a convolutional neural network together, which they called "Segmentation Neural Network" for the classification of blood cells. When these studies are examined, it is seen that the U-Net model is preferred more than other CNN architectures due to its overall superior performance in segmentation of biomedical images.

U-Net models have many hyper-parameters that need to be optimized. On the other hand, many optimization techniques are focused at finding network architectures that should be efficient in terms of both segmentation performance and computational complexity. These techniques, called Neural Architecture Search (NAS), have been applied to various biomedical imaging problems such as tumor segmentation and vessel segmentation [4,36].

This study presents a NAS technique for automated cell segmentation using two metaheuristics including Differential Evolution (DE) and Local Search (LS). Optimizing the architectures of deep neural networks with DE and LS in NAS studies is not novel; however, optimizing U-Net architecture with these methods in NAS for cell segmentation is novel. NAS is used for the optimization of the U-Net backbone which is a well-known convolutional neural network for medical image segmentation. To the best of our knowledge, this is the one of initial applications of NAS using metaheuristics for cell segmentation. The proposed approach combines the advantages of different NAS techniques proposed in [16, 40] to maintain diversity in the search space. We use the cell structure which is represented as a directed acyclic graph (DAG) consisting of N intermediate, one input, and one output vertice. Parameter settings presented in the NAS-Bench-101 [40] is used to develop the cell structures. In addition, we consider the operations in the cell proposed in the Differentiable Architecture Search (DARTS) algorithm [16]. Gaussian Error Linear Unit (GELU) has been used as the activation function in all operations instead of Rectified Linear Unit (ReLU) [17]. Two different cell segmentation data sets (see Sect. 3.1) have been used to evaluate the performance of different U-Net architectures. The results reveal that DE performs better than LS for both data sets.

2   Methodologies

Neural Architecture Search (NAS) is one of the most basic and common techniques in the field of Automated Machine Learning [11]. In this field, several approaches have been developed in recent years, most of them based on graphs. Graphs are used to represent each part, which is called cell. Network architectures are created by combining multiple cells. Each cell can contain a different graph structure, or the same graph structure, usually called repeated cell, which can be used to reduce the search space [40]. Unlike state-of-the-art CNNs, cells with widely different structures are formed since there is no repeated pattern rule between operations in the cell.

Evolutionary Algorithms (EAs) are commonly used approaches in most NAS studies [2,24,35]. In these algorithms, each individual represents an architecture. New architectures are created using evolutionary operators such as crossover, mutation, and population converges towards the best architectures. One of the first studies in the field of NAS with evolutionary algorithms is the EvoNAS study [25]. Genetic algorithm is used to find the best CNN architecture in this study and the proposed method is evaluated on CIFAR10 and CIFAR100 datasets.

Real et al. [24] showed that architectures generated by Evolutionary Algorithms outperform hand-crafted architectures. They used Regularized Evolution for the NAS-Bench-101 benchmark data set and, obtained competitive results with fewer parameters compared to the NASNet-A [42] method. The tournament selection method is used and two mutation methods named hidden state mutation and op mutation are proposed. One of the connections between the states is randomly selected and mutated, i.e. add or remove a connection, in the hidden state mutation method. Similarly, one of the existing operations is randomly selected and replaced by another operation in the op mutation method.

Qiang et al. [23] proposed a Particle Swarm Optimization based NAS method. They presented NAS framework on the deep belief network for unsupervised learning. Variable mutation rate is used for better exploitation and exploration in this study. High mutation rates are selected in the early stages of this mutation process, whereas lower mutation rates are selected in the later stages.

In another study, Awad et al. [2] focused on Differential Evolution (DE) method and showed that this method yields state-of-the-art performance for different NAS benchmarks. DE achieved better and more robust results for 4 NAS benchmarks: NAS-Bench-101 [40], NAS-Bench-1Shot1 [41], NAS-Bench-201 [9], and NAS-HPO [15]. They used binomial crossover and rand1 mutation operation for DE steps.

In this study, we propose a Neural Architecture Search approach, referred to as UNAS-Net, for the optimization of the U-Net backbone. UNAS-Net consists of two main parts: metaheuristics and search space. As metaheuristics, we select Differential Evolution and Local Search metaheuristics. We present the detail of employed metaheuristics and search space configuration in the following subsections.

2.1   Search Space

As in the U-Net structure, UNAS-Net consists of consecutive encoder and decoder blocks. Encoder blocks are used for feature extraction and reducing the spatial dimensions. A lower-dimensional representation of the input image is learned through the encoder blocks. On the other hand, decoder blocks are used for reconstruction and increasing the spatial dimension. Image is reconstructed with the learned features in the decoder blocks. In this study, all of these blocks are called encoder and decoder cells. Each cell is represented by a directed acyclic graph (DAG) containing V vertices and E edges. Encoder cell takes the 2D Cell image as input and applies a series of operations inside the encoder cell to this image.

Encoder cells are followed by the max pooling operation with stride 2, which is used for reducing the spatial dimensions and called down sampling. After the consecutive encoder cells and down sampling operations, the bottleneck layer is used. At the next step, images are reconstructed with decoder cells using the extracted features. In the decoder cell, the transpose convolution is applied for the up sampling of the images. Finally, the sigmoid activation function is applied to the output from the decoder cell in the last layer, resulting in a binary image. Figure 1 shows the general structure of UNAS-Net. Each vertex, $v \in V$, has one $l \in L$ (Table 1) label which indicates the operation. On the other hand, each edge, $e \in E$, represents the connection between two vertices. The first and last vertices in the cell are fixed and are called $INPUT$ and $OUTPUT$, respectively. The search space grows exponentially as the size of V and the size of E increase. Therefore, we limit the total number of vertices and edges. We consider the following parameter settings used in [40]:

- 11 different operations presented in Table 1 ($|L| = 11$),
- The maximum size of E is set to 9 ($|E| \leq 9$), and
- The maximum size of V is set to 7 ($|V| \leq 7$)

There are two parameters that directly affect to depth and performance of the network in NAS: the number of cells ($N_c \in \{2, 3, 4, 5\}$) [4] and the number of feature maps of the first cell ($N_f^{init} \in \{8, 16, 32\}$) [4]. The number of feature maps is doubled after each down sampling operation.

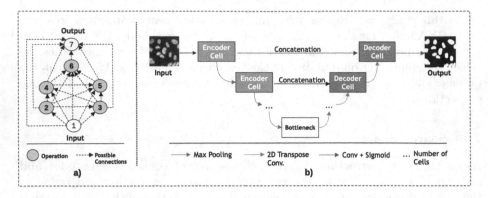

Fig. 1. The general structure of UNAS-Net. **(a)** The structure of cells that can be created. Each cell can have a maximum of 5 intermediate vertices and 9 edges. Each operation can be linked to the operations that come after it. **(b)** the UNAS-Net structure, which is consists of several encoder and decoder cells. The same type of cell structure is used in each encoder and decoder cell.

2.2   Architecture Encoding

Graphs can be represented in several ways. Binary matrices are one of the most commonly used representation for graph encoding. In this study, we use matrix encoding to represent the *DAGs* within each cell. We consider a 7×7 binary upper triangular matrix to encode the connections between the vertices since the maximum number of operations is set to 7. The labels of 5 intermediate vertices, without including the input and output vertices, are represented by a list of 5 elements. 337 billion distinct graphs can be produced using the proposed encoding method ($2^{21} * 11^5 = 337B$). However, this number will decrease due to the maximum edge limit, isomorphic networks, and infeasible graphs in which there is no path from input to output [40].

Table 1. Possible operations that can be selected for cell. Operations are selected based on the DARTS [16]

Operation name
• 1×1 2D convolution (1×1 conv)
• 3×3 2D convolution (3×3 conv)
• 5×5 2D convolution (5×5 conv)
• 7×7 2D convolution (7×7 conv)
• 3×3 2D depthwise separable convolution (3×3 depconv)
• 5×5 2D depthwise separable convolution (5×5 depconv)
• 7×7 2D depthwise separable convolution (7×7 depconv)
• 3×3 2D dilated convolution (3×3 dilconv)
• 5×5 2D dilated convolution (5×5 dilconv)
• 7×7 2D dilated convolution (7×7 dilconv)
• Skip-connection (skip)

Discrete and continuous search spaces have been used in the literature for NAS. Based on the results, it is clear that using a continuous search space achieves better results [2]. Therefore, in this study, solutions are represented in continuous space (except the local search) to better investigate the search space. We use a vector of length 28 to represent the solutions. The first 21 items in this vector indicate the connections of the vertices; the next 5 elements indicate the selected operations, and the last two elements indicate the number of cells (n_c) and the number of feature maps of the first cell (n_f^{init}), respectively. All of these values in the vector are scaled to $[0, 1]$. These vectors can not be used to create a network architecture. Therefore, a mapping between continuous values and network architecture is required [2].

At the mapping phase, the continuous values are divided into bins that are equal to the number of values that can be selected [2]. The first 21 items in this vector represented the connections of the vertices. These values can only be binary: if the value is 1, there is a connection between two vertices; otherwise, there is no connection. For the discrete values $\{0, 1\}$, each interval corresponds to one state; i.e. the values in $[0, 0.5)$ is decoded as 0, and the values in $[0.5, 1)$ is decoded as 1. Mapping is performed similarly for operations. For discrete values $L \in \{1 \times 1\ conv,\ 3 \times 3\ conv,\ 7 \times 7\ conv,\ 3 \times 3\ depconv,\ 5 \times 5\ depconv,\ 7 \times 7\ depconv,\ 3 \times 3\ dilconv,\ 5 \times 5\ dilconv,\ 7 \times 7\ dilconv,\ skip\}$, the interval $[0, 1/11)$ corresponds to $1 \times 1\ conv$, $[1/11, 2/11)$ to $3 \times 3\ conv$, $[2/11, 3/11)$ to $7 \times 7\ conv$, and so on. Figure 2 shows the architecture encoding from the candidate solution to UNAS-Net.

2.3 Metaheuristics

Differential Evolution (DE) and Steepest Descent Local Search (LS) are two metaheuristics used in this study to build distinct neural networks. DE, which is categorized as population-based evolutionary algorithm, has achieved successful results in NAS studies especially in continuous space [2]. Therefore, we consider DE algorithm in this study. We consider the following parameter settings of DE recommended in [2,6]: Population size = 20; Mutation factor = 0.5; Crossover probability = 0.5; Mutation method = rand1; Crossover method = Binomial; Boundary handling method = Random. The implementation of DE which is publicly available[1] is adapted for cell segmentation.

Local Search (LS) is used as a basis approach in NAS studies [21,27,37]. Therefore, we consider LS as the second metaheuristic. The steps of the LS algorithm used in this study are as follows: (1) Generate 10 random solutions. Initial solutions consist of continuous values. At the mapping phase, these solutions, which consist of continuous values, are mapped to discrete values and their performance is evaluated. (2) Choose the best among these solutions, called incumbent. (3) Generate neighbors of the incumbent. All neighbors of the incumbent that are within one hamming distance are generated. (4) If any of the neighbors of this solution is better than the incumbent solution, use this solution in the

---

[1] https://github.com/automl/DE-NAS.

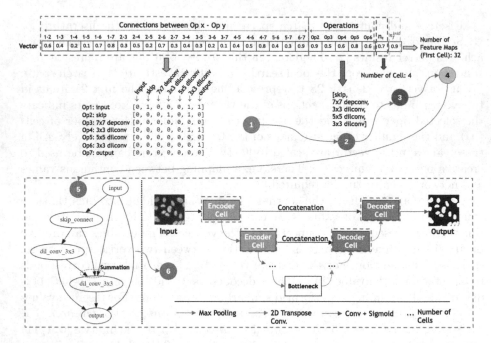

Fig. 2. Architecture encoding from solution vector to UNAS-Net. Circles from 1 to 6 indicate the sequence of operations. The following steps are applied sequentially while creating an architecture from a generated solution vector: 1) The adjacency matrix is generated; 2) The selected operations are determined; 3) The number of cell is determined; 4) The number of feature maps is defined; 5) The graph is generated with a adjacency matrix and operations, and it is used in each cell; 6) UNAS-Net is created with the parameters obtained in the previous steps.

next step as incumbent and repeat the same steps until the termination condition is satisfied. (5) The open source software[2] is used and developed for the implementation of Local Search.

3   Experimental Setup

To assess the performance of the proposed methods, we conduct experiments on two different data sets in this study. Selected data sets are presented in Sect. 3.1. For experimental studies, the tests are performed on the computer with the following configurations: Ryzen 5600X processor, 12 GB RTX 3060 GPU, 16 GB RAM. The Pytorch library is used to create different UNAS-Net models and the following hyper-parameters are chosen for all models: Optimizer: Adam, Loss Function: Dice loss, Evaluation Metrics: F1-Score and Intersection over Union (IoU) [4], Learning Rate: 0.001 and Batch Size: 8.

---

[2] https://github.com/naszilla/naszilla.

Each generated solution should be trained over long training epochs with the whole training data set, and its quality should be evaluated on the validation data set. However, applying these steps for every solution is inefficient. Therefore, half of the training and validation sets are used to train and evaluate the networks produced by the metaheuristics [12]. The images in the test set are never used during the training steps. The test set is only used to measure the actual performance of the best solutions obtained at the end of the metaheuristic. Moreover, the Early Stopping method is used to evaluate more solutions. Each solution is trained with half of the training set during the maximum of 36 training epochs [40]. The training is interrupted in the corresponding epoch if the validation loss value obtained during the training epochs does not improve for three consecutive epoch (Early Stopping). The best validation F1-Score value obtained by the trained solution at the end of the training epochs is used to evaluate the quality of that solution. Both metaheuristics are terminated when 500 solutions are trained. The best 3 models obtained at the end of the runs are trained with the whole training data set over long training epochs (200 epochs) and tested with the whole test set. The results acquired for the test set are used as the actual performance of the model.

3.1   Data Set

The segmentation performance of the architectures is evaluated using two different cell data sets in this study: Optofil and Cell Nuclei [3]. This section describes these data sets. In this study, we do not apply data augmentation for both data sets.

Optofil Data set: This data set consists of A2780 cell culture images obtained using a commercial brightfield automated cell counting device (Quantacell, Optofil). The data set contains 4914 grayscale images with a size of 128×128. We divide the data set into three parts: train, test, and validation. These parts contain the following number of images respectively: 3539, 982, and 393. The ground truth images (segmented images) are created through FiJi software [26]. The original and ground truth image pairs for five exemplary images of A2780 cells are presented in Fig. 3.

Cell Nuclei Data set: Cell Nuclei data set contains segmented nuclei images. The images are obtained under various conditions [1]. There are 670 colored images in this data set, which is publicly available. Also, it contains segmented images corresponding to each colored image, and these images are named ground truths. We need to apply pre-processing step since each image is a different size and data set contains the segmented masks of each nucleus. Therefore, we use the pre-processed data set created in [3]. We divide the data set into three parts: train, test, and validation. These parts contain the following number of images respectively: 483, 134, and 53. Figure 4 shows a set of image pairs that include both original and segmented images.

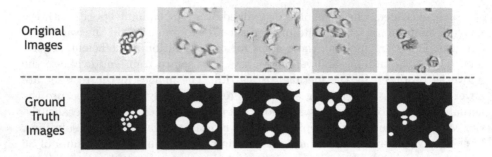

Fig. 3. Original and ground truth image pairs from the Optofil data set.

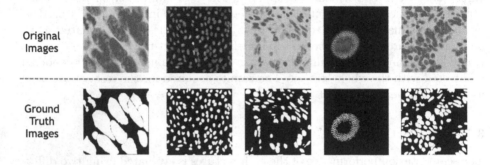

Fig. 4. Original and ground truth image pairs from the Cell Nuclei data set.

4   Results and Discussion

In this section, the results of experimental studies are reported and discussed. Two metaheuristics, namely DE and LS, are utilized in NAS for the optimization of the U-Net backbone. When biomedical image segmentation studies are examined, it is clear that the U-Net model which is state-of-the-art for the biomedical image segmentation outperforms alternative CNN designs in terms of overall performance [29] Therefore, we chose U-Net as the basis and compared our approaches to U-Net with the ResNet-34 Encoder (URes34-Net) proposed in [22,38].

The models developed for the biomedical image segmentation problem should have high segmentation performance and short inference time. Therefore, it's important to compare the models in terms of segmentation performance and computational complexity. In this study, the performance of all algorithms are compared in terms of two performance metrics, namely F1-Score and IoU, and the number of floating point operations (FLOPs).

Table 2 shows the average F1-score, IoU, and FLOPs values obtained for five different seed values. The results show that LS provides the best F1-Score value for the Optofil data set. However, in terms of the IoU metric, DE gives the best results and produces model with the least computational complexity.

Table 2. Comparison of different methods in terms of segmentation performance and computational complexity (measured in GigaFLOPs)

Method	Optofil (Mean ± Stdev)			Cell nuclei (Mean ± Stdev)		
	F1 score	IoU	FLOPs (G)	F1 score	IoU	FLOPs (G)
LS	**83.22 ± 0.91**	70.57 ± 0.90	22.80	91.16 ± 0.15	83.95 ± 0.27	82.99
DE	83.08 ± 0.41	**70.93 ± 0.77**	**4.18**	**91.53 ± 0.12**	**84.48 ± 0.20**	**35.03**
URes34-Net	82.69 ± 0.29	70.01 ± 0.14	30.85	91.11 ± 0.25	83.83 ± 0.37	62.51

DE has 0.14 worse F1-Score than LS while 5.45 times better in computational complexity. The metaheuristics outperform URes34-Net in terms of both segmentation performance and computational complexity for the Optofil data set. Considering this result, we can conclude that metaheuristics improve the segmentation performance of U-Net with less computational complexity. Based on the results obtained for the Cell Nuclei data set, it is seen that DE is better than the other two methods in terms of both segmentation performance and computational complexity. Besides, there is no significant difference between the different methods in terms of F1-Score and IoU values. However, in terms of computational complexity, DE is 2.37 times better than LS; 1.78 times better than URes34-Net.

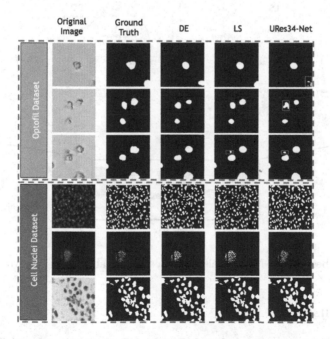

Fig. 5. Segmented images obtained for different cell images as a result of experiments with three different methods. The red squares represent regions where the segmentation is incorrect. (Color figure online)

Three exemplary images for each data set are used to illustrate the performance of our approaches and URes34-Net. Figure 5 shows the segmented images obtained by three algorithms as well as the ground truths for these six exemplary images. Networks produced by metaheuristics perform more accurate segmentation. Red boxes shown in Fig. 5 indicate the false positive samples for Optofil data set. DE yields good segmentation performance for these three images.

Figure 6 illustrates the best UNAS-Net models obtained by each metaheuristic for each data set. When the best models are examined, it is seen that operations with bigger filter size are generally selected. The initial filter numbers are usually 8 and 16, and the cell numbers are usually 2 and 3. Moreover, it is observed that most produced models have a maximum of 8 connections.

Based on the results, DE outperforms the other methods in terms of IoU metric and the number of FLOPs. This is somewhat expected, since DE is a population-based approach that maintains diversity via genetic operators during the search. However, LS is a single-point search method. In this study, a new solution with Hamming distance 1 is generated from the current one. Therefore, LS could get stuck in local optima. To escape the local optimum, Iterated Local Search can be used in NAS.

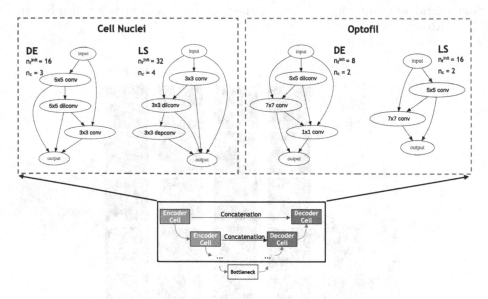

Fig. 6. The best UNAS-Net models generated by metaheuristics for two data sets.

5   Conclusion

In this study, two metaheuristics (Differential Evolution and Local Search) are utilized in NAS for automated cell segmentation. We consider U-Net architecture that is a successful and well-known convolutional neural network (CNN) for medical image segmentation in NAS. The performances of metaheuristic approaches addressed in this study are compared with URes34-Net. We conduct the experiments on two data sets, namely Optofil and Cell Nuclei. The results reveal that DE both improves the segmentation performance of U-Net and reduces the computational complexity for both data sets.

As the future works, the search algorithm will be extended with estimators. We aim to spend less GPU time this way. Parameter tuning can be performed for parameters that are kept constant in order to improve the segmentation performance. Study can be addressed as a bi-objective optimization problem with the following objective functions: segmentation performance vs complexity. Another future work can be utilizing the UNAS-Net structure, which gives successful results for cell segmentation, for 3D vessel segmentation problem.

References

1. 2018 data science bowl. https://www.kaggle.com/c/data-science-bowl-2018/data
2. Awad, N.H., Mallik, N., Hutter, F.: Differential evolution for neural architecture search. arXiv preprint arXiv:2012.06400 (2020)
3. Aydın, M., et al.: A deep learning model for automated segmentation of fluorescence cell images. J. Phys: Conf. Ser. **2191**(1), 012003 (2022)
4. Baldeon Calisto, M., Lai-Yuen, S.K.: EMONAS-Net: efficient multiobjective neural architecture search using surrogate-assisted evolutionary algorithm for 3D medical image segmentation. Artif. Intell. Med. **119**, 102154 (2021)
5. Chaffer, C.L., Weinberg, R.A.: A perspective on cancer cell metastasis. Science **331**(6024), 1559–1564 (2011)
6. de-la Cruz-Martínez, S.J., Mezura-Montes, E.: Boundary constraint-handling methods in differential evolution for mechanical design optimization. In: 2020 IEEE Congress on Evolutionary Computation (CEC), pp. 1–8 (2020)
7. Delgado-Ortet, M., Molina, A., Alférez, S., Rodellar, J., Merino, A.: A deep learning approach for segmentation of red blood cell images and malaria detection. Entropy **22**(6), 657 (2020)
8. Deshmukh, B.S., Mankar, V.H.: Segmentation of microscopic images: a survey. In: 2014 International Conference on Electronic Systems, Signal Processing and Computing Technologies, pp. 362–364 (2014)
9. Dong, X., Yang, Y.: NAS-Bench-201: extending the scope of reproducible neural architecture search. arXiv preprint arXiv:2001.00326 (2020)
10. Dzyubachyk, O., van Cappellen, W.A., Essers, J., Niessen, W.J., Meijering, E.: Advanced level-set-based cell tracking in time-lapse fluorescence microscopy. IEEE Trans. Med. Imaging **29**(3), 852–867 (2010)
11. Elsken, T., Metzen, J.H., Hutter, F.: Neural architecture search: a survey. J. Mach. Learn. Res. **20**(1), 1997–2017 (2019)
12. Gülcü, A., Kuş, Z.: Hyper-parameter selection in convolutional neural networks using microcanonical optimization algorithm. IEEE Access **8**, 52528–52540 (2020)
13. Habibzadeh, M., Jannesari, M., Rezaei, Z., Baharvand, H., Totonchi, M.: Automatic white blood cell classification using pre-trained deep learning models: ResNet and Inception. In: Verikas, A., Radeva, P., Nikolaev, D., Zhou, J. (eds.) Tenth International Conference on Machine Vision (ICMV 2017), vol. 10696, pp. 274–281. International Society for Optics and Photonics, SPIE (2018)
14. Hollandi, R., et al.: nucleAizer: a parameter-free deep learning framework for nucleus segmentation using image style transfer. Cell Syst. **10**(5), 453-458.e6 (2020)

15. Klein, A., Hutter, F.: Tabular benchmarks for joint architecture and hyperparameter optimization. arXiv preprint arXiv:1905.04970 (2019)
16. Liu, H., Simonyan, K., Yang, Y.: Darts: differentiable architecture search (2019)
17. Liu, Z., Mao, H., Wu, C.Y., Feichtenhofer, C., Darrell, T., Xie, S.: A convnet for the 2020s. arXiv preprint arXiv:2201.03545 (2022)
18. Meijering, E.: Cell segmentation: 50 years down the road [life sciences]. IEEE Signal Process. Mag. **29**(5), 140–145 (2012)
19. Mookiah, M.R.K., et al.: A review of machine learning methods for retinal blood vessel segmentation and artery/vein classification. Med. Image Anal. **68**, 101905 (2021)
20. Nath, S.K., Palaniappan, K., Bunyak, F.: Cell segmentation using coupled level sets and graph-vertex coloring. In: Larsen, R., Nielsen, M., Sporring, J. (eds.) MICCAI 2006. LNCS, vol. 4190, pp. 101–108. Springer, Heidelberg (2006). https://doi.org/10.1007/11866565_13
21. Den Ottelander, T., Dushatskiy, A., Virgolin, M., Bosman, P.A.N.: Local search is a remarkably strong baseline for neural architecture search. In: Ishibuchi, H., et al. (eds.) EMO 2021. LNCS, vol. 12654, pp. 465–479. Springer, Cham (2021). https://doi.org/10.1007/978-3-030-72062-9_37
22. Pi, J., et al.: FS-UNet: mass segmentation in mammograms using an encoder-decoder architecture with feature strengthening. Comput. Biol. Med. **137**, 104800 (2021)
23. Qiang, N., Ge, B., Dong, Q., Ge, F., Liu, T.: Neural architecture search for optimizing deep belief network models of fMRI data. In: Li, Q., Leahy, R., Dong, B., Li, X. (eds.) MMMI 2019. LNCS, vol. 11977, pp. 26–34. Springer, Cham (2020). https://doi.org/10.1007/978-3-030-37969-8_4
24. Real, E., Aggarwal, A., Huang, Y., Le, Q.V.: Regularized evolution for image classifier architecture search. In: Proceedings of the AAAI Conference on Artificial Intelligence, vol. 33, pp. 4780–4789 (2019)
25. Real, E., et al.: Large-scale evolution of image classifiers. In: International Conference on Machine Learning, pp. 2902–2911. PMLR (2017)
26. Schindelin, J., et al.: Fiji: an open-source platform for biological-image analysis. Nat. Methods **9**(7), 676–682 (2012)
27. Schneider, L., Pfisterer, F., Binder, M., Bischl, B.: Mutation is all you need. arXiv preprint arXiv:2107.07343 (2021)
28. Sharif, J.M., Miswan, M.F., Ngadi, M.A., Salam, M.S.H., bin Abdul Jamil, M.M.: Red blood cell segmentation using masking and watershed algorithm: a preliminary study. In: 2012 International Conference on Biomedical Engineering (ICoBE), pp. 258–262 (2012)
29. Siddique, N., Paheding, S., Elkin, C.P., Devabhaktuni, V.: U-net and its variants for medical image segmentation: a review of theory and applications. IEEE Access **9**, 82031–82057 (2021)
30. Song, Y., et al.: A deep learning based framework for accurate segmentation of cervical cytoplasm and nuclei. In: 2014 36th Annual International Conference of the IEEE Engineering in Medicine and Biology Society, pp. 2903–2906 (2014)
31. Su, H., Yin, Z., Huh, S., Kanade, T.: Cell segmentation in phase contrast microscopy images via semi-supervised classification over optics-related features. Med. Image Anal. **17**(7), 746–765 (2013). Special Issue on the 2012 Conference on Medical Image Computing and Computer Assisted Intervention
32. Vonesch, C., Aguet, F., Vonesch, J.L., Unser, M.: The colored revolution of bioimaging. IEEE Signal Process. Mag. **23**(3), 20–31 (2006)

33. Wang, M., Zhou, X., Li, F., Huckins, J., King, R.W., Wong, S.T.: Novel cell segmentation and online SVM for cell cycle phase identification in automated microscopy. Bioinformatics **24**(1), 94–101 (2007)
34. Wang, X., He, W., Metaxas, D., Mathew, R., White, E.: Cell segmentation and tracking using texture-adaptive snakes. In: 2007 4th IEEE International Symposium on Biomedical Imaging: From Nano to Macro, pp. 101–104 (2007)
35. Wei, C., Niu, C., Tang, Y., Wang, Y., Hu, H., Liang, J.: NPENAS: neural predictor guided evolution for neural architecture search. IEEE Transactions on Neural Networks and Learning Systems, pp. 1–15 (2022)
36. Weng, Y., Zhou, T., Li, Y., Qiu, X.: NAS-Unet: Neural architecture search for medical image segmentation. IEEE Access **7**, 44247–44257 (2019)
37. White, C., Nolen, S., Savani, Y.: Exploring the loss landscape in neural architecture search. In: de Campos, C., Maathuis, M.H. (eds.) Proceedings of the Thirty-Seventh Conference on Uncertainty in Artificial Intelligence. Proceedings of Machine Learning Research, vol. 161, pp. 654–664. PMLR (2021)
38. Yakubovskiy, P.: Segmentation models pytorch (2020). https://github.com/qubvel/segmentation_models.pytorch
39. Yin, Z., Bise, R., Chen, M., Kanade, T.: Cell segmentation in microscopy imagery using a bag of local Bayesian classifiers. In: 2010 IEEE International Symposium on Biomedical Imaging: From Nano to Macro, pp. 125–128 (2010)
40. Ying, C., Klein, A., Christiansen, E., Real, E., Murphy, K., Hutter, F.: NAS-Bench-101: towards reproducible neural architecture search. In: Chaudhuri, K., Salakhutdinov, R. (eds.) Proceedings of the 36th International Conference on Machine Learning, vol. 97, pp. 7105–7114. PMLR (2019)
41. Zela, A., Siems, J., Hutter, F.: NAS-Bench-1shot1: benchmarking and dissecting one-shot neural architecture search. arXiv preprint arXiv:2001.10422 (2020)
42. Zoph, B., Vasudevan, V., Shlens, J., Le, Q.V.: Learning transferable architectures for scalable image recognition. In: Proceedings of the IEEE Conference on Computer Vision and Pattern Recognition, pp. 8697–8710 (2018)

Analytical Methods to Separately Evaluate Convergence and Diversity for Multi-objective Optimization

Takato Kinoshita[1] , Naoki Masuyama[2]([✉]) , Yusuke Nojima[2] ,
and Hisao Ishibuchi[3]

[1] Graduate School of Engineering, Osaka Prefecture University,
1-1 Gakuen-cho Naka-ku, Sakai-Shi, Osaka 599-8531, Japan
`sbb01065@st.osakafu-u.ac.jp`
[2] Graduate School of Informatics, Osaka Metropolitan University,
1-1 Gakuen-cho Naka-ku, Sakai-Shi, Osaka 599-8531, Japan
`{masuyama,nojima}@omu.ac.jp`
[3] Guangdong Provincial Key Laboratory of Brain-inspired Intelligent Computation,
Department of Computer Science and Engineering, Southern University of Science
and Technology, Shenzhen 518055, China
`hisao@sustech.edu.cn`

Abstract. This paper proposes two analytical methods which completely separate the search performance of multi-objective evolutionary algorithms (MOEAs) into convergence and diversity for quantitatively comparing MOEAs. Specifically, Convergence-Diversity Pair (C-D Pair) is proposed to statistically compare the convergence and diversity of two MOEAs. C-D Pair provides analytical information on the overall experimental results. In addition, Convergence-Diversity Diagram (C-D Diagram) is also proposed to visualize a pair of convergence and diversity of a solution set as a single point in a two-dimensional space. C-D Diagram enables a detailed and intuitive comparison of the search performance trends of multiple MOEAs. Moreover, this paper introduces two diversity indicators. These indicators are designed to evaluate only the diversity of the population in an MOEA by completely eliminating the effect of the convergence. Computational experiments demonstrate the analytical capability and validity of the proposed analytical methods by using various test problems.

Keywords: Multiobjective optimization · Performance analysis · Visualization

This work was supported by Japan Society for the Promotion of Science (JSPS) KAKENHI Grant Number JP19K20358 and 22H03664. National Natural Science Foundation of China (Grant No. 61876075), Guangdong Provincial Key Laboratory (Grant No. 2020B121201001), the Program for Guangdong Introducing Innovative and Enterpreneurial Teams (Grant No. 2017ZT07X386), The Stable Support Plan Program of Shenzhen Natural Science Fund (Grant No. 20200925174447003), Shenzhen Science and Technology Program (Grant No. KQTD2016112514355531).

L. Di Gaspero et al. (Eds.): MIC 2022, LNCS 13838, pp. 172–186, 2023.
https://doi.org/10.1007/978-3-031-26504-4_13

1   Introduction

In many real-world situations, one task has multiple objectives to be optimized simultaneously. Such optimization problems are called Multi-objective Optimization Problems (MOPs). Since the objective functions usually have tradeoff relationships, MOPs have the Pareto-optimal solution set (PS) and the Pareto optimal front (PF) which is the image of the PS in the objective space. One popular approach is the use of population-based search algorithms like Multiobjective Evolutionary Algorithms (MOEAs) which can obtain a number of non-dominated solutions approximating the PF in a single run. The search performance of each MOEA is examined with respect to the convergence of the population toward the PF and the population diversity over the entire PF.

To simultaneously evaluate both convergence and diversity, Hypervolume (HV) [23] and Inverted Generational Distance (IGD) [5] have been popularly used as performance indicators. Because HV and IGD aggregate the information of one solution set into a single scalar value, we can quantitatively evaluate the overall search performance of MOEAs. Meanwhile, it is difficult to analyze factors that contribute to the improvement of the search performance of MOEAs from the HV and IGD values. Therefore, we have often qualitatively discussed convergence and diversity based on comparisons of visualized information, such as scatter plots and parallel coordinates of solution sets. If users can interpret the actual states of the solution sets from the visualized information, the visualized information gives a more intuitive analysis than performance indicators that separately evaluate convergence or diversity, such as Generational Distance (GD) [19] and Spread (Δ) [8].

MOPs with four or more objectives are called Many-objective Optimization Problems (MaOPs) and have attracted much attention in recent years due to their difficulty in search and analysis [10]. Although it is possible to directly visualize the solution set by visualization methods such as scatter plot matrices and parallel coordinates, the larger the number of objectives, the more complicated the visualized information. Therefore, it is not easy to interpret the actual states of the solution sets for MaOPs from the visualized information. To analyze the search performance of MOEAs on MaOPs, a quantitative and intuitive analytical method that ensures interpretability is desired.

One approach to address these issues is to devise performance indicators that separately evaluate convergence or diversity and integrate these indicator values. In this paper, we propose Convergence-Diversity Pair (C-D Pair) and Convergence-Diversity Diagram (C-D Diagram) as analytical methods for simultaneously comparing convergence and diversity indicator values of multiple algorithms. C-D Pair statistically compares two series of indicator values for two MOEAs. C-D Diagram visualizes a pair of two indicator values as a point on a two-dimensional space. In addition, we introduce two diversity indicators and a single convergence indicator.

The remainder of this paper is organized as follows. In Sect. 2, we review conventional performance indicators. In Sect. 3, we explain two diversity indicators, C-D Pair, and C-D Diagram. Then, in Sect. 4, we verify the validity and analytical capability of the proposed analytical methods through computational experiments using solution sets obtained by some MOEAs. Finally, this paper is concluded in Sect. 5 where some future topics are also suggested.

2 Conventional Performance Indicators

In this section, we explain the three conventional performance indicators: HV, IGD, and GD. They have several undesirable characteristics, which may cause difficulties in practicality, fairness, or interpretability.

At first, we give here the preliminaries in this paper: We can generally formulate an MOP as follows:

$$\text{Minimize}\quad \boldsymbol{f}(\boldsymbol{x}) = (f_1(\boldsymbol{x}), \ldots, f_m(\boldsymbol{x}), \ldots, f_M(\boldsymbol{x})),$$
$$\text{subject to}\quad \boldsymbol{x} \in S \subset \boldsymbol{R}^D, \tag{1}$$

where $\boldsymbol{x}$ is a D-dimensional decision vector in the search space S, f_m is the mth objective function ($m = 1, \ldots, M$), and M is the number of objectives.

In addition, the solution set $X = \{\boldsymbol{x}_1, \boldsymbol{x}_2, \ldots\} \subset \boldsymbol{R}^D$ denotes a non-dominated solution set obtained from the search of an MOEA, and the solution distribution $Y = \boldsymbol{f}[X] = \{\boldsymbol{f}(\boldsymbol{x}_1), \boldsymbol{f}(\boldsymbol{x}_2), \ldots\} \subset \boldsymbol{R}^M$ denotes the image of the solution set X under an objective function vector $\boldsymbol{f}$. Moreover, $\boldsymbol{q} \in \boldsymbol{R}^M$ denotes the reference point used in HV, and $R \subset \boldsymbol{R}^M$ denotes the reference point set on the PF used in IGD and GD.

2.1 Hypervolume

Hypervolume (HV) [23] is a Pareto-compliant performance indicator [21]. In addition, in contrast to the reference point set of IGD and GD, HV needs only one reference point. Therefore, HV is one of the most used performance indicators [17]. The HV value for the solution set X is defined by the measure of the area in the objective space that is dominated by the objective vectors in the solution distribution Y and dominates the reference point $\boldsymbol{q}$ as follows:

$$\text{HV}(X) = \mathcal{L}\left(\bigcup_{\boldsymbol{y} \in Y}\left(\prod_{m=1}^{M}(y_m, q_m]\right)\right), \tag{2}$$

where $\mathcal{L}(\cdot)$ denotes the Lebesgue measure. Larger HV values indicate that Y more closely approximates PF in terms of both convergence and diversity.

While HV has excellent characteristics, it also has two practical drawbacks. One is the enormous computational complexity, which increases exponentially with the number of objectives [2]. For this reason, it is considered difficult to apply HV to MaOPs with more than ten objectives [16]. The other is the difficulty of appropriately setting a reference point in MaOPs, where the reference point setting significantly affects the solution distribution to optimize HV [11]. Therefore, depending on the PF shapes and reference point settings, HV may overestimate or underestimate the quality of solution distributions, preventing a fair comparison of the search performance of MOEAs.

2.2 Inverted Generational Distance

Inverted Generational Distance (IGD) [5], like HV, evaluates the search performance of MOEAs in terms of both convergence and diversity. The advantages of IGD over HV are its low computational complexity and easy interpretation due to its simple definition [12]. IGD value for the solution set X is defined by the average of the minimum distance from each reference point $r \in R$ to the solution distribution Y as follows:

$$\text{IGD}(X) = \frac{1}{|R|} \left(\sum_{r \in R} \min_{y \in Y} \{d(y, r)\} \right), \tag{3}$$

where $d(y, r)$ denotes the distance between vector y and vector r. In this paper, we use the Euclidean distance. Intuitively, IGD evaluates the degree of approximation of the solution distribution to the reference point set by the average distance. Therefore, when $|X| = |R|$, clearly the solution set X with solution distribution $Y = R$ is optimal in terms of IGD.

Fair comparisons of the search performance of MOEAs using IGD require R consisting of a large number of reference points uniformly distributed over the entire PF [12]. Therefore, the true PF must be known. Moreover, when $|X| \ll |R|$, it is pointed out that the solution distribution Y, which is much less diverse than the true PF, optimizes IGD, even if the ideal reference point set R is given [12]. Considering this issue and that the solution distribution becomes sparser in the objective space as the number of objectives increases, it is possible to obtain counterintuitive evaluations from IGD for MaOPs.

2.3 Generational Distance

Unlike HV and IGD, Generational Distance (GD) [19] evaluates only convergence. Intuitively, using reference point set R, the GD value for the solution set X is defined by the average of the approximate distances from the objective vectors in Y to the PF as follows:

$$\text{GD}(X) = \frac{1}{|Y|} \left(\sum_{y \in Y} \min_{r \in R} \{d(y, r)\} \right). \tag{4}$$

Comparing (4) with (3), we can understand the similarity between GD and IGD. Therefore, GD can be easily interpreted as IGD. The computational complexity is also equivalent.

While HV and IGD are difficult to analyze theoretically [12,16], GD is relatively easy. The approximate distance to the PF given by GD is always larger than or equal to the true distance. That is, GD may underestimate but never overestimate the closeness of any solutions to the PF. This characteristic is described as follows:

$$\forall R, \forall y, 0 \le \delta(y) \le \tilde{\delta}_R(y) = \min_{r \in R} \{d(y, r)\}, \tag{5}$$

where $\delta\left(\boldsymbol{y}\right)$ denotes the true distance from objective vector $\boldsymbol{y}$ to the PF, and $\tilde{\delta}_R\left(\boldsymbol{y}\right)$ denotes the distance from objective vector $\boldsymbol{y}$ to the approximate PF by the reference point set R. The accuracy of the approximate distances given by GD increases monotonically for the inclusion relation among reference point sets. That is, for any extended reference point set $R' \supset R$, the following holds,

$$\forall\boldsymbol{y}, 0 \leq \delta\left(\boldsymbol{y}\right) \leq \tilde{\delta}_{R'}\left(\boldsymbol{y}\right) \leq \tilde{\delta}_R\left(\boldsymbol{y}\right). \tag{6}$$

From (5), conservation of solution optimality holds for the inclusion relation of the reference point set. That is, providing that $\tilde{\delta}_R\left(\boldsymbol{y}\right) = 0$ holds, the following holds,

$$0 = \delta\left(\boldsymbol{y}\right) = \tilde{\delta}_{R'}\left(\boldsymbol{y}\right) = \tilde{\delta}_R\left(\boldsymbol{y}\right). \tag{7}$$

From these characteristics, we can say that the reference point set of GD allows for as many additional reference points as possible, providing that the inclusion relationship is conserved.

For a simple understanding of this discussion, consider an extreme example in Fig. 1. First, let $|X| = |R|$ and assume that R is uniformly distributed over the entire PF, as shown in Fig. 1a. Such a distribution of R provides a fair comparison for GD as well as IGD because approximation accuracy for distance is uniform on average. In this case, as with IGD, the solution set X is optimal for GD such that the solution distribution Y holds $Y = R$.

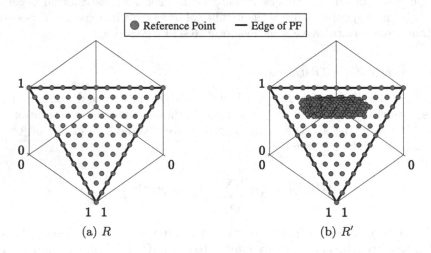

(a) R (b) R'

Fig. 1. Example of uniform reference point set R and unevenly extended reference point set $R' \supset R$

Then, let $R' \supset R$ be the unevenly extended reference point set by adding countless reference points to a local region of the PF, as in Fig. 1b. From $R' \supset R$, considering the rightmost "$\leq$" of (6), the inequality holds in the region where the reference points are added and in its neighborhood, while equality holds in other regions. Thus, the effect of each reference point on the approximation accuracy is limited within its neighborhood. Also, when the reference point set is set to R' and the accuracy of the approximation is checked on average, it is uneven, as is the distribution. Therefore, the bias in the distribution of the reference point set can be interpreted as a bias in the approximation accuracy. On the other hand, from (7), the optimality of the solution distribution $Y = R$ is conserved even if the reference point set is R'. In this respect, GD differs significantly from IGD. If the reference point set is R', the solution distribution that optimizes IGD value should be concentrated in the region where the reference points are added.

From the above, GD is considered easier to handle and interpret as a performance indicator compared to HV and IGD. Therefore, we use GD as the convergence indicator in this paper.

3   Proposed Analytical Methods

3.1   Diversity Indicators

Currently, as far as we know, there is no promising indicator to evaluate diversity only. Therefore, this paper proposes two indicators that aim to evaluate only diversity.

Ratio of Missing PFs. Some MOPs have disconnected PF shapes, i.e., the PF is composed of several sub-regions. In such MOPs, a necessary condition for diversity of MOEAs is to obtain at least one solution in each sub-region of the PF. The first proposed diversity indicator is the ratio of sub-regions of the PF that the used MOEA fails to obtain at least one solution. We call this indicator Ratio of Missing PFs.

In this paper, we define it for MaF7 [4] (DTLZ7 [9]). The objective function of MaF7 is defined as follows:

$$\min \begin{cases} f_1(\boldsymbol{x}) = x_1 \\ f_2(\boldsymbol{x}) = x_2 \\ \dots \\ f_{M-1}(\boldsymbol{x}) = x_{M-1} \\ f_M(\boldsymbol{x}) = h\left(f_1, \dots, f_{M-1}, g\right)\left(1 + g\left(\boldsymbol{x}_M\right)\right) \end{cases}, \tag{8}$$

where $f_1, \dots, f_{M-1}$ are positional functions, and f_M is a distance function. MaF7 has 2^{M-1} sub-regions of the PF. Thus, it would be quite difficult for MOEAs to obtain all sub-regions for MaOPs with more than ten objectives. The proposed indicator compares the obtained solution distribution Y with the reference point set R by the positional functions $f_1, \dots, f_{M-1}$ and calculates the ratio of sub-regions that no solution exists nearby them.

The specific procedure is as follows:

1. For the reference point set R, generate labels by applying k-means++ [1] to $R|_{1,\dots,M-1}$.

2. For the reference point set R, construct a classifier using $R|_{1,...,M-1}$ and the labels obtained in step 1. In this paper, we use k-nearest neighbor method as the classifier.
3. For the solution distribution Y, classify $Y|_{1,...,M-1}$ using the classifier obtained in step 2.

Here, $R|_{1,...,M-1}$ and $Y|_{1,...,M-1}$ are restricted sets of objective vectors defined as follows:

$$R|_{1,...,M-1} = \{r|_{1,...,M-1} = (r_1,\ldots,r_{M-1}) : r \in R\} \subset R^{M-1}, \tag{9}$$

$$Y|_{1,...,M-1} = \{y|_{1,...,M-1} = (y_1,\ldots,y_{M-1}) : y \in Y\} \subset R^{M-1}. \tag{10}$$

For k-means++ in step 1, parameter k is set to 2^{M-1}, and for k-nearest neighbor method in step 2, k is set to 1.

Hypervolume of Pareto-Optimal Solutions. The second diversity indicator depends on the structure of the test problems. In DTLZ test suite [9], many test problems are separated into positional variables and distance variables. For those problems, the optimal value is known for each distance variable. Therefore, by specifying the distance variables of the obtained solution set X as the optimal values, we can obtain a Pareto optimal solution set X_{P*} that preserves the solution distribution of the solution set X on the PF.

To eliminate the influence of convergence and obtain the pure diversity evaluation for the solution set X, we propose to evaluate this optimized solution set X_{P*} with an existing performance indicator such as HV and IGD. In this paper, we use HV and call this indicator Hypervolume of Pareto-optimal Solutions (HV_{P*}).

3.2   Proposed Methods for Comparison and Visualization

In the case of performance indicators that simultaneously evaluate convergence and diversity, such as HV and IGD, comparisons of the search performance of MOEAs are based on comparisons of large and small. Because MOEA is based on stochastic search, different results are obtained for each run. Therefore, in general, we compare representative values such as the mean or median, considering differences among runs. In addition, significant differences in representative values are confirmed by hypothesis testing.

In this paper, we evaluate the search performance of MOEAs using two indicators. When comparing convergence and diversity individually, we can use the conventional performance comparison method described above. However, to compare convergence and diversity simultaneously, it is required to use a performance comparison method that combines the two evaluation values. The following two approaches can be used to achieve this:

Convergence-Diversity Pair (C-D Pair)
A pair of statistical test results in terms of convergence and diversity indicator values for each problem is listed in a table together with the conventional

indicator value like HV or IGD. C-D Pair gives us the additional information of the conventional indicator value.

Convergence-Diversity Diagram (C-D Diagram)
For each run, two indicator values are combined as a pair and visualized in a scatter plot on a two-dimensional space. In addition, to aid in understanding, the marginal distribution for each indicator estimated by kernel density estimation are displayed outside the axes of C-D Diagram.

C-D Pair can provide a clear overview of comparisons for the performance of MOEAs, and C-D Diagram can provide a visual information that help us to intuitively understand the features for the performance of MOEAs.

4   Computational Experiments

4.1   Analysis for Many-Objective MaF7 Problems

First, we focus on the MaF7 problem with various specifications on the number of objectives. The MOEAs used in the experiments are RVEA-CA [14], RVEA-iGNG [15], and RVEA [3]. RVEA is one of the representative decomposition-based MOEAs and is the base algorithm for the others. RVEA-iGNG is an adaptive decomposition-based MOEA that adjusts the distribution of the reference vector set using GNG, a clustering algorithm. It is a promising MOEA for MaOP and shows higher search performance than other methods on various test problems. In addition, RVEA-CA is an improved version of RVEA-iGNG. All the algorithms and problems are implemented on the PlatEMO[1] platform [18].

In this experiment, we evaluate the experimental results using HV together with GD and Ratio of Missing PFs proposed in Sect. 3.1. Then, we analyze these indicator values with C-D Pair and C-D Diagram, respectively.

Parameter Settings. We use 5-, 8-, and 10-objective MaF7 problems. The population size and the number of evaluations are set to 210, 240, and 230 and 50,190, 80,160, and 100,050, for each number of objectives, respectively. For RVEA-CA, λ is set to 100. For RVEA-iGNG, we set parameters as $\varepsilon_b = 0.2$, $\varepsilon_n = 0.006$, $\alpha = 0.5$, $\alpha_{\max} = 50$, $d = 0.995$, $\lambda = 50$, $f_r = 0.1$. For RVEA, α is set to 2, and f_r is set to 0.1.

In this experiment, we use Simulated Binary Crossover (SBX) [6] and Polynomial Mutation (PM) [7]. For SBX, the distribution index η_c is set to 20, and the crossover probability p_c is set to 1. For PM, the distribution index η_m is set to 20, and the mutation probability p_m is set to $\frac{1}{D}$, where D is the number of dimensions of the decision space.

For the HV calculation, considering the number of objectives, the reference point q is set to $(1.5, \ldots, 1.5)^\mathsf{T}$.

For GD and Ratio of Missing PFs, we use the reference point set provided by PlatEMO. The sizes of the reference point set are set to 5×10^6 for GD and 1×10^5 for Ratio of Missing PFs.

---

[1] https://github.com/BIMK/PlatEMO.

Table 1. Median HV value and its standard deviation of 31 runs on many-objective MaF7 problems and C-D Pair

Problem M	RVEA-CA	RVEA-iGNG		RVEA	
	Median (SD)	Median (SD)	(C, D)	Median (SD)	(C, D)
MaF7 5	3.7163e+0 (1.00e−1)	4.5843e+0 (3.25e−2) −	(−, −)	4.2465e+0 (1.75e−2) −	(+, −)
8	1.4810e+1 (1.53e−1)	1.4850e+1 (1.24e−1) ≈	(−, ≈)	1.2813e+1 (1.77e−1) +	(+, +)
10	3.1804e+1 (1.84e−1)	3.1307e+1 (2.25e−1) +	(−, +)	2.6096e+1 (7.26e−1) +	(+, +)
+/≈/−		1/1/1	(0/0/3, 1/1/1)	2/0/1	(3/0/0, 2/0/1)

The algorithm with the best median value for each row is highlighted.
The result of statistical comparison with RVEA-CA at a 0.05 level by Wilcoxon's rank-sum test is shown on the right end of each left column in each row except for RVEA-CA.
"+", "≈", and "−" mean that RVEA-CA performs significantly better than, non-significantly better/ worse than, and significantly worse than each algorithm, respectively.

For k-means++, the number of reassignments of the initial cluster center of gravity is set to 1,000, and the maximum number of iterations is set to 1×10^{18}.

Experimental Results and Analysis for MaF7. Table 1 and Fig. 2 show the HV values with C-D Pair and C-D Diagram, respectively. Each column in Table 1 represents each algorithm, and each row in Table 1 represents each test instance. Each right column in Table 1 except for RVEA-CA shows the results of the comparison by C-D Pair.

Table 1 shows that RVEA-iGNG is the best in terms of the HV value for 5- and 8-objective MaF7 problems, while RVEA-CA outperforms the others on 10-objective MaF7 problem.

From Fig. 2, it can be seen that the convergence trends are very different between RVEA and the others. In addition, Table 1 also shows that RVEA-CA significantly outperforms RVEA in terms of convergence, i.e., the GD value on 5-, 8-, and 10-objective MaF7 problems. The reference vector adaptations based on clustering algorithms in RVEA-CA and RVEA-iGNG may contribute to this trend by concentrating the reference vectors in local regions in the PF.

Table 1 shows that RVEA-CA is outperformed by both RVEA-iGNG and RVEA in terms of the HV values on the 5-objective MaF7 problem. From Fig. 2a, the reason for this result may be that RVEA-CA misses many sub-regions of the PF on the 5-objective MaF7 problem.

When we focus on the comparison between RVEA-CA and RVEA-iGNG for 8- and 10-objective problems, Table 1 shows that no significant difference in terms of the HV value is observed for the 8-objective problem, while RVEA-CA outperforms RVEA-CA for the 10-objective with a significant difference in terms of the HV value. The results for diversity of C-D Pair in Table 1 show the same pattern, and we can observe the detailed distributions for diversity that support these comparison results in Fig. 2b and Fig. 2c. Therefore, the difference

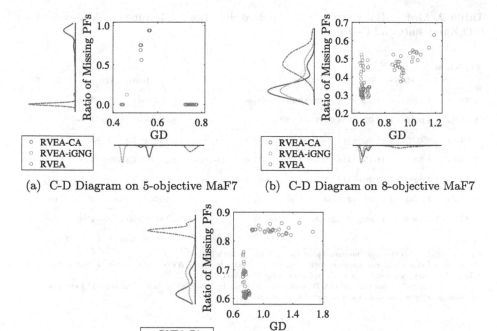

(a) C-D Diagram on 5-objective MaF7   (b) C-D Diagram on 8-objective MaF7

(c) C-D Diagram on 10-objective MaF7

Fig. 2. Comparisons by C-D Diagrams on many-objective MaF7 problems

in diversity seems to be the reason for the results that RVEA-CA outperforms the others on the 10-objective MaF7 problem.

We can easily extend the conventional summary of experimental results with the additional information from C-D Pair and briefly check the features of the search performance for the whole experiment from C-D Pair. In addition, we can understand detailed trends for a test instance from C-D Diagram. C-D Pair and C-D Diagram provide us complementary information to understand the features of MOEAs.

4.2   Analysis for a Three-Objective DTLZ Test Suite

In this section, we focus on the three-objective DTLZ test suite. The MOEAs used in the experiments are NSGA-II [8], MOEA/D [20], and IBEA [22]. They are a representative dominance-, decomposition-, and indicator-based MOEA, respectively.

In this experiment, we evaluate the experimental results using HV together with GD and HV_{P*} proposed in Sect. 3.1. Then, we analyze these indicator values with C-D Pair and C-D Diagram, respectively.

Table 2. Median HV value and its standard deviation of 31 runs on three-objective DTLZ test suite and C-D Pair

Problem M	NSGA-II Median (SD)	MOEA/D Median (SD)	(C, D)	IBEA Median (SD)	(C, D)
DTLZ1 3	1.1567e+0 (4.52e−3)	1.1729e+0 (4.03e−3) −	($\approx$, −)	8.6323e−1 (5.02e−2) +	(−, +)
DTLZ2 3	8.6959e−1 (4.07e−3)	8.9661e−1 (6.55e−5) −	(−, −)	8.9378e−1 (1.25e−3) −	(−, −)
DTLZ3 3	6.4609e−1 (3.00e−1)	3.6906e−1 (3.92e−1) +	(+, −)	4.7963e−1 (1.64e−1) +	(−, +)
DTLZ4 3	8.5062e−1 (1.15e−1)	6.5875e−1 (2.32e−1) $\approx$	(−, $\approx$)	8.9377e−1 (1.23e−3) −	(−, −)
DTLZ5 3	4.0002e−1 (2.19e−4)	3.8073e−1 (4.49e−5) +	(+, +)	3.9975e−1 (2.29e−4) +	(−, +)
DTLZ6 3	4.0042e−1 (1.59e−4)	3.8065e−1 (6.18e−5) +	(+, +)	3.9657e−1 (8.80e−4) +	(−, +)
DTLZ7 3	4.7863e−1 (1.00e−2)	4.6921e−1 (1.71e−3) +	(+, +)	4.8812e−1 (1.61e−2) −	(−, −)
+/$\approx$/−		4/1/2	(4/1/2, 3/1/3)	4/0/3	(0/0/4, 4/0/3)

The algorithm with the best median value for each row is highlighted.
The result of statistical comparison with NSGA-II at a 0.05 level by Wilcoxon's rank-sum test is shown on the right end of each left column in each row except for NSGA-II.
"+", "$\approx$", and "−" mean that NSGA-II performs significantly better than, non-significantly better/worse than, and significantly worse than each algorithm, respectively.

Parameter Settings. We use three-objective DTLZ test suite. The population size and the number of evaluations are set to 120 and 30,000, respectively.

In this experiment, we use SBX and PM with the same settings as Sect. 4.1.

For the HV and HV_{P*} calculation, the reference point q is set to $(1.1, \ldots, 1.1)^{\mathsf{T}}$.

Experimental Results and Analysis for DTLZ Test Suite. Table 2 and Fig. 3 show the HV values with C-D Pair and C-D Diagram, respectively. Each column in Table 2 represents each algorithm, and each row in Table 2 represents each test instance. Each right column in Table 2 except for NSGA-II shows the results of the comparison by C-D Pair.

Since HV_{P*} is an HV-based indicator, the direction of optimization is maximization. Thus, for visual consistency, the vertical axis is set in the opposite direction (i.e., smaller at the top).

Table 2 shows that MOEA/D outperforms the others in terms of the HV value on DTLZ1 and DTLZ2 problems, and C-D Pair in Table 2 also shows that MOEA/D almost outperforms NSGA-II in terms of both convergence and diversity on these test problems. In addition, Fig. 3a and Fig. 3b support these results, and we can observe that the variance of MOEA/D per run is small from Fig. 3a and Fig. 3b. Since DTLZ1 and DTLZ2 have the PFs that cover the entire objective space, MOEA/D seems to achieve the high search performance and robustness of the search performance as shown in Fig. 3a and Fig. 3b.

Figure 3e and Fig. 3f show that MOEA/D is outperformed by the others in terms of both convergence and diversity on DTLZ5 and DTLZ6 test problems. MOEA/D is a conventional decomposition-based MOEA and uses the reference

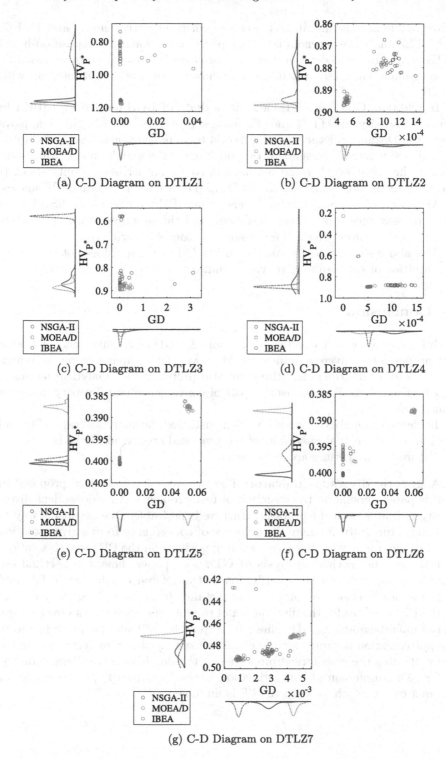

Fig. 3. Comparisons by C-D Diagrams on 3-objective DTLZ test suite

vector set uniformly distributed over the entire objective space, and DTLZ5 and DTLZ6 have the degenerated PF shapes. Conventional decomposition-based MOEAs are known to deteriorate their search performance on degenerated PFs because many uniformly distributed reference vector sets do not intersect with the PFs [15].

In addition, C-D Pair in Table 2 shows that NSGA-II outperforms IBEA in terms of diversity on DTLZ5 and DTLZ6 test problems. Since NSGA-II employs Crowding Distance for the second criterion to rank solutions, NSGA-II emphasizes diversity over convergence at the end stage of the search process when solutions in the population become non-dominated each other. As a result, NSGA-II outperforms the others on DTLZ5 and DTLZ6 that have degenerated PF shapes.

Moreover, C-D Pair in Table 2 shows that IBEA outperforms NSGA-II in terms of convergence on all test instances, and the overview of Fig. 3 supports its remarkable convergence. These results are consistent with [13].

The above shows that the analysis using C-D Diagram does not contradict our intuition or the results of previous studies.

5   Conclusion

In this paper, we gave theoretical discussions for GD as a convergence indicator and proposed two diversity indicators. Moreover, for comparisons on the search performance of multiple algorithms, we also proposed two analytical methods, C-D Pair and C-D diagram using a pair of convergence and diversity indicator values.

In computational experiments, we demonstrated the analytical capability and the validity of the analysis results of the proposed analytical methods.

Future research issues are as follows:

1. **A generic diversity indicator:** The two diversity indicators proposed in this paper depend on the structure of test problems. An independent diversity indicator from the problem structure is desirable. One promising way to achieve this is to eliminate the influence of convergence from a given solution set by approximating the corresponding position on the PF for each solution.

2. **Further theoretical analysis of GD:** This paper showed in Sect. 2.3 the relationship between the distribution of the reference point set in GD and the approximation accuracy for the distance from each objective vector to the PF. The bias for the distribution of the reference point set seems to cause the unfairness among GD values. We expect to estimate the lower bound of approximation accuracy for the distance from an objective vector to the PF by utilizing the geometrical properties of the dominance relationship on the PF. The discussion of this estimation is considered an important topic for practical use, such as when the PF is unknown.

References

1. Arthur, D., Vassilvitskii, S.: K-means++: the advantages of careful seeding. In: Proceedings of the Eighteenth Annual ACM-SIAM Symposium on Discrete Algorithms, SODA 2007, pp. 1027–1035. Society for Industrial and Applied Mathematics, USA (2007)
2. Bringmann, K., Friedrich, T.: Approximating the volume of unions and intersections of high-dimensional geometric objects. In: Hong, S.-H., Nagamochi, H., Fukunaga, T. (eds.) ISAAC 2008. LNCS, vol. 5369, pp. 436–447. Springer, Heidelberg (2008). https://doi.org/10.1007/978-3-540-92182-0_40
3. Cheng, R., Jin, Y., Olhofer, M., Sendhoff, B.: A reference vector guided evolutionary algorithm for many-objective optimization. IEEE Trans. Evol. Comput. **20**(5), 773–791 (2016). https://doi.org/10.1109/TEVC.2016.2519378
4. Cheng, R., et al.: A benchmark test suite for evolutionary many-objective optimization. Complex Intell. Syst. **3**(1), 67–81 (2017). https://doi.org/10.1007/s40747-017-0039-7
5. Coello, C.A.C., Cortés, N.C.: Solving multiobjective optimization problems using an artificial immune system. Genet. Program Evolvable Mach. **6**(2), 163–190 (2005). https://doi.org/10.1007/s10710-005-6164-x
6. Deb, K., Agrawal, R.B.: Simulated binary crossover for continuous search space. Complex Syst. **9**(2), 115–148 (1995)
7. Deb, K., Goyal, M.: A combined genetic adaptive search (GeneAS) for engineering design. Comput. Sci. Inform. **26**, 30–45 (1996)
8. Deb, K., Pratap, A., Agarwal, S., Meyarivan, T.: A fast and elitist multiobjective genetic algorithm: NSGA-II. IEEE Trans. Evol. Comput. **6**(2), 182–197 (2002). https://doi.org/10.1109/4235.996017
9. Deb, K., Thiele, L., Laumanns, M., Zitzler, E.: Scalable test problems for evolutionary multiobjective optimization. In: Abraham, A., Jain, L., Goldberg, R. (eds.) Evolutionary Multiobjective Optimization. Advanced Information and Knowledge Processing, pp. 105–145. Springer, London (2005). https://doi.org/10.1007/1-84628-137-7_6
10. Fleming, P.J., Purshouse, R.C., Lygoe, R.J.: Many-objective optimization: an engineering design perspective. In: Coello Coello, C.A., Hernández Aguirre, A., Zitzler, E. (eds.) EMO 2005. LNCS, vol. 3410, pp. 14–32. Springer, Heidelberg (2005). https://doi.org/10.1007/978-3-540-31880-4_2
11. Ishibuchi, H., Imada, R., Setoguchi, Y., Nojima, Y.: How to specify a reference point in hypervolume calculation for fair performance comparison. Evol. Comput. **26**(3), 411–440 (2018). https://doi.org/10.1162/evco_a_00226
12. Ishibuchi, H., Imada, R., Setoguchi, Y., Nojima, Y.: Reference point specification in inverted generational distance for triangular linear Pareto front. IEEE Trans. Evol. Comput. **22**(6), 961–975 (2018). https://doi.org/10.1109/TEVC.2017.2776226
13. Ishikawa, T., Fukumoto, H., Oyama, A., Nishida, H.: Improved binary additive epsilon indicator for obtaining uniformly distributed solutions in multi-objective optimization. In: Proceedings of the Genetic and Evolutionary Computation Conference Companion, GECCO 2019, pp. 209–210 (2019). https://doi.org/10.1145/3319619.3322025
14. Kinoshita, T., Masuyama, N., Liu, Y., Nojima, Y., Ishibuchi, H.: Reference vector adaptation and mating selection strategy via adaptive resonance theory-based clustering for many-objective optimization (2022). https://doi.org/10.48550/ARXIV.2204.10756

15. Liu, Q., Jin, Y., Heiderich, M., Rodemann, T., Yu, G.: An adaptive reference vector-guided evolutionary algorithm using growing neural gas for many-objective optimization of irregular problems. IEEE Trans. Cybern. **52**(5), 2698–2711 (2022). https://doi.org/10.1109/TCYB.2020.3020630

16. Shang, K., Ishibuchi, H., He, L., Pang, L.M.: A survey on the hypervolume indicator in evolutionary multiobjective optimization. IEEE Trans. Evol. Comput. **25**(1), 1–20 (2021). https://doi.org/10.1109/TEVC.2020.3013290

17. Tanabe, R., Ishibuchi, H.: An analysis of quality indicators using approximated optimal distributions in a 3-D objective space. IEEE Trans. Evol. Comput. **24**(5), 853–867 (2020). https://doi.org/10.1109/TEVC.2020.2966014

18. Tian, Y., Cheng, R., Zhang, X., Jin, Y.: PlatEMO: a MATLAB platform for evolutionary multi-objective optimization. IEEE Comput. Intell. Mag. **12**(4), 73–87 (2017). https://doi.org/10.1109/MCI.2017.2742868

19. Van Veldhuizen, D.A.: Multiobjective evolutionary algorithms: classifications, analyses, and new innovations. Ph.D. thesis, USA (1999). aAI9928483

20. Zhang, Q., Li, H.: MOEA/D: a multiobjective evolutionary algorithm based on decomposition. IEEE Trans. Evol. Comput. **11**(6), 712–731 (2007). https://doi.org/10.1109/TEVC.2007.892759

21. Zitzler, E., Brockhoff, D., Thiele, L.: The hypervolume indicator revisited: on the design of pareto-compliant indicators via weighted integration. In: Obayashi, S., Deb, K., Poloni, C., Hiroyasu, T., Murata, T. (eds.) EMO 2007. LNCS, vol. 4403, pp. 862–876. Springer, Heidelberg (2007). https://doi.org/10.1007/978-3-540-70928-2_64

22. Zitzler, E., Künzli, S.: Indicator-based selection in multiobjective search. In: Yao, X., et al. (eds.) PPSN 2004. LNCS, vol. 3242, pp. 832–842. Springer, Heidelberg (2004). https://doi.org/10.1007/978-3-540-30217-9_84

23. Zitzler, E., Thiele, L.: Multiobjective evolutionary algorithms: a comparative case study and the strength Pareto approach. IEEE Trans. Evol. Comput. **3**(4), 257–271 (1999). https://doi.org/10.1109/4235.797969

How a Different Ant Behavior Affects on the Performance of the Whole Colony

Carolina Crespi(✉), Rocco A. Scollo(ID), Georgia Fargetta(ID),
and Mario Pavone(ID)

Department of Mathematics and Computer Science, University of Catania,
Viale A. Doria 6, 95125 Catania, Italy
{carolina.crespi,rocco.scollo,georgia.fargetta}@phd.unict.it,
mpavone@dmi.unict.it

Abstract. This paper presents an experimental analysis of how different behavior performed by a group of ants affects the optimization efficiency of the entire colony. Two different interaction ways of the ants with each other and with the environment, that is a weighted network, have been considered: (*i*) *Low Performing Ants* (LPA), which destroy nodes and links of the network making it then dynamic; and (*ii*) *High Performing Ants* (HPA), which, instead, repair the destroyed nodes or links encountered on their way. The purpose of both ant types is simply to find the exit of the network, starting from a given entrance, whilst, due to the uncertainty and dynamism of the network, the main goal of the entire colony is maximize the number of ants that reach the exit, and minimize the path cost and the resolution time. From the analysis of the experimental outcomes, it is clear that the presence of the LPAs is advantageous for the entire colony in improving its performances, and then in carrying out a better and more careful optimization of the environment.

Keywords: Ant Colony Optimization · Metaheuristic · Optimization · Dynamic networks · Uncertain optimization environments

1 Introduction

Ant Colony Optimization (ACO) is a well-known optimization procedure and represents nowadays the most representative methodology into the Swarm Intelligence family as it was successfully applied in many hard combinatorial optimization problems [4,8]. ACO is a metaheuristic that takes inspiration from observing foraging behavior of natural ant colonies since they can find exactly the shortest path from their nest to source of food, and they communicate with each other through chemical signals called pheromones. Thanks to these properties, it has become powerful optimization techniques for solving different kinds of complex combinatorial optimization problems [14], such as scheduling and routing problems [7,10], coloring [3,9], the robot path planning to patrol areas where humans cannot get there [1,2,16], transportation problems [11], and feature selection [13]. Ant colonies are also recognized to be the best organized and

© The Author(s), under exclusive license to Springer Nature Switzerland AG 2023
L. Di Gaspero et al. (Eds.): MIC 2022, LNCS 13838, pp. 187–199, 2023.
https://doi.org/10.1007/978-3-031-26504-4_14

cooperative system, able to make its community work at its best, and able to perform complex tasks [12]: any action of each ant is related only to its local environment, local interactions with other ants, simple social rules, and in the total absence of centralized decisions. It is known that it is not the single ant that finds the best solution but its cooperation and interaction with the environment and the rest of the colony that produces the desired result. These features have been implemented in ACO algorithms to solve not only the previously mentioned problems but also to evaluate how they affect the efficiency of the algorithm [6] and to investigate and analyze crowds' behavior [5]. This research paper proposes an analysis of what happens if in an ACO algorithm some ants act in a different way from the rest of the colony. In particular, the presented study consists of analyse two different kinds of ants, which act in different way: *Low Performing Ants* (LPAs) that can accidentally destroy some nodes or links of the network, therefore making them not crossable; and *High Performing Ants* (HPAs) that instead repair them. These different actions performed by the ants make the network dynamics in the sense that both actions (destroy or repair) change instantaneously the environment, modifying consequently the network topology. This means that a node or a link can be not crossable in the timestep t, but becoming crossable just after $(t + 1)$. Both kinds of ants must find the exit point of the network, starting from a given entrance, with the overall goal to maximize the number of ants that reach the exit, and minimize the path cost and the resolution time. The problem studied is a general path problem, however, the shortest path, in this case, is not a good evaluation metric due to the dynamism of the network. Moreover, thinking about a possible application of this study in the field of swarm robotics, it is desirable that if there are some robots exploring unknown environments, the same robots will be able to come out from them in the maximum possible number. Two different complex networks have been considered to analyze how the presence of LPA affects the performance of the entire colony at different levels of available information: *high trace*, i.e. high amount of pheromone released, and *low trace*, low amount of pheromone released. Analyzing the investigation conducted on entire colony from an optimization point of view, emerges that the presence of a group of LPA helps and stimulates the rest of the colony to work better, especially when the amount of trace shared is high. Indeed, the disturbing actions performed by LPAs force the rest of the colony to change its behavior, and, consequently, to explore new paths.

2   The Model

The presented model has been realized using the software NetLogo [15], and the environment in which the ants move is a weighted network defined mathematically as a graph $G = (V, E, w)$, where V is the set of vertices, E is the set of edges and $w \colon V \times V \to \mathbb{R}^+$ is the *weighted* function that assigns a positive cost to each edge of the graph. The weight indicates how difficult is crossing a particular edge. The starting point is a node randomly chosen in one side of the graph (e.g. left side), whilst the exit point is another node randomly chosen in the opposite side to the starting one (e.g. right side). Every link is crossable in both directions. The colony is composed of two kinds of ants:

- *Low Performing Ants* (LPAs): they are always low performing in the sense that they do not work properly and so they can destroy, with a certain probability $0 \leq \rho_e \leq 1$, some edges of the network or, with a probability $0 \leq \rho_v \leq 1$ some nodes of the network. They do not leave any amount of pheromone after crossing an edge (i, j) of their path;
- *High Performing Ants* (HPAs): they always are high performing, in fact, if they find they find a destroyed node they can repair it with a probability $0 \leq \rho_v \leq 1$ and, if they find a destroyed edge they can repair it with a probability $0 \leq \rho_e \leq 1$. Moreover, they release two different kinds of information about the path: the classic pheromone information after they have crossed ad edge (i, j), and a more sophisticate information named $\eta_{ij}(t) = 1/w_{ij}(t)$, where $w_{ij}(t)$ is the weight of the edge (i, j) at a time t and so $\eta_{ij}(t)$ indicate to the rest of the colony how difficult is that path.

It is important to highlight that the action of destroy an edge or a node means that this becomes impracticable, i.e. uncrossable. Instead, repair an edge or a node means that it is practicable again. Both actions, therefore, make the network dynamic. The number of HPAs in the colony is determined by the *performing factor* $p_f \in [0, 1]$, and therefore, once it is set, the remaining ants (i.e. $1 - p_f$) will be LPAs. Note that when $p_f = 1$, i.e. all ants are HPA, the ACO classical version is obtained.

Let be $A_i = \{j \in V : (i, j) \in E\}$ the set of vertices adjacent to vertex i and $\pi^k(t) = (\pi_1, \pi_2, \ldots, \pi_t)$ the set of vertices visited by an ant k at a certain time t, where $(\pi_i, \pi_{i+1}) \in E$ for $i = 1, \ldots, t - 1$. Due to the action of the HPAs that can repair damaged nodes and/or links, the path $\pi^k(t)$ is not just a simple path, because an ant can visit again a vertex due to a back-tracking operation. The probability $p_{ij}^k(t)$ with which an ant k placed on a vertex i chooses as destination one of its neighbor vertices j at the time t is defined according to the Ant Colony Optimization *proportional transition rule*:

$$
p_{ij}^k(t) = \begin{cases} \dfrac{\tau_{ij}(t)^\alpha \cdot \eta_{ij}^\beta}{\sum_{l \in J_i^k} \tau_{il}(t)^\alpha \cdot \eta_{il}^\beta} & \text{if } j \in J_i^k \\ 0 & \text{otherwise,} \end{cases}
\tag{1}
$$

where $J_i^k = A_i \backslash \{\pi_t^k\}$ are all the possible displacements of the ant k from vertex i, $\tau_{ij}(t)$ is the pheromone intensity on the edge (i, j) and $\eta_{ij}(t)$ is the desirability of the edge (i, j) at a given time t, while α and β are two parameters that determine the importance of pheromone intensity with respect to the desirability of an edge. For contextualization reasons with the environment/scenario tackled, from now on the term pheromone will be replaced with the term *trace*. The amount of trace released by the k ant after crossing an edge (i, j) at a time t is constant and it is defined as:

$$
\Delta \tau_{ij}^k(t) = K.
\tag{2}
$$

The desirability $\eta_{ij}(t)$ at a given time t, establish how much an edge (i, j) is promising. In particular and it is defined as $\eta_{ji}(t) = \frac{1}{w_{ij}(t)}$. This information is released by each ant as the trace, however it does not depend on the ant itself, but only on the edge (i, j). Each link is updated asynchronously with two kinds of updating rules based on the ticks T of the software used for the simulations[1]. A *local updating rule* that updates the trace levels at the end of each tour of the winning ant, according to the follow rule:

$$\tau_{ij}(t+1) = \tau_{ij}(t) + K,\tag{3}$$

where K represents the trace that every ant leave after crossing an edge (i, j) and $\tau_{ij}(t)$ is the amount of trace on the link at time t. A *global updating rule* that update the amount of trace on all the links of the network every T ticks:

$$\tau_{ij}(t+1) = (1 - \rho) \cdot \tau_{ij}(t),\tag{4}$$

where $\tau_{ij}(t)$ is the amount of trace on the edge (i, j) at a time t and ρ is the evaporation decay parameter. The aim of the ants is to explore the graph and find in the shortest time, the cheapest path from the starting point to the end point, orienting themselves using the amount of trace on the paths and the information exchanged about their desirability. At the same time, they must maximize their number at the end point, that is the exit. Mathematically, this means the one have to optimize three different objective functions: minimize the *path cost function* and the *time cost function*, and maximize the *exit function* that represent how many ants have reached the end point. Since the path cost function and the time cost function must both be minimized, they have been put together into the following unified objective function:

$$\min \sum_{i=1}^{t-1} w(\pi_i^k, \pi_{i+1}^k) + |\pi^k|.\tag{5}$$

It represents both the minimization of the cost of the path and the resolution time, where the first term represents the path made by an ant k, while the second term represents the number of steps made by the same ant k. It can be used as a time term because each unit of time corresponds to an ant displacement, i.e. the number of the nodes visited by an ant corresponds to the resolution time.

Finally, the exit function is defined as:

$$\max \sum_{g \in G} \sum_{k \in N} k_g.\tag{6}$$

It represents the maximization of the number of ants that must reach the exit, where G is the total number of groups, g is the index of the group to which the ant k belongs, k_g is the ant k that belongs to g group and N is the set of ants.

---

[1] Each tick correspond to an ant displacement and movement.

3   Experiments and Results

The simulations have been realized using two different kinds of scenarios, that correspond to different networks with increasing complexity. Within each scenario, two parameters of the model have been varied. In particular, the amount K of trace deposited by each ant on the links and the value of the parameter β that measures the importance of the information with respect to the amount of trace itself. This choice was meant to study and understand if and how the values of the model affect the performances of the colony when it is composed of two different kinds of ants.

The two scenarios are:

- *Scenario B1*: a network with $|V| = 225$ nodes and $|E| = 348$ links.
- *Scenario B2*: a network with $|V| = 225$ nodes but $|E| = 495$ links.

The general experimental setup is the following. For each scenario, $N = 1000$ ants divided into $G = 10$ groups have been considered. This means that each group is composed of $N_g = 100$ ants that start their journey from the starting point at regular intervals computed multiplying the values of rows and columns, so $T_l = 225$ ticks. As said previously, the colony is composed of two different kinds of ants: high performing ants (HPAs) that always work at their best, and low performing ants (LPAs) that may destroy some nodes or links of their path. The number of HPAs and LPAs is regulated by a performing factor p_f that establishes the fraction of the first respect to the second. It goes from $p_f = 0.0$ (that defines a colony of just LPAs) to $p_f = 1.0$ (that defines a colony of just HPAs and correspond to the ACO classic version) with steps of $p_f = 0.10$. For instance, in a colony of 100 ants a value of $p_f = 0.30$ means that 30 ants are HPAs while the other 70 are LPAs.

Due to limited time resources, the ants must find the exit in a maximum time, which depends on the number of groups and the complexity of the network. This time is set to $T_{max} = 2 \times G \times T_l$, where G is the number of groups, T_l is the launch interval and 2 is just a corrective factor. The initial trace intensity on the links is set to 1.0 and it decreases over time according to the trace evaporation interval, $T_d = 50$ (i.e. every 50 ticks the amount of trace evaporate with the evaporation rate $\rho = 0.10$). For the scenarios defined as High Trace, the parameters α and β are both set to 1 and the amount of trace deposited by each ant on the links of its path is set to $K = 0.1$. For the scenarios defined as Low Trace, the parameter α is set to 1, the parameter β is set to 0.5 and the amount of trace deposited by each ant on the links of its path is set to $K = 0.001$. Since the parameter β regulates the influence of the information with respect to the amount of trace, one can expects that decreasing both β and K the colony will act taking more into account the information acquired about the path and less the information released with the trace. Finally, the edge destruction-repair probability and vertex destruction-repair probability are for both configurations $\rho_e = 0.02$ and $\rho_v = 0.02$. With these configurations of the parameters, 10 independent simulations have been performed, starting from the value $p_f = 0.0$ of the performing factor to $f = 1.0$, with steps of 0.1. Two different kinds of analysis have been done: (*i*) a *group analysis* to understand how many ants have reached the exit, considering both the value of the performing factor and the number of groups; and (*ii*) an *overall analysis* considering the (1) path cost found by the colony, (2) how much time the ants have used to find it, and (3) how many of them have reached the exit in time. In the following results, the label *High Trace* refers to a value of $K = 0.1$ and a value of $\beta = 1.0$, while the label *Low Trace* refers to a value of $K = 0.001$ and a value of $\beta = 0.5$. It is worth emphasizing once again that, due to the dynamism of the network produced by the actions of the two types of ants, it is not possible to consider the shortest path as evaluation metric, and therefore the number of ants that reach the exit (to be maximized), the cost of the paths and the resolution time (both to be minimized) were considered as the investigation measure.

3.1   Group Analysis

As said, in this first kind of analysis both the performing factor and the number of groups have been considered to evaluate the number of ants that have reached the exit. A heat map has been used to plot the results and by looking at the legend on the right of each plot one can easily understand that the lighter the blue is, the higher the value of the number of ants is. On the contrary, the darker the blue is, the lower the same number is. The absence of color implies that no ants have reached the exit for that value of the performing factor or for that value of the group. In Fig. 1 are shown the results obtained for the simulation performed in scenario B1 and in particular in Fig. 1a are plotted the number of ants that have reached the exit per ticks when there is a high-level trace. In Fig. 1b is plotted the same quantity but when there is a low-level trace.

Comparing Fig. 1a and Fig. 1b one can easily see that the best results, that is the maximum number of ants that reach the exit, are obtained not only for different values of the performing factor, but also for different values of the groups. In particular, when there is a high-level trace the best performances of the colony are obtained by the last groups and when the performing factor is round $p_f = 0.9$. On the contrary, comparing these results with the ones obtained in the same scenario with a low-level trace, one can see that in this case, the best results are obtained from the last groups not only when the performing factor in equal to $p_f = 0.9$ but also when it is equal to $p_f = 1.0$. This indicates that the presence of LPAs is much more important when the trace level is high. In fact, it is noted that the number of ants per ticks exiting is greater for values of the performing factor equal to $f = 0.9$ or, a little bit lower, at $f = 0.8$. This behavior is similar for all the groups as the performing factor varies. These results are justified by the fact that when the trace is high, the ants are mistakenly affected and tend to follow incorrect paths. Furthermore, from the plots in Fig. 1 it is observed that in the case in which there is a high-level trace, the number of ants that reach the exit is higher respect to the one obtained when there is a low-level trace. This indicates that the action of the LPAs is crucial to maximize the ants when there is a high-level trace, since for $p_f = 1.0$ the performances of

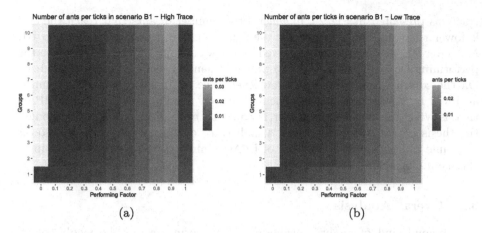

(a)   (b)

Fig. 1. Heat map representing the number of ants that have reached the exit per ticks in scenario B1. The performances of the colony change depending on the amount of the trace released by the ants: in (a) one can see that they reach their best for the last group $g = 10$ and when the performing factor is equal to $p_f = 0.9$, if there is a high level trace. The trend is similar in (b), that is when there is a low level trace released by the ants even if in this case the good performances continue to the value of the performing factor $p_f = 1.0$

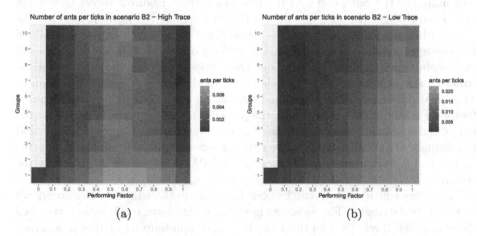

(a)   (b)

Fig. 2. Heat map representing the number of ants that have reached the exit per ticks in scenario B2. In (a) one can see that the colony reaches its best for the first groups and when the performing factor is round $p_f = 0.5$, if there is a high level trace. On the contrary, when there is a low level trace released, as in (b), the best performances are obtained for higher values of the performing factor, grater then $p_f > 0.7$ and for more groups following the firsts.

the colony are worst. Figure 2 shows the same analysis carried out for scenario B2. The trend is similar to the one presented in the previous heat maps for scenario B1, but with some differences. In this case, when there is a high-level trace, as presented in Fig. 2a, in general the number of ants that reach the exit is lower than the one obtained for the same configuration but in scenario B1. Moreover, the best results are achieved by the first groups of the colony when the performing factor is round $p_f = 0.5$ that is, when it is composed of some LPAs. On the contrary, when there is a low-level trace, as in Fig. 2b, the optimum is achieved by the first group and when the colony is composed by mainly HPAs, that is when the performing factor is equal to $p_f = 1.0$. This makes stronger the thesis of this paper, for which a high trace confuses the colony and so, at the same time, a small percentage of LPAs stimulates the rest of the group to change its behavior.

3.2   Overall Analysis

This second kind of analysis evaluates the performances of the whole colony considering only how they vary with respect to the performing factor, not considering the number of groups in which are divided the ants. The analysis is carried out considering, as in the Sect. 3.1, the different performances of the colony when there is a high-level trace and a low-level trace. The quantities analyzed are the number of ants that have reached the exit, the path cost, and the resolution time. The aim of the experiments was to maximize the number of ants and minimize the path cost and the resolution time. Figure 3 shows how many ants have reached the exit in scenario B1. In particular, Fig. 3a represents the results obtained for high-level trace, while Fig. 3b represents the ones obtained for low-level trace. As one can see, the actions of the LPAs are more powerful and useful when there is an excess of trace released by all the ants, since the colony reaches better results when there is a small percentage of LPAs within it, as is clear from Fig. 3a for which the best value is obtained when the performing factor $p_f = 0.9$. On the other hand, the presence of LPAs seems to not boost the performances of the colony when there is a low-level trace. Figure 3b shows that the number of ants that reached the exit is approximately the same for $p_f = 0.9$ and $p_f = 1.0$, indicating that the presence of LPAs does not affect positively the colony.

The same considerations can be done for scenario B2, whose results are shown in Fig. 4. In particular, Fig. 4a shows how many ants have reached the exit when there is a high-level trace. In this case, the maximum number of ants is obtained when the performing factor is $p_f = 0.5$. Figure 4b, on the other hand, shows the same quantity when there is a low-level trace, and here the best performances of the colony are obtained when the performing factor is $p_f = 1.0$. As in the previous case, the presence of LPAs seems to be more important and helpful when there is an excess of trace release along the path since in this case, the colony has better performances when it is not composed of just HPAs. A note of interest is that scenario B2 has been obtained lower average values of the number of ants with respect to the ones obtained for scenario B1. This may depend on the complexity of the network: the higher it is the worst the performances of the colony will be.

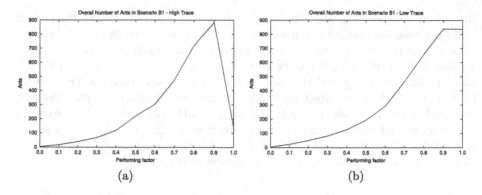

(a)   (b)

Fig. 3. Overall number of ants that have reached the exit in scenario B1. In (a) the values obtained for a high level trace; in (b) the ones obtained for a low level trace. The presence of LPAs is much more important and useful when there is a high-level trace, leading the colony to better performances. The best values are obtained for $p_f = 0.9$ when there is a high-level trace and for $p_f = 1.0$ when there is a low-level trace.

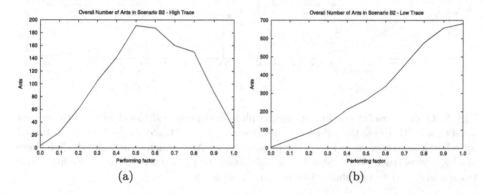

(a)   (b)

Fig. 4. Overall number of ants that have reached the exit in scenario B1. In (a) the values obtained for a high level trace; in (b) the ones obtained for a low level trace. As in Fig. 3, the presence of LPAs is much more helpful when there is a high-level trace. The best values are obtained for $p_f = 0.5$ when there is a high-level trace and for $p_f = 1.0$ when there is a low-level trace.

The path cost and the resolution time are both quantities to be minimized so they have been put together in the same plot. In particular, the principal plot represents the resolution time, the inset one the path cost. This has been done both for scenario B1, in Fig. 5, and for scenario B2, in Fig. 6. In particular, Fig. 5a represents how the resolution time and path cost vary with respect to the performing factor in scenario B1 with high-level trace; Fig. 5b shows the same quantities in the same scenario with a high-level trace. In this case, the best values are the lowest ones because they correspond to the best performances of the colony. Comparing these results with the ones regarding the number of ants in Fig. 3a, one can realize that in scenario B1, when there is a high-level trace, the colony has better performances when it is composed of a small fraction of LPAs because the maximum number of ants that reaches the exit, the minimum value of the resolution time and the minimum of the path cost is obtained for a value of the performing factor equal to $p_f = 0.9$. Doing the same with Fig. 5b and Fig. 3b, one can see that when there is a low-level trace the presence of LPAs not only does not affect positively the number of ants that reach the exit but neither on the resolution time and on the path cost find by the colony. In this case, indeed, the best values are obtained when the colony is composed of just HPAs, reinforcing the hypothesis for which the presence of LPAs is useful to regulate the actions when there is an excess of trace.

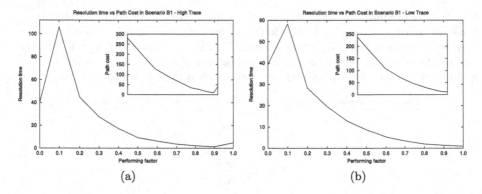

Fig. 5. Overall resolution time (principal plot) and path cost (inset plot) of the colony for scenario B1. In (a) the values obtained for a high-level trace; in (b) the ones obtained for a low level trace. As in Fig. 3, the presence of LPAs is much more helpful when there is a high-level trace. The best values are obtained for $p_f = 0.9$ when there is a high-level trace and for $p_f = 1.0$ when there is a low-level trace.

It is not surprising that the same results have been obtained also for scenario B2, as represented in Fig. 6. In this case, the importance of the presence of LPAs is clear especially looking at the values obtained when there is a high-level trace. Indeed, there is a lot of difference between the path cost found by the colony (in the inset plot) when the performing factor is equal to $p_f = 1.0$ and the one obtained when the performing factor is equal to $p_f = 0.9$. The second value is much better than the first and it is obvious that the same worst performances are present also considering the resolution time in the principal plot, and the number of ants, as shown in Fig. 4a. On the contrary, but as previously shown, when there is a low-level trace, the presence of LPAs does not help the colony to boost its performances, which are better when it is composed of just HPAs. Figure 6, indeed shows that the best values of the resolution time and the path cost are obtained when the performing factor is equal to $p_f = 1.0$.

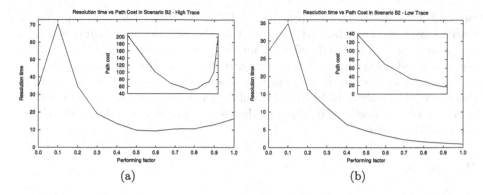

Fig. 6. Overall resolution time (principal plot) and path cost (inset plot) of the colony for scenario B1. In (a) the values obtained for a high-level trace; in (b) the ones obtained for a low level trace. As in Fig. 4, the presence of LPAs is much more helpful when there is a high-level trace. When there is a high-level trace, the best value of the resolution time is for $p_f = 0.6$ and the one for the path cost is for $p_f = 0.5$. When there is a low-level trace the same bests are obtained for $p_f = 1.0$.

4   Conclusions

This paper aims to investigate how different behaviors of the ants in the Ant Colony Optimization algorithm affect the global performances of the colony. To do this, two different kinds of ants have been considered: (1) *low performing ants* (LPAs), which can damage with certain probability nodes and links of their paths, and which do not help the rest of the colony sharing their information about the cost of each link; and (2) *high performing ants* (HPAs) which, on the contrary, may repair with a certain probability the damaged nodes and links and share their information about the cost of the links. The model has been tested on two networks with increasing complexity and has been investigated if and how the presence of LPA affects the performances of the group when different levels of information are present. Two different kinds of analysis have been carried out: (*i*) a *group analysis*, to analyze how the number of ants that reach the endpoint of the network varies with respect to the performing factor and the group of the colony considered; and (*ii*) an *overall analysis* to analyze how the number of ants of the colony, the path cost find by it and its resolution time of the network vary taking into account only the performing factor. Both kinds of analysis have been realized naming High Trace the configuration for which the amount of pheromone released by the ants is $K = 0.1$ and the parameter $\beta = 1.0$ (i.e. more information available), and Low Trace the configuration for which the amount of pheromone release by the ants is $K = 0.001$ and the parameter $\beta = 0.5$ (i.e. less information available). From the group analysis, emerges that the presence of LPA helps the rest of the colony especially when there is a condition of high-level trace because the disturbing actions performed by the LPAs stimulate the others to search for other paths and to share their information among the other groups of the colony. This seems to be true even for the overall analysis which considers the objective functions. The presence of LPAs is crucial when the amount of trace shared by the ants, and so present in the environment, is too high. An excess of information is self-defeating for the group because, since their actions are calibrated according to this quantity, it does not allow the ants to explore the rest of the network, letting them choose the same path over and over. In this sense, the presence of LPAs is helpful for the rest of the group because their actions force the rest of the colony to change its behavior in order to search for more fruitful paths.

References

1. Akka, K., Khaber, F.: Mobile robot path planning using an improved ant colony optimization. Int. J. Adv. Robot. Syst. **15**(3) (2018). https://doi.org/10.1177/1729881418774673
2. Brand, M., Masuda, M., Wehner, N., Yu, X.: Ant colony optimization algorithm for robot path planning. In: 2010 International Conference On Computer Design and Applications, vol. 3, pp. V3-436–V3-440 (2010). https://doi.org/10.1109/ICCDA.2010.5541300
3. Consoli, P., Collerà, A., Pavone, M.: Swarm intelligence heuristics for graph coloring problem. In: 2013 IEEE Congress on Evolutionary Computation, pp. 1909–1916 (2013). https://doi.org/10.1109/CEC.2013.6557792
4. Consoli, P., Pavone, M.: O-BEE-COL: optimal BEEs for COLoring graphs. In: Legrand, P., Corsini, M.-M., Hao, J.-K., Monmarché, N., Lutton, E., Schoenauer, M. (eds.) EA 2013. LNCS, vol. 8752, pp. 243–255. Springer, Cham (2014). https://doi.org/10.1007/978-3-319-11683-9_19
5. Crespi, C., Fargetta, G., Pavone, M., Scollo, R.A., Scrimali, L.: A game theory approach for crowd evacuation modelling. In: Filipič, B., Minisci, E., Vasile, M. (eds.) BIOMA 2020. LNCS, vol. 12438, pp. 228–239. Springer, Cham (2020). https://doi.org/10.1007/978-3-030-63710-1_18
6. Crespi, C., Scollo, R.A., Pavone, M.: Effects of different dynamics in an ant colony optimization algorithm. In: 2020 7th International Conference on Soft Computing Machine Intelligence (ISCMI2020), pp. 8–11. IEEE (2020). https://doi.org/10.1109/ISCMI51676.2020.9311553
7. Deng, W., Xu, J., Zhao, H.: An improved ant colony optimization algorithm based on hybrid strategies for scheduling problem. IEEE Access **7**, 20281–20292 (2019). https://doi.org/10.1109/ACCESS.2019.2897580
8. Dorigo, M., Stützle, T.: Ant colony optimization: overview and recent advances. In: Gendreau, M., Potvin, J.-Y. (eds.) Handbook of Metaheuristics. ISORMS, vol. 272, pp. 311–351. Springer, Cham (2019). https://doi.org/10.1007/978-3-319-91086-4_10
9. Fidanova, S., Pop, P.: An improved hybrid ant-local search algorithm for the partition graph coloring problem. J. Comput. Appl. Math. **293**, 55–61 (2016). https://doi.org/10.1016/j.cam.2015.04.030
10. Jia, Y.H., Mei, Y., Zhang, M.: A bilevel ant colony optimization algorithm for capacitated electric vehicle routing problem. IEEE Trans. Cybern. 1–14 (2021). https://doi.org/10.1109/TCYB.2021.3069942
11. Jovanovic, R., Tuba, M., Voß, S.: An efficient ant colony optimization algorithm for the blocks relocation problem. Eur. J. Oper. Res. **274**(1), 78–90 (2019). https://doi.org/10.1016/j.ejor.2018.09.038

12. O'Shea-Wheller, T., Sendova-Franks, A., Franks, N.: Differentiated anti-predation responses in a superorganism. PLoS One **10**(11), e0141012 (2015). https://doi.org/10.1371/journal.pone.0141012
13. Peng, H., Ying, C., Tan, S., Hu, B., Sun, Z.: An improved feature selection algorithm based on ant colony optimization. IEEE Access **6**, 69203–69209 (2018). https://doi.org/10.1109/ACCESS.2018.2879583
14. Pintea, C.-M., Matei, O., Ramadan, R.A., Pavone, M., Niazi, M., Azar, A.T.: A fuzzy approach of sensitivity for multiple colonies on ant colony optimization. In: Balas, V.E., Jain, L.C., Balas, M.M. (eds.) SOFA 2016. AISC, vol. 634, pp. 87–95. Springer, Cham (2018). https://doi.org/10.1007/978-3-319-62524-9_8
15. Wilensky, U.: NetLogo. Center for Connected Learning and Computer-Based Modeling, Northwestern University, Evanston, IL (1999). http://ccl.northwestern.edu/netlogo/
16. Zhang, D., You, X., Liu, S., Pan, H.: Dynamic multi-role adaptive collaborative ant colony optimization for robot path planning. IEEE Access **8**, 129958–129974 (2020). https://doi.org/10.1109/ACCESS.2020.3009399

Evaluating the Effects of Chaos in Variable Neighbourhood Search

Sergio Consoli[1]([✉]) and José Andrés Moreno Pérez[2]

[1] European Commission, Joint Research Centre (JRC), Ispra, VA, Italy
`sergio.consoli@ec.europa.eu`
[2] Instituto Universitario de Desarrollo Regional, Universidad de La Laguna,
Tenerife, Spain
`jamoreno@ull.edu.es`

Abstract. Metaheuristics are problem-solving methods which try to find near-optimal solutions to very hard optimization problems within an acceptable computational timeframe, where classical approaches usually fail, or cannot even been applied. Random mechanisms are an integral part of metaheuristics, given randomness has a role in dealing with algorithmic issues such as parameters tuning, adaptation, and combination of existing optimization techniques. In this paper, it is explored whether deterministic chaos can be suitably used instead of random processes within Variable Neighbourhood Search (VNS), a popular metaheuristic for combinatorial optimization. As a use case, in particular, the paper focuses on labelling graph problems, where VNS has been already used with success. These problems are formulated on an undirected labelled graph and consist on selecting the subset of labels such that the subgraph generated by these labels has, respectively, an optimal spanning tree or forest. The effects of using chaotic sequences in the VNS metaheuristic are investigated during several numerical tests. Different one-dimensional chaotic maps are applied to VNS in order to compare the performance of each map in finding the best solutions for this class of graph problems.

Keywords: Deterministic chaos · Metaheuristics · Variable neighbourhood search · Labelling graph problems · Algorithm dynamics

1 Introduction

The term "chaos" covers a rather broad class of phenomena showing random-like behaviors at a first glance, even if they are generated by deterministic systems. This kind of processes is used to denote phenomena which are of a purely stochastic nature, such as the behavior of a waft of smoke or ocean turbulence, or the dynamic of molecules inside a vessel filled with gas, among many others [25]. However, chaotic system behaviors are easily mistaken for random noises given they share the property of long term unpredictable irregular behavior and broad band spectrum.

© The Author(s) 2023, corrected publication 2023
L. Di Gaspero et al. (Eds.): MIC 2022, LNCS 13838, pp. 200–214, 2023.
https://doi.org/10.1007/978-3-031-26504-4_15

A classical topic in studying real world phenomena is to distinguish then between chaotic and random dynamics [18]. Deterministic chaotic systems are necessarily nonlinear, and conventional statistical procedures are insufficient for their analysis [39]. If the output of a deterministic chaotic system is analysed with these approaches, it will be erroneously recognised as the result of a random process. Therefore, characterizing the irregular behavior that can be caused either by deterministic chaos or by randon processes is challenging because of the surprising similarity that deterministic chaotic and random signals often show. Thus, it is still an open problem to distinguish among these two types of phenomena [25].

Deterministic chaos and its applications can be observed in control theory, computer science, physics, biology, and many other fields [18]. The interest in studying chaotic systems arises indeed when the theme of chaos reaches a high interdisciplinary level involving not only mathematicians, physicians and engineers but also biologists, economists and scientists from different areas. Several research works have shown that order could arise from disorder in various fields, from biological systems to condensed matter, from neuroscience to artificial neural networks [1]. In these cases, disorder often indicates both non-organized patterns and irregular behavior, whereas order is the result of self-organization and evolution, and often arises from a disorder condition or from the presence of dissymmetries. Gros [19] discusses the origin of self-organization where, leveraging from various key points from evolutionary theory and biology, it emphasizes the idea that life exists at the edge of chaos. Other examples in which the concept of stochastic driven procedures leads to ordered results are, e.g., Monte Carlo and evolutionary optimization [39], together with stochastic resonance in which the presence of noise improves the transmission of information [14].

The discovery of the phenomenon of deterministic chaos has brought about the need to identify manifestations of this phenomenon also in experimental data. Research on this line has focused so far on exploring the properties of cause and effect of chaotic phenomena, and also on using deterministic chaotic processes as instruments to improve other systems. This article focuses on the latter, and in particular on exploiting chaos for the improvement of heuristic optimization [32]. The goal consists on evaluating to performance between chaotic and random dynamics within a metaheuristic algorithm, showing the use of chaos in the inner optimization process, and focussing the attention on how chaos supports the birth of order from disorder also in this field [38]. This means to investigate the effects of the introduction of either deterministic chaotic or random sequences in a complex optimization routine. For this purpose, in particular, in this work we focus on Variable Neighbourhood Search (VNS), a popular explorative metaheuristic for combinatorial optimization problems based on dynamic changes of the neighbourhood structure in the solution space during the search process [21]. To compare the performance between a VNS procedure that runs using chaotic signals and that of a traditional random-based VNS, we consider as use case a set of labelling graph problems, i.e. the labelled spanning tree and forest problems. These problems are formulated on an undirected labelled graph, and consist on selecting the subset of labels such that the subgraph generated by these labels has an optimal spanning tree or forest. This family of problems has many real-world applications in different fields, such as in data compression, telecommunications network design, and multimodal transportation systems [9,10,12,13]. For example, in multimodal transportation systems there are often circumstances where it is needed to guarantee a complete service between the terminal nodes of the network by using the minimum number of provider companies. This situation can be modelled as a labelling graph problem, where each edge of the input graph is assigned a label, denoting a different company managing that link, and one wants to obtain a spanning tree of the network using the minimum number of labels. This spanning tree will reduce the construction cost and the overall complexity of the network.

The effects of using chaos in VNS on this family of combinatorial optimization problems are evaluated, aiming at disentangling the improvement in the optimization power due to the inclusion of a deterministic chaotic map within the VNS approach, one of the most popular metaheuristic used for tackling this class of problems. For the task, as it will be shown next, different popular one-dimensional chaotic maps are considered. The rest of the paper is structured as follows. Section 2 provides an overview of the background literature, while Sect. 3 presents the considered labelling graph problems used as test-bench. Section 4 describes the VNS methodology implemented for this family of problems. Section 5 describes how we used chaos in VNS and the deterministic chaotic maps considered in our experiments. Section 6 shows the obtained empirical results and findings, while in Sect. 7 we provide our main conclusions.

2   Related Work

The active use of chaos has been recently widely investigated in the literature [18,25]. The link between chaos and randomness has been largely investigated in several works (see e.g. [20,26,31] among others). Particularly interesting results have arisen in computer systems and algorithms, where chaos has been observed in the dynamics of algorithmic routines [24] and evolutionary algorithms [38,39]. The latter is a topic of great interest, linked to the work presented in this paper. Chaos indeed has been used to substitute pseudo-random number generators in a variety of heuristic optimization procedures. The use of chaos inside evolutionary optimization is discussed in [27,38], where it is thoroughly evaluated whether pure chaotic sequences improve the performance of evolutionary strategies. Davendra et al. [15] use with success a chaos driven evolutionary algorithm for PID control, while El-Shorbagy et al. [17] propose a chaos-based evolutionary algorithm for nonlinear programming problems. Hong et al. [22] propose a chaotic Genetic Algorithm for cyclic electric load forecasting; for the same problem, Dong et al. [16] introduce a hybrid seasonal approach using a chaotic Cuckoo Search algorithm together with a Support Vector Regression model. Another example on the use of chaos in a Genetic Algorithm is present in [28], with an application for the solution of a chip mapping problem. Senkerik et al. [33,34]

discuss the impact of chaos on Differential Evolution, powering the algorithm by a multi-chaotic framework used for parent selection and population diversity. Pluhacek et al. [29,30] has widely explored the use of deterministic chaos inside Particle Swarm Optimization. Hong et al. [23] introduce a novel chaotic Bat Algorithm for forecasting complex motion of floating platforms. Chen et al. [6] propose a Whale Optimization Algorithm with a chaos-based mechanism relying on quasi-opposition for global optimization problems. In [40], instead, an improved Artificial Fish Swarm Algorithm based on chaotic search and feedback strategy has been described. Wang et al. [36] recently present an improved Grasshopper Optimization Algorithm using an adaptive chaotic strategy to further improve the comprehensive ability of grasshopper swarms in the early exploration and later development, and apply the algorithm to pattern synthesis of linear array in RF antenna design.

We do not attempt to hide the fact that, in certain ways, the field has been progressing in a way that seems to us less useful, and sometimes even harmful, to the development of the field in general. For example, many of the contributions that appear in the new literature, in our opinion do appear rather marginal additions to a list of relevant and widely accepted metaheuristics [35]. Nevertheless, it can be stated that, based on the listed and further other research papers in the literature, several contributions have shown the value that chaos appears to provide as an additional tool for heuristic optimization routines. It is evident the increasingly rising attention of the research community towards the hybridization of modern optimization algorithms and chaotic dynamics.

To the best of our knowledge, however, no attempts have been made on the use of chaos within the Variable Neighbourhood Search algorithm. We want to fill this gap, and, therefore, in this paper we use chaos to try to improve the VNS metaheuristic, testing it through different chaotic functions. As shown next, we evaluate the performance of the impact of a chaotic version of VNS on a set of labelling graph problems, used as testbench, to a non-chaotic version of the same algorithm.

3   The Labelled Spanning Tree and Forest Problems

In this paper we scratch a chaotic version of VNS, aimed to achieve further improvements to a classic, random-based VNS implementation tackling two classical labelling graph problems, namely the Minimum Labelling Spanning Tree (MLST) [4] and the k-Labelled Spanning Forest (kLSF) [3] problems. Variants exist (see e.g. [8,10]), but these two problems are maybe the most prominent and general of this family. They are defined on a labelled graph, that is an undirected graph, $G = (V, E, L)$, where V is its set of nodes and E is the set of edges that are labelled on the set L of labels.

The MLST problem [4] consists on, given a labelled input graph $G = (V, E, L)$, getting a spanning tree with the minimum number of labels; i.e., the aim is to find the labelled spanning tree $T^* = (V, E^*, L^*)$ of the input graph that minimizes the size of label set $|L^*|$.

Instead, the kLSF problem [3] is defined as follows. Given a labelled input graph $G = (V, E, L)$ and an integer positive value $\bar{k}$, find a labelled spanning forest $F^* = (V, E^*, L^*)$ of the input graph having the minimum number of connected components with the upper bound $\bar{k}$ for the number of labels to use, i.e. $\min |Comp(G^*)|$ with $|L^*| \leq \bar{k}$.

Therefore in both problems, the matter is to find an optimal set of labels L^*. Since a solution to the MLST problem would be a solution also to the kLSF problem if the obtained solution tree would not violate the limit $\bar{k}$ on the used number of labels, it is easily deductable that the two problems are deeply correlated. Given the subset of labels $L^* \subseteq L$, the labelled subgraph $G^* = (V, E^*, L^*)$ may contain cycles, but each of them can be arbitrarily break by eliminating edges in polynomial time until a forest, or a tree, is obtained.

The NP-hardness of the MLST and kLSF problems has been proved in [4] and in [3], respectively. Therefore any practical solution approach to both problems requires heuristics. Several optimization algorithms to the MLST problem have been approached in the literature [2,37], showing in several cases the particular suitability of the VNS heuristic [9,11,12]. For the kLSF problem, in [3] a Genetic Algorithm and the Pilot Method metaheuristics have been proposed. In particular, in [7,13], some metaheuristics based on Greedy Randomized Adaptive Search Procedure and Variable Neighbourhood Search have been designed, obtaining high-quality results in most cases and showing the effectivenes of the VNS approach [13]. Given VNS has demonstrated to be a promising strategy for this class of problems, we have chosen it as a benchmark for testing the use of chaos inside the VNS metaheuristic. Nevertheless, note that the approach can be easily adapted and generalised to other optimization problems where the solution space consists of subsets of a reference set, such as feature subset selection problems or a variety of location problems.

4   Variable Neighbourhood Search

Variable Neighbourhood Search (VNS) is an explorative metaheuristic for combinatorial optimization problems based on dynamic changes of the neighbourhood structure of the solution space during the search process [21]. The guiding principle of VNS is that a local optimum with respect to a given neighbourhood may not be locally optimal with respect to another neighbourhood. Therefore VNS looks for new solutions in increasingly distant neighbourhoods of the current solution, jumping only if a better solution than the current best solution is found [21]. The process of changing neighbourhoods when no improvement occurs is aimed at producing a progressive diversification.

Given a labelled graph $G = (V, E, L)$ with n vertices, m edges, and ℓ labels, each solution is encoded by a binary string [9], i.e. $C = (c_1, c_2, \ldots, c_\ell)$ where

$$c_i = \begin{cases} 1 & \text{if label } i \text{ is in solution } C \\ 0 & \text{otherwise} \end{cases} \qquad (\forall i = 1, \ldots, \ell). \qquad (1)$$

Denote with $N_k(C)$ the neighbourhood space of the solution C, and with $k_{\max}$ the maximum size of the neighbourhood space. In order to impose a neighbourhood structure on the solution space S, comprising all possible solutions, the distance considered between any two such solutions $C_1, C_2 \in S$, is the Hamming distance [9, 12]:

$$\rho(C_1, C_2) = |C_1 - C_2| = \sum_{i=1}^{\ell} \lambda_i \qquad (2)$$

where $\lambda_i = 1$ if label i is included in one of the solutions but not in the other, and 0 otherwise, $\forall i = 1, ..., \ell$. Then, given a solution C, its kth neighbourhood, $N_k(C)$, is considered as all the different sets having a Hamming distance from C equal to k labels, where $k = 1, 2, \ldots, k_{\max}$, and $k_{\max}$ is the maximum dimension of the shaking. In a more formal way, the kth neighbourhood of a solution C is defined as $N_k(C) = \{S \subset L : \rho(C, S) = k\}$, where $k = 1, ..., k_{\max}$.

Algorithm 1: Variable Neighbourhood Search for the MLST problem

---

Input: A labelled, undirected, connected graph $G = (V, E, L)$ with n vertices, m edges, ℓ labels;

Output: A spanning tree T;

Initialisation:

- Let $C \leftarrow \emptyset$ be the global set of used labels;
- Let $H = (V, E(C))$ be the subgraph of G restricted to V and edges with labels in C, where $E(C) = \{e \in E : L(e) \in C\}$;
- Let C' be a set of labels;
- Let $H' = (V, E(C'))$ be the subgraph of G restricted to V and edges with labels in C', where $E(C') = \{e \in E : L(e) \in C'\}$;
- Let $Comp(C')$ be the number of connected components of $H' = (V, E(C'))$;

begin

    $C \leftarrow Generate\text{-}Initial\text{-}Solution()$;

    repeat

        Set $k \leftarrow 1$ and $k_{\max} \leftarrow (|C| + |C|/3)$;

        while $k < k_{max}$ **do**

            $C' \leftarrow Shaking\ phase(C,\ k)$;

            Local search(C');

            if $|C'| < |C|$ **then**

                Move $C \leftarrow C'$;

                Restart with the first neighbour: $k \leftarrow 1$;

            else

                Increase the size of the neighbourhood structure: $k \leftarrow k + 1$;

            end

        end

    until *termination conditions*;

    Update $H = (V, E(C))$;

    $\Rightarrow$ Take any arbitrary spanning tree T of $H = (V, E(C))$.

end

Algorithm 2: *Shaking phase* procedure

Procedure Shaking phase(C, k):
Set $C' \leftarrow C$;
for $i \leftarrow 1$ *to* k **do**
    Select at random a number between 0 and 1: $rnd \leftarrow random[0,1]$;
    if $rnd \leq 0.5$ **then**
        Delete at random a label $c' \in C'$ from C', i.e. $C' \leftarrow C' - \{c'\}$;
    else
        Add at random a label $c' \in (L - C)$ to C', i.e. $C' \leftarrow C' \cup \{c'\}$;
    end
    Update $H' = (V, E(C'))$ and $Comp(C')$;
end

For illustrative purpose and a better comprehension, in Algorithm 1 is described the VNS implementation for the MLST problem [9,12]. The VNS solution approach for the kLSF problem is very akin [7,13] and only differ from the fact that an upper bound $\bar{k}$ for the number of labels has to be imposed, and that a forest instead of a spanning tree has to be considered for halting the algorithm. Note that given a subset of labels $L^* \subseteq L$, the labelled subgraph $G^* = (V, E^*, L^*)$ may contain cycles, but they can arbitrarily break each of them by eliminating edges in polynomial time until a forest or a tree is obtained.

Algorithm 1 starts from an initial feasible solution C generated at random and lets parameter k vary during the execution. In the successive shaking phase (*Shaking phase($N_k(C)$)* procedure, see Algorithm 2) a random solution C' is selected within the neighbourhood $N_k(C)$ of the current solution C. This is done by randomly adding further labels to C, or removing labels from C, until the resulting solution has a Hamming distance equal to k with respect to C [9]. Addition and deletion of labels at this stage have the same probability of being chosen. For this purpose, a random number is selected between 0 and 1 ($rnd \leftarrow random[0,1]$). If this number is smaller than 0.5, the algorithm proceeds with the deletion of a label from C. Otherwise, an additional label is included at random in C from the set of unused labels $(L - C)$. The procedure is iterated until the number of addition/deletion operations is exactly equal to k [12].

The successive local search (*Local search(C')* procedure, see Algorithm 3) consists of two steps [9]. In the first step, since deletion of labels often gives an infeasible incomplete solution, additional labels may be added in order to restore feasibility. In this case, addition of labels follows the criterion of adding the label with the minimum number of connected components. Note that in case of ties in the minimum number of connected components, a label not yet included in the partial solution is chosen at random within the set of labels producing the minimum number of components (i.e. $u \in S$ where $S = \{\ell \in (L - C') : \min Comp(C' \cup \{\ell\})\}$). Then, the second step of the local search tries to delete labels one by one from the specific solution, whilst maintaining feasibility [9,12].

Algorithm 3: *Local search* procedure

Procedure Local search(C'):

while $Comp(C') > 1$ **do**

  Let S be the set of unused labels which minimize the number of connected
    components, i.e. $S = \{\ell \in (L - C') : \min Comp(C' \cup \{\ell\})\}$;

  Select at random a label $u \in S$;

  Add label u to the set of used labels: $C' \leftarrow C' \cup \{u\}$;

  Update $H' = (V, E(C'))$ and $Comp(C')$;

end

for $i \leftarrow 1$ *to* $|C'|$ **do**

  Delete label i from the set C', i.e. $C' \leftarrow C' - \{i\}$;

  Update $H' = (V, E(C'))$ and $Comp(C')$;

  if $Comp(C') > 1$ **then**

    Add label i to the set C', i.e. $C' \leftarrow C' \cup \{i\}$;

  end

  Update $H' = (V, E(C'))$ and $Comp(C')$;

end

After the local search phase, if no improvements are obtained ($|C'| \geq |C|$), the neighbourhood structure is increased ($k \leftarrow k+1$) giving a progressive diversification ($|N_1(C)| < |N_2(C)| < ... < |N_{k_{\max}}(C)|$), where $k_{\max} \leftarrow (|C| + |C|/3)$ from [9,12]. Otherwise, the algorithm moves to the improved solution ($C \leftarrow C'$) and sets the first neighbourhood structure ($k \leftarrow 1$). Then the procedure restarts with the shaking and local search phases, continuing iteratively until the user termination conditions are met.

5   Using Chaos in VNS

Chaos is a non-periodic, long-term behavior in a deterministic system that exhibits sensitive dependence on initial conditions, and is a common nonlinear phenomenon in our lives [25]. The dynamic properties of chaos can be summarised as following [40]:

1. *"Sensitive dependence to Initial Conditions (SIC)"*: Chaos is highly sensitive to the initial value.
2. *"Certainty"*: Chaos is produced by a certain iterative formula.
3. *"Ergodicity"*: Chaos can go through all states in certain ranges without repetition.

In general, the most important defining property of chaotic variables is the first one, which requires that trajectories originating from very nearly identical initial conditions diverge at an exponential rate [28]. Pseudorandomness and ergodicity are other important dynamic characteristics of a chaotic structure, which ensure that the track of a chaotic variable can travel ergodically over the whole space of interest.

Chaos is similar to randomness. The variation of the chaotic variable has indeed an inherent property in spite of the fact that it looks like a disorder. Edward Lorenz irregularity in a toy model of the weather displayed first chaotic or strange attractor in 1963. It was mathematically defined as randomness generated by simple deterministic systems. A deterministic structure can have no stochastic (probabilistic) parameters. Therefore chaotic systems are not at all equal to noisy systems driven by random processes. The irregular behavior of the chaotic systems arises from intrinsic nonlinearities rather than noise [25].

Several experimental studies have shown already the benefits of using chaotic signals rather than random signals [18], although a general rule can not be drawn yet [32]. Chaos has been used as a novel addition to optimization techniques to help escaping from local optima, and chaos-based searching algorithms have aroused intense interests [32,39].

As from the second property of chaos just listed above, one-dimensional non-invertible maps are the simplest systems with capability of generating chaotic dynamics. They are capable of providing simple deterministic chaotic signals, that we can use inside our VNS procedure (Algorithm 1) in place of the pseudo-random number generation occurring in the shaking phase (Algorithm 2). Here the chaotic mapping of a shaking $N_k(\cdot)$ to an incumbent solution, C, includes the following major steps:

1. Variable C in the solution space is mapped to the chaotic space, by using a deterministic chaotic map chosen by the user.
2. Using the selected chaotic dynamics, select the kth chaotic variable from the generating map.
3. The chaotic variable is then mapped back to the solution space, producing the next solution C'.

Please note that after this step, in case of an infeasible incomplete solution is obtained, additional labels may be added in order to restore feasibility, following the criterion of adding the label with the minimum number of connected components with respect to the incumber solution C'.

In the following we briefly depict some well-known one-dimensional chaotic maps that we employ in our experiments. For more in-depth descriptions, the reader in referred to [5,32].

Logistic map

The logistic map is a chaotic polynomial map. It is often cited as an example of how complex behavior can arise from a very simple nonlinear dynamical equation. This map is defined by:

$$x_{n+1} = f(\mu, x_n) = \mu x_n (1 - x_n), \qquad 0 < \mu \leq 4 \tag{3}$$

in which μ is a control parameter, and x is the variable. Since Eq. (3) represents a deterministic dynamic system, it seems that its long-term behavior can be predicted.

Tent Map

In mathematics, the tent map is an iterated function, in the shape of a tent, forming a discrete-time dynamical system. It takes a point x_n on the real line and maps it to another point, according to:

$$x_{n+1} = \begin{cases} \mu x_n & \text{if } x_n < 1/2 \\ \mu(1 - x_n) & \text{otherwise,} \end{cases} \tag{4}$$

where μ is a positive real constant. The tent map and the logistic map are topologically conjugate, and thus the behavior of the two maps is in this sense identical under iteration.

Bernoulli Shift Map

The Bernoulli shift (Bshift) map belongs to class of piecewise linear maps which consist of a number of piecewise linear segments. This map is a particularly simple example, consisting of two linear segments to model the active and passive states of the source. It is defined as follows:

$$x_{n+1} = \begin{cases} \dfrac{x_n}{1 - \lambda} & \text{if } 0 < x_n < (1 - \lambda) \\ \dfrac{x_n - (1 - \lambda)}{\lambda} & \text{otherwise.} \end{cases} \tag{5}$$

Sine Map

The sine map is described by the following equation:

$$x_{n+1} = \frac{\alpha}{4} \sin(\pi x_n), \tag{6}$$

where $0 < \alpha \le 4$. This map has qualitatively the same shape as the logistic map. Such maps are also called unimodal chaotic maps.

Circle map

The circle map [22] is represented by

$$x_{n+1} = x_n + b - \frac{a}{2\pi} \sin(2\pi x_n), \tag{7}$$

where a can be interpreted as a strength of nonlinearity, and b as an externally applied frequency. The circle map exhibits very unexpected behavior as a function of these parameters; with $a = 0.5$ and $b = 0.2$, it generates chaotic sequences in $(0, 1)$.

Iterative Chaotic Map with Infinite Collapses

The iterative chaotic map with infinite collapses (ICMIC) is defined by:

$$x_{n+1} = \sin(\alpha/x_n), \tag{8}$$

where $\alpha \in (0, \infty)$ is an adjustable parameter.

6   Experimental Results

We show here our computational experience on the use of chaos within the VNS methodology for the considered labelling graph problems. We examine the VNS implementation using pseudo-random number generation in the shaking phase (*Rand*), and the same VNS model including the different deterministic chaotic maps in the shaking step, denoted respectively with: *Logistic*, *Tent*, *Bshift*, *Sine*, *Circle*, and *ICMIC*. To test the performance and the efficiency of the algorithms, we run an experimental evaluation on a set of labelled graphs having numbers of vertices $|V| = 100, 200, 300, 400, 500, 1000$, labels $|L| = 0.25\,V|, 0.5\,V|, |V|,$ $1.25\,V|$, and edges $|E| = (|V| - 1)/|V|$. These are the well-known benchmark instances in the literature taken from [9,11,12] for the MLST literature, and from [7,13] for the kLSF problem. All the considered instances are available upon request from the authors. For each combination of $|V|$ and $|L|$, ten different problem instances are generated; the parameter $\bar{k}$ for the kLSF is determined experimentally as $\lfloor |V|/2^j \rfloor$, where j is the smallest value such that the generated instances do not report a single connected solution when solved with maximum vertex covering [3]. The algorithms have been implemented in C++ under the Microsoft Visual Studio 2015 framework, and all the computations run on an Intel Quad-Core i7 64-bit microprocessor at 2.30 GHz with 32 GB RAM.

For each dataset, solution quality is evaluated as the average objective function value (i.e. the number of labels of the solution for the MLST problem, or the number of connected components for the kLSF problem) among the 10 problem instances. As in [9,12,13], a maximum allowed CPU time of 1 hour has been chosen as stopping condition for all the VNS algorithms. Selection of the maximum allowed CPU time as the stopping criterion is made in order to have a fair and direct comparison between the different VNS implementations with respect to the quality of their solutions. All the algorithms run until the maximum CPU time is reached and, in each case, the best solution is recorded, along with the total number of iterations required to obtain such best solution.

We show in Fig. 1 the bar chart of the sum of the objective function values obtained by the different VNS variants for the *MLST* problem instances (left) and for the *kLSF* problem instances (right). The results show that the deterministic chaotic maps perform well in the considered instances for both problems, with the exception of *Sine* and *Circle* that appear to not bring a real improvement over that classical VNS with pseudo-random number generation. The best results are obtained when using the *ICMIC* map, reaching the best solutions in both problems. Fine results are also reached, respectively, by the *Tent*, *Logistic*, and *Bshift* maps, which follow immediately after *ICMIC* and clearly outperform *Rand*.

We also compare the total number of iterations at which the best solutions are obtained when executing VNS with each of the discussed chaotic maps, and show the relative bar chart in Fig. 2. We see a consistent drop with respect to the number of iterations required by all the VNS variants using the chaotic maps, meaning that they are able to converge earlier with respect to *Rand*. Looking at the figure, the best performance in terms of total number of iterations is

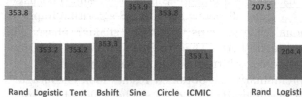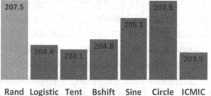

Fig. 1. Bar chart of the objective function values obtained by the different VNS variants for solving the *MLST* problem instances (left) and the *kLSF* problem instances (right).

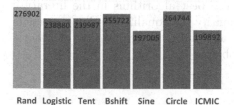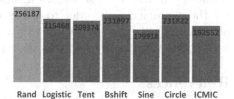

Fig. 2. Bar chart of the total number of iterations required by the different VNS variants for solving the *MLST* problem instances (left) and the *kLSF* problem instances (right).

obtained by *Sine* for both problems, immediately followed by the *ICMIC* map, and by the *Logistic* and *Tent* maps, afterwards. However, although *Sine* is faster than the other chaotic maps, it does not show top performance with respect to the objective function values (Fig. 1), meaning it is more prone to get stuck into local optima. Instead, looking at *Bshift* and *Circle*, they appear to be sometimes slower than the other employed maps. Nevertheless, summarizing, as seen already with respect to solution quality, it is again evident the value of using the chaotic maps in VNS, given all the chaotic VNS variants always outperform *Rand* with respect to the required number of iterations.

7   Conclusions

This paper introduces the novel idea of combining the two concepts of chaotic sequences and Variable Neighbourhood Search (VNS). Different popular one-dimensional chaotic maps have been considered, and they have been injected into the shaking phase of the VNS algorithm in place of classical pseudo-random number generation. The chaotic VNS variants have been tested on a family of labelling graph problems, namely the Minimum Labelling Spanning Tree (MLST) problem and the k-Labelled Spanning Forest (kLSF) problem. In order to evaluate the effectiveness of the chaotic maps in reaching the best solution for the considered problems, objective function values and total number of iterations required by the different VNS implementations have been computed upon a set of problem instances commonly used in the literature. Simulation results on this set of benchmarks indicate that searching efficiency of the VNS algorithm improves when using the one-dimensional chaotic maps within the shaking phase. The proposed chaotic variants work quite better than the classical VNS algorithms using randomness for the two problems introduced in previous works.

Summarizing, although preliminary, the obtained results look encouraging, showing an overall validity of the employed methodology. The achieved VNS optimization strategy using chaos seems to be highly promising for both labelling graph problems. The experiments carried out confirm the efficiency, lower number of iterations, and scalability of the chaotic VNS implementations. Ongoing investigation will consist in performing a thorough statistical analysis of the resulting chaotic VNS strategies against the best algorithms in the literature for these problems, in order to better quantify and qualify the improvements obtained. Further investigation will deal also with the application of chaotic variants of VNS to other optimization problems.

References

1. Abel, D., Trevors, J.: Self-organization vs. self-ordering events in life-origin models. Phys. Life Rev. **3**(4), 211–228 (2006)
2. Cerulli, R., Fink, A., Gentili, M., Voß, S.: Metaheuristics comparison for the minimum labelling spanning tree problem. In: Golden, B.L., Raghavan, S., Wasil, E.A. (eds.) The Next Wave on Computing. Optimization, and Decision Technologies, pp. 93–106. Springer-Verlag, New York (2005). https://doi.org/10.1007/0-387-23529-9_7
3. Cerulli, R., Fink, A., Gentili, M., Raiconi, A.: The k-labeled spanning forest problem. Procedia. Soc. Behav. Sci. **108**, 153–163 (2014)
4. Chang, R.S., Leu, S.J.: The minimum labelling spanning trees. Inf. Process. Lett. **63**(5), 277–282 (1997)
5. Chen, G., Huang, Y.: Chaotic maps: dynamics, fractals, and rapid fluctuations (synthesis lectures on mathematics and statistics). Morgan Claypool Publishers (2011). https://doi.org/10.2200/S00373ED1V01Y201107MAS011
6. Chen, H., Li, W., Yang, X.: A whale optimization algorithm with chaos mechanism based on quasi-opposition for global optimization problems. Expert Syst. Appl. **158**, 113612 (2020)
7. Consoli, S., Moreno-Pérez, J.A.: Variable neighbourhood search for the k-labelled spanning forest problem. Electr. Notes Discrete Math. **47**, 29–36 (2015)
8. Consoli, S., Moreno-Pérez, J.A., Darby-Dowman, K., Mladenović, N.: Discrete particle swarm optimization for the minimum labelling Steiner tree problem. In: Krasnogor, N., Nicosia, G., Pavone, M., Pelta, D. (eds.) Nature Inspired Cooperative Strategies for Optimization. Studies in Computational Intelligence, vol. 129, pp. 313–322. Springer-Verlag, New York (2008). https://doi.org/10.1007/s11047-009-9137-9
9. Consoli, S., Darby-Dowman, K., Mladenović, N., Moreno-Pérez, J.A.: Greedy randomized adaptive search and variable neighbourhood search for the minimum labelling spanning tree problem. Eur. J. Oper. Res. **196**(2), 440–449 (2009)
10. Consoli, S., Darby-Dowman, K., Mladenović, N., Moreno-Pérez, J.A.: Variable neighbourhood search for the minimum labelling Steiner tree problem. Ann. Oper. Res. Accepted Publ. **172**(1), 71–96 (2009)

11. Consoli, S., Moreno-Pérez, J.A., Mladenović, N.: Intelligent variable neighbourhood search for the minimum labelling spanning tree problem. Electron. Notes Discrete Math. **41**, 399–406 (2013)

12. Consoli, S., Mladenović, N., Moreno-Pérez, J.A.: Solving the minimum labelling spanning tree problem by intelligent optimization. Appl. Soft Comput. **28**, 440–452 (2015)

13. Consoli, S., Moreno-Pérez, J.A., Mladenović, N.: Comparison of metaheuristics for the k-labeled spanning forest problem. Int. Trans. Oper. Res. **24**(3), 559–582 (2017)

14. Cui, L., Yang, J., Wang, L., Liu, H.: Theory and application of weak signal detection based on stochastic resonance mechanism. Secur. Commun. Netw. **2021**, 5553490 (2021)

15. Davendra, D., Zelinka, I., Senkerik, R.: Chaos driven evolutionary algorithms for the task of PID control. Comput. Math. Appl. **60**(4), 1088–1104 (2010)

16. Dong, Y., Zhang, Z., Hong, W.-C.: A hybrid seasonal mechanism with a chaotic cuckoo search algorithm with a support vector regression model for electric load forecasting. Energies **11**(4), 1009 (2018)

17. El-Shorbagy, M., Mousa, A., Nasr, S.: A chaos-based evolutionary algorithm for general nonlinear programming problems. Chaos, Solitons Fractals **85**, 8–21 (2016)

18. Etkin, D.: 5 - disasters and complexity. In: Etkin, D. (ed.) Disaster Theory, pp. 151–192. Butterworth-Heinemann, Boston (2016)

19. Gros, C.: Complex and Adaptive Dynamical Systems: A Primer. Springer, Cham (2008). https://doi.org/10.1007/978-3-642-04706-0

20. Hamza, R.: A novel pseudo random sequence generator for image-cryptographic applications. J. Inf. Secur. Appl. **35**, 119–127 (2017)

21. Hansen, P., Mladenović, N., Moreno-Pérez, J.A.: Variable neighbourhood search: methods and applications. Ann. Oper. Res. **175**(1), 367–407 (2010). https://doi.org/10.1007/s10479-009-0657-6

22. Hong, W.-C., Dong, Y., Zhang, W., Chen, L.-Y., Panigrahi, B.K.: Cyclic electric load forecasting by seasonal SVR with chaotic genetic algorithm. Int. J. Electr. Power Energy Syst. **44**(1), 604–614 (2013)

23. Hong, W.-C., Li, M.-W., Geng, J., Zhang, Y.: Novel chaotic bat algorithm for forecasting complex motion of floating platforms. Appl. Math. Model. **72**, 425–443 (2019)

24. Hoyle, A., Bowers, R., White, A.: Evolutionary behaviour, trade-offs and cyclic and chaotic population dynamics. Bull. Math. Biol. **73**(5), 1154–1169 (2011). https://doi.org/10.1007/s11538-010-9567-7

25. Jørgensen, S.: Chaos. In: Jørgensen, S.E., Fath, B.D. (eds.) Encyclopedia of Ecology, pp. 550–551. Academic Press, Oxford (2008)

26. Lozi, R.: Emergence of randomness from chaos. Int. J. Bifurcat. Chaos **22**(2), 1250021 (2012)

27. Lu, Y., Zhoun, J., Qin, H., Wang, Y., Zhang, Y.: Chaotic differential evolution methods for dynamic economic dispatch with valve-point effects. Eng. Appl. Artif. Intell. **24**(2), 378–387 (2011)

28. Moein-darbari, F., Khademzadeh, A., Gharooni-fard, G.: Evaluating the performance of a chaos genetic algorithm for solving the network on chip mapping problem. In: Proceedings - 12th IEEE International Conference on Computational Science and Engineering, CSE 2009, vol. 2, pp. 366–373 (2009)

29. Pluhacek, M., Senkerik, R., Zelinka, I.: Particle swarm optimization algorithm driven by multichaotic number generator. Soft. Comput. **18**(4), 631–639 (2014). https://doi.org/10.1007/s00500-014-1222-z

30. Pluhacek, M., Senkerik, R., Viktorin, A., Kadavy, T.: Chaos-enhanced multiple-choice strategy for particle swarm optimisation. Int. J. Parallel Emergent Distrib. Syst. **35**(6), 603–616 (2020)
31. Sahari, M., Boukemara, I.: A pseudo-random numbers generator based on a novel 3d chaotic map with an application to color image encryption. Nonlinear Dyn. **94**(1), 723–744 (2018). https://doi.org/10.1007/s11071-018-4390-z
32. Salcedo-Sanz, S.: Modern meta-heuristics based on nonlinear physics processes: A review of models and design procedures. Phys. Rep. **655**, 1–70 (2016)
33. Senkerik, R., Viktorin, A., Pluhacek, M., Kadavy, T.: On the population diversity for the chaotic differential evolution. In 2018 IEEE Congress on Evolutionary Computation, CEC 2018 - Proceedings, 8477741 (2018)
34. Senkerik, R., et al.: Differential evolution and deterministic chaotic series: a detailed study. Mendel **24**(2), 61–68 (2018)
35. Sörensen, K., Sevaux, M., Glover, F.: A history of metaheuristics. In: Martí, R., Pardalos, P.M., Resende, M.G.C. (eds.) Handbook of Heuristics, pp. 791–808. Springer, Cham (2018). https://doi.org/10.1007/978-3-319-07124-4_4
36. Wang, H., Kang, Y., Li, B.: Synthesis for sidelobe suppression of linear array based on improved grasshopper optimization algorithm with adaptive chaotic strategy. Int. J. RF Microwave Comput. Aided Eng. **32**(4), e23048 (2022)
37. Xiong, Y., Golden, B., Wasil, E.: Improved heuristics for the minimum labelling spanning tree problem. IEEE Trans. Evol. Comput. **10**(6), 700–703 (2006)
38. Zelinka, I.: A survey on evolutionary algorithms dynamics and its complexity - mutual relations, past, present and future. Swarm Evol. Comput. **25**, 2–14 (2015)
39. Zelinka, I., et al.: Impact of chaotic dynamics on the performance of metaheuristic optimization algorithms: An experimental analysis. Inf. Sci. **587**, 692–719 (2022)
40. Zhu, K., Jiang, M.: An improved artificial fish swarm algorithm based on chaotic search and feedback strategy. In: Proceedings - 2009 International Conference on Computational Intelligence and Software Engineering, CiSE'09, p. 5366958 (2009)

Investigating Fractal Decomposition Based Algorithm on Low-Dimensional Continuous Optimization Problems

Arcadi Llanza[1,2]([✉]) [ID], Nadiya Shvai[1] [ID], and Amir Nakib[1,2] [ID]

[1] Cyclope.ai, Paris, France
{arcadi.llanza,nadiya.shvai}@cyclope.ai
[2] University Paris Est Créteil, Laboratoire LISSI, 94400 Vitry sur Seine, France
nakib@u-pec.fr

Abstract. This paper analyzes the performance of the Fractal Decomposition Algorithm (FDA) metaheuristic applied to low-dimensional continuous optimization problems. This algorithm was originally developed specifically to deal efficiently with high-dimensional continuous optimization problems by building a fractal-based search tree with a branching factor linearly proportional to the number of dimensions. Here, we aim to answer the question of whether FDA could be equally effective for low-dimensional problems. For this purpose, we evaluate the performance of FDA on the Black Box Optimization Benchmark (BBOB) for dimensions 2, 3, 5, 10, 20, and 40. The experimental results show that overall the FDA in its current form does not perform well enough. Among different function groups, FDA shows its best performance on Misc. moderate and Weak structure functions.

Keywords: Continuous optimization · Metaheuristics · Fractal decomposition · Black Box Optimization Benchmark

1 Introduction

The general form of an optimization problem considered in this paper is defined by Eq. 1:

$$Min f(\boldsymbol{x}), s.t. \boldsymbol{B}_l \leq \boldsymbol{x} \leq \boldsymbol{B}_u \qquad (1)$$

where $f(\boldsymbol{x})$ denotes the function to be minimize. It is assumed to be continuous. $\boldsymbol{x} = (x_1, x_2, ..., x_D)$ is the variable vector in $\mathbb{R}^D$. Here, $\boldsymbol{x}$ is a given parameter. Moreover, the function is constrained by $\boldsymbol{B}_l = (B_{l1}, B_{l2}, ..., B_{lD})$ as the lower boundary and $\boldsymbol{B}_u = (B_{u1}, B_{u2}, ..., B_{uD})$ as the upper boundary.

The Fractal Decomposition Algorithm (FDA) is a deterministic metaheuristic method that has been shown to solve large scale (50 up to 1000 dimensions) continuous optimization problems with high-performance [11,12]. This research aims to benchmark, for the first time, FDA in a low-dimensional (5 up to 40 dimensions) constrained continuous optimization problem such as Black Box

L. Di Gaspero et al. (Eds.): MIC 2022, LNCS 13838, pp. 215–229, 2023.
https://doi.org/10.1007/978-3-031-26504-4_16

Optimization Benchmark (BBOB) [6]. In this paper, the Black Box optimization refers to the design and analysis of algorithms for problems where the structure of the objective function is unknown and unexploitable. The rest of the paper is organized as follows. First, Sect. 2 reviews the related work. Then, Sect. 3 examines the foundations of the proposed method DFDA. Afterwards, Sect. 4 describes the used benchmark. Section 5 presents the experiments and results. Finally, in Sect. 6 the conclusion and further research directions are presented.

2   Related Work

To our knowledge, no other study has been previously done on FDA performance for low-dimensional continuous optimization problems. In the original FDA paper [11], the benchmark considered was SOCO2011. The problem dimension was set to a range of values from 50 to 1000. FDA ranked first for each considered.

An overview of state-of-the-art (SOTA) methods for the BBOB benchmark is provided. The methods are reported in order of their average performance from best to worst. In the case of noiseless BBOB, generally, it is Evolutionary Algorithms (EAs) that perform better. Nevertheless, Local Searches (LS) and other hybrid methods are competitive as well. *Hansen et al.* proposed in [5] a multistart BI-population CMA-ES with equal budgets for two interlaced restart strategies, one with increasing population size and one with varying small population sizes. In [2], *Bosman et al.* introduced the Adapted Maximum-Likelihood Gaussian Model Iterated Density-Estimation Evolutionary Algorithm (AMaLGaM-IDEA). AMaLGaM-IDEA is a parameter-free algorithm with incremental model building (iaMaLGaM IDEA). MA-LS-Chain [10] was proposed by *Molina et al.*. It uses a memetic algorithm with continuous local search. The Variable Neighbourhood Search (VNS) was suggested by *Garcia et al.* in [3]. IPOP-SEP-CMA-ES [15] is an algorithm with a multistart strategy with increasing population size introduced by *Ros et al.*

The Age-Layered Population Structure (ALPS) Evolutionary Algorithm (EA) is a method presented by *Hornby et al.* in [7]. ALPS claims to avoid premature convergence than other EAs methods.

The Prototype Optimization with Evolved Improvement Steps (POEMS) was introduced by *Kubalik et al.* in [9]. POEMS is a stochastic local search-based algorithm.

The restarted estimation of distribution algorithm (EDA) with Cauchy distribution (Cauchy-EDA) probabilistic model was suggested by *Povsik et al.* in [13]. Cauchy-EDA claims to be usable for many fitness landscapes. On the contrary, EDA with Gaussian distribution tends to converge prematurely.

The Differential Ant-Stigmergy Algorithm (DASA) was presented in [8] by *Korovsec et al.* DASA is a stigmergy-based algorithm for solving optimization problems with continuous variables.

Hansen et al. analysed the Nelder-Mead downhill simplex method [5]. Nelder-Mead is a multistart strategy applied on local and global levels. On the one hand, at the local level, ten restarts are conducted with a small number of iterations and reshaped simplex. On other hand, at the global level, independent restarts are launched until 105D function evaluations are exceeded.

3   The Fractal Decomposition Based Algorithm

Nakib *et al.* introduced a fractal decomposition [11] based on hyperspheres to solve high dimensional continuous optimization problems with low complexity. FDA [11] uses two components to find the optima in the landscape: the First component, called fractal decomposition, is used as an exploration technique. Then, the second component, called Intensive Local Search (ILS), is used as the exploitation technique to search in the local regions previously identified as promising regions. The basic pattern used in the fractal decomposition is a hypersphere because when scaled into a high dimensional space, its computational complexity is low. FDA covers the space with few hyperspheres allowing it to obtain a good performance in its exploratory phase. An inflation procedure is applied to the hyperspheres (see Fig. 1) to ensure that all space is covered.

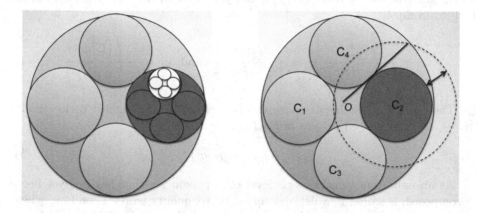

Fig. 1. Hypersphere fractal decomposition at 4 levels of depth [11].

The following subsection is dedicated to the different FDA components.

3.1   Exploration Component

At this phase, promising regions are searched for by conveniently subdividing the search space into smaller regions that might contain a good solution. The partition of space is modeled after a form of a hypersphere. This shape is a suitable representation that allows FDA to be extremely competitive in large-scale spaces.

Given a center C_k of a hypersphere with radius r the centers of its decomposition can be obtained as in Eq. 2:

$$C_{k+1}^i = C_k + (-1)^i \cdot ((r - r')e_{\lfloor \frac{i+1}{2} \rfloor}) \tag{2}$$

Then, the quality of each generated hypersphere is evaluated based on two points $\vec{s_1}$ and $\vec{s_2}$ originated based on Eq. 3.

$$\vec{s_1} = \vec{C^l} + \alpha \frac{r_l}{\sqrt{D}} \times \vec{e}_d, \quad \text{for} \quad d = 1, 2, \ldots, D$$

$$\vec{s_2} = \vec{C^l} - \alpha \frac{r_l}{\sqrt{D}} \times \vec{e}_d, \quad \text{for} \quad d = 1, 2, \ldots, D \tag{3}$$

Subsequently, for the aforementioned positions $(\vec{s_1}, \vec{s_2})$ and the center of the hypersphere $\vec{C_l}$, their fitnesses, f_1, f_2, and f_c, respectively, are evaluated. Furthermore, the distances to the best position found so far (BSF) are also computed via the $L^2 - norm$.

During this process, it is important to point out that the best solutions and their coordinates are saved. At this point, the slope g_1, g_2, and g_c is calculated on the three positions' fitness (f_1, f_2, and f_c) based on the $L_2\,norm$ distance to BSF as in Eq. 4.

$$g_1 = \frac{f(\vec{s_1})}{\|\vec{s_1} - BSF\|}, g_2 = \frac{f(\vec{s_2})}{\|\vec{s_2} - BSF\|} \quad \text{and} \quad g_c = \frac{f(\vec{C^l})}{\|\vec{C^l} - BSF\|} \tag{4}$$

The quality of the hypersphere will be defined by the highest ratio among g_1, g_2, and g_c, denoted by q as in Eq. 5:

$$q = \max\{g_1, g_2, g_c\} \tag{5}$$

As an important remark, each level of hyperspheres that has not yet been decomposed, is stored in a list and sorted by its quality score. Therefore, when all the hyperspheres in a level have been explored, the next level of hyperspheres unlocks until the stopping criterion is reached. By default the stopping criterion is defined on one of the following conditions:

- The maximum number of evaluations given by the benchmark is reached.
- The maximum decomposition level k is reached.

3.2   Exploitation Component

In the intensification phase, ILS is used for its simplicity and efficiency to find a local/global optimum at the end of its execution.

This technique uses two candidate solutions (x^{s_1} and x^{s_2}) that are evaluated per dimension. Each candidate is located one step size (ω) in the opposite directions $w.r.t.$ the current solution x^s based on Eq. 6:

$$x^{s_1} = x^s + \omega \times e_i$$
$$x^{s_2} = x^s - \omega \times e_i \tag{6}$$

where e_i is the unit vector in which the i-th element is set to 1 and the other elements to 0.

Then the best solution among x^s, x^{s_1}, and x^{s_2} is selected to be the next current solution x^s. The factor ω is halved whenever ILS cannot find a better solution in any of the dimensions until a stopping criterium is reached. This can happen in any of the following cases:

- The maximum number of evaluations given by the benchmark is reached.
- ω keeps decreasing until a given value ω_{min} that denotes the tolerance or the minimum precision needed by the benchmark.

4 Benchmark

BBOB is a continuous optimization problem with mixed-integer domains which consists of a set of 6 suits (*bbob, bbob-noisy, bbob-biobj, bbob-largescale, bbob-mixint, bbob-biobj-mixint*). In this study, we analyze the *bbob* suit that consists of a set of functions (f1 - f24) divided into five groups (Separable, Misc. moderate, Ill-conditioned, Multi-modal, and Weak structure) which are scalable to the dimension. The functions are defined within the hypercube $[-5, 5]^D$, where D is the dimensionality of the search space. Furthermore, each function has 15 instances to ensure results are statistically significant when reported. Generally, the difficulty in BBOB increases from the first to the last group. Groups are used to aggregate the obtained results, into more meaningful reports on the performance of functions with particular characteristics.

To compare real-parameter global optimizers, we used the Comparing Continuous Optimizers (COCO) framework [4]. COCO provides benchmark function testbeds, easy-to-parallelize experimentation templates, and tools for processing and visualization tools for data generated by one or more optimizers.

4.1 Performance Metrics

Average execution time (aRT) was introduced in [14] under the name as ENES and afterwards referred to in [1,5] as success performance and ERT correspondingly. This metric estimates the expected execution time of the restart algorithm. Typically, the average over all the trials is taken by varying only the reference instantiation parameters θ_i.

The execution time of the restart algorithm is given in Eq. 7. Here, we imply that the instance of the optimization problem $p = (n, f_\theta, \theta_i)$ is given by triple of search space dimension, function, and instantiation parameters. The subscripts us and s denote unsuccessful and successful trials, ΔI is the precision, and $J \sim BN(1, 1 - p_s)$ is a random variable with negative binomial distribution that models the number of unsuccessful runs until a success is observed given $p_s > 0$ the success probability of the algorithm.

$$\mathbf{RT}(n, f_\theta, \Delta I) = \sum_{j=1}^{J} RT_j^{us}(n, f_\theta, \Delta I) + RT^s(n, f_\theta, \Delta I) \tag{7}$$

Therefore, the expected runtime of the restart algorithm is represented in Eq. 8:

$$\mathbb{E}(\mathbf{RT})\& = \&\mathbb{E}(RT^s) + \frac{1 - p_s}{p_s} \mathbb{E}(RT^{us}). \tag{8}$$

Given a dataset that succeeds at least once ($n_s \geq 1$) with runtimes RT_i^s, and n_{us} unsuccessful runs with RT_j^{us} evaluations, the average runtime is expressed as in Eq. 9:

$$\begin{aligned}
aRT &= \frac{1}{n_s} \sum_i RT_i^s + \frac{1 - p_s}{p_s} \frac{1}{n_{us}} \sum_j RT_j^{us} \\
&= \frac{\sum_i RT_i^s + \sum_j RT_j^{us}}{n_s} \\
&= \frac{\#FEs}{n_s}, \tag{9}
\end{aligned}$$

where #FEs denotes the total number of function evaluations made in all trials before reaching the target precision.

5 Experiments and Discussion

In the following subsections, the results of the 24 BBOB functions are analyzed. FDA has been benchmarked on the dimensions $D = 2, 3, 5, 10, 20, 40$. The maximum number of function evaluations (maximum budget) is chosen as $1000 \times D$. Experimental results show that FDA performs best on the separable functions. However, it works also quite well on Misc. moderate functions with lower dimensions. On the other hand, the algorithm fails to solve optimization problems with functions based on a multi-modal structure. The summary of FDA performance comparing to benchmark SOTA is provided in Fig. 2.

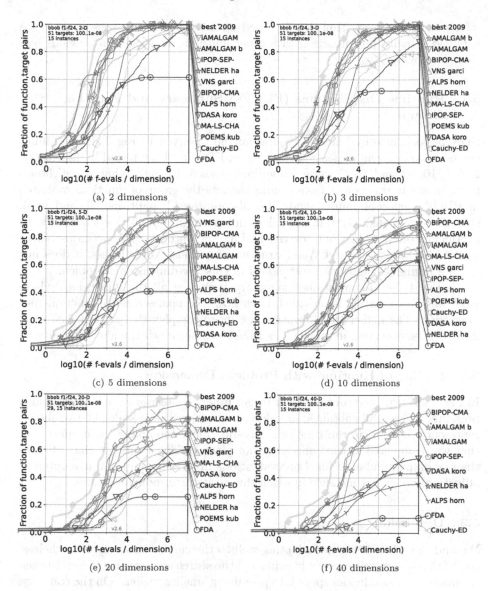

Fig. 2. Comparison of 12 SOTA methods per dimensions 2, 3, 5, 10, 20, and 40. The x-axis represents a particular budget in time (evaluations). The y-axis represents the performance of the 24 bbob functions that the given method has managed to solve (the higher the value the better). Each function is represented by 15 instances.

The rest of the experiments section is organized as follows. First, function results are broken down by dimension and function group to better understand FDA performance. Then, an aggregated graph is offered to summarize the previously mentioned information and expand the view of the dimensions.

Afterward, target precision details based on FDA evaluation consumption are supplied. Finally, a comparison with other methods mentioned in previous sections is provided.

5.1   Runtime Distributions (ECDFs) Summary and Function Groups

FDA performance by function groups can be observed in Fig. 3. Each column represents a different dimension complexity. FDA has been tested in dimensions 2, 3, 5, 10, 20, and 40. However, only dimensions 5, 10, and 40 have been chosen to be shown in this graph. Each column depicts the group of functions available in BBOB (separable, misc. moderate, ill-conditioned, multi-modal, and weak structure functions). Overall, FDA does not perform well on the benchmark. In particular, often it does not succeed finding the optima. FDA performs better in the Misc. moderate and Weak structure functions. Nevertheless, it does not reach a minimum performance standard in this low-dimensional problem.

In Fig. 4 the aforementioned summary including all the dimensions where FDA was benchmarked is provided. Each graph compares a set of functions per dimension. The higher the dimension the more complex the problem becomes. As it can be observed, in many cases FDA does not reach the global optimum.

5.2   Scaling of Runtime with Problem Dimension

In Fig. 5 the expected runtime 8 per target function precision $w.r.t.$ dimension is presented. The values obtained are plotted in a logarithmic scale. The symbol + represents the median run time of successful runs to complete the hardest goal that was completed at least once (but not always). The symbol × characterizes the maximum number of function evaluations in a trial. The FDA attempted to adjust only once per instance function due to its deterministic nature.

5.3   Discussion

We find that the FDA is not adapting well to the current problem. Nonetheless, the FDA has a promising start in exploring the search space with its fractal-based technique that subdivides space into promising smaller regions. On the contrary, ILS turns out to be too slow in the intensification phase harming the good initial performance. In particular, the main disadvantage of the FDA intensification step is not being able to abandon an unpromising solution trajectory.

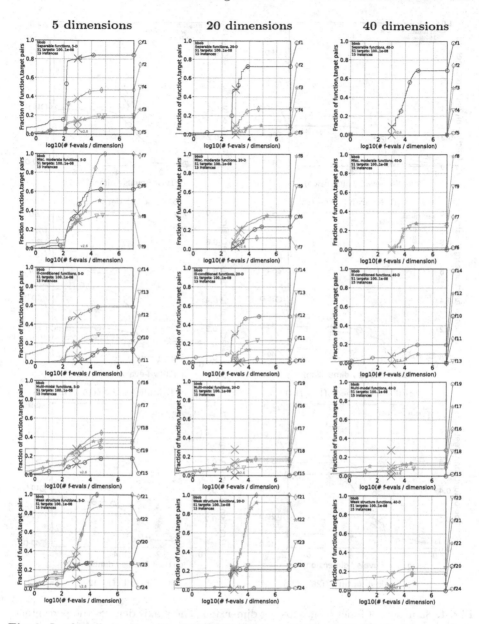

Fig. 3. In this illustration the 24 BBOB functions are represented. Each row refers respectively to separable, misc. moderate, ill-conditioned, multi-modal, and weak structure functions. Moreover, each column refers to the functions computed on 5, 20, and 40 dimensions. Particularly, the y-axis denotes the percentage of times a function has located the optimal target (Every independent function being run 15 times). The x-axis represents the number of evaluations divided by the problem dimension.

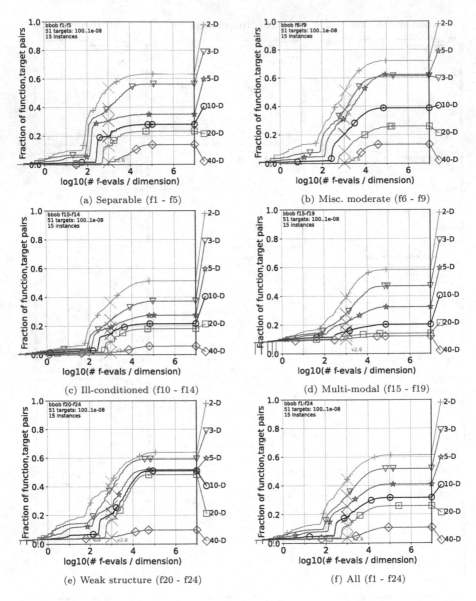

Fig. 4. Summary of function groups per dimension. The y-axis denotes the percentage of times a function has located the optimal target (Every independent function being run 15 times). The x-axis represents the number of evaluations divided by the problem dimension.

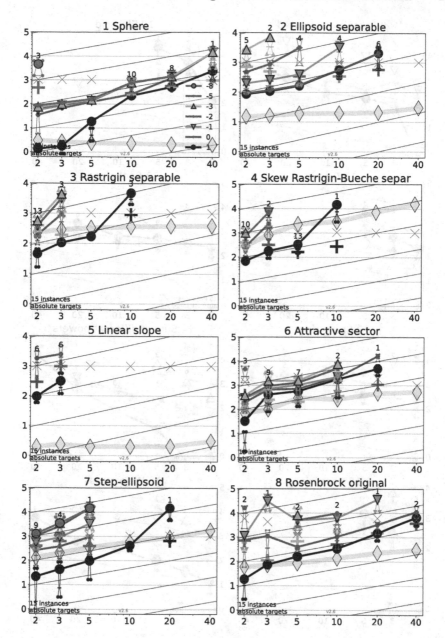

Fig. 5. In this figure the 24 BBOB functions are displayed at different target precision. The diamond shape in the graphs presents the best method in 2009 at the most complex target (red line). The y-axis denotes the expected runtime (ERT; the lower the better). The x-axis shows the problem dimension. Finally, as a legend is used the ERT for fixed target precision values of 10k. (Color figure online)

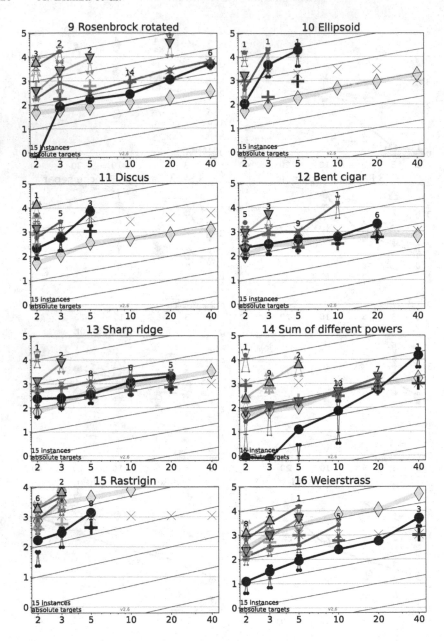

Fig. 5. (*continued*)

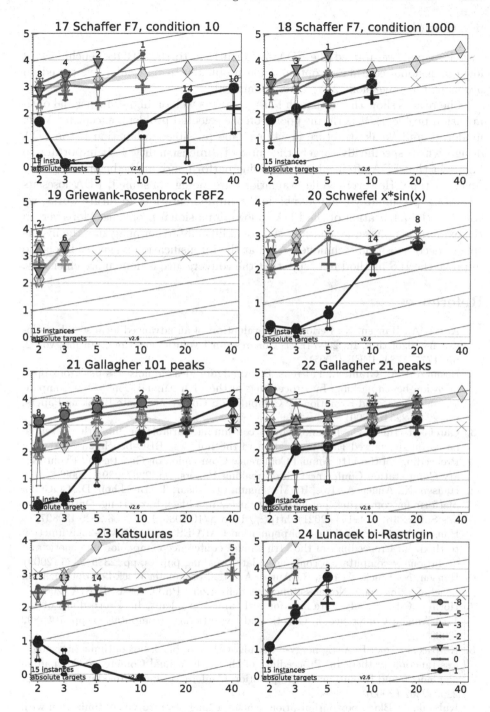

Fig. 5. (*continued*)

6 Conclusions

In this paper, we applied the FDA metaheuristic to the BBOB benchmark of low-dimensional continuous optimization problems. In particular, we tested on *bbob* suit for dimensions 2, 3, 5, 10, 20, and 40.

The results show that although FDA performs well on high-dimensional optimization problems, it performs poorly on the selected problem, except for Misc. moderate and Weak structure functions. One possible explanation is the FDA design which specifically targets the curse of dimensionality. It is important to highlight that FDA often exhibits a leading performance at the beginning of any function type. However, while evaluations are being consumed, FDA loses its advantage presumably due to its basic ILS implementation. In future works, we consider the adaptation of the FDA to low-dimensional problems. Moreover, in the exploration phase, we consider the possibility of further analyzing the hypersphere decomposition techniques. Also, at the intensification stage, we intend to improve the ILS method to ensure that the convergence is done more efficiently.

References

1. Auger, A., Hansen, N.: Performance evaluation of an advanced local search evolutionary algorithm. In: 2005 IEEE Congress on Evolutionary Computation, vol. 2, pp. 1777–1784. IEEE (2005)
2. Bosman, P.A., Grahl, J., Thierens, D.: Amalgam ideas in noiseless black-box optimization benchmarking. In: Proceedings of the 11th Annual Conference Companion on Genetic and Evolutionary Computation Conference: Late Breaking Papers, pp. 2247–2254 (2009)
3. García-Martínez, C., Lozano, M.: A continuous variable neighbourhood search based on specialised EAs: application to the noiseless BBO-benchmark 2009. In: Proceedings of the 11th Annual Conference Companion on Genetic and Evolutionary Computation Conference: Late Breaking Papers, pp. 2287–2294 (2009)
4. Hansen, N., Auger, A., Ros, R., Mersmann, O., Tušar, T., Brockhoff, D.: COCO: a platform for comparing continuous optimizers in a black-box setting. Optim. Methods Softw. **36**, 114–144 (2021). https://doi.org/10.1080/10556788.2020.1808977
5. Hansen, N.: Benchmarking a bi-population CMA-ES on the BBOB-2009 function testbed. In: Proceedings of the 11th annual conference companion on genetic and evolutionary computation conference: late breaking papers, pp. 2389–2396 (2009)
6. Hansen, N., Finck, S., Ros, R., Auger, A.: Real-parameter black-box optimization benchmarking 2009: Noiseless functions definitions. Ph.D. thesis, INRIA (2009)
7. Hornby, G.S.: Steady-state alps for real-valued problems. In: Proceedings of the 11th Annual Conference on Genetic and Evolutionary Computation, pp. 795–802 (2009)
8. Korošec, P., Šilc, J.: A stigmergy-based algorithm for black-box optimization: noiseless function testbed. In: Proceedings of the 11th Annual Conference Companion on Genetic and Evolutionary Computation Conference: Late Breaking Papers, pp. 2295–2302 (2009)
9. Kubalik, J.: Black-box optimization benchmarking of prototype optimization with evolved improvement steps for noiseless function testbed. In: Proceedings of the 11th Annual Conference Companion on Genetic and Evolutionary Computation Conference: Late Breaking Papers, pp. 2303–2308 (2009)

10. Molina, D., Lozano, M., Herrera, F.: A memetic algorithm using local search chaining for black-box optimization benchmarking 2009 for noisy functions. In: Proceedings of the 11th Annual Conference Companion on Genetic and Evolutionary Computation Conference: Late Breaking Papers, pp. 2359–2366 (2009)
11. Nakib, A., Ouchraa, S., Shvai, N., Souquet, L., Talbi, E.G.: Deterministic metaheuristic based on fractal decomposition for large-scale optimization. Appl. Soft Comput. **61**, 468–485 (2017)
12. Nakib, A., Souquet, L., Talbi, E.G.: Parallel fractal decomposition based algorithm for big continuous optimization problems. J. Parallel Distrib. Comput. **133**, 297–306 (2019)
13. Pošík, P.: Bbob-benchmarking a simple estimation of distribution algorithm with cauchy distribution. In: Proceedings of the 11th Annual Conference Companion on Genetic and Evolutionary Computation Conference: Late Breaking Papers, pp. 2309–2314 (2009)
14. Price, K.V.: Differential evolution vs. the functions of the 2/sup nd/ICEO. In: Proceedings of 1997 IEEE International Conference on Evolutionary Computation (ICEC'97), pp. 153–157. IEEE (1997)
15. Ros, R.: Benchmarking sep-CMA-ES on the BBOB-2009 function testbed. In: Proceedings of the 11th Annual Conference Companion on Genetic and Evolutionary Computation Conference: Late Breaking Papers, pp. 2435–2440 (2009)

A Comparative Analysis of Different Multilevel Approaches for Community Detection

Guido Bordonaro, Rocco A. Scollo[✉][iD], Vincenzo Cutello[iD], and Mario Pavone[iD]

Department of Mathematics and Computer Science, University of Catania, v.le A. Doria 6, 95125 Catania, Italy
rocco.scollo@phd.unict.it, cutello@unict.it, mpavone@dmi.unict.it

Abstract. Community Detection is one of the most investigated problems as it finds application in many real-life areas. However, detecting communities and analysing community structure are very computationally expensive tasks, especially on large networks. In light of this, to better manage large networks, two new Multi-Level models are proposed in order to reduced and simplify the original graph via aggregation of groups of nodes. Both models have been applied on two variants of an immune-inspired algorithm, the first one based on a fully random-search process, and the second based on a hybrid approach. From the experimental analysis it clearly appears that the two proposed models help the random-search and the hybrid immune-inspired algorithms to significantly improve their performances from both computational and quality of found solutions point of view. In particular, the hybrid variant appears to be very competitive and efficient.

Keywords: Community detection · Multi-level search · Metaheuristics · Hybrid metaheuristics · Immune-inspired computation · Random search

1 Introduction

Community detection is one of the most important and relevant problems in network sciences and graph analysis as it finds application in many industrial and research areas, as a consequence many approaches to the problem have been developed in the last years. The ability to discover communities in a complex network allows to gain crucial and important information, such as, for instance: identify the interactions between different entities; reveal social relationships among people; detect proteins that have same specific function within a cell; identify different web pages dealing with the same or related topics; and many others. From these simple examples, we can see that community detection plays a key role in many research and application areas, such as biology, medicine, economics, social sciences, engineering, ecology, etc. Informally, a community

L. Di Gaspero et al. (Eds.): MIC 2022, LNCS 13838, pp. 230–245, 2023.
https://doi.org/10.1007/978-3-031-26504-4_17

(or cluster) is defined as a set of network's elements that are strongly connected inside the community but weakly linked with all other ones. In a nutshell, each community represents a set of vertices (or entities) that share common properties, or play similar roles, or show similar interests within the network. For all these reasons, therefore, the study of community structures inspires intense research activities to visualize and understand the dynamics of a network at different scales [10,14,16].

As a consequence, several iterative and exact search algorithms for clustering problems have been proposed, and they have been proved to be robust in finding as cohesive as possible communities in complex networks [18,19]. However, discovering communities and analysing their structure on large complex networks (with thousands and/or millions of nodes) is very hard and extremely computationally expensive, which makes most of the algorithms proposed in literature unsuitable to be applied. Thus, to efficiently handle large networks and find reliable partitions it is usually useful to apply multi-level approaches, in which the main optimization algorithm is applied to the reduced graph obtained from the best partition found up to that point, where nodes represent communities, and the edges the connections between them. This iterative process allows for a more exhaustive exploration of the search space.

In this research work, we present some multi-level approaches and we study how they affect the performances of two different immune algorithms (IA): a first algorithm based on a fully random-search [7,25], and a second based on a hybrid approach [7]. The goal of the developed multi-level approaches is therefore to help the two versions of IA in efficiently handling large networks to detect community structures. The two main multi-level approaches proposed have been designed, and then adapted to the two variants of IA, as they explore the search space in different way. The first approach uses a backtracking technique to previous levels if the aggregations made so far do not lead to improvements for a given number of levels. The second one, on the other hand, introduces a quality-based aggregation, which simply merges those nodes in the same community having a high ratio between internal degree and total degree (see Eq. 4, Sect. 4.1). The obtained outcomes show the usefulness of the proposed multi-level approaches, proving their remarkably good influence in improving the quality of the found solutions for both variants, particularly on larger networks.

2   The Modularity as Evaluation Metric

As we already said, a community in a network is defined as a set of elements that are highly linked within it but weakly outside. The goal of community detection is to divide (or partition) the network nodes into groups, so that the connections are strong inside and weak outside. In a nutshell, the aim of community detection is to identify the modules or communities in networks and their hierarchical organization.

To assess the quality of the detected communities, the *modularity* function is certainly the most used quality metric [20]. Modularity is based on the idea that a random network is not expected to have a community structure, therefore the possible existence of communities can be revealed by the difference of density between vertices of the network and vertices of a random network having the same size and degree distribution. Formally it can be defined as follow: given an undirected graph $G = (V, E)$, with V the set of vertices ($|V| = N$), and E the set of edges ($|E| = M$), the modularity Q of a community is given by

$$Q = \frac{1}{2M} \left[\sum_{i=1}^{N} \sum_{j=1}^{N} \left(A_{ij} - \frac{d_i d_j}{2M} \right) \delta(i, j) \right],$$ (1)

where A_{ij} is the adjacency matrix of G, d_i and d_j are the degrees of nodes i and j respectively, and $\delta(i, j) = 1$ if i, j belong to the same community, 0 otherwise. As asserted in [3], the modularity value for unweighted and undirected graphs lies in the range $[-0.5, 1]$. Hence, low Q values reflect a bad nodes partitioning and imply the absence of real communities; high values instead identify good partitions and imply then the presence of highly cohesive communities. However, although modularity is the most used evaluation metric, it fails to reveal relatively small communities as it tends to produce larger communities [11]. Considering the modularity as objective function the community detection can be tackled as a combinatorial optimization problem where the goal becomes to find a clustering that maximize Q. In view of this, we must consider the fact that community detection has been proven to be a $\mathcal{NP}$-complete problem [3].

3   Immune Metaheuristics

Immune Algorithms (IA), which are one of the most successfully population-based metaheuristics in search and optimization tasks, are based on the principles and dynamics through which the immune system protects a living organism [5]. What makes the immune system a source of inspiration is its ability to detect, distinguish, learn and remember all foreign entities to the body (called Antigens - Ag) [12], defending consequently the organism from potential, dangerous diseases. There are several ways for the immune system to fastly recognize and attack foreign entities. One of these is the clonal selection principle that works in the following way: all cells that recognize the antigen start a cloning phase of themselves, i.e. duplication, and the number of copies is proportional to their recognition quality. Afterwards, each clone undergoes a hypermutation process, and the number of mutations (greater than one) is inversely proportional way to its recognition quality. From a computational point of view, the antigen is the problem to be solved, while the cells belonging to the organism represent the solutions to the problem. All those algorithms that simulate this principle belong to the special class *Clonal Selection Algorithms* (CSA) [6,8,22], and have in the (i) cloning, (ii) hypermutation and (iii) aging operators their strength. The two immune algorithms proposed in this research work, RANDOM-IA and HYBRID-IA, are both based on this principle.

As in all population-based metaheuristics, both algorithms maintain a population of d candidate solutions at each time step t. In this work, each solution is a subdivision of the vertices of the graph $G = (V, E)$ in communities; specifically, a solution x is an integers sequence of $N = |V|$ integers belonging to the range $[1, N]$, where $x_i = j$ indicates that the vertex i has been added to the cluster j. For the initial population, at time step $t = 0$, each vertex i is randomly assigned to a j. Once initial population is generated, and, in general, just after a new offspring population is created, the procedure ComputeFitness($P^{(*)}$) is called to compute the fitness function (Eq. 1), that is to evaluate the quality (poor or good) of each solution. Finally, the algorithms will stop once a given termination criterion is reached, which, in this work, is the maximum number of allowed generations ($MaxGen$). The pseudo-code and main parameters are reported in Algorithm 1, where the boolean variable LS enables (or not) the Local Search procedure [24]. Thus, when $LS = true$, we will have HYBRID-IA, otherwise RANDOM-IA.

Algorithm 1. Pseudo-code of RANDOM-IA and HYBRID-IA.

1: **procedure** IMMUNEALGORITHM(d, dup, ρ, τ_B, LS)
2:    $t \leftarrow 0$
3:    $P^{(t)} \leftarrow$ InitializePopulation(d)
4:    ComputeFitness($P^{(t)}$)
5:    **while** ¬StopCriterion **do**
6:       $P^{(clo)} \leftarrow$ Cloning($P^{(t)}, dup$)
7:       $P^{(hyp)} \leftarrow$ Hypermutation($P^{(clo)}, \rho$)
8:       ComputeFitness($P^{(hyp)}$)
9:       $(P_a^{(t)}, P_a^{(hyp)}) \leftarrow$ Aging($P^{(t)}, P^{(hyp)}, \tau_B$)
10:      $P^{(select)} \leftarrow (\mu + \lambda)-$Selection($P_a^{(t)}, P_a^{(hyp)}$)
11:      **if** (LS) **then**
12:        $P^{(t+1)} \leftarrow$ LocalSearch($P^{(select)}$)
13:        ComputeFitness($P^{(t+1)}$)
14:      **else**
15:        $P^{(t+1)} \leftarrow P^{(select)}$
16:      **end if**
17:      $t \leftarrow t + 1;$
18:    **end while**
19: **end procedure**

After the phase of fitness computation (line 4), the *cloning* procedure takes place (line 6), each solution is copied dup times, producing an intermediate population $P^{(clo)}$ of size $d \times dup$. Inside the cloning operator, an age is assigned to each clone, which represents the number of generations each clone has been in the population. Thus, the range from the assigned age to the maximum age allowed (user-defined parameter τ_B) represents how long the solution can live/remain in the population. Such a lifespan affects the efficiency of the exploration of the search space, as well as on the exploitation phase. Specifically, a random age chosen in the range $[0 : \frac{2}{3}\tau_B]$ is assigned to each clone, so to guarantee to the solution a minimal number of generations ($\frac{1}{3}\tau_B$ in the worst case). It is worth emphasizing that age assignment together with the aging operator play a crucial role on the performance of the algorithm as their combination helps the algorithms to avoid to get trapped into local optima thanks to a proper solutions' diversity produced [9].

The *hypermutation operator* (line 7) aims at carefully exploring the neighborhood of each solution so to find better solutions iteration by iteration. A predetermined number of mutations - more than one - are carried out on each solution, whose mutation rate is *inversely proportional* to its fitness value, i.e. the higher the fitness value of the element, the smaller its mutation rate. Specifically, let x be a solution, the *mutation rate* $\alpha = e^{-\rho \hat{f}(x)}$ is defined as the probability to move a node from one community to another, where ρ (user-defined parameter) determines the shape of the mutation rate, and $\hat{f}(x)$ is the fitness function normalized in the range $[0, 1]$. Formally, for each element in the population we randomly select two different communities c_i and c_j ($c_i \neq c_j$) such that c_i is chosen among all existing ones, and the second one in the range $[1, N]$. Then, all vertices in c_i are moved to c_j with probability given by α. However, since c_j is chosen randomly in the range $[1, N]$, it can happen that it doesn't match any existing one; in such a case, a new community c_j is created and added to the existing ones. This approach allows to create and discover new communities by moving a variable percentage of nodes from existing communities, and, further, in HYBRID-IA version, it balances the effects of local search by allowing the algorithm to avoid premature convergences towards local optima.

Once the fitness values are computed for all the elements in the population (line 8) the *Static aging operator* (line 9) is applied in both versions, and it removes older elements from both populations $P^{(t)}$ and $P^{(hyp)}$. Let $\tau_B{}^1$ the maximum number of generations allowed within which the solution can stay into the population; then, starting from the age assigned in the cloning operator, which is increased by one at each iteration, as soon as the solution exceeds the τ_B age (i.e. age $= \tau_B + 1$), it is removed from the population, regardless of its fitness value. Only an exception occurs for the overall best current solution, which is instead kept even if its age is older than τ_B. Such a variant is called *elitist static aging operator*. The best d survivors from both populations $P_a^{(t)}$ and $P_a^{(hyp)}$ are selected for the new (temporary) population $P^{(select)}$ (line 10), on which the local search (line 12) is carried out if LS is true. For this selection, we developed the $(\mu + \lambda)$-*Selection operator*, where $\mu = d$ and $\lambda = (d \times dup)$, which ensures monotonicity in the evolution dynamics.

The main and most important difference between RANDOM-IA and HYBRID-IA versions is the use of the *Local Search* (LS) in the last. Local Search aims at properly speeding up the convergence of the algorithm, driving it towards more promising regions [24]. The *Move Vertex* approach (MV) [17] was used for intensifying the exploration of the neighborhood of each solution. The LS basic idea is to assess deterministically if a node can be moved from its community to another one. MV approach considers the *move gain* that is the variation in modularity produced when a node is moved from a community to another. However, before formally defining the move gain, it is important to point out that the modularity Q (Eq. 1) can be rewritten as:

---

[1] A user-defined parameter.

$$Q(c) = \sum_{i=1}^{k} \left[\frac{\ell_i}{M} - \left(\frac{d_i}{2M} \right)^2 \right], \tag{2}$$

where k is the number of communities detected; $c = \{c_1, \ldots, c_i, \ldots c_k\}$ is the set of communities; ℓ_i and d_i are, respectively, the number of links inside the community i, and the sum of the degrees of vertices belonging to the community i. Let u a vertex of the community c_i; the *move gain* is, therefore, defined as the modularity variation produced by moving u from c_i to c_j. Formally:

$$\Delta Q_u(c_i, c_j) = \frac{l_{c_j}(u) - l_{c_i}(u)}{M} + d_V(u) \left[\frac{d_{c_i} - d_V(u) - d_{c_j}}{2M^2} \right], \tag{3}$$

where $l_{c_i}(u)$ and $l_{c_j}(u)$ are respectively the number of links of u with nodes in c_i and c_j, while $d_V(u)$ is the degree of u when considering all vertices in V. If $\Delta Q_u(c_i, c_j) > 0$ then moving u from c_i to c_j produces an increment in modularity, and then a possible improvement. It follows, that the goal of MV is then finding a node u to move so to maximize $\max_{v \in Adj(u)} \Delta Q_u(i, j)$, where $u \in C_i$, $v \in C_j$ and $Adj(u)$ is the adjacency list of node u. For each solution in $P^{(select)}$, the Local Search begins by sorting the communities in increasing order with respect to the ratio between the sum of inside links and the sum of the node degrees in the community. In this way, poorly formed communities are identified. After that, MV acts on each community of the solution, starting from nodes that lie on the border of the community, that is those that have at least an outgoing link. Also for communities, the nodes are sorted with respect to the ratio between the links inside and node degree. The key idea behind LS is to deterministically repair the solutions which were produced by the hypermutation operator, by discovering then new partitions with higher modularity value. Equation 3 can be calculated efficiently because M and $d_V(u)$ are constants, the terms l_{c_i} and d_{c_i} can be stored and updated using appropriate data structures, while the terms $l_{c_i}(u)$ can be calculated during the exploration of all adjacent nodes of u. Therefore, the complexity of the move vertex operator is linear on the dimension of the neighborhood of node u.

4   Multi-level Approaches

The multi-level approach is an optimization technique used to improve a community detection algorithm, both in terms of objective function and computational cost. This approach consists in creating a new graph in which the nodes are the communities of the partial solution found by the base algorithm, while edges between communities are merged together with a weight given by the sum of the edges between nodes in the corresponding two communities. Edges between nodes in the same community are translated in self-loops in the new graph. In this way the modularity value for the partition does not change in the new graph [1]. In Fig. 1 is shown the creation of a new level starting from the partition found on the current graph. The reduced graph is then passed as input

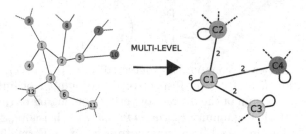

Fig. 1. Creation of the community network by the multi-level optimization. Communities will be translated in nodes in the next level, while edges are merged together with a weight that is the number of edges that those communities share. Self-loops identify internal edges.

to the base algorithm to compute the next solution. These steps are repeated until no further improvement can be achieved or for a certain amount of time. At each iteration the size of the reduced graph decreases and consequently the efficiency of the base algorithm is greatly improved.

In a metaheuristic algorithm, the implementation of multi-level optimization could lead to wrong solutions, because some parts of the solution would be *locked* and any further improvement to the solution would be done only by fusing together the remaining nodes. After few iterations, the graph will be reduced to a small number of nodes where any combination between them would not result in a modularity improvement, but the solution would remain of low quality. In light of this, in the following sections we propose two multi-level approaches: the first one uses a backtracking mechanism to give the underlying base metaheuristic algorithm a chance to improve specific parts of the global solution; the second one uses a heuristic to merge nodes together.

4.1   Random and Smart Explosion

The first approach proposed consists of the classical multi-level optimization with a backtracking mechanism that brings the algorithm back to a previous level when there is no improvement of the modularity value for a certain number of levels. Then to the base algorithm we provide the graph of the level with the best partition found in which some nodes, that represent communities, are disaggregated to the original graph. This allows to free nodes that had been blocked in an earlier wrong solution trying to repair communities not well-formed. The communities which *explode* are randomly selected from those in the best partition and the number is given by a user-defined parameter N_e. Usually the number of communities to disaggregate is kept low in order to avoid degrading too much the current solution and letting the underlying base algorithm to focus mainly on those nodes that are now free. After the roll-back to a previous level, the multi-level approach continue in the classical way until a new stagnation of the modularity value occurs. Then the back-tracking mechanism is applied again and this process is repeated for a certain number of times. Finally, the algorithm stops and returns the best solution found.

A complete disaggregation of one or more communities affects the performance of the base algorithm, disrupting correct parts of the current solution and increasing the number of nodes to evaluate. A further improvement of this approach consists in a *smart explosion* of the communities, in which only a subset of nodes is disaggregated. In this way the method only critical nodes, that is the nodes that lie on the boundary of the community, are disaggregated. Critical nodes are identified using the internal-total degree ratio:

$$\frac{k_i^{int}(C)}{k_i} < T_e \tag{4}$$

where $k_i^{int}(C)$ is the sum of the weights of edges that node i shares with other nodes belonging to the same community C, k_i is the sum of weights of all incident edges of node i and T_e is a user-defined threshold. In Fig. 2 is shown the application of the *smart explosion* approach.

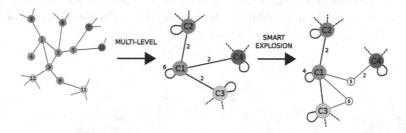

Fig. 2. Example of smart explosion approach. In this case nodes 5 and 6 have an internal-total degree ratio less or equal than 0.5 and the method disaggregates them from their own community in order to let the base algorithm relocate them to a better community.

4.2 Smart Merge

A naive multi-level approach, that blindly merges all nodes in their respective supposed communities, could lead to wrong associations node-community, as described before. The second proposed approach modifies the multi-level optimization introducing a *quality-based aggregation*. In particular during the aggregation phase, only those nodes belonging to the same community and with a high internal-total degree ratio (Eq. 4) will be merged together. In this way, the nodes that are supposed to be already associated with the correct community and that will not change in subsequent iterations, will be merged together, reducing the size of the graph and the complexity of the base algorithm. On the other hand, critical nodes are kept free and can be moved to the correct community by the underlying base algorithm. In Fig. 3 is shown how the multi-level with smart merge mechanism works. This approach is useful and efficient with a base

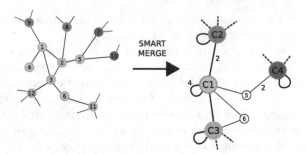

Fig. 3. Creation of the network of the next level by the multi-level optimization with smart merge mechanism. In this case nodes 5 and 6 are kept free because they share a number of links with other communities greater than or equal to those they share with their own community.

algorithm that finds good partition in a relatively small time. Algorithms that tend to converge slowly starting from low quality solutions, do not receive a significant improvement by this approach because the aggregation heuristic used in the smart merge (that depends on the threshold T_m) decreases the graph size slowly, affecting the overall computational time.

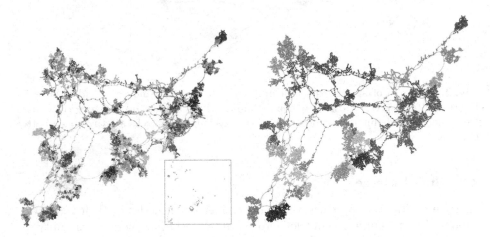

Fig. 4. Communities detected by the smart merge approach (left plot), and by the smart merge considering the connected components (right plots).

However, although this approach allows to reach high values of modularity, by inspecting the graphical representation of the detected communities (left plot in Fig. 4), it is possible to note how a single community is composed by elements disconnected from each other (see inset plot in Fig. 4). This happens because these disconnections are disregarded by *smart merge* approach, as it asserts the goodness of a vertex by checking only if its links are inside or outside. In light of this, to overcome this limitation, it was enough to add a control on the communities detected by the basic version of the algorithm (HYBRID-IA), which divides the clusters into their connected components. Through this simple check, the detected communities appear to be more compact graphically (right plot in Fig. 4), as well as reaching higher modularity values (see Table 3, Sect. 5). This variant is called *smart-merge+check-connect*.

5   Experimental Results

To assess the robustness and efficiency of the proposed multilevel approaches, three well-known benchmark networks were used, which are reported in Table 1. Obviously, the comparison with the relative basic versions is also presented in this section so to check the improvements produced by the proposed approaches. In particular, RANDOM-IA has been considered as the basic algorithm for the *random* and *smart explosion* approaches due to its stochastic nature; in this way, it can repair a small region of the network disaggregated by the explosion mechanism. On the other hand, as described in Sect. 4.2, HYBRID-IA has been used as basic underlying metaheuristic for the *smart merge* approach, because this algorithm reaches in just few iterations solutions with high modularity value. Consequently, the backtracking approaches developed in *random* and *smart explosion*, if applied on HYBRID-IA should disaggregate a high number of communities at each stagnation of modularity, and then correct/repair all communities, increasing however the network size, and therefore considerably slowing down the convergence of the entire algorithm.

For all the experiments, both versions use the same parameter configurations, and specifically a population of $d = 100$ solutions; a duplication factor $dup = 2$; $\tau_B = 20$ as the maximum age allowed; and a mutation shape $\rho = 1.0$. Due to the different algorithmic structure of the two versions, a different number of iterations $MaxGen$ was considered. In particular, for RANDOM-IA we set $MaxGen = 1000$ for each level, while in HYBRID-IA the number of iterations is related to the size of the network: $MaxGen$ starts from 50 iterations and progressively decreases proportionally to the size of the network, to a minimum of 10 iterations.

Table 1. The benchmark networks used in the experiments.

Name	Description	$|V|$	$|E|$
E-mail [15]	University e-mail network	1133	5451
Yeast [4]	Protein-protein interaction network in budding yeast	2284	6646
Power [26]	Topology of the Western States Power Grid of the US	4941	6594

For the multi-level optimization process, the *random explosion* reverts just $N_e = 1$ community to the original network, while the *smart explosion* approach disaggregate $N_e = 2$ communities using a threshold $T_e = 0.5$. The multi-level optimization with the *smart merge* mechanism instead uses a $T_m = 0.5$ to construct the network for the next level. Although multi-level optimization can stop its execution when it detects a modularity stagnation, for an easier comparison all algorithms were run 30 times for each instance and for a prefixed CPU time. In particular, in *random explosion* and *smart explosion*, which use RANDOM-IA as basic algorithm, the CPU time limit was fixed, respectively, to 1200 s for *E-mail*, 2400 s for *Yeast* and 3600 s for *Power*. In *smart merge* approach, which uses HYBRID-IA as underlying basic algorithm, the CPU time limit was fixed to 120 s for *E-mail*, 900 s for *Yeast* and 3600 s for *Power*.

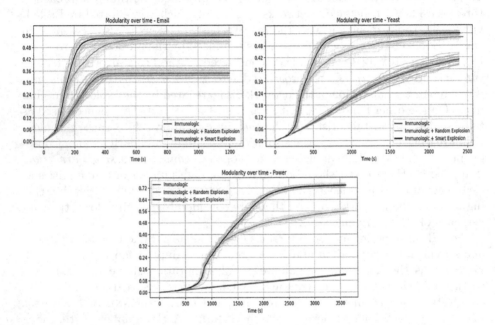

Fig. 5. Convergence analysis over time of RANDOM-IA; RANDOM-IA with *random explosion*; and RANDOM-IA with *smart explosion*, on the benchmark networks *Email* (top left), *Yeast* (top right) and *Power* (bottom).

The first analysis of this research work focused on investigating the impact that the two *random* and *smart explosion* approaches have on the basic version (RANDOM-IA), and how much they positively affect its overall performances. Figure 5 therefore shows the convergence behaviour of the proposed multi-level approaches compared with RANDOM-IA. In particular, the three convergence curves of (1) RANDOM-IA, (2) RANDOM-IA with *random explosion*, and (3) RANDOM-IA with *smart explosion* are displayed, from which it is possible to analyze how much improvement the two proposed multi-level approaches produce compared to the basic version. With regard to the larger benchmark networks, it can be clearly seen how the improvements produced by the two multi-level approaches are remarkably reaching significantly higher modularity values. Inspecting only the comparison between the two multi-level approaches it is possible to assert: (*a*) on the *Email* network the *random explosion* shows an initially slower convergence than *smart explosion*, whilst, afterwards, the two curves join showing the same convergence behaviour. However, towards the end of the run, *random explosion* is able to improve and reach slightly higher modularity value than *smart explosion*; (*b*) on the *Yeast* and *Power* networks, instead, *smart explosion* clearly outperforms *random explosion*, especially on the larger network (*Power*), where the distance between the curves is quite significant and clear in favour of *smart explosion*.

Table 2. *Random* and *Smart Explosion* versus basic algorithm (Random-IA). Best modularity found, mean and standard deviation (σ) as comparison measures.

	Email		Yeast		Power	
	best	*mean ± σ*	*best*	*mean ± σ*	*best*	*mean ± σ*
Random-IA	0.3841	0.3465 ± 0.0186	0.4411	0.4089 ± 0.0171	0.1260	0.1200 ± 0.0026
Random-IA+RE	0.5627	0.5416 ± 0.0142	0.5388	0.5210 ± 0.0091	0.5791	0.5568 ± 0.0124
Random-IA+SE	0.5539	0.5282 ± 0.0160	0.5538	0.5404 ± 0.0092	0.7532	0.7364 ± 0.0093

In Table 2 we can see, respectively, the best modularity found, the mean of the best, and the standard deviation (*mean* $\pm\ \sigma$), for all three benchmark networks considered. The outcomes showed in the table confirm what asserted from the convergence plots, that the *random explosion* works better on the smaller networks (i.e. *Email*), whilst *smart explosion* on the other two. With regard to the *Power* network, which is the larger and then the most significant from the multi-level approach perspective, the modularity value found by *smart explosion* is way better than the others, especially with respect to the basic version that instead finds low values of modularity (0.1260). This points out, then, how multi-level approach designed in smart explosion helps the random-search algorithm (Random-IA) in revealing good community structures.

The same analysis was conducted to understand how the *smart merge* approach affects the performance of Hybrid-IA, that is the basic version on which it is applied. In Fig. 6 is therefore shown the convergence behaviour of the multi-level approach compared to the basic one. By inspecting the three plots, it can be seen how *smart merge* and the *smart-merge+check-connect* variant are similar on the *Email* network, whilst in the *Yeast* one the connected-components version is shown to be slightly better than the *smart merge* version alone. It is important to point out that both multi-level versions improve in any case the performance of the basic algorithm, although such improvements are moderate. The improvements produced by the *smart merge* and *smart-merge+check-connect* approaches are best seen on the larger network *Power*, where the gap

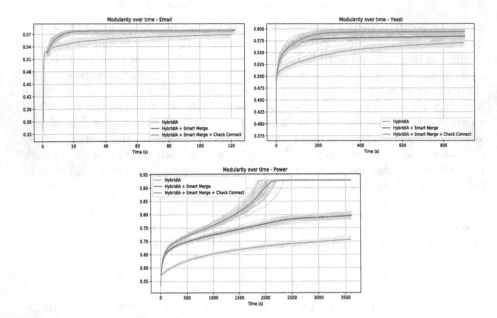

Fig. 6. Convergence analysis over time of HYBRID-IA; HYBRID-IA with *smart merge*; and HYBRID-IA with smart merge and *check connection*, on the benchmark networks *Email* (top left), *Yeast* (top right) and *Power* (bottom).

between the three curves is clear and marked. In particular the variant *smart-merge+check-connect* produces sharply best performance, reaching considerably higher modularity values to the basic version, and the *smart merge* one.

Table 3. *Smart Merge* versus basic algorithm (HYBRID-IA). Best modularity found, mean and standard deviation (σ) as comparison measures.

	Email		Yeast		Power	
	best	mean ± σ	best	mean ± σ	best	mean ± σ
HYBRID-IA	0.5782	0.5690 ± 0.0049	0.5858	0.5710 ± 0.0057	0.7202	0.7065 ± 0.0063
HYBRID-IA+SM	0.5824	0.5801 ± 0.0018	0.5929	0.5845 ± 0.0045	0.8125	0.7964 ± 0.0099
HYBRID-IA+SM+C	0.5813	0.5782 ± 0.0019	0.5998	0.5940 ± 0.0033	0.9321	0.9294 ± 0.0015

These improvements are also confirmed by the results reported in Table 3, both in term of best modularity value found, and in the mean. Indeed, by inspecting the table, it clearly appears that, due to the high quality solutions produced by HYBRID-IA on networks not excessively large, the effects and improvements produced by the multi-level approach are limited, while instead on the large one, where the basic algorithm struggles to reach high modularity values, the improvement contribution given by the multi-level approach is notable and mainly relevant (0.7202 vs 0.9321).

Table 4. Comparison with state-of-the-art algorithms.

Network	MSG-VM	SS+ML	Louvain	CNTS	CNTS-ML	HYBRID-IA+SM+C
Email	0.5746	0.5813	0.5758	0.5820	0.5815	0.5813
Yeast	0.5948	0.6068	0.5962	0.6053	0.6055	0.5998
Power	0.9381	0.9392	0.9371	0.9380	0.9392	0.9321

Finally, Table 4 reports the comparisons between the *smart-merge+check-connect* variant (being the best approach) and the state-of-the-art: SS+ML, a multi-level algorithm based on a single-step greedy coarsening and fast greedy refinement [21]; MSG-VM, a multistep greedy algorithm with vertex mover [23]; Louvain, a fast multi-level greedy algorithm [2]; CNTS, a combined neighborhood tabu search [13]; and CNTS-ML, the multi-level version of the CNTS algorithm [13]. It is possible to see how the proposed multilevel approach is competitive with the community detection state-of-the-art on the first two benchmark networks, a little less on the *Power* one. However, on this last network, the results obtained by HYBRID-IA with *smart-merge+check-connect* are not so far from the compared ones.

6 Conclusions and Future Work

The multi-level models we propose for community detection on quite large networks and which are based on two variants of an immune-inspired algorithm, were experimentally shown to be very competitive and efficient. Yet, still trailing some state of the art methodologies, especially on extremely large networks. Given such promising initial results, as future work we plan to tackle even larger networks, in particular biological and online social networks. We will focus our research direction on implementing mechanisms, such as reinforcement and probabilistic learning, to better guide the level construction phase of the multi-level approaches to further improve both the objective function and convergence.

References

1. Arenas, A., Duch, J., Fernández, A., Gómez, S.: Size reduction of complex networks preserving modularity. New J. Phys. **9**(6), 176–176 (2007). https://doi.org/10.1088/1367-2630/9/6/176

2. Blondel, V.D., Guillaume, J.L., Lambiotte, R., Lefebvre, E.: Fast unfolding of communities in large networks. J. Stat. Mech: Theory Exp. **10**, 10008–10019 (2008). https://doi.org/10.1088/1742-5468/2008/10/P10008

3. Brandes, U., et al.: On modularity clustering. IEEE Trans. Knowl. Data Eng. **20**(2), 172–188 (2008). https://doi.org/10.1109/TKDE.2007.190689

4. Bu, D., et al.: Topological structure analysis of the protein-protein interaction network in budding yeast. Nucleic Acids Res. **31**(9), 2443–2450 (2003). https://doi.org/10.1093/nar/gkg340

5. Coello Coello, C.A., Cutello, V., Lee, D., Pavone, M.: Recent advances in immuno-logical inspired computation. Eng. Appl. Artif. Intell. **62**, 302–303 (2017)
6. Cutello, V., Oliva, M., Pavone, M., Scollo, R.A.: An immune metaheuristics for large instances of the weighted feedback vertex set problem. In: 2019 IEEE Symposium Series on Computational Intelligence (SSCI), pp. 1928–1936 (2019).https://doi.org/10.1109/SSCI44817.2019.9002988
7. Cutello, V., Fargetta, G., Pavone, M., Scollo, R.A.: Optimization algorithms for detection of social interactions. Algorithms **13**(6), 139–153 (2020). https://doi.org/10.3390/a13060139
8. Cutello, V., Nicosia, G., Pavone, M., Stracquadanio, G.: An information-theoretic approach for clonal selection algorithms. In: Hart, E., McEwan, C., Timmis, J., Hone, A. (eds.) Artificial Immune Systems, pp. 144–157. Springer, Berlin (2010). https://doi.org/10.1007/978-3-642-14547-6
9. Di Stefano, A., Vitale, A., Cutello, V., Pavone, M.: How long should offspring lifespan be in order to obtain a proper exploration? In: 2016 IEEE Symposium Series on Computational Intelligence (SSCI), pp. 1–8 (2016). https://doi.org/10.1109/SSCI.2016.7850270
10. Didimo, W., Montecchiani, F.: Fast layout computation of clustered networks: algorithmic advances and experimental analysis. Inf. Sci. **260**, 185–199 (2014). https://doi.org/10.1016/j.ins.2013.09.048
11. Fortunato, S., Barthélemy, M.: Resolution limit in community detection. Proc. Natl. Acad. Sci. **104**(1), 36–41 (2007). https://doi.org/10.1073/pnas.0605965104
12. Fouladvand, S., Osareh, A., Shadgar, B., Pavone, M., Sharafi, S.: DENSA: an effective negative selection algorithm with flexible boundaries for self-space and dynamic number of detectors. Eng. Appl. Artif. Intell. **62**, 359–372 (2017). https://doi.org/10.1016/j.engappai.2016.08.014
13. Gach, O., Hao, J.K.: Combined neighborhood tabu search for community detection in complex networks. RAIRO-Oper. Res. **50**(2), 269–283 (2016). https://doi.org/10.1051/ro/2015046
14. Girvan, M., Newman, M.E.J.: Community structure in social and biological networks. Proc. Natl. Acad. Sci. **99**(12), 7821–7826 (2002). https://doi.org/10.1073/pnas.122653799
15. Guimerà, R., Danon, L., Díaz-Guilera, A., Giralt, F., Arenas, A.: Self-similar community structure in a network of human interactions. Phys. Rev. E **68**, 065103 (2003). https://doi.org/10.1103/PhysRevE.68.065103
16. Gulbahce, N., Lehmann, S.: The art of community detection. BioEssays **30**(10), 934–938 (2008). https://doi.org/10.1002/bies.20820
17. Kernighan, B.W., Lin, S.: An efficient heuristic procedure for partitioning graphs. Bell Syst. Tech. J. **49**(2), 291–307 (1970). https://doi.org/10.1002/j.1538-7305.1970.tb01770.x
18. Newman, M.E.J.: Fast algorithm for detecting community structure in networks. Phys. Rev. E **69**, 066133 (2004). https://doi.org/10.1103/PhysRevE.69.066133
19. Newman, M.E.J.: Finding community structure in networks using the eigenvectors of matrices. Phys. Rev. E **74**, 036104 (2006). https://doi.org/10.1103/PhysRevE.74.036104
20. Newman, M.E.J., Girvan, M.: Finding and evaluating community structure in networks. Phys. Rev. E **69**, 026113 (2004). https://doi.org/10.1103/PhysRevE.69.026113
21. Noack, A., Rotta, R.: Multi-level algorithms for modularity clustering. In: Vahrenhold, J. (ed.) SEA 2009. LNCS, vol. 5526, pp. 257–268. Springer, Heidelberg (2009). https://doi.org/10.1007/978-3-642-02011-7_24

22. Pavone, M., Narzisi, G., Nicosia, G.: Clonal selection: an immunological algorithm for global optimization over continuous spaces. J. Global Optim. **53**, 769–808 (2012). https://doi.org/10.1007/s10898-011-9736-8
23. Schuetz, P., Caflisch, A.: Efficient modularity optimization by multistep greedy algorithm and vertex mover refinement. Phys. Rev. E **77**, 046112 (2008). https://doi.org/10.1103/PhysRevE.77.046112
24. Scollo, R.A., Cutello, V., Pavone, M.: Where the local search affects best in an immune algorithm. In: Baldoni, M., Bandini, S. (eds.) AIxIA 2020. LNCS (LNAI), vol. 12414, pp. 99–114. Springer, Cham (2021). https://doi.org/10.1007/978-3-030-77091-4_7
25. Spampinato, A.G., Scollo, R.A., Cavallaro, S., Pavone, M., Cutello, V.: An immunological algorithm for graph modularity optimization. In: Ju, Z., Yang, L., Yang, C., Gegov, A., Zhou, D. (eds.) UKCI 2019. AISC, vol. 1043, pp. 235–247. Springer, Cham (2020). https://doi.org/10.1007/978-3-030-29933-0_20
26. Watts, D.J., Strogatz, S.H.: Collective dynamics of "small-world" networks. Nature **393**(6684), 440–442 (1998). https://doi.org/10.1038/30918

Tchebycheff Fractal Decomposition Algorithm for Bi-objective Optimization Problems

N. Aslimani[1]([⊠]) [ID], E-G. Talbi[1] [ID], and R. Ellaia[2] [ID]

[1] University of Lille, Lille, France
n.aslimani@yahoo.fr, el-ghazali.talbi@univ-lille.fr
[2] LERMA EMI, Mohammed V University in Rabat, Rabat, Morocco
ellaia@emi.ac.ma

Abstract. In most of the existing multi-objective metaheuristics based on decomposition, the reference points and the subspaces are statically defined. In this paper, a new adaptive strategy based on Tchebycheff fractals is proposed. A fractal decomposition of the objective space based on Tchebycheff functions, and adaptive strategies for updating the reference points are performed. The proposed algorithm outperforms popular multi-objective evolutionary algorithms both in terms of the quality of the obtained Pareto fronts (convergence, cardinality, diversity) and the search time.

Keywords: Bi-objective optimization · Fractal decomposition · Tchebycheff scalarization · Adaptive reference points

1 Introduction

Many real world problems require optimizing multiple conflicting objectives. Pareto optimality is generally used in the context of multi-objective optimization problems (MOPs). Indeed, while single objective optimization problems involves a unique optimal solution, MOPs present a set of compromised solutions, known as the Pareto optimal set [12]. These solutions are optimal in the sense that no single objective can be improved without decreasing at least one of the others[1]. We consider a MOP of the form:

$$\min_{X \in S} \mathbf{F}(X) = (f_1(X), \cdots, f_m(X))^{\mathrm{T}} \tag{1}$$

where: $f_k : \mathbb{R}^n \longrightarrow \mathbb{R}$, for $k \in \{1, \cdots, m\}$, denotes the objective functions, S is the decision space: $S = \prod_{i=1}^{n} [l_i, u_i]$, X is the decision vector with n decision variables: $X = (x_1, \cdots, x_n) \in \mathbb{R}^n$,. Let $X = (x_1, x_2, ..., x_n)$ and $Y = (y_1, y_2, ..., y_n)$

---

[1] Without loss of generality, we assume that all objectives are to be minimized.

The ELSAT2020 project is co-financed by the European Union with the European Regional Development Fund, the French state and the Hauts de France Region Council.

be decision vectors (solutions). Solution X is said to dominate solution Y, denoted as $X \preceq Y$, if and only if:

$$\forall i \in [1, m] \ : \ f_i(X) \leq f_i(Y)) \wedge (\exists j \in [1, m] \ : \ f_j(X) < f_j(Y) \qquad (2)$$

A solution X is Pareto optimal if it is not dominated by any other solution which means there is no other solution $Y \in S$ such that $Y \preceq X$. The set of all Pareto optimal solutions is called Pareto set (PS). The corresponding set of Pareto optimal objective vectors is called Pareto front (PF). The main goal in solving MOPs is to find a "good" approximation of the Pareto front in terms of convergence, cardinality and diversity. In the last two decades many metaheuristics (e.g. evolutionary algorithms, swarm intelligence, local search) have been designed for solving MOPs [19,25]. Multi-objective metaheuristics can be classified in three main categories:

- **Dominance-based approaches:** dominance-based approaches[2] use the concept of dominance and Pareto optimality to guide the search process. The main differences between the various proposed approaches arise in the following search components: fitness assignment, diversity management, and elitism [8].
- **Indicator-based-based approaches:** Those approaches optimize a multi-objective performance indicator (e.g. hypervolume in IBEA [27]). The quality of a solution is measured according to its contribution to the performance indicator.
- **Decomposition-based approaches:** the objective space is decomposed into subspaces [23]. Independent or cooperative search is carried out in those subspaces.

In this paper, we propose a new decomposition approach based on adaptive reference points and Tchebycheff fractal decomposition of the objective space for solving bi-objective optimization problems. Compared to existing decomposition-based multi-objective metaheuristics, the main characteristics of the proposed algorithm are the following:

- **Fast and accurate convergence**: in most of multi-objective metaheuristics (e.g. evolutionary algorithms, swarm intelligence), the initial solutions are generated randomly which are generally of poor quality [15]. Based on the anchor points, our approach starts from a couple of Pareto reference points.
- **Parallel independent decomposition of the objective space**: in most of the proposed decompositions strategies of the literature (e.g. MOEA/D [23]), the generated sub-problems are not independent, and correspond to single objective optimization problems using some scalarization strategies (e.g. weighted metrics, Tchebycheff) [12,17]. There is a need of cooperation in solving the sub-problems. In our approach, all the generated sub-problems are independent. A parallel scalable implementation of the approach is straightforward. Moreover, our fractal decomposition procedure can be used as an *intensification* for any approximation found by a metaheuristic.

---

[2] Also named Pareto approaches.

- **No need to manage diversity during search**: diversity management in multi-objective search is handled by complex procedures based on density estimation procedures (e.g. kernel methods such as fitness sharing, nearest-neighbor techniques such as crowding, and histograms) [18]. In our approach, diversity is ensured by the fractal decomposition step. We never generate solutions in the same region of the objective space.
- **Suitability for interactive optimization**: preferences in interactive optimization are generally defined in the objective space [1,11]. Our approach can handle efficiently those preferences to focus the search into a reduced part of the objective space.

The paper is organized as follows. Section 2 presents our positioning in relation to the literature concerning decomposition-based metaheuristics for multi-objective optimization. In Sect. 3 we detail the main components of our algorithm. In Sect. 4, the experimental settings and computational results against competing methods are detailed and analyzed. Finally, the Sect. 5 enumerates some perspectives for the proposed approach.

2   Reference Points in Decomposition-Based Algorithms

A general framework for decomposition-based multi-objective optimization consists in partitioning the objective space into a number of single objective problems ((e.g. MOEA/D [23]). Using some predefined weights and/or reference points, various scalarization strategies may be applied (e.g. weighted sum, Tchebycheff, Penalty-based boundary intersection) [7]. Each sub-problem corresponds to a multi-objective weight vector and reference point, and each sub-problem optimal solution is a Pareto solution. The Tchebycheff function introduces the concept of ideal point or reference point z_i^* as follows [13]:

$$Minimize \max_{i=1,...,k} [\omega_i(f_i(x) - z_i^*)]$$
$$Subject\ to\ \ x \in X \tag{3}$$

where $z^* = (z_1^*, ..., z_k^*)$ is the reference point, and $\omega = (\omega_1, ..., \omega_k)$ is the weight vector.

There have been numerous studies of decomposition approaches to use different types of reference points for providing evolutionary search directions. According to the position of reference point relative to the true PF in the objective space (Fig. 1).

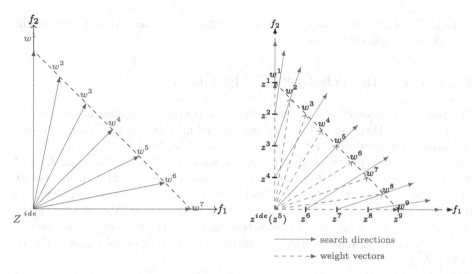

(a) Tchebychev approach with simple reference points

(b) Tchebychev approach with multiple utopian reference points

Fig. 1. Illustration of Tchebycheff decomposition according to the selection of the reference points.

The generation of reference points has a great impact on the distribution of the obtained Pareto solutions. In a *static generation*, the reference points might not be well distributed given that the geometric shape of the PF is unknown apriori. Indeed, uniform reference points do not result in evenly distributed Pareto optimal solutions, especially, on irregular PFs due to the constraints and non-linear objective functions (e.g. disconnected, concave). Hence, Static generation of reference points can lead to the loss of diversity of the obtained PF and waste of computing efforts. In *adaptive strategies*, the reference points are ajusted dynamically according to the shape of the PF. Based on the search knowledge used, adaptive strategies can be classified into:

- **Global knowledge:** is based on knowledge on all subproblems. It consists to approximate the whole PF geometric shape, where periodically interpolate/fit the PF shape and then uniformly samples a set of points on the estimated PF for the generation of subproblems. Various interplolation methods have been investigated to approximated the PF using current non-dominated solutions: piecewise linear interpolation [6], cubic spline interpolation [14] (limited to bi-objective), piecewise hyperplane [5], Gaussian process [21], self-organizing mapping (SOM) network [4].
- **Local knowledge:** instead of using global information (e.g. population or archive), this class of strategy use local information to update the reference points [9,10,20,22,24,26]. For instance, a region with few non-dominated solutions is decomposed into multiple ones by adding new search directions [20].

In [16], the weight vectors in the dense regions are removed and new weight vectors are generate in the sparse regions.

3   The Fractal Tchebycheff Algorithm

The fractal Tchebycheff algorithm (TFA) operates recursively on the two anchor points (or any other pair of Pareto solutions) by introducing a geometric fractal decomposition composed of several levels. The initial level seeks to find a given number of solutions between the two anchor solutions by relying on Tchebycheff scalarization approach that consider the two axes connecting the anchor points to the utopia point. For the next level, the algorithm operates recursively on the set of pairs of solutions from the previous level and consider instead of the utopia point a secondary utopia point resulting from the intersection of the two local vertical and horizontal axes generated from the pair of solutions as shown in Fig. 2.

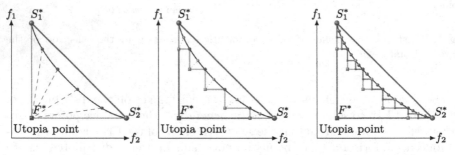

a) First fractal decomposition   b) Second fractal decomposition   c) Third fractal decomposition
   level   level   level

Fig. 2. Illustration of the Tchebycheff fractal algorithm (TFA) with three levels

3.1   Regularisation of the Tchebycheff Fractal

In order to be suitably adapted to the geometry of the PF, we are considering an alternative scheme which aim to regularize the distribution of the solutions generated by the different fractal levels. The idea is to deform the various utopian axes by transforming them into sides of an equilateral triangle whose basis is formed by the segment connecting the two solutions treated.

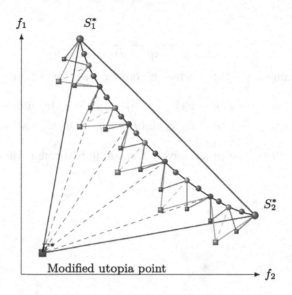

Fig. 3. Illustration of the regularized Tchebycheff fractal (RTFA) algorithm.

To obtain the positions of the different modified utopian points, we use the following complex transformations:

$$\omega = a + (b - a) \cdot r \exp(i\theta) = b + (a - b) \cdot r \exp(-i\theta) \tag{4}$$

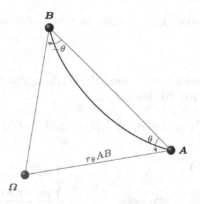

Fig. 4. Illustration of the isocele triangle based on the pair solutions $\{A, B\}$.

where a and b denotes the affix of the pair of solutions, and ω denotes the affix of the modified utopia point associated to the solutions A, B such that $AB\Omega$ is isosceles. Thus the radius r is related to the angle θ through the equation:

$$a + (b - a) \cdot r \exp(i\theta) = b + (a - b) \cdot r \exp(-i\theta) \tag{5}$$

which lead to:

$$r_\theta = \frac{1}{\exp(i\theta) + \exp(-i\theta)} = \frac{1}{2\cos(\theta)} \qquad (6)$$

By transforming Eq. (4) to cartesian coordinates, we obtain:

$$x_\omega = x_A + (x_B - x_A)r_\theta \cos(\theta) - (y_B - y_A)r_\theta \sin(\theta) \qquad (7)$$
$$y_\omega = y_A + (x_B - x_A)r_\theta \sin(\theta) + (y_B - y_A)r_\theta \cos(\theta) \qquad (8)$$

In fact, we will reduce progressively the angle θ through the fractal levels: $\theta_0 > \theta_1 > \theta_2$.

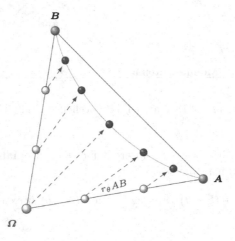

Fig. 5. Generation of solutions using regular reference points on the triangular utopian axis.

We generates in a uniform way along these triangular utopian axis N_t targets points $(T_i)_{1 \leqslant i \leqslant N_t}$, (see Fig. 5) defined as:

$$T_i = \Omega + \alpha_i(A - \Omega) + \beta_i(B - \Omega), \qquad (9)$$

such as:

$$\alpha_i = \begin{cases} \dfrac{\max(1 + n - i, 0)}{2 + n}, & \text{if } N_t = 2 \\ \dfrac{\max(1 + n - i, 0)}{1 + n}, & \text{otherwise} \end{cases} \qquad \beta_i = \begin{cases} \dfrac{\max(i + n - N_t, 0)}{2 + n}, & \text{if } N_t = 2 \\ \dfrac{\max(i + n - N_t, 0)}{1 + n}, & \text{otherwise} \end{cases},$$

with: $n = \lfloor N_t/2 \rfloor$

3.2   Dynamic Management of the Fractal Distribution

Since each fractal has its own subspace, the generation of a static number of solutions for each fractal will induce an irregularity in the final distribution of the solutions on the PF. An alternative consists in generating a dynamic number of solutions in each fractal in order to adapt to changes in scope induced by the local geometry of the PF. To adapt the number of solutions in each fractal, the following steps are carried out:

- Firstly we compute the spacing between each pair of the N_f solutions generated by the fundamental fractal (level $j = 0$). We note S_j the spacing vector at the jth level ($j = 1, 2$).
- We compute the minimal and the maximal spacing in the set:

$$\alpha_j = \min S_j, \quad \beta_j = \max S_j$$

- Set the reference spacing: $dx = \lfloor \frac{\beta_j}{N_f} \rfloor$
- Then we reduce the spacing vector by α: $S_j = S_j / \alpha_j$
- Finally we get the number of solutions for the pth fractal (corresponding to the pth pair of solutions) at the jth fractal level ($j = 1, 2$) using the formula:

$$N_j(p) = \begin{cases} \lfloor \dfrac{S_j(p)}{dx} \rfloor, & \text{if} \quad dx > 0 \\ \lfloor S_j(p) \rfloor - j + 1, & otherwise \end{cases} \tag{10}$$

The algorithm will not consider the fractals without any points.

4   Computational Experiments

In order to evaluate the performance of the proposed TFA algorithm, seven test problems are selected from the literature: Zdt1, Zdt2, Zdt3, Zdt4, Zdt6, Pol and Kur. These problems are covering different type of difficulties and are selected to illustrate the capacity of the algorithm to handle diverse type of Pareto fronts. In fact, all these test problems have different levels of complexity in terms of convexity and continuity. For instance, the test problems Kur and Zdt3 have disconnected Pareto fronts; Zdt4 has too many local optimal Pareto solutions, whereas Zdt6 has non convex Pareto optimal front with low density of solutions near Pareto front.

Three performance measures were adopted in this study: the generational distance (GD) to evaluate the convergence, the Spacing (S) and the Spread (Δ) to evaluate the diversity and cardinality. The convergence metric (GD) measure the extent of convergence to the true Pareto front. It is defined as:

$$GD = \frac{1}{N} \sum_{i=1}^{N} d_i, \tag{11}$$

where N is the number of solutions found and d_i is the Euclidean distance between each solution and its nearest point in the true Pareto front. The Spread Δ, beside measuring the regularity of the obtained solutions, also quantifies the extent of spread in relation to the true Pareto front. The Spread is defined as:

$$\Delta = \frac{d_f + d_l + \sum_{i=1}^{N} |d_i - \bar{d}|}{d_f + d_l + d_i + (N-1)\bar{d}}. \tag{12}$$

where d_i is the Euclidean distance between two consecutive solutions in the obtained set, d_f and d_l denotes the distance between the boundary solutions of the true Pareto front and the extreme solutions in the set of obtained solutions, $\bar{d}$ denotes the average of all distances d_i, $i = 1, 2, \cdots, N-1$ under assumption of N obtained non-dominated solutions. The Spacing metric S indicates how the solutions of an obtained Pareto front are spaced with respect to each other. It is defined as:

$$S = \sqrt{\frac{1}{N} \sum_{i=1}^{N} (d_i - \bar{d})^2} \tag{13}$$

4.1   Numerical Results

The proposed two algorithms FTA and RTFA (including with regularisation of the Tchebycheff fractal) are compared with three popular evolutionary algorithms: a decomposition-based evolutionary algorithm MOEA/D [23], and two Pareto-based evolutionary algorithms: NSGA-II [3] and PESA-II [2][3] The obtained computational results are summarized in Table 1 in term of the mean and the standard deviation (Std) of the used metrics (GD, S, Δ) for 10 independent experiments, the average number of Pareto solutions found (NS), the average number of function evaluations (FEs), the average execution time in seconds (Time).

By analyzing the obtained results of Table 1, it is clear that the RTFA algorithm has the best performance in terms of convergence to the front as well as a better distribution of solutions. Moreover, we observe that for the same number of evaluation functions our approach is able to generate a much larger number of solutions compared to the other approaches. The reason is that evolutionary algorithms imposes in general a restriction on the size of the archive of non-dominated points (here 100 points) since the procedure of archiving is very expensive at each iteration of the algorithm. The capture of the front by the

---

[3] MATLAB implementation obtained for the yarpiz library available at www. yarpiz.com.

RTFA algorithm is continuous and precise, as illustrated by the Fig. 6. In addition, the RTFA algorithm is much faster (less expensive in CPU time) compared to other algorithms.

Table 1. Comparison of MOEA/D, NSGA-II, PESA-II and RFTA for some considered test problems.

Problem	Method	GD		S		Δ		NS	CPU(s)
		Mean	Std	Mean	Std	Mean	Std		
Zdt1	MOEA/D	2,81E−02	2,92E−02	2,42E−02	6,83E−03	9,95E−01	7,98E−02	100	86,67
	NSGA-II	9,17E−02	1,03E−02	2,17E−02	1,56E−03	7,95E−01	2,86E−02	100	105,08
	PESA-II	5,38E−02	5,31E−03	3,73E−01	2,68E−02	8,82E−01	6,01E−02	100	40,97
	RFTA	**2,91E−03**	**5,66E−11**	**7,27E−03**	**1,04E−10**	**7,09E−01**	**1,04E−10**	**37**	**38,72**
Zdt2	MOEA/D	1,32E−01	4,55E−02	2,66E−02	3,87E−03	1,13E+00	8,99E−03	100	61,61
	NSGA-II	1,49E−01	1,74E−02	1,65E−02	2,31E−03	9,10E−01	1,69E−02	100	113,18
	PESA-II	9,04E−02	3,42E−03	5,40E−01	4,61E−02	8,32E−01	2,73E−02	100	32,82
	RFTA	**2,15E−03**	**5,69E−11**	**8,30E−03**	**3,97E−10**	**7,14E−01**	**3,97E−10**	**39**	**40,29**
Zdt3	MOEA/D	2,08E−02	8,00E−03	6,01E−02	1,52E−02	1,19E+00	3,43E−02	100	82,50
	NSGA-II	6,27E−02	4,74E−03	3,75E−02	1,23E−02	8,25E−01	2,12E−02	100	112,91
	PESA-II	4,45E−02	3,42E−03	3,54E−01	3,22E−02	8,48E−01	1,04E−01	100	33,98
	RFTA	**3,20E−03**	**1,09E−03**	**6,53E−02**	**2,73E−03**	**9,10E−01**	**2,73E−03**	**42**	**53,90**
Zdt4	MOEA/D	8,28E−01	5,75E−01	1,70E−01	1,45E−01	1,09E+00	5,77E−02	100	62,65
	NSGA-II	4,46E−01	1,33E−01	3,03E−01	1,79E−01	8,85E−01	9,56E−02	100	129,75
	PESA-II	1,12E+01	5,16E−01	2,72E+01	4,86E+00	1,11E+00	5,16E−02	100	18,00
	RFTA	**2,90E−03**	**1,74E−04**	**8,31E−03**	**2,67E−03**	**7,12E−01**	**2,67E−03**	**37**	**54,54**
Pol	MOEA/D	5,40E−01	1,08E−01	1,35E+00	1,01E+00	1,25E+00	1,54E−01	100	91,11
	NSGA-II	2,48E+00	5,84E−02	1,75E+00	2,09E−03	9,72E−01	3,55E−03	100	123,86
	PESA-II	1,52E+01	6,02E+00	1,01E+01	1,16E+00	9,77E−01	9,85E−02	100	45,64
	RFTA	**1,25E−02**	**8,17E−04**	**4,28E+00**	**5,59E−02**	**9,00E−01**	**5,59E−02**	**15**	**26,19**
Zdt6	MOEA/D	4,22E−01	1,18E−01	4,63E−02	2,02E−02	1,09E+00	5,98E−02	100	57,86
	NSGA-II	3,08E−01	2,38E−02	1,57E−01	4,15E−02	8,73E−01	2,93E−02	100	121,89
	PESA-II	4,42E−01	4,69E−03	2,00E+00	6,31E−01	1,03E+00	4,97E−02	100	29,170
	RFTA	**1,38E−02**	**2,32E−03**	**1,06E−01**	**2,63E−02**	**9,25E−01**	**2,63E−02**	**54**	**57,94**
Kur	MOEA/D	5,51E−03	5,51E−03	6,76E−01	4,35E−01	1,42E+00	1,80E−01	100	80,83
	NSGA-II	5,31E−04	2,64E−05	1,22E−01	4,15E−03	4,10E−01	3,10E−02	100	117,86
	PESA-II	1,78E−01	1,28E−02	4,15E+00	4,86E−01	9,04E−01	5,75E−02	100	33,76
	RFTA	**1,25E−03**	**1,53E−05**	**4,06E−01**	**2,54E−02**	**8,45E−01**	**2,54E−02**	**37**	**53,59**

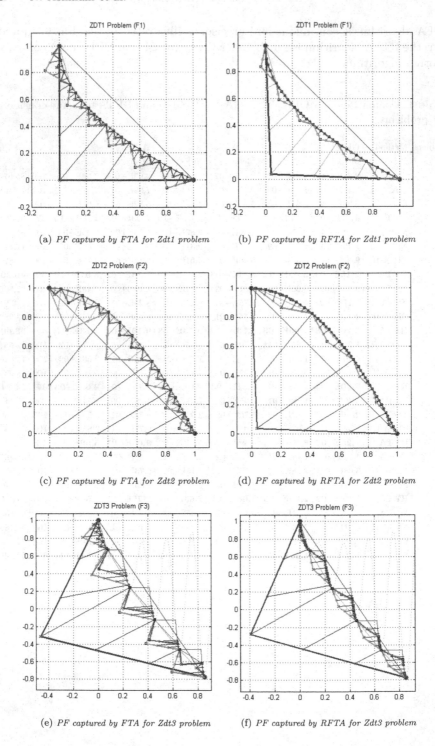

(a) *PF captured by FTA for Zdt1 problem*   (b) *PF captured by RFTA for Zdt1 problem*

(c) *PF captured by FTA for Zdt2 problem*   (d) *PF captured by RFTA for Zdt2 problem*

(e) *PF captured by FTA for Zdt3 problem*   (f) *PF captured by RFTA for Zdt3 problem*

Fig. 6. PF captured by RFTA for Zdt1, Zdt2, Zdt3, Kur and Msc problems

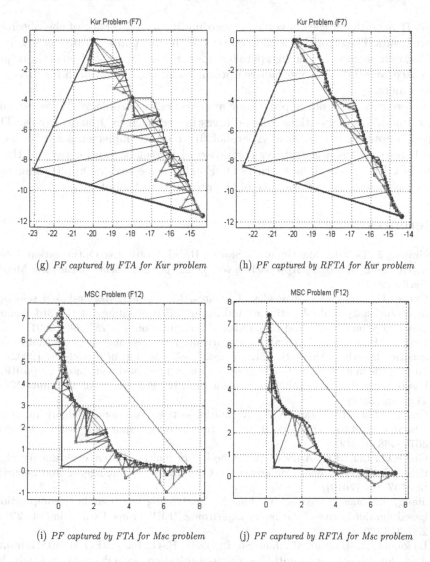

(g) *PF captured by FTA for Kur problem*   (h) *PF captured by RFTA for Kur problem*

(i) *PF captured by FTA for Msc problem*   (j) *PF captured by RFTA for Msc problem*

Fig. 6. (*continued*)

5   Conclusion and Perspectives

In this paper, we have successfully developed the RFTA algorithm which is based on the Tchebycheff fractal decomposition of the objective space, and adaptive multiple reference points. The proposed algorithm was tested on various benchmark problems with different features and complexity levels. The results obtained show that the approach is efficient in converging to the true Pareto fronts and finding a diverse set of solutions along the Pareto front. Our approach outperforms some popular evolutionary algorithms such as MOEA/D, NSGA-II, and

PESA-II in terms of the convergence, cardinality and diversity of the obtained Pareto fronts. The algorithm is characterized by its fast and accurate convergence, very low complexity of archiving, parallel independent decomposition of the objective space, suitability for interactive optimization, and the no need for diversity management.

We are extending our approach for many objective problems. Moreover, different fractal decompositions will be investigated such as spherical ones. The propose approach is intrinsically parallel. Its parallel implementation on large scale heterogeneous architectures composed of dozens of multi-cores and thousands of GPU cores will be carried out. Finally, some real-life applications will be used to assess the performance of the proposed methodology.

References

1. Branke, J., Deb, K., Miettinen, K., Slowiń, R.: Multiobjective Optimization: Interactive and Evolutionary Approaches, vol. 5252. Springer Science & Business Media, Berlin (2008)
2. Corne, D., Jerram, N., Knowles, J., Oates, M.: Pesa-II: Region-based selection in evolutionary multiobjective optimization. In: Proceedings of the 3rd Annual Conference on Genetic and Evolutionary Computation, pp. 283–290 (2001)
3. Deb, K., Pratap, A., Agarwal, S., Meyarivan, T.: A fast and elitist multiobjective genetic algorithm: NSGA-II. IEEE Trans. Evol. Comput. $6(2)$, 182–197 (2002)
4. Gu, F., Cheung, Y.M.: Self-organizing map-based weight design for decomposition-based many-objective evolutionary algorithm. IEEE Trans. Evol. Comput. $22(2)$, 211–225 (2017)
5. Gu, F., Liu, H.L., Tan, K.C.: A multiobjective evolutionary algorithm using dynamic weight design method. Int. J. Innovative Comput. Inf. Control $8(5(B))$, 3677–3688 (2012)
6. Gu, F.Q., Liu, H.L.: A novel weight design in multi-objective evolutionary algorithm. In: International Conference on Computational Intelligence and Security, pp. 137–141 (2010)
7. Jiang, S., Yang, S., Wang, Y., Liu, X.: Scalarizing functions in decomposition-based multiobjective evolutionary algorithms. IEEE Trans. Evol. Comput. $22(2)$, 296–313 (2017)
8. Liefooghe, A., Basseur, M., Jourdan, L., Talbi, E.-G.: ParadisEO-MOEO: a framework for evolutionary multi-objective optimization. In: Obayashi, S., Deb, K., Poloni, C., Hiroyasu, T., Murata, T. (eds.) EMO 2007. LNCS, vol. 4403, pp. 386–400. Springer, Heidelberg (2007). https://doi.org/10.1007/978-3-540-70928-2_31
9. Liu, H.L., Chen, L., Zhang, Q., Deb, K.: An evolutionary many-objective optimisation algorithm with adaptive region decomposition. In: IEEE Congress on Evolutionary Computation (CEC), pp. 4763–4769 (2016)
10. Liu, H.L., Chen, L., Zhang, Q., Deb, K.: Adaptively allocating search effort in challenging many-objective optimization problems. IEEE Trans. Evol. Comput. $22(3)$, 433–448 (2017)
11. Luque, M., Ruiz, F., Miettinen, K.: Global formulation for interactive multiobjective optimization. OR Spectr. $33(1)$, 27–48 (2011). https://doi.org/10.1007/s00291-008-0154-3
12. Miettinen, K.: Nonlinear Multiobjective Optimization, vol. 12. Springer Science & Business Media, Berlin (2012)

13. Miettinen, K., Mäkelä, M.: On scalarizing functions in multiobjective optimization. OR Spectr. **24**(2), 193–213 (2002). https://doi.org/10.1007/s00291-001-0092-9
14. Pilát, M., Neruda, R.: General tuning of weights in MOEA/D. In: IEEE Congress on Evolutionary Computation (CEC), pp. 965–972 (2016)
15. Poles, S., Fu, Y., Rigoni, E.: The effect of initial population sampling on the convergence of multi-objective genetic algorithms. In: Barichard, V., Ehrgott, M., Gandibleux, X., T'Kindt, V. (eds.) Multiobjective Programming and Goal Programming, pp. 123–133. Springer, Berlin (2009). https://doi.org/10.1007/978-3-540-85646-7_12
16. Qi, Y., Ma, X., Liu, F., Jiao, L., Sun, J., Wu, J.: MOEA/D with adaptive weight adjustment. Evol. Comput. **22**(2), 231–264 (2014)
17. Santiago, A., et al.: A survey of decomposition methods for multi-objective optimization. In: Castillo, O., Melin, P., Pedrycz, W., Kacprzyk, J. (eds.) Recent Advances on Hybrid Approaches for Designing Intelligent Systems. SCI, vol. 547, pp. 453–465. Springer, Cham (2014). https://doi.org/10.1007/978-3-319-05170-3_31
18. Silverman, B.W.: Density Estimation for Statistics and Data Analysis, vol. 26. CRC Press, Boca Raton (2018)
19. Talbi, E.G.: Metaheuristics: from Design to Implementation, vol. 74. John Wiley & Sons, Hoboken (2009)
20. Wang, M., Wang, Y., Wang, X.: A space division multiobjective evolutionary algorithm based on adaptive multiple fitness functions. Int. J. Pattern Recogn. Artif. Intell. **30**(03), 1659005 (2016)
21. Wu, M., Kwong, S., Jia, Y., Li, K., Zhang, Q.: Adaptive weights generation for decomposition-based multi-objective optimization using gaussian process regression. In: Proceedings of the Genetic and Evolutionary Computation Conference, pp. 641–648 (2017)
22. Xiang, Y., Zhou, Y., Li, M., Chen, Z.: A vector angle-based evolutionary algorithm for unconstrained many-objective optimization. IEEE Trans. Evol. Comput. **21**(1), 131–152 (2016)
23. Zhang, Q., Li, H.: MOEA/D: a multiobjective evolutionary algorithm based on decomposition. IEEE Trans. Evol. Comput. **11**(6), 712–731 (2007)
24. Zhao, H., Zhang, C., Zhang, B., Duan, P., Yang, Y.: Decomposition-based subproblem optimal solution updating direction-guided evolutionary many-objective algorithm. Inf. Sci. **448**, 91–111 (2018)
25. Zhou, A., Qu, B.Y., Li, H., Zhao, S.Z., Suganthan, P., Zhang, Q.: Multiobjective evolutionary algorithms: a survey of the state of the art. Swarm Evol. Comput. **1**(1), 32–49 (2011)
26. Zhou, C., Dai, G., Zhang, C., Li, X., Ma, K.: Entropy based evolutionary algorithm with adaptive reference points for many-objective optimization problems. Inf. Sci. **465**, 232–247 (2018)
27. Zitzler, E., Künzli, S.: Indicator-based selection in multiobjective search. In: Yao, X. (ed.) PPSN 2004. LNCS, vol. 3242, pp. 832–842. Springer, Heidelberg (2004). https://doi.org/10.1007/978-3-540-30217-9_84

Local Search for Integrated Predictive Maintenance and Scheduling in Flow-Shop

Andrea Ecoretti, Sara Ceschia[ID], and Andrea Schaerf[✉][ID]

DPIA, University of Udine, via delle Scienze 206, 33100 Udine, Italy
ecoretti.andrea@spes.uniud.it, {sara.ceschia,andrea.schaerf}@uniud.it

Abstract. We address the Permutation Flow-Shop Scheduling Problem with Predictive Maintenance presented by Varnier and Zerhouni (2012), that consists in finding the integrated schedule for production and maintenance tasks such that the total production time and the advance of maintenance services are minimized. Predictive maintenance services are scheduled based on a prognostics system that is able to provide the remaining useful life of a machine. To solve this problem, we propose a local search method with neighborhoods specifically tailored for maintenance interventions. Computational experiments performed on generated benchmarks demonstrate the effectiveness and scalability of our method with respect to an exact technique based on the mathematical model proposed by Varnier and Zerhouni (2012).

Keywords: Predictive maintenance · Local search · Flow-shop · Scheduling

1 Introduction

In manufacturing, production scheduling and maintenance planning are strongly correlated. The former consists in sequencing and assigning jobs to machines in order to maximize productivity, while the latter concerns the scheduling of maintenance services necessary to retain the machines in operating conditions. These two activities are naturally in trade-off because a maintenance service makes the machine unavailable for production, however an excessive delay may cause a failure, requiring then longer and expensive maintenance interventions to repair or replace the broken machine. As a consequence, an integrated planning of these two activities could improve both the reliability and the productivity of a manufacturing system [16].

We can generally distinguish three policies for maintenance scheduling [20]: *corrective* maintenance, *preventive* maintenance and *predictive* maintenance. The *corrective* maintenance (*run-to-break*) allows a machine to run until the failure of the system, so that the maintenance is done after the breakdown. *Preventive* maintenance consists in interventions planned to keep machines in a specified condition. It includes periodic inspections, critical item replacement, calibration, and lubrication. Predictive maintenance requires a monitoring system (sensors, vibration analysis, infrared scanning, ...) able to estimate the time before the failure of the machine such that maintenance is performed only when it is needed. Both *preventive* and *predictive* maintenance schedule interventions in advance to prevent unexpected system breakdowns, however in the first case, decisions are made using reliability theory and historical data about failures of similar machines, while for the *predictive* policy, they are dependent on the current health state of the machine detected by monitoring devices.

Given its wide range of real applications in industries, the joint optimization of scheduling production and maintenance has attracted the interest also of the research community in the last decades. Ma *et al.* [17] provide a review of the most recent results on scheduling problems with deterministic availability constraints motivated by preventive maintenance. However, for this class of problems maintenance periods are fixed in advance, such that they are not part of the decision process.

Lee *et al.* [14] studied the problem of processing a set of jobs on parallel machines where each machine requires maintenance once during the planning period; the final objective is to minimize the total weighted completion time of jobs. For this problem, the authors give some theoretic results about optimality and complexity, and then propose a Column Generation approach able to solve instances up to 40 jobs and 8 machines. The parallel machine scheduling problem has been tackled with a multi-objective perspective by Berrichi *et al.* [2], that propose a Multi-Objective Ant Colony Optimization approach. In this case, the aim is to simultaneously minimize the makespan and the unavailability of machines, which is estimated through reliability models using a time increasing function dependent on the failure rate.

More recently, many papers have addressed the integrated production and maintenance scheduling problem for the single machine case [5,9,13,19].

We tackle the Permutation Flow-Shop Scheduling Problem with Predictive Maintenance (PFSP-PM) defined in the work by Varnier and Zerhouni [24] where the production system is modeled as a permutation flow-shop and a predictive maintenance policy is adopted. The authors assume that machines are continuously monitored and a data-driven prognostic system is able to evaluate the Remaining Useful Life (RUL) for each machine, such that a maintenance service must be planned within the RUL [18]. In addition, each machine can run in two modes: the *nominal* mode (the default one) and the *degraded* mode with longer processing times for jobs (the machine runs slower), but RUL greater than the nominal one. The aim of the PFSP-PM is to define the joint scheduling for production and maintenance, such that the makespan and the advance of maintenance services are minimized. The authors developed a mixed integer linear programming model which was tested on random generated instances. Subsequently, the problem has been extended with probabilistic values for RULs and degradation levels that are modeled through fuzzy logic [10]. The authors present a heuristic solution method based on Variable Neighborhood Search that employs three neighborhood structures: *swap jobs, insert job, maintenance move* and report comparative results between different variants of the method on some Taillard benchmarks [22], conveniently completed by the maintenance data. The same problem has been solved also using a Genetic Algorithm [12].

Ladj *et al.* [11] studied the PFSP-PM with various RULs and degradation values for each machine depending on the job being processed. For the solution of this problem, they implemented a Genetic Algorithm, which was tested on the same dataset used in the work by Ladj *et al.* [10] with instances up to 200×20 ($n \times m$) size.

Finally, other examples of integration of maintenance and production scheduling in multi-product process plant are presented by Biondi *et al.* [3] and Aguirre *et al.* [1].

The paper is organized as follows. Section 2 introduces the problem. Section 3 provides the mathematical model. The proposed search method is described in Sect. 4. Section 5 presents the experimental analysis. Finally, conclusions and future work are outlined in Sect. 6.

2   Problem Definition

The problem consists in n jobs that must be scheduled on m machines, as each one is split into exactly m tasks that need to be performed in the same order, i.e. the i−th task must be executed on the i−th machine. Each machine can perform a single task at a time, and no preemption is allowed.

It is assumed that there is a prognostic system that estimates the RUL of each machine $i \in \mathcal{M} = \{1, \ldots, m\}$ for the nominal running mode (RUL_i^n) and for the degraded running mode (RUL_i^d), where a machine reduces its processing power so as to avoid early failures and increase its RUL. For each job $j \in \mathcal{J} = \{1, \ldots, n\}$, p_{ij} denotes its processing time when machine i runs in nominal mode, while p'_{ij} when it is set in degraded mode.

Throughout the planning period, each machine i can be exposed to at most one maintenance service of duration d_i that must be scheduled before its RUL_i. At the beginning of the planning period, a machine is set in a running mode (nominal or degraded) and it can change its status only if a maintenance is performed, after which the machine will always run in nominal mode.

The problem consists in defining a feasible schedule for jobs and maintenance services that minimizes the weighted sum of two objectives: (i) the makespan; (ii) the earliness of maintenance services. The two weights, called α and β respectively, can be set independently for each instance.

The Permutation Flowshop Scheduling Problem has been proved to be NP-hard for more than two machines [15], thus also the PFSP-PM is NP-hard.

Figure 1 shows a file containing a small exemplary instance written in OPL, the IBM ILOG Optimization Programming Language [7]. The matrices P_n and P_d contain the processing times of jobs on machines running in nominal and degraded mode, respectively. The vector d stores the duration of the maintenance service on each of the two machines; finally, the two RUL vectors represent the

```
Machines = 2;
Jobs = 5;
alpha = 10;
beta = 1;
P_n = [[3 4 4 3 4]
       [3 2 3 2 3]];
P_d = [[4 5 5 4 5]
       [4 3 4 3 4]];
d = [2 3];
RUL_n = [8 5];
RUL_d = [14 9];
```

Fig. 1. Example of input file in OPL format.

remaining useful life of each machine in nominal and degraded mode. A visual representation of the corresponding optimal solution is reported in Fig. 2 as a Gantt chart. The figure represents the actual processing times of tasks and the machine modes selected: m_1 runs in nominal mode whereas m_2 in degraded. The jobs sequence is [3 2 1 5 4]. The maintenance tasks start exactly at the respective RULs so that there is no penalty for earliness. As a consequence, for this solution the value of the objective function is 220, corresponding to the makespan (22) multiplied by its weight $\alpha = 10$.

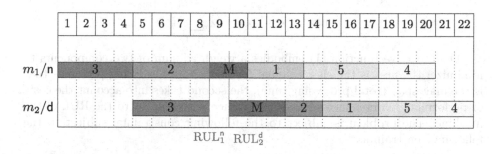

Fig. 2. An optimal solution of the input file in Fig. 1 represented as a Gantt chart.

3   Mathematical Model

As proposed by Varnier and Zerhouni [24], the problem can be formulated as a multi-objective mixed integer linear programming model. We present here our model, which is based on the one by Varnier and Zerhouni [24], correcting some minor inaccuracies and revising the objective function.

The formulation relies on three main decision variables: (i) the completion time c_{ij} of job j on machine i, i.e. the completion time of the i-th task of job

j, (ii) the completion time c_i^m of the maintenance service on machine i, (iii) a binary variable Z_i taking value 1 if machine i runs in degraded mode at the beginning of the planning period. All the decision variables used in the model are reported on Table 1.

Table 1. Decision variables

Decision variables	
c_{ij}	Completion time of job $j \in \mathcal{J}$ on machine $i \in \mathcal{M}$
c_i^m	Completion time of the maintenance service on machine $i \in \mathcal{M}$
Z_i	1 If machine $i \in \mathcal{M}$ runs in degraded mode at the beginning of the planning period, 0 if it is set in nominal mode
P_{ij}	Actual processing time of job j on machine i, depending on the running mode of the machine
X_{jk}	1 if job $j \in \mathcal{J}$ precedes job $k \in \mathcal{J}$ in the schedule, 0 otherwise
Y_{ij}	1 if the maintenance service on machine $i \in \mathcal{M}$ is planned before job $j \in \mathcal{J}$, 0 otherwise
C_{MAX}	Makespan
e_i	Earliness of completion time of maintenance service on machine $i \in \mathcal{M}$

$$\min z = \alpha\, C_{\mathrm{MAX}} + \beta \sum_{i \in \mathcal{M}} e_i \tag{1}$$

The objective function (1) embodies a weighted sum of two components to minimize: the first is the classical goal for flow shop scheduling problems that is the makespan (total processing time); the second takes into account the costs of performing maintenance services in advance with respect to the RUL of the machine. The problem is subject to the following constraints: subject to the following constraints:

$$P_{ij} = p_{ij}(1 - Z_i) + \left[p_{ij}Y_{ij} + p'_{ij}(1 - Y_{ij})\right]Z_i \quad \forall i \in \mathcal{M}, j \in \mathcal{J} \tag{2}$$

$$c_{i+1,j} - P_{i+1,j} \geq c_{ij} \quad \forall i \in 1,\dots,m-1, \forall j \in \mathcal{J} \tag{3}$$

$$c_{1j} \geq P_{1j} \quad \forall j \in \mathcal{J} \tag{4}$$

$$c_{ij} \geq c_{ik} + P_{ij} - M X_{jk} \quad \forall j \in 1,\dots,n-1, \forall i \in 1,\dots,m$$
$$\forall k \in j+1,\dots,n$$

$$c_{ik} \geq c_{ij} + P_{ik} - M(1 - X_{jk}) \quad \forall j \in 1,\dots,n-1, \forall i \in 1,\dots,m$$
$$\forall k \in j+1,\dots,n \tag{5}$$

$$c_{ij} \leq c_i^m - d_i + M Y_{ij} \quad \forall i \in \mathcal{M}, \forall j \in \mathcal{J}$$

$$c_i^m \leq c_{ij} - P_{ij} + M(1 - Y_{ij}) \quad \forall i \in \mathcal{M}, \forall j \in \mathcal{J}$$

$$c_i^m \geq d_i \quad \forall i \in \mathcal{M} \tag{6}$$

$$c_i^m - d_i \leq \mathrm{RUL}_i^n(1 - Z_i) + \mathrm{RUL}_i^d Z_i \qquad \forall i \in \mathcal{M} \qquad (7)$$

$$c_{mj} \leq C_{\mathrm{MAX}} \qquad \forall j \in \mathcal{J} \qquad (8)$$

$$e_i \geq \mathrm{RUL}_i^n(1 - Z_i) + \mathrm{RUL}_i^d Z_i - (c_i^m - d_i) \qquad \forall i \in \mathcal{M} \qquad (9)$$

$$c_{ij} \geq 0 \qquad \forall i \in \mathcal{M}, \forall j \in \mathcal{J}$$

$$c_i^m \geq 0 \qquad \forall i \in \mathcal{M}$$

$$Z_i \in \{0, 1\} \qquad \forall i \in \mathcal{M}$$

$$X_{jk} \in \{0, 1\} \qquad \forall j \in \mathcal{J}, \forall k \in \mathcal{J}$$

$$Y_{ij} \in \{0, 1\} \qquad \forall i \in \mathcal{M}, \forall k \in \mathcal{J}$$

$$Z_{ij} \in \{0, 1\} \qquad \forall i \in \mathcal{M}, \forall k \in \mathcal{J}$$

$$e_1 \geq 0 \qquad \forall i \in \mathcal{M}$$

$$C_{\mathrm{MAX}} \geq 0$$

Constraints (2) set the actual processing time for jobs considering three possible situations: (i) the machine runs in nominal mode from the beginning of the planning period, so that $P_{ij} = p_{ij}$; (ii) the machine runs in degraded mode from the beginning of the planning period and job j is scheduled before the maintenance service on machine i, so that $P_{ij} = p'_{ij}$; (ii) the machine runs in degraded mode at the beginning of the planning period but job j is scheduled after the maintenance service on machine i, so that $P_{ij} = p_{ij}$. Equations (2) can be linearized introducing the binary variables $Z_{ij} = Y_{ij} Z_i$ as follows:

$$P_{ij} = p_{ij} + (p'_{ij} - p_{ij})Z_i + (p_{ij} - p'_{ij})Z_{ij} \qquad \forall i \in \mathcal{M}, j \in \mathcal{J}$$

$$Z_{ij} \leq Y_{ij} \qquad \forall i \in \mathcal{M}, j \in \mathcal{J}$$

$$Z_{ij} \leq Z_i \qquad \forall i \in \mathcal{M}, j \in \mathcal{J}$$

$$1 - Y_{ij} - Z_i + Z_{ij} \geq 0 \qquad \forall i \in \mathcal{M}, j \in \mathcal{J} \qquad (10)$$

Constraints (3–4) ensure that each task of a job is performed in the correct order that represents the flow on the production line. Constraints (5) guarantee that only one task can be assigned to a machine simultaneously (M is a big number). Similarly, Constraints (6) indicate that a maintenance service cannot be performed when the machine is already occupied by a production task. Constraints (7) impose that the maintenance service starts before the RUL of the machine. Constraints (8) define the makespan as the total production time. Constraints (9) calculate the advanced time of a maintenance service with respect to the machine RUL. The latter is different from the model of Varnier and Zerhouni [24] where the earliness is computed respect to the completion time of the maintenance service. Indeed, we believe that considering the RUL as the time in which the maintenance must be started, rather than ended, is more intuitive and independent of the duration of the maintenance task.

4   Solution Method

The implementation with IBM ILOG CPLEX of the MIP model described in Sect. 3 is able to optimally resolve only small instances (see comparative results on Sect. 5.3), thus for the solution of the PFSP-PM we resort to a metaheuristic approach, in particular we developed a local search method guided by a Simulated Annealing (SA) procedure.

4.1   Local Search

In order to implement our local search method, we need to define the search space, the initial solution, the cost function, and the neighborhood relations.

A solution is coded by three main data structures:

- an integer-valued vector $\mathbf{S} = [j_1, \ldots, j_n]$ $(j_k \in \mathcal{J})$ that represents the jobs' permutation, such that job j_k is the k-th job to be executed on each machine;
- an integer-valued vector $\mathbf{P} = [p_1, \ldots, p_m]$ $(p_k \in \{1, \ldots, n+1\})$ that stores for each machine the position of the maintenance task in the machine sequence;
- a boolean-valued vector $\mathbf{O} = [o_1, \ldots, o_m]$ such that if $o_k = 1$ machine k runs in degraded mode from the beginning of the planning period, otherwise it runs in nominal mode.

For example, for the instance shown in Fig. 1, the optimal solution of Fig. 2 is encoded as $\mathbf{S} = [3\ 2\ 1\ 5\ 4]$, $\mathbf{P} = [3\ 2]$ and $\mathbf{O} = [0\ 1]$.

Notice that there are $n + 1$ potential different positions of the maintenance task, as it can also be done before the first job or after the last one.

The actual schedule is deterministically obtained from the above mentioned data structures in the following way: first, we schedule the tasks in chronological order one machine at a time at the earliest time and we compute the makespan. At this point, in order to minimize the earliness of the maintenance interventions, all tasks are processed in reverse order and rescheduled at the latest time, keeping the makespan fixed and respecting the RUL.

The initial solution is generated totally at random: first a random permutation of jobs is created, then the position of the maintenance service is randomly selected between 1 and $n + 1$, finally the run mode is drawn. This means that Constraints (7) imposing that the maintenance intervention must be executed before the RUL of the machine can be violated. The cost function is thus the weighted sum of the two objectives (makespan and earliness) and the violation of the Constraints (7) multiplied by a suitable hard weight, such that a single violation of a hard constraint is never preferred to the objectives.

The typical neighborhood relations in permutation flow shop scheduling are based on *swap* and *insert* moves:

SwapJobs (SJ) Given two positions k_1 and k_2 (with $k_1 \neq k_2$) in the job sequence
S, the move $\text{SJ}\langle k_1, k_2 \rangle$ swaps the corresponding jobs j_{k_1} and j_{k_2}.

InsertJob (IJ) The move $\text{IJ}\langle k_1, k_2 \rangle$ (with $k_1 \neq k_2$) removes job j_{k_1} from its
current position k_1 in **S** and relocates it at position k_2, shifting the jobs
inbetween accordingly (either forward or backward).

In addition, we devised two new neighborhood relations tailored for managing
maintenance tasks:

FlipMode (FM) The move $\text{FM}\langle o_k \rangle$ changes the current run mode of machine k
from degraded to nominal (if $o_k = 1$) or *vice versa* (if $o_k = 0$).

ShiftMaintenance (SM) The move $\text{SM}\langle m, k \rangle$ shifts the maintenance task of
machine m from its current position p_m to position k. This means that the
maintenance task is executed on machine m before job j_k in the sequence **S**.

The local search method uses a neighborhood composed of the union of these
four atomic neighborhoods: SJ, IJ, FM, and SM.

4.2 Simulated Annealing

We use Simulated Annealing [8] to guide the local search. For a review of the
different variants of SA, we refer the interested reader to the work by Franzin
and Stützle [6].

The SA procedure starts from the initial solution built as described in
Sect. 4.1 and then, at each iteration, randomly selects a move in the composite neighborhood $\text{SJ} \cup \text{IJ} \cup \text{FM} \cup \text{SM}$.

The selection of the move works as follows. First we select the atomic neighborhood and then we draw the specific move inside it. The neighborhood selection is based on fixed probabilities: We add three real-valued parameters called
σ_{IJ}, σ_{FM} and σ_{SM}, such that at each step neighborhoods IJ, FM, and SM are
selected with probability σ_{IJ}, σ_{FM} and σ_{SM}. Consequently, the SJ probability is
equal to $1 - \sigma_{\text{IJ}} - \sigma_{\text{FM}} - \sigma_{\text{SM}}$. Within the single neighborhood, the specific move
is selected uniformly.

The move is always accepted if the difference of cost Δ induced is null or
negative (i.e. the value of the objective function improves o remains equal),
whereas if $\Delta > 0$ it is accepted with probability $\exp^{-\Delta/T}$, where T is a control
parameter called *temperature*. Indeed, SA starts with an initial temperature T_0,
which is decreased according to the standard geometric cooling scheme ($T_i = c \cdot T_{i-1}$) after a fixed number of samples n_s. To the basic SA procedure, we add
the *cut-off* mechanism such that the temperature decreases also if a maximum
number of moves has been accepted. This is expressed as a fraction ρ of the
number of iterations n_s (with $0 \leq \rho \leq 1$). In order to guarantee the same
running time to all configurations of SA, we use the total number of iterations $\mathcal{I}$
as stop criterion. To keep $\mathcal{I}$ fixed, we recompute n_s from $n_s = \mathcal{I} \Big/ \left(\frac{\log(T_f/T_0)}{\log c} \right)$,
where T_f is the final temperature.

5   Experimental Analysis

The code implementing the SA algorithm is written in C++ and compiled using GNU g++ v. 9.3.0. We implemented the mathematical model in IBM ILOG CPLEX (v. 12.10) [7], using its C++ interface CONCERT. All experiments ran on an Ubuntu Linux 20.04.2 LTS machine with 64 GB of RAM and 32 AMD Ryzen Threadripper PRO 3975WX processors (3.50 GHz), hyper-threaded to 64 virtual cores. A single virtual core has been dedicated to each SA experiment, whereas no limits have been imposed to the use of multiple CPUs or memory for CPLEX.

5.1   Instance Generator

No instances have been made available for this problem. Therefore, partly following [24] (see Sect. IV.B), we implemented a parametrized instance generator able to create instances of any size, providing as input n and m. The processing time p_{ij} is randomly selected from a uniform distribution $U[20; 50]$; if the machine runs in degraded mode, its processing time is computed as $p'_{ij} = p_{ij} \cdot \gamma$, with $\gamma \in U[1; 1.5]$. The duration of a maintenance service is generated from $U[10; 30]$.

Finally the RUL_i^n for a machine running in nominal mode is drawn from a distribution $U[\vartheta, \vartheta + \sum_{j \in \mathcal{J}} p_{ij}]$, where $\vartheta = 0$ for the first machine, while $\vartheta = \sum_{k=1}^{i} \sum_{j \in \mathcal{J}} p_{i-1,j}/n$ for $i = \{2, \ldots, m\}$. The corresponding RUL_i^d is set as $\text{RUL}_i^n \cdot \lambda$ with $\lambda \in U[1.5; 2]$.

The generation of RUL is different from Varnier and Zerhouni [24], because their method that uses the distribution $U[0, \sum p_{i,j}]$ does not take into account the fact that the last machines are inevitably idle in the initial part of the schedule. As a consequence, the last machines end up starting always with the maintenance task. In order to create more realistic situations, we propose a different procedure, where the RUL of a machine is extracted from the distribution described above. Thereby the starting time is the sum of the average time spent on all previous machines, that is when we expect that the machine starts to work.

We decided to set the weights α and β for all instances to 10 and 1 respectively, in order to give priority to the minimization of the makespan upon the earliness of the maintenance.

We generated three datasets of instances depending on the size $n \times m$, as summarized in Table 2. The sizes of the Small instances are the ones used by Varnier and Zerhouni [24], and the Medium ones are their extension. The Large instances follow the size of the classical benchmarks of Taillard for flow-shop scheduling [22].

In addition to these instances, we generated a *training* dataset composed by 60 instances of *large* type to be used for the tuning of the SA parameters and the neighborhood rates.

The name of each instance reported on Tables 4 and 5 follows this pattern: fspm-D-m-n, where D is the dataset, m is the number of machines and n the

Table 2. Datasets.

Name	#instances	n	m
Small	28	$\{4, 6, 8, 10\}$	3–9
Medium	35	$\{15, 20, 25, 30, 35\}$	3–9
Large	12	$\{20, 50, 100, 200\}$	$\{5, 10, 20\}$

number of jobs. All validation instances and their best solutions are available for inspection and future comparisons at https://bitbucket.org/satt/pfsp-pm-data.

5.2 Parameter Tuning

The parameters of SA and the move probabilities σ_* have been tuned using the tool JSON2RUN [23] that implements the F-Race procedure [4]. The cooling rate c has been fixed to 0.99 given that results are not sensitive to its variations. The tuning process has been performed on the training dataset in two stages: a first stage dedicated to SA parameters and a second stage for move probabilities. In the first stage, all move probabilities have been set according to preliminary experiments.

Table 3. Parameter settings.

Name	Description	Value	Range
T_0	Initial temperature	450	100–500
T_f	Final temperature	1.06	0.5–1.5
ρ	Accepted moves ratio	0.19	0.1–0.2
σ_{IJ}	Probability of a IJ move	0.42	0.0–0.7
σ_{FM}	Probability of a FM move	0.17	0.0–0.2
σ_{SM}	Probability of a SM move	0.09	0.0–0.1

The resulting best configuration is shown in Table 3, which reports also the initial ranges. The maximum number of iterations $\mathcal{I}$ was fixed to 10^7.

5.3 Experimental Results

For the validation experiments, we impose a time limit of one or two hours for the MIP solver and $\mathcal{I} = 10^8$ for SA which corresponds on average to about 300 and 600 s for Medium and Large instances, respectively.

Experiments on the Small dataset have not been reported since both the MIP solver and SA are able to find consistently the optimal solution for all instances in short computational times (from milliseconds to 240 s for MIP and an average time of about 18 s for SA).

Table 4. Comparative results on the Medium dataset.

Instance	SA					MIP(1 h)	Δ
	min	avg	max	stdev	time	z	
fspm-M-3-15	5840	5840.0	5840	0.00	138.2	5840	0.00%
fspm-M-3-20	7670	7670.0	7670	0.00	151.8	7670	0.00%
fspm-M-3-25	9930	9930.0	9930	0.00	175.6	9930	0.00%
fspm-M-3-30	11800	11800.0	11800	0.00	201.2	–	–
fspm-M-3-35	13185	13186.2	13206	4.63	225.4	13204	−0.13%
fspm-M-4-15	6423	6426.1	6431	3.88	160.5	6467	−0.63%
fspm-M-4-20	8896	8896.0	8896	0.00	210.8	8896	0.00%
fspm-M-4-25	9524	9524.0	9524	0.00	220.1	9597	−0.76%
fspm-M-4-30	11778	11778.0	11778	0.00	243.3	11815	−0.31%
fspm-M-4-35	14530	14530.0	14530	0.00	299.4	14531	−0.01%
fspm-M-5-15	6878	6878.0	6878	0.00	188.6	6878	0.00%
fspm-M-5-20	9091	9097.1	9114	10.34	222.1	9123	−0.28%
fspm-M-5-25	11100	11100.0	11100	0.00	257.6	11106	−0.05%
fspm-M-5-30	12220	12220.0	12220	0.00	280.0	12357	−1.11%
fspm-M-5-35	13680	13690.4	13700	8.10	330.5	13922	−1.66%
fspm-M-6-15	7294	7294.7	7316	4.02	203.1	7294	0.01%
fspm-M-6-20	9080	9088.1	9107	10.18	252.8	9369	−3.00%
fspm-M-6-25	10995	11009.3	11011	4.50	280.2	11125	−1.04%
fspm-M-6-30	12521	12562.7	12650	36.91	322.1	12770	−1.62%
fspm-M-6-35	14120	14120.3	14130	1.83	399.4	14250	−0.91%
fspm-M-7-15	7950	7953.5	8003	13.45	238.6	8077	−1.53%
fspm-M-7-20	9562	9586.1	9847	50.81	274.7	9715	−1.33%
fspm-M-7-25	11570	11575.9	11620	14.35	322.5	11684	−0.92%
fspm-M-7-30	12584	12584.0	12584	0.00	364.8	12875	−2.26%
fspm-M-7-35	14440	14513.4	14579	42.96	426.3	14842	−2.21%
fspm-M-8-15	8351	8369.5	8451	27.53	262.3	8402	−0.39%
fspm-M-8-20	10086	10105.3	10220	30.81	311.5	10318	−2.06%
fspm-M-8-25	11903	11912.0	12044	27.56	363.6	12022	−0.91%
fspm-M-8-30	13630	13636.1	13683	13.79	413.5	13927	−2.09%
fspm-M-8-35	14893	14894.2	14899	2.17	474.4	15530	−4.09%
fspm-M-9-15	8739	8747.2	8780	16.68	289.8	8787	−0.45%
fspm-M-9-20	10198	10239.7	10257	21.90	338.6	10423	−1.76%
fspm-M-9-25	11704	11730.4	11829	26.00	394.5	12239	−4.16%
fspm-M-9-30	13710	13777.1	14570	168.23	436.7	14003	−1.61%
fspm-M-9-35	15795	15859.3	16485	121.92	532.7	16850	−5.88%
Avg.	10904.9	10917.9	10990.9	18.93	291.6	11054.1	−1.27%

Table 5. Results on the Large dataset.

Instance	SA					MIP(2 h)
	min	avg	max	stdev	time	z
fspm-L-5-20	9116	9129.4	9130	2.56	78.1	9130
fspm-L-5-50	18780	18780.0	18780	0.00	164.1	19360
fspm-L-5-100	37010	37010.0	37010	0.00	301.5	1038740
fspm-L-5-200	72590	72590.0	72590	0.00	627.5	–
fspm-L-10-20	10835	10862.3	11026	62.15	134.6	10931
fspm-L-10-50	21201	21246.3	21300	33.40	296.4	22840
fspm-L-10-100	39690	39769.4	40079	120.51	567.8	50980
fspm-L-10-200	74180	74798.8	75630	692.34	1105.2	–
fspm-L-20-20	14814	14968.8	15342	113.39	255.4	15283
fspm-L-20-50	25548	25695.1	26200	169.10	579.1	28625
fspm-L-20-100	42784	43159.9	44511	465.86	1076.5	–
fspm-L-20-200	77900	79939.8	84818	2402.04	2155.1	–
Avg.	37037.3	37329.1	38034.7	338.44	611.8	

Table 4 presents the results obtained by SA in comparison with the MIP solver on the Medium dataset. For SA we show the average running times in seconds. The MIP was not able to find any proven optimal solution within the timeout of one hour. The last column reports the gap between SA and MIP, computed as $\Delta = (\text{avg}_{SA} - z_{MIP})/z_{MIP}$. The MIP solver produced a memory error for instance fspm-M-3-30, thus no value is available for this instance.

We also experimented a *warm start* approach, where first SA is run for 10^6 iterations (about 3 s) and then the MIP solver is invoked for the remaining time up to one hour, using as initial solution the one found by SA in the first stage. However, the outcome is that MIP is not able to improve the initial solution for the 75% of the cases, and for the remaining ones the average improvement is only 0.23%. We thus decided to not report these experimental results.

We see that SA is able to outperform the MIP solver in most instances, being equal on all the others, with an average improvement of 1.27%. This result is obtained using much shorter time (about 300 s on average) and on a single core, whereas the MIP solver is allowed to use all cores and 1 h of computational time.

For the Large dataset (Table 5), we decided to grant the MIP solver two hours, in order to give it more possibilities to find a solution. Unsurprisingly, it has not been able to find a feasible solution for the largest instances. We also notice a particularly high value on instance fspm-L-5-100; this is probably due to the timeout that evidently stopped the MIP solver in a very early stage of the search. Generally, the average value of the SA solver is constantly better than the result of the MIP solver.

Looking at the values reported on Table 5, we can conclude that the SA results are rather stable, and actually for three out of four small (five machines) instances the standard deviation is null. Finally, it can be seen that the computational time increases significantly with the size of the instance.

6   Conclusions

We proposed a multi-neighborhood SA approach for a integrated predictive maintenance and flow shop scheduling problem introduced in the literature by Varnier and Zerhouni [24]. We also formulated and implemented a MIP model so as to assess the quality of the solution produced by our SA. The solver turned out to be able to find always the proven optimal solution for the small instances, and to outperform the MIP solver for the medium and large ones. The performances on the Large dataset exhibit good scalability and robustness, although with longer running times. In order to reduce the computational time, we are going to investigate the use of massively parallel computing using GPUs.

The future work includes the extension of the problem formulation to the case of different RULs dependent on the job sequence being processed for each machine. We will also test our solution method on the well known benchmarks from Taillard [22], which however would need to be completed with suitable maintenance data. Finally, we plan to explore different solution techniques, for example by adapting the iterated greedy algorithm by Ruiz and Stützle [21] designed for the basic flow shop scheduling problem.

Acknowledgement. We thank Hildarahi Luz Orihuela Lino for developing the preliminary version of the mathematical model.

This work has been co-funded by the ERDF-ROP (2014–2020), Friuli Venezia Giulia (Italy), Axis 1, Action 1.3.

References

1. Aguirre, A.M., Papageorgiou, L.G.: Medium-term optimization-based approach for the integration of production planning, scheduling and maintenance. Comput. Chem. Eng. **116**, 191–211 (2018)
2. Berrichi, A., Yalaoui, F., Amodeo, L., Mezghiche, M.: Bi-objective ant colony optimization approach to optimize production and maintenance scheduling. Comput. Oper. Res. **37**(9), 1584–1596 (2010)
3. Biondi, M., Sand, G., Harjunkoski, I.: Optimization of multipurpose process plant operations: a multi-time-scale maintenance and production scheduling approach. Comput. Chem. Eng. **99**, 325–339 (2017)
4. Birattari, M., Yuan, Z., Balaprakash, P., Stützle, T.: F-Race and iterated F-Race: an overview. In: Bartz-Beielstein, T., Chiarandini, M., Paquete, L., Preuss, M. (eds.) Experimental Methods for the Analysis of Optimization Algorithms, pp. 311–336. Springer, Heidelberg (2010). https://doi.org/10.1007/978-3-642-02538-9_13
5. Bougacha, O., Varnier, C., Zerhouni, N., Hajri-Gabouj, S.: Integrated production and predictive maintenance planning based on prognostic information. In: 2019 International Conference on Advanced Systems and Emergent Technologies (IC_ASET), pp. 363–368. IEEE (2019)

6. Franzin, A., Stützle, T.: Revisiting simulated annealing: a component-based analysis. Comput. Oper. Res. **104**, 191–206 (2019)
7. ILOG: CPLEX Optimizer (2019). https://www.ibm.com/products/ilog-cplex-optimization-studio, v. 12.10
8. Kirkpatrick, S., Gelatt, D., Vecchi, M.: Optimization by simulated annealing. Science **220**, 671–680 (1983)
9. Ladj, A., Varnier, C., Tayeb, F.B.S.: IPro-GA: an integrated prognostic based GA for scheduling jobs and predictive maintenance in a single multifunctional machine. IFAC-PapersOnLine **49**(12), 1821–1826 (2016)
10. Ladj, A., Benbouzid-Si Tayeb, F., Varnier, C., Dridi, A.A., Selmane, N.: A hybrid of variable neighbor search and fuzzy logic for the permutation flowshop scheduling problem with predictive maintenance. Procedia Comput. Sci. **112**, 663–672 (2017)
11. Ladj, A., Tayeb, F.B.S., Varnier, C.: Tailored genetic algorithm for scheduling jobs and predictive maintenance in a permutation flowshop. In: 2018 IEEE 23rd International Conference on Emerging Technologies and Factory Automation (ETFA), vol. 1, pp. 524–531. IEEE (2018)
12. Ladj, A., Tayeb, F.B.S., Varnier, C.: Hybrid of metaheuristic approaches and fuzzy logic for the integrated flowshop scheduling with predictive maintenance problem under uncertainties. Eur. J. Ind. Eng. **15**(5), 675–710 (2021)
13. Ladj, A., Varnier, C., Tayeb, F.B.S., Zerhouni, N.: Exact and heuristic algorithms for post prognostic decision in a single multifunctional machine. Int. J. Prognostics Health Manag. **8**(2) (2017)
14. Lee, C.Y., Chen, Z.L.: Scheduling jobs and maintenance activities on parallel machines. Nav. Res. Logist. (NRL) **47**(2), 145–165 (2000)
15. Lenstra, J.K., Kan, A.R., Brucker, P.: Complexity of machine scheduling problems. In: Annals of Discrete Mathematics, vol. 1, pp. 343–362. Elsevier (1977)
16. Liu, Q., Dong, M., Chen, F., Lv, W., Ye, C.: Single-machine-based joint optimization of predictive maintenance planning and production scheduling. Robot. Comput.-Integr. Manuf. **55**, 173–182 (2019)
17. Ma, Y., Chu, C., Zuo, C.: A survey of scheduling with deterministic machine availability constraints. Comput. Ind. Eng. **58**(2), 199–211 (2010)
18. Medjaher, K., Zerhouni, N., Gouriveau, R.: From Prognostics and Health Systems Management to Predictive Maintenance 1: Monitoring and Prognostics. Wiley, Hoboken (2016)
19. Pan, E., Liao, W., Xi, L.: A joint model of production scheduling and predictive maintenance for minimizing job tardiness. Int. J. Adv. Manuf. Technol. **60**(9), 1049–1061 (2012)
20. Paz, N.M., Leigh, W.: Maintenance scheduling: issues, results and research needs. Int. J. Oper. Prod. Manag. **14**, 47–69 (1994)
21. Ruiz, R., Stützle, T.: A simple and effective iterated greedy algorithm for the permutation flowshop scheduling problem. Eur. J. Oper. Res. **177**(3), 2033–2049 (2007)
22. Taillard, E.: Benchmarks for basic scheduling problems. Eur. J. Oper. Res. **64**(2), 278–285 (1993)
23. Urli, T.: Json2run: a tool for experiment design & analysis. CoRR abs/1305.1112 (2013)
24. Varnier, C., Zerhouni, N.: Scheduling predictive maintenance in flow-shop. In: Proceedings of the IEEE 2012 Prognostics and System Health Management Conference (PHM-2012 Beijing), pp. 1–6. IEEE (2012)

An Investigation of Hyper-Heuristic Approaches for Teeth Scheduling

Felix Winter[✉] and Nysret Musliu

Christian Doppler Laboratory for Artificial Intelligence and Optimization
for Planning and Scheduling, DBAI, TU Wien, Vienna, Austria
{winter,musliu}@dbai.tuwien.ac.at

Abstract. Modern day production sites for teeth manufacturing often utilize a high-level of automation and sophisticated machinery. Finding efficient machine schedules in such a production environment is a challenging task, as complex constraints need to be fulfilled and multiple cost objectives should be minimized.

This paper investigates a hyper-heuristic solution approach for the artificial teeth scheduling problem which originates from real-life production sites of the teeth manufacturing industry. We propose a collection of innovative low-level heuristic strategies which can be utilized by state-of-the-art selection-based hyper-heuristic strategies to efficiently solve practical problem instances. Furthermore, the paper introduces eight novel large-scale scheduling scenarios from the industry, which are included in the experimental evaluation of the proposed techniques.

An extensive set of experiments with well-known hyper-heuristics on benchmark instances shows that our methods improve state-of-the-art results for the large majority of the instances.

Keywords: Hyper-heuristics · Scheduling · Metaheuristics · Low-level heuristics

1 Introduction

In the artificial teeth manufacturing industry a large magnitude of synthetic teeth is produced daily by utilizing a high-level of automation. Thereby, various different shapes and colors are processed using automated machinery that is able to manufacture many products simultaneously. Due to the large quantities of teeth that need to be produced each day, efficient scheduling methods are required that consider all the complex constraints raised by the automated environment and aim to minimize several cost factors.

Recently, we introduced the artificial teeth scheduling problem (ATSP) in [17], which captures the requirements and aims of real-life production environments in the teeth manufacturing area. Besides the formal specification of the problem, a collection of real-life benchmark instances was provided to evaluate exact and heuristic solution approaches. In [17] we further proposed an exact constraint-modeling approach that achieved optimal results for some small

© The Author(s), under exclusive license to Springer Nature Switzerland AG 2023
L. Di Gaspero et al. (Eds.): MIC 2022, LNCS 13838, pp. 274–289, 2023.
https://doi.org/10.1007/978-3-031-26504-4_20

problem instances but could not produce feasible solutions for practically-sized instances within reasonable runtime. Therefore, a construction heuristic as well as a metaheuristic approach using local search were proposed that could successfully provide upper bounds to all real-life instances. However, optimal solutions are still unknown for realistically-sized instances and thus there is still a potential to improve the quality of solutions.

The existing local search based approach for the ATSP belongs to the category of traditional metaheuristics that efficiently utilize domain-specific search neighborhoods and thereby are capable of producing good solutions also to large-scale instances. In this paper, we investigate a novel hyper-heuristic approach for the ATSP which is able to automatically utilize and integrate several heuristics including the existing metaheuristic to achieve high-quality solutions.

The main aim of hyper-heuristics is to develop heuristic techniques that generalize well on different problem domains. Over the recent decade this topic has been the subject of intensive study and a large amount of work has been reviewed in surveys (e.g. [4,6]). Furthermore, several frameworks that allow researchers to compare and evaluate hyper-heuristic approaches have been proposed (e.g. [2,14,15,19]). A group of techniques from this area, known under the term selection-based hyper-heuristics, defines problem-independent methods that utilize a set of low-level heuristics. Such low-level operators consist of domain specific strategies which are applied by a hyper-heuristic in an iterative search process.

One of the main advantages of a hyper-heuristic approach is that strategies that have been successful on related problems can be efficiently reused on novel problem domains such as the ATSP. Selection-based hyper-heuristics were used to efficiently tackle several large realistic scheduling problems in the literature (e.g. [3,10,12]). Thus, there is a potential to utilize successful strategies from this area also as an effective approach for teeth scheduling. However, using selection-based hyper-heuristics requires carefully designed low-level operators, which to the best of our knowledge have not been proposed for teeth scheduling previously.

This paper proposes a set of innovative low-level heuristics including mutational- and local search operators as well as a crossover operator for the ATSP. We experimentally show that these operators can be efficiently used with hyper-heuristic strategies from the literature, like e.g. the winner of a popular hyper-heuristic competition [14], to solve large-scale instances. The work provides an extensive experimental evaluation of these hyper-heuristics and the proposed methods on publicly available benchmark instances. Furthermore, we introduce 8 novel real-life instances that we gathered from our industry partners and include them in our evaluation. The results show that a hyper-heuristic approach for the ATSP can produce competitive results compared to the state-of-the-art solution methods. Using the innovative low-level heuristics together with the hyper-heuristic approaches we achieve several novel upper bounds and improve results compared to the existing approaches for the large majority of the realistic instances.

This work is an extended full paper version of a recently published poster [18]. Besides many more details on the hyper-heuristic approach and the empirical evaluation, this work includes an additional crossover operator, 8 novel real-life instances, and additional experiments that are not included in the poster submission. Further, we draw new conclusions based on the extended results.

The remainder is structured as follows: First, we review the specification of the ATSP in Sect. 2. Afterwards, we propose the low-level heuristics used in the hyper-heuristic approach in Sect. 3. In Sect. 4, we describe the new realistic instances and report experimental results. At the end of the paper we give concluding remarks and mention possible future work.

2   The Artificial Teeth Scheduling Problem

In teeth manufacturing, large quantities of teeth are usually processed in batches, where each job in the schedule uses multiple moulds to produce teeth. Furthermore, each mould is used for a certain tooth shape which is associated to a product line. A job can consist of moulds that belong to different lines and determines the color that should be applied to each mould. Thus, the final product type produced in each mould is given by its product line and color. Each job is further configured by a length- and program parameter, where the length defines the number of the job's production cycles and thereby determines the total number of produced teeth. The program parameter determines how many moulds are simultaneously processed by the job, which mould types are compatible, and the processing time of a single cycle. As every program requires a fixed amount of moulds to be processed per cycle, it can happen that more teeth than necessary are produced. Usually this cannot be completely avoided, and therefore one of the problem's goals is to minimize the amount of waste caused by excessive teeth. Consecutively scheduled jobs may either use different programs or share the same program with a different set of mould- and or color assignments. In any case a setup time is required between jobs, however if different programs are used a longer setup time is required.

Finally, the aim of the ATSP is to create a schedule fulfilling all customer demands in a way that makespan, total tardiness, and waste are minimized. The ATSP can be viewed as a single machine batch scheduling problem variant which has several unique properties and constraints compared to other machine scheduling variants. For example, while traditional machine scheduling problems usually provide a predetermined set of jobs as part of the input, instances of the ATSP include customer demands but do not specify any job information. Thus, solution methods which are able to create and schedule efficient jobs are needed to solve the ATSP.

Figure 1 visualizes a schedule with three jobs for an example instance.

The figure shows three jobs J1–J3 being scheduled on the horizontal time line. Time points t1, t3, and t5 indicate the starting times of each job, whereas t2, t4, and t6 denote the job end times. J1 and J2 both use the production program P1, whereas J3 uses P2. Note that the setup time between J1 and J2 (visualized by the horizontal space between jobs) is much smaller than it is between J2 and J3, as J1 and J2 both use program P1, but J3 uses P2. Furthermore, the horizontal length of the jobs indicates the number of assigned production cycles.

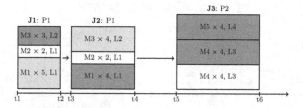

Fig. 1. A small example schedule for the ATSP (see also [17])

Table 1. Input parameters of the ATSP

Description	Parameter
Set of colors	C
Set of programs	P
Set of mould types	M
Set of product lines	L
Set of demands	D
Setup time between identical programs	$sj \in \mathbb{N}$
Setup time between different programs	$sp \in \mathbb{N}$
Max product types per job	$w \in \mathbb{N}_{>0}$
Min cycles per job	$c_{\min} \in \mathbb{N}_{>0}$
Max cycles per job	$c_{\max} \in \mathbb{N}_{>0}$
Number of available moulds per type	$a_m \in \mathbb{N} \; \forall m \in M$
Number of mould slots per program	$am_p \in \mathbb{N} \; \forall p \in P$
Cycle time per program	$t_p \in \mathbb{N}_{>0} \; \forall p \in P$
Admissible program per mould type	$p_m \in P \; \forall m \in M$
Product line of each mould type	$l_m \in L \; \forall m \in M$
Requested mould type per demand	$dm_d \in M \; \forall d \in D$
Requested mould quantity per demand	$dq_d \in \mathbb{N}_{>0} \; \forall d \in D$
Due date of each demand	$dd_d \in \mathbb{N} \; \forall d \in D$
Requested color for each demand	$dc_d \in C \; \forall d \in D$
Set of compatible colors per color	$comp_c \in 2^C \; \forall c \in C$

As the program defines the total number of assigned moulds, we see in Fig. 1 that J1 and J2 both use 10 moulds, whereas J3 uses 12 moulds. Mould types M1, M2, and M3 are compatible only with program P1, and mould types M4 and M5 are associated to P2. We can further see in the figure, that each mould type is associated to a certain product line (e.g. M3 corresponds to line L2), and that the same mould type may be used with different colorings within the same job. Note that any two colors may only be used within the same job if they are compatible (color compatibility is given as part of the instance parameters).

Using the instance parameters shown in Table 1 and the variables shown in Table 2 the constraints of the ATSP are defined as follows:[1]

---

[1] We make use of the Iverson bracket notation: $[P] = 1$, if $P = true$ and $[P] = 0$ if $P = false$.

Table 2. Variables of the ATSP

Description	Variables
Number of assigned jobs	$j \in N \quad J = \{1, \ldots, j\}$
Program assigned to each job	$jp_i \in P \quad \forall i \in J$
Length of each job (i.e. the number of cycles)	$jl_i \in N_{>0} \quad \forall i \in J$
The number of mould types (with color) assigned to each job	$jm_{i,m,c} \in N \quad \forall i \in J, m \in M, c \in C$
The total number of mould types (with color) produced by each job	$totaljm_{i,m,c} = jm_{i,m,c} \cdot jl_i \quad \forall i \in J, m \in M, c \in C$
The processing time for each job	$jt_i \in N_{>0} \quad \forall i \in J$
The finishing time for each job	$je_i \in N_{>0} \quad \forall i \in J$
The finishing job for each demand (after completion the demand is fulfilled)	$dj_d \in J \quad \forall d \in D$

- The number of assigned moulds to each job must be equal to the number of mould slots of the job's program:

$$\sum_{m \in M} \sum_{c \in C} jm_{i,m,c} = am_{(jp_i)} \quad \forall i \in J$$

- The number of scheduled moulds per job must not exceed mould availability:

$$\sum_{c \in C} jm_{i,m,c} \leq a_m \quad \forall i \in J, m \in M$$

- The number of different product types within a single job must be less than or equal to the allowed maximum:

$$\sum_{c \in C} \sum_{l \in L} \sum_{m \in M} ([l_m = l] \cdot [jm_{i,m,c} > 0]) \leq w \quad \forall i \in J$$

- All demands need to be fulfilled:

$$\sum_{d \in D} [dm_d = m \wedge dc_d = c] \cdot dq_d \leq \sum_{i \in J} totaljm_{i,m,c} \ \forall m \in M, c \in C$$

- Job moulds must be compatible with the job's program:

$$\sum_{c \in C} jm_{i,m,c} \cdot [jp_i \neq p_m] = 0 \quad \forall i \in J, m \in M$$

- A single job must not use incompatible colors:

$$\left[\sum_{m \in M} jm_{i,m,c_1} > 0 \right] \leq \left[\sum_{m \in M} jm_{i,m,c_2} = 0 \right]$$
$$\forall i \in J, c_1 \in C, c_2 \in (C \setminus comp_{c_1})$$

- Set the job processing times: $jt_i = jl_i \cdot t_{(jp_i)} \quad \forall i \in J$
- Set job finishing times:

$$je_i = jt_1 + \sum_{k=2}^{i} (jt_k + sj + [jp_k \neq jp_{k-1}] \cdot (sp - sj)) \ \forall i \in J$$

- Set demand finishing jobs:

$$\sum_{i=1}^{dj_d} totaljm_{i,m,c} \geq \sum_{d' \in D'} dq_{d'} \quad \forall d \in D \text{ where } m = dm_d,$$

$$c = dc_d, D' = \{d' \in D | dd_{d'} \leq dd_d \wedge dm_{d'} = m \wedge dc_{d'} = c\}$$

The objective function of the ATSP aims to minimize three solution objectives:

1. Minimize the total **makespan** of the schedule: $ms = je_j$
2. **Waste** (i.e. the number of excessively produced moulds) should be minimized:

$$waste = \sum_{i \in J} \sum_{m \in M} \sum_{c \in C} totaljm_{i,m,c} - \sum_{d \in D} dq_d$$

3. The total **tardiness** of all demands in the schedule should be mimized:

$$tard = \sum_{d \in D} \max(0, je_{(dj_d)} - dd_d)$$

The objectives are aggregated in a normalized weighted sum where objectives marked with * denote the costs of a given reference solution:

$$minimize \quad \frac{ms}{ms*} + \frac{waste}{waste*} + \frac{tard}{tard*}$$

This particular normalization scheme was chosen together with expert practitioners from the industry to capture the most important aims in the real-life environment. For all instances we evaluate, reference costs are generated using the construction heuristic from [17], which serves as a baseline approach.

Previously, we proposed seven local search neighborhood operators for the ATSP in [17]. As some of the low-level heuristics in this paper build upon these operators, we shortly review them here:

1. **Swap two jobs:** Swaps the positions of two existing jobs.
2. **Increment length:** Increments the cycles of a job by 1.
3. **Decrement length:** Decrements the cycles of a job by 1.
4. **Change single mould assignment:** Changes a single assigned mould type and/or color to a different mould type and/or color within the same job.
5. **Delete last job:** Deletes the last job in the schedule.
6. **Append new job:** Appends a new job at the end of the current schedule. Move parameters define the job program, as well as the mould quantities that should be used in the newly created job.
7. **Swap mould assignments between two jobs:** Swaps a single mould type and/or color assignment from a job with a single mould and/or color assignment from another job.

3 Low-Level Heuristics for the ATSP

In this section we propose several low-level heuristics for the ATSP, which we categorize into *mutational*, *local search*, and *crossover* operators.

Mutational Heuristics: The idea behind the *mutational* low-level heuristics is to perform small perturbative changes to the components of a candidate solution.

We thus propose 7 mutation moves corresponding to the search neighborhoods which were described earlier: *Swap Two Jobs, Increment Job Length, Decrement Job Length, Change Single Mould Assignment, Delete Last Job, Append new Job, Swap Two Mould Assignments.* The mutational low-level operator works as follows for all 7 move types: Given is a candidate solution that is mutated by iteratively applying uniformly random selected moves of the selected move type.

For example, a mutation low-level heuristic which uses the *Swap Two Jobs* move, randomly selects one out of all possible swap moves for the given candidate solution. This move is then performed to modify the solution, and the process continues for a number of iterations which can be controlled by setting a parameter α that lies within $[0.0, 1.0]$ and a parameter k that defines the maximum number of iterations. The iteration limit is then $\lfloor \alpha \cdot k \rfloor$ so that k moves will be performed in any mutational low-level heuristic move if $\alpha = 1.0$.

Local Search Heuristics: In the following we propose several *local search* based low-level heuristics for the ATSP. These low-level operators conduct a local search for a number of iterations to find improved solutions. The iteration limit for all local search operators is configured by a parameter β that lies within $[0.0, 1.0]$.

Stochastic Hill Climber: The first local search heuristic we propose randomly generates moves in each search iteration in a similar way as it has been described for the simulated annealing [7] based approach from [17]. After a move has been randomly selected, this heuristic evaluates the change in solution costs that would be caused by the move. Then, an acceptance function is used to decide whether the move should be accepted depending on a potential cost improvement and a given temperature parameter τ. Note that for this heuristic the temperature value that is given as a parameter will not change but is fixed.

The iteration limit for this low-level heuristic is determined by $\lfloor \beta \cdot k \rfloor$, where k is another given integer parameter. Additionally, a temperature parameter τ and a time limit t configure the fixed temperature value and a timeout. The heuristic stops if the iteration limit or time limit is reached.

This low-level heuristic can end on solutions of reduced quality. However, the heuristic always returns the overall best solution found (if no improvement was found, the operator simply returns the initial solution).

Simulated Annealing: We propose another low-level local search operator that directly implements a simulated annealing approach.

Similarly as with the stochastic hill climber operator, the initial temperature is given as parameter τ, but here we use an additional cooling rate parameter γ to configure the geometrical cooling schedule. Again we determine the iteration limit as $\lfloor \beta \cdot k \rfloor$ with another integer parameter k, and the time limit for the simulated annealing low-level heuristic is set by parameter t.

Full Neighborhood Move Heuristics: In addition to the stochastic hill climber and simulated annealing operators that both generate random moves in each

---

Algorithm 1: Crossover Low-Level Heuristic for the ATSP

---

fn *Crossover* (S_1, S_2)

    $randomJob1Position = random(1, length(S_1))$
    $result =$ new empty job sequence
    for $i = 1$ *to* $(randomJob1Position - 1)$ **do**
        $result.append(S_1[i])$

    $bestJob2Position = null$
    $bestNumProducts = -1$
    for $i = 1$ *to* $length(S_2)$ **do**
        $numProducts = CountProducts(result, S_2[i])$
        if $numProducts > bestNumProducts$ **then**
            $bestNumProducts = numProducts$
            $bestJob2Position = i$

    $result.append(S_2[bestJob2Position])$
    for $i = (randomJob1Position + 1)$ *to* $length(S_1)$ **do**
        $result.append(S_1[i])$

    return *result*

---

search iteration, we further propose a set of local search low-level heuristics that consider the full search neighborhood for particular move operators.

For example, a full neighborhood move heuristic for the *Swap Two Jobs* move type generates moves for all possible pairs of jobs in the current candidate solution. Afterwards, the change in solution quality for all generated moves is evaluated and the full neighborhood move heuristic applies the move leading to the best solution quality (ties are broken by the order of the generated moves). In case none of the generated moves leads to an improvement no move is performed.

Based on this idea, we propose four full neighborhood move heuristics:

1. *Full Change Mould Neighborhood:* Considers all possible instantiations of the *Change Single Mould Assignment* neighborhood operator on the given candidate solution.
2. *Full Job Length Neighborhood:* Considers all length modifying neighborhood operators (i.e. *Increment/Decrement Job Length*) for all jobs in the given candidate solution.
3. *Full Swap Job Neighborhood:* Considers *Swap Two Jobs* moves for all possible job pairs in the given candidate solution.
4. *Full Swap Mould Neighborhood:* Selects the best of all possible *Swap Two Mould Assignments* moves.

Crossover Heuristic: We now propose a crossover low-level heuristic that combines two given candidate solutions for the ATSP. The main idea behind this operator is to exchange a single job between two given candidate solutions S_1 and S_2. To achieve this, first a random job is selected from S_1. Then, the heuristic exchanges this job with a job from S_2 by deleting the original job in S_1 and injecting the new job at the same position. To select a job from S_2 that is a good replacement for the original job, the algorithm calculates for each job in S_2 the effect on the number of fulfilled demands in S_1 that would occur if this job was chosen to replace the randomly selected job from S_1. Thereby, the heuristic selects the job which produces the most moulds that can be used to satisfy demands in the resulting solution. The main idea behind this strategy is to inject efficiently designed jobs from other solutions into a good candidate solution, without causing a lot of overproduction or many unfulfilled demand violations. Algorithm 1 describes the detailed procedure where S_1 and S_2 are the given candidate solutions.

The algorithm starts by randomly selecting a job position from S_1. Afterwards, a partial solution *result* is built by copying all jobs that are scheduled before the selected job position from S_1 to *result*. Then, the algorithm iterates over the jobs from S_2 to determine the number of demanded products that could be fulfilled if any single job was appended at the end of the current partial solution *result* (the function *CountProducts* represents the calculation of this number in the pseudocode). Finally, after the best fitting job is added to the resulting solution, the remaining jobs from $S1$ are appended.

4   Experimental Evaluation

We evaluated the proposed hyper-heuristic approach on the benchmark instances from [17] which consist of 6 small instances (I1–I6) that were created by shrinking realistic scheduling scenarios as well as 6 large real-life instances (I7–I12). In this paper, we further introduce 8 additional large real-life instances (I13–I20) that we received from our industry partners. All newly introduced real-life benchmark instances are publicly available online.[2]

Evaluated Hyper-Heuristics. For the evaluation of the proposed techniques we implemented the low-level heuristics within the latest version of the HyFlex framework [14]), as it is a popular and widely used selection-based hyperheuristic framework which was also used in two hyper-heuristic competitions. In addition to the proposed methods, we also implemented the construction heuristic from [17] within framework so that the evaluated hyper-heuristics can select it to generate initial solutions. We further contacted the author that achieved the first place of the CHeSC competition [14] and received an up-to-date implementation of GIHH [13] which was the winning algorithm. Additionally, we evaluated an improved 'lean' version of GIHH called Lean-GIHH (LGIHH) from [1] and further received an implementation of the HAHA hyper-heuristic [9] which scored sixth place in the competition.

Besides the algorithms that participated in CHeSC, we evaluated a recent hyper-heuristic approach from [5] (CHU) as well as two hyper-heuristics based on self-adaptive large neighborhood search strategies from [8] (ALNS) and [16] (ALNS2) and were shown to deliver high-quality results on scheduling problems. These three approaches have recently been adapted and implemented for use within the HyFlex framework by [11], and we thus included them in our evaluation.

---

[2] https://cdlab-artis.dbai.tuwien.ac.at/papers/atsp_mic/.

Table 3. Overview on overall best results per instance.

Inst.	SA	GIHH	LGIHH	HAHA	ALNS	ALNS2	CHU
I 1	**2.53**	2.95	**2.53**	**2.53**	**2.53**	2.54	**2.53**
I 2	1.96	1.95	**1.94**	**1.94**	1.95	1.95	**1.94**
I 3	**2.23**	**2.23**	**2.23**	**2.23**	**2.23**	**2.23**	**2.23**
I 4	**2.54**	**2.54**	**2.54**	**2.54**	**2.54**	**2.54**	**2.54**
I 5	2.13	2.12	**2.11**	2.12	**2.11**	2.13	2.12
I 6	**3**	**3**	**3**	**3**	**3**	**3**	**3**
I 7	2.95	**2.85**	**2.85**	**2.85**	2.91	2.87	2.9
I 8	**2.38**	2.45	2.45	2.47	2.52	2.45	2.49
I 9	2.99	2.86	2.86	**2.85**	2.91	2.91	2.89
I 10	2.67	2.66	**2.64**	2.66	2.74	2.67	2.68
I 11	**2.76**	2.77	2.77	2.78	2.81	2.79	2.8
I 12	2.97	**2.90**	2.95	2.91	2.95	2.96	2.91
I 13	2.97	2.83	**2.81**	2.85	2.89	2.88	2.87
I 14	2.99	**2.83**	**2.83**	2.84	2.93	2.91	2.88
I 15	2.99	**2.81**	**2.81**	**2.81**	2.88	2.89	2.84
I 16	2.98	**2.66**	2.69	2.67	2.82	2.73	2.77
I 17	2.94	**2.79**	**2.79**	**2.79**	2.83	**2.79**	**2.79**
I 18	2.98	**2.72**	**2.72**	**2.72**	2.85	**2.72**	2.77
I 19	2.94	**2.7**	**2.7**	**2.7**	2.72	2.71	**2.7**
I 20	2.98	**2.78**	**2.78**	**2.78**	2.84	2.82	2.82

Experimental Environment. All experiments were run on a cluster with 10 identical nodes, each having 24 cores, an Intel(R) Xeon(R) CPU E5-2650 v4 @ 2.20GHz and 252 GB RAM. In addition to the hyper-heuristic approach proposed in this paper, we reevaluated the state-of-the-art metaheuristic for the ATSP (i.e. simulated annealing (SA) starting from an initial solution produced by a construction heuristic) to allow a direct comparison to the hyper-heuristic results. We received the source code and parameter configuration of these methods from the authors of [17].

Based on manual tuning trials we further set the parameters of the low-level heuristics as follows: For the mutational operators, we set the iteration limit $k = 100$. Additionally, we set the default intensity of mutation to $\alpha = 0.1$. However, note that this is only the initial value of α and this parameter can be controlled dynamically during search by each individual hyper-heuristic.

For the *Stochastic Hill Climber* heuristic we set the time limit t to 60 s. Further, we determine the iteration limit k dependent on instance size parameters by calculating the product of the number of colors, the number of mould types, the number of demands, and the total number of demanded moulds. Additionally, we set the temperature $\tau = 0.01$ and set the default depth of search $\beta = 0.1$

For the *Simulated Annealing* heuristic, we use the same parameters as for the *Stochastic Hill Climber* heuristic, except for τ, which we set to 0.4735. Further, we set the cooling rate parameter to 0.9274. These two parameter values correspond to the simulated annealing configuration that was determined in [17].

Table 4. Overview on mean objective costs per instance.

Inst.	SA	GIHII	LGIHH	HAHA	ALNS	ALNS2	CHU
I 1	**2.62**	2.95	2.91	**2.62**	2.74	2.92	2.7
I 2	2.02	1.99	1.97	**1.96**	2	2.25	1.99
I 3	**2.23**	**2.23**	**2.23**	**2.23**	**2.23**	**2.23**	**2.23**
I 4	**2.54**	**2.54**	**2.54**	**2.54**	**2.54**	**2.54**	**2.54**
I 5	2.2	2.27	2.20	**2.14**	2.2	2.36	2.26
I 6	**3**	**3**	**3**	**3**	**3**	**3**	**3**
I 7	2.99	2.87	**2.86**	**2.86**	2.95	2.92	2.92
I 8	**2.42**	2.48	2.48	2.49	2.56	2.63	2.52
I 9	3	**2.89**	**2.89**	**2.89**	2.95	2.94	2.92
I 10	2.83	**2.69**	**2.69**	2.7	2.78	2.75	2.74
I 11	**2.77**	2.78	2.78	2.78	2.84	2.85	2.81
I 12	2.99	**2.93**	2.95	2.95	2.96	2.96	2.95
I 13	2.99	2.84	**2.83**	2.85	2.94	2.93	2.9
I 14	3	**2.86**	2.87	2.87	2.97	2.96	2.92
I 15	3	**2.82**	**2.82**	**2.82**	2.93	2.96	2.87
I 16	3	2.72	2.72	**2.71**	2.9	2.87	2.84
I 17	2.99	**2.79**	**2.79**	**2.79**	2.87	2.9	2.85
I 18	2.99	2.79	2.74	**2.73**	2.9	2.89	2.82
I 19	2.98	**2.7**	**2.7**	**2.7**	2.76	2.85	2.74
I 20	3	**2.78**	**2.78**	**2.78**	2.91	2.95	2.85

As some full neighborhood move heuristics can require a large processing time when computing all possible moves, we additionally imposed a maximum runtime limit of 10 min to each low-level heuristics in this category. Using this parameter configuration we performed 10 repeated runs for each instance and evaluated approach, where every single run was given a time limit of 1 h.

Computational Results: The best results over 10 runs for each instance and approach are shown in Table 3. Columns 2–8 display the best solution quality per instance produced with SA and the evaluated hyper-heuristic approaches. Note that we rounded all results to two digits after the decimal point and that best results for each line are formatted in boldface. For the remainder of this section, we apply a similar formatting to all tables unless stated otherwise. The results displayed in Table 3 show that LGIHH is able to produce the best costs for 15 of the 20 instances and therefore seems to be the overall best performing method in this comparison. However, GIHH and HAHA come close as these approaches produced best results for 12 out of 20 instances. Furthermore, we note that for instances 8, 11 SA produced slightly better results than GIHH, LGIHH and HAHA.

Table 4 shows the mean objective costs over 10 repeated runs per instance. This time, we see that HAHA seems to be the overall best performing approach as it produced the best mean objective costs for 14 out of 20 instances, whereas GIHH and LGIHH produced best mean objective costs for 11 out of 20 instances.

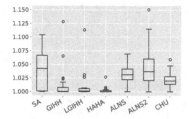

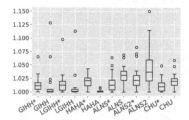

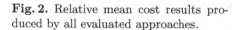

Fig. 2. Relative mean cost results produced by all evaluated approaches.

Fig. 3. Results produced without and with full neighborhood operators.

Therefore, the results indicate that HAHA produced robust results over the 10 repeated runs more often than the other evaluated approaches. However, we note that for instances 8, and 10–14 SA, GIHH and LGIHH reached better mean objective costs than HAHA. The standard deviation of the objective costs is not shown in the table, but it was close to 0 for most approaches and instances with only a few exceptions. This indicates that all evaluated approaches are able to produce reasonably robust results regarding different random seeds.

Figure 2 further provides a visual comparison of the relative mean cost results over all instances in form of box plots. To calculate the relative mean cost result for each instance, we divide the actual mean cost result for each instance by the lowest mean cost result achieved for that instance considering all methods which are included in the plot. Thus, a value of 1.0 denotes the best mean cost for an instance regarding the plotted methods. All boxplots in the remainder of this section visualize the relative mean cost results in this way.

We see in Fig. 2 that results produced by HAHA lead to the lowest median and lowest inter-quartile range, indicating that it performed best over all evaluated instances. However, GIHH and LGIHH produce a median value of similar quality.

As the full neighborhood search heuristics may need large amounts of execution time, the evaluation of the full neighborhood can require up to several minutes for a single low-level heuristic call for large instances, we conducted a second set of experiments without using the full neighborhood search heuristics to investigate whether this has a negative effect on the overall performance of the hyper-heuristics. To mark results produced without the full neighborhood low-level heuristics, we add a * after the identifiers of the evaluated methods.

Figure 3 summarizes the overall mean cost results for all evaluated hyper-heuristic strategies without and with the full neighborhood low-level operators. We can see in the results shown in Fig. 3 that for GIHH, LGIHH and HAHA, the inclusion of the full neighborhood operators leads to improved results. On the other hand, for ALNS, ALNS2 and CHU, results are improved without these low-level heuristics. This indicates that not all hyper-heuristic approaches benefit from the computationally costly full exploration of the neighborhoods. However, if these operators are utilized efficiently they actually produce the best mean cost results in our experiments together with GIHH, LGIHH, and HAHA.

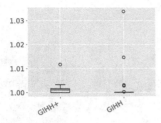

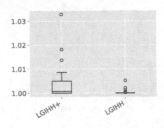

Fig. 4. Results for GIHH without and with the crossover low-level heuristic.

Fig. 5. Results for LGIHH without and with the crossover low-level heuristic.

Table 5. Comparison of results achieved with SA and HAHA.

Inst.	SA	S̄A	SA σ	HAHA	H̄AHA	HAHA σ
I 1	2.53	**2.62**	0.17	2.53	**2.62**	0.18
I 2	1.96	2.02	0.08	1.94	**1.96**	0.01
I 3	2.23	**2.23**	0	**2.23**	2.23	0
I 4	2.54	**2.54**	0	2.54	**2.54**	0
I 5	2.13	2.2	0.12	2.12	**2.14**	0.01
I 6	3	**3**	0	3	**3**	0
I 7	2.95	2.99	0.01	2.85	**2.86**	0.01
I 8	2.38	**2.42**	0.02	2.47	2.49	0.01
I 9	2.99	3	0	2.85	**2.89**	0.01
I 10	2.67	2.83	0.09	2.66	**2.7**	0.01
I 11	2.76	**2.77**	0.01	2.78	2.78	0
I 12	2.97	2.99	0.01	2.91	**2.95**	0.01
I 13	2.97	2.99	0.01	2.85	**2.85**	0
I 14	2.99	3	0	2.84	**2.87**	0.02
I 15	2.99	3	0	2.81	**2.82**	0.01
I 16	2.98	3	0.01	2.67	**2.71**	0.02
I 17	2.94	2.99	0.02	2.79	**2.79**	0
I 18	2.98	2.99	0.01	2.72	**2.73**	0.01
I 19	2.94	2.98	0.02	2.7	**2.7**	0
I 20	2.98	3	0.01	2.78	**2.78**	0

Table 6. The best bounds achieved by exact methods, SA, and HAHA.

Inst.	LB	Exact	SA	HAHA
I 1	2.08	**2.53**	**2.53**	2.53
I 2	1.25	1.96	**1.94**	**1.94**
I 3	2.23	**2.23**	**2.23**	**2.23**
I 4	2.54	**2.54**	**2.54**	**2.54**
I 5	1.63	**2.1**	2.13	2.12
I 6	3	3	**3**	**3**
I 7	0.5	–	2.95	**2.85**
I 8	0.15	–	**2.38**	2.47
I 9	0.59	–	2.97	**2.85**
I 10	0.53	–	2.67	**2.66**
I 11	0.34	–	**2.76**	2.78
I 12	1.02	–	**2.89**	2.91
I 13	0.60	–	2.97	**2.85**
I 14	0.46	–	2.99	**2.84**
I 15	0.56	–	2.99	**2.81**
I 16	0.37	–	2.98	**2.67**
I 17	0.20	–	2.94	**2.79**
I 18	0.39	–	2.98	**2.72**
I 19	0.18	–	2.94	**2.7**
I 20	0.18	–	2.98	**2.78**

We conclude that the inclusion of the full neighborhood low-level heuristics has a positive impact on the performance of the hyper-heuristics and thereby can improve results. A possible explanation could be that hyper-heuristic strategies such as GIHH and HAHA perform an online selection and performance evaluation of the given low-level heuristics allowing them to automatically detect and remove inefficient heuristics during the search process.

To investigate the crossover heuristic, we ran additional experiments with LGIHH and GIHH using all low-level operators except the crossover operator (the other evaluated hyper-heuristics do not support crossover low-level heuristics). Figures 4 and 5 compare relative mean cost results achieved without and with the crossover operator (a + indicates that no crossover operator was used).

We can see in both figures that the median and inter-quartile range is closer to the value 1.0 when the crossover heuristic was included. This indicates that the operator can lead to improved results, especially with LGIHH where a signed wilcoxon signed-rank test showed a statistical significant improvement. Further, the results displayed in Table 3 show that the overall best results for most of the largest instances (I12–I120) could be reached with GIHH and LGIHH. This indicates that the utilization of the crossover operator can improve results for large-scale instances, even though HAHA (which does not support crossover operators) produces the best mean cost results for most instances in Table 4.

In Table 5, we further present a detailed comparison between the results from HAHA, which produced the best mean costs in our experiments and SA, which is the state-of-the-art heuristic for large instances. The table displays the best cost, mean cost, and standard deviation over all 10 individual runs per instance for SA and HAHA. Best mean cost results per instance are formatted in boldface.

Looking at the mean cost results shown in Table 5, we see that HAHA improves results for instances 2, 5, 7, 9, 10, 12–20 compared to SA and a wilcoxon signed-rank test shows an overall statistical significant cost improvement with HAHA. However, the SA approach reaches better mean costs for instances 8, and 11. Comparing the best cost results per instance, we further see that HAHA is able to reach equal or better results than SA for all instances except instance 8 and 11. The standard deviation is for both approaches reasonable low regarding all instances. These results show that the hyper-heuristic approach can improve results for the large majority of the benchmark instances compared to SA.

Finally, Table 6 summarizes the best bounds achieved by the state-of-the-art exact methods, SA and HAHA. The table includes the currently best known upper bounds for each instance regarding the SA approach and thus combines the best results given in [17] together with the best results achieved in the reevaluation of the SA approach conducted in this work. Furthermore, we ran experiments with the exact approach from [17] to find lower bounds for instances 13–20, however, no bound could be produced within 1 h (indicated by a -).

Table 6 displays in Columns 2–5 from left to right, the best lower bounds achieved by exact methods (LB) followed by the best upper bounds achieved by exact methods (Exact), SA, and HAHA. The results show that the hyper-heuristic approach produced the best solutions for the large majority of the instances. Only for instances 5, 8, and 11–12 exact methods or SA found better bounds.

5   Conclusion

This work provides a hyper-heuristic approach for the ATSP, by proposing a set of innovative low-level operators for selection-based hyper-heuristics.

In an extensive experimental evaluation using publicly available instances and newly introduced real-life instances, we demonstrated that the proposed methods can be used as an efficient solution approach with hyper-heuristics such as GIHH and HAHA. Computational results show that such an approach could produce high-quality solutions for all instances and improves results obtained by the state-of-the-art heuristic on the majority of the instances. The proposed methods further provide improved upper bounds for many realistic instances.

Future work could extend the collection of low-level operators with a destroy-and-repair method by utilizing exact methods as a repair mechanism.

Acknowledgments. The financial support by the Austrian Federal Ministry for Digital and Economic Affairs, the National Foundation for Research, Technology and Development and the Christian Doppler Research Association is gratefully acknowledged.

References

1. Adriaensen, S., Now'e, A.: Case study: an analysis of accidental complexity in a state-of-the-art hyper-heuristic for hyflex. In: 2016 IEEE Congress on Evolutionary Computation (CEC). IEEE (2016)
2. Asta, S., Özcan, E., Parkes, A.J.: Batched mode hyper-heuristics. In: Nicosia, G., Pardalos, P. (eds.) LION 2013. LNCS, vol. 7997, pp. 404–409. Springer, Heidelberg (2013). https://doi.org/10.1007/978-3-642-44973-4_43
3. Bouazza, W., Sallez, Y., Trentesaux, D.: Dynamic scheduling of manufacturing systems: a product-driven approach using hyper-heuristics. Int. J. Comput. Integr. Manuf. **34**(6), 641–665 (2021)
4. Burke, E.K., et al.: Hyper-heuristics: a survey of the state of the art. J. Oper. Res. Soc. **64**(12), 1695–1724 (2013)
5. Chuang, C.: Combining multiple heuristics: studies on neighborhood-base heuristics and sampling-based heuristics. Thesis, Carnegie Mellon University (2020)
6. Drake, J.H., Kheiri, A., Özcan, E., Burke, E.K.: Recent advances in selection hyper-heuristics. Eur. J. Oper. Res. **285**(2), 405–428 (2020)
7. Kirkpatrick, S., Gelatt, C.D., Vecchi, M.P.: Optimization by simulated annealing. Science **220**(4598), 671–680 (1983)
8. Laborie, P., Godard, D.: Self-adapting Large Neighborhood Search: Application to Single-Mode Scheduling Problems (2007)
9. Lehrbaum, A., Musliu, N.: A new hyperheuristic algorithm for cross-domain search problems. In: Hamadi, Y., Schoenauer, M. (eds.) LION 2012. LNCS, pp. 437–442. Springer, Heidelberg (2012). https://doi.org/10.1007/978-3-642-34413-8_41
10. Li, W., Özcan, E., John, R.: Multi-objective evolutionary algorithms and hyper-heuristics for wind farm layout optimisation. Renew. Energy **105**, 473–482 (2017)
11. Mischek, F., Musliu, N.: A collection of hyper-heuristics for the hyflex framework. Technical report, TU Wien, CD-TR, 2021/2 (2021)
12. Mısır, M., Smet, P., Vanden Berghe, G.: An analysis of generalised heuristics for vehicle routing and personnel rostering problems. J. Oper. Res. Soc. **66**(5), 858–870 (2015)
13. Mısır, M., Verbeeck, K., De Causmaecker, P., Vanden Berghe, G.: An intelligent hyper-heuristic framework for CHeSC 2011. In: Hamadi, Y., Schoenauer, M. (eds.) LION 2012. LNCS, pp. 461–466. Springer, Heidelberg (2012). https://doi.org/10.1007/978-3-642-34413-8_45

14. Ochoa, G., et al.: HyFlex: a benchmark framework for cross-domain heuristic search. In: Hao, J.-K., Middendorf, M. (eds.) EvoCOP 2012. LNCS, vol. 7245, pp. 136–147. Springer, Heidelberg (2012). https://doi.org/10.1007/978-3-642-29124-1_12
15. Pillay, N., Beckedahl, D.: EvoHyp - a Java toolkit for evolutionary algorithm hyper-heuristics. In: 2017 IEEE Congress on Evolutionary Computation (CEC), pp. 2706–2713 (2017)
16. Thomas, C., Schaus, P.: Revisiting the self-adaptive large neighborhood search. In: van Hoeve, W.-J. (ed.) CPAIOR 2018. LNCS, vol. 10848, pp. 557–566. Springer, Cham (2018). https://doi.org/10.1007/978-3-319-93031-2_40
17. Winter, F., Mrkvicka, C., Musliu, N., Preininger, J.: Automated production scheduling for artificial teeth manufacturing. In: Proceedings of the International Conference on Automated Planning and Scheduling, vol. 31, pp. 500–508 (2021)
18. Winter, F., Musliu, N.: A hyper-heuristic approach for artificial teeth scheduling. In: Genetic and Evolutionary Computation Conference, Companion Volume, GECCO 2022, Boston, MA, USA, 9–13 July 2022. ACM (2022)
19. Zhang, Y., Bai, R., Qu, R., Tu, C., Jin, J.: A deep reinforcement learning based hyper-heuristic for combinatorial optimisation with uncertainties. Eur. J. Oper. Res. **300**, 418–427 (2021)

A Mixed-Integer Programming Formulation and Heuristics for an Integrated Production Planning and Scheduling Problem

D. M. Silva[1(✉)] and G. R. Mateus[2]

[1] Instituto Federal de Minas Gerais (IFMG), Formiga, MG, Brazil
diego.silva@ifmg.edu.br
[2] Universidade Federal de Minas Gerais (UFMG), Belo Horizonte, MG, Brazil
mateus@dcc.ufmg.br

Abstract. This paper proposes a new mixed-integer programming formulation for an integrated multiproduct, multiperiod, and multistage capacitated lot sizing with hybrid flow shop problem (CLSP-HFS). Heuristics that combine relax-and-fix with fix-and-optimize are also proposed to solve it, using strategies for decomposing the set of variables by product, period and stage. A relax-and-fix heuristic takes an initial feasible solution, and a fix-and-optimize heuristic tries to improve it. In order to evaluate the performance of the combined strategy, some experiments were done considering seven datasets as a benchmark, each one composed of ten randomly generated instances with 5, 10, 15, 20, 25, 30, and 40 products. They are processed in parallel machines during three stages along a planning horizon of eight periods. Experimental results suggest that period-based strategies achieve a percentage deviation close to zero from the optimum, while product-based strategies offer a compromise between solution quality and computational time.

Keywords: CLSP · HFS · RFO-heuristic

1 Introduction

Production planning consists of defining how much to produce and store in each period of a planning horizon to meet customer demand with minimal production and inventory costs. It is related to the acquisition of resources and raw materials and the production activities necessary to transform them into finished products as economically or efficiently as possible [15,20]. Transformation occurs by executing a set of jobs on one or more machines, in one or more stages, to be sequenced to meet customer demand and previously established constraints and objectives. Scheduling determines the most suitable time to execute each planned operation and on which machine, taking into account conditions such as order

Supported by Fapemig and CNPq Brazil.

delivery date, minimum makespan, balanced use of machines, etc., in order to increase production productivity [8]. Although some practical techniques can be used as a scheduling procedure, they do not guarantee that the finished product batches will be delivered on time [5]. In addition, scheduling results also affect production planning when capacity limits restrict lot sizing and scheduling from being treated hierarchically and separately [4]. An integrated approach can overcome these drawbacks.

Scheduling problems, in general, are highly complex and depend on environmental characteristics, process constraints, objectives, or performance criteria. Some classic variants are well known in the literature, such as the job shop and the flow shop problems [17]. These can also follow extensions like permutational and hybrid. Some exploit a single machine or several, in series or parallel, identical, uniform, or unrelated. More specifically, the hybrid flowshop (HFS) consists of a series of stages, each with a set of parallel machines and a set of tasks to be processed in series passing through each stage [14]. Therefore, the problem is to determine, for each task, on which machine of each stage it should be processed and at what time. In this paper, we propose an integrated procedure considering multiproducts, multiperiods, and multistages, in a combination and variants of two classic problems called capacitated lot sizing (CLSP) [11] and hybrid flow shop (HFS). To the best of our knowledge, few works in the literature on production planning and scheduling address the integration of the two problems. We also present a mixed-integer programming formulation and heuristics used to relax-and-fix and fix-and-optimize techniques to solve the proposed problem.

This paper is organized as follows. In the Sect. 2 we present some recent works on the integration of production planning and scheduling, with a focus on those who approach the problem through the use of MIP-heuristics relax-and-fix and fix-and-optimize. Then, in Sect. 3 we present a mixed-integer programming formulation to solve the integration of capacitated lot sizing problem with hybrid flow shop (CLSP-HFS). The Sect. 4 provides the main features of the relax-and-fix and fix-and-optimize algorithms, and the Sect. 5 describes the decomposition strategies adopted to solve the CLSP-HFS using the proposed formulation. Next, Sect. 6 presents a comparative performance analysis of the decomposition strategies implemented considering the percentage deviation from the optimum (or best-known solution) and runtime for seven sets of instances of the integrated problem. Finally, Sect. 7 discusses the final remarks and the future of this research.

2   Related Works

In the general case, both the capacitated lot sizing and the hybrid flow shop problems are NP-Hard [2,14]. Consequently, most works in the literature propose to solve real-world instances or considered difficult to solve, using matheuristics and hybridization with metaheuristics. This section presents a brief review of works that employed relax-and-fix (RF) and fix-and-optimize (FO) decomposition heuristics [1,20] to solve similar problems, integrated or not. In common, they share the choice to decompose binary variables related to decisions about the setup of a machine, carryover of setup between periods, or both. Some of these problems address the integration of production planning and scheduling, but none resemble CLSP-HFS. RF and FO methods solve a MIP problem heuristically by iteratively decomposing it into smaller MIP subproblems whose integer and binary variables must sometimes be continuous (relaxed), sometimes fixed, and sometimes discrete. At each iteration, a subproblem is solved by keeping decisions done in past iterations fixed, optimizing the decisions that belong to the current iteration, and relaxing future decisions. When combined, RF constructs a feasible solution, and FO improves it.

One of the first works that applied these techniques to solve production planning problems heuristically was presented by [20]. Since then, many works have applied RF, FO, or combined relax-and-fix-and-optimize (RFO). Fix-and-optimize heuristic solved the *multilevel capacitated lot sizing problem with multi-period setup carryover* (MLCLSP-L) and variants decomposing the original problem using combinations of machines, processes and products to split setup and carryover decision variables per product and period [9, 22]. The *capacitated lot sizing with sequence-dependent setups and substitution* (CLSD-S) was solved by [12] using relax-and-fix, fix-and-optimize, and relax-and-fix-and-optimize methods to split variables per period, product and substitute. In [25] the authors used a *rolling horizon window strategy* to solve MLCLSP and a real-world production scheduling problem decomposing variables per period, product, or both.

Integration of production planning and scheduling was also approached using these decomposition techniques. For example, in [18] we found a relax-and-fix method to solve a *lot sizing problem with permutation flow shop and setups dependent on the sequence*. In [10] the *parallel machine capacitated lot sizing and scheduling with sequence-dependent setup* (CLSD-PM) was solved using MIP-based search with neighborhood heuristics, that includes combining relax-and-fix with other improvement heuristics or local search techniques. It decomposes integer variables of the model per period with re-optimization of part of them.

Hybridization is also used with decomposition techniques. In [6] a hybrid method using fix-and-optimize and bees algorithm (BA) solved a *multilevel capacitated lot sizing problem with carryover* by substituting the well-defined partitioning in [22] by a bee algorithm to decompose setup and carryover variables. The CLSP *with carryover* was solved in [7] using a hybridization of fix-and-optimize and genetic algorithm. Finally, [13] used fix-and-optimize with variable neighborhood search (VNS) based on k-degree connection neighborhood to solve a *stochastic multiproduct capacitated lot sizing with setup, backlogging, and carryover* (S-MICLSP-L) decomposing variables per product, period, resource and demand.

Recently, more complex decomposition strategies have emerged. An enhanced relax-and-fix and RFO were presented in [24] to solve a *two-stage multiproduct lot sizing and scheduling problem with parallel machines* using six different strategies that combine machines, stages, products and periods. A *generalized lot sizing and scheduling* (GLSP) variant that uses a process configuration selection is solved using four different variations of relax-and-fix: forward, backward, with overlapping and minimizing backlog. In [1] the authors proposed new decomposition strategies that use information about the chronological order of periods, demands per product and period, flexibility to produce products, the discrepancy of production times, and efficiency and criticality of machines to decompose setup variables. Finally, [4] explores the hybridization of relax-and-fix and fix-and-optimize with intensification metaheuristics path-relinking (PR) and kernel search (KS) using decomposition per period with re-optimization of part of the variables in an *integrated lot sizing and scheduling problem on parallel machines with non-triangular sequence-dependent setup costs and times with carryover* (ILSSP-NT).

3   Mixed-Integer Programming Formulation

This section describes a mixed-integer program (MIP) formulation to integrate the capacitated lot sizing problem with the hybrid flow shop (CLSP-HFS). Sets, parameters, and decision variables are present as follows. In this integrated problem, we decide the lot sizing of $|P|$ products along a planning horizon divided in $|T|$ periods. To produce them are available a set of machines M^k that operates in $|K|$ different stages of production. Subscripts p, t, m and k refers, respectively, to products, periods, machines and stages. It is allowed to produce more than a product per period but with no preemption (i.e., if the production of p starts in the period t, it must be finished within t without interruption). To produce a product p in a given period t and stage k are available $|M^k|$ machines in parallel, each one with capacity given by w_{pt}^{mk} units of time and a processing time e_p^{mk}, where $\hat{e}_p^k = \max_{m \in M^k} \{e_p^{mk}\}$ denotes the worst processing time to produce p on stage k. The schedule to produce p is feasible if it fits within this capacity. A setup time to produce p is required to prepare machines between different products and stages and takes es_p^{mk} units of time before starting processing a new batch of a product if reconfiguration is required. Each period also have a total capacity of w_t^{mk} units of time, and must be greater than all w_{pt}^{mk} that belong the same period t. The production of a product p in a stage k only starts after the production of p be finished in the stage $(k-1)$. Products may be stored to meet future demands with a inventory cost cs_{pt}, and must respect the capacity limit of warehousing S_t. The initial inventory level is given by S_0^p. Producing a product p in the period t costs cf_{pt} per unit of product plus a setup cost of cu_p^m per unit of time to prepare the machine, so the final setup cost is given by $\left(cu_p^m \cdot es_p^{mk}\right)$. The total costs involve production, inventory, and setup costs and must be minimized. Let the decision variables be x_{pt}, integer, which decides how many units of a product p must be produced in the period t; s_{pt}, integer, which decides the number of products p that must be stored in the warehouse on period t; h_{pt}^k, continuous, that decides the start time of processing of a lot for the product p in the stage k of the period t; r_{pt}^{mk}, binary, that assumes value 1 if the machine m of stage k is assigned to produce product p on the period t, and 0 otherwise; and $q_{pp't}^{mk}$ that assumes 1 if the product p is scheduled before que product p' on machine m of stage k on the period t, and 0 otherwise.

This work is based on the CLSP formulation of [16], whose model embedding capacity information due to scheduling constraints in a classical lot sizing model, and the HFS multistage formulation of [23] that uses precedence variables to model scheduling and sequencing between stages of a unique period. Furthermore, it extends these works by proposing a multiperiod, multiproduct, and multistage formulation for integrated lot sizing and scheduling with setup costs whose scheduling decisions directly impact the total cost. Finally, integrating both problems is also interesting because it permits generating feasible lot sizing respecting the factory floor constraints. The CLSP-HFS problem is formulated as follows:

$$\min \sum_p \sum_t cf_{pt}x_{pt} + cs_{pt}s_{pt} + \sum_p \sum_t \sum_k \sum_m r_{pt}^{mk} \cdot es_p^{mk} \cdot cu_p^m \tag{1}$$

s.t.

$$s_{p(t-1)} - s_{pt} + x_{pt} = d_{pt} \qquad\qquad \forall p,t \tag{2}$$

$$s_{p0} = S_0^p \qquad\qquad \forall p \tag{3}$$

$$\sum_{p \in P} s_{pt} \leq S_t \qquad\qquad \forall t \tag{4}$$

$$\hat{e}_p^k \cdot x_{pt} - \sum_m w_{pt}^{mk} \cdot r_{pt}^{mk} \leq 0 \qquad\qquad \forall p,t,k \tag{5}$$

$$\sum_p \left(w_{pt}^{mk} + es_p^{mk} \right) \cdot r_{pt}^{mk} \leq w_t^{mk} \qquad\qquad \forall t,m,k \tag{6}$$

$$\sum_m r_{pt}^{mk} \leq 1 \qquad\qquad \forall p,t,k \tag{7}$$

$$\left(2 - r_{pt}^{mk} - r_{p't}^{mk} \right) \cdot N + \left(1 - q_{pp't}^{mk} \right) \cdot N +$$
$$h_{p't}^k - h_{pt}^k - e_p^{mk}x_{pt} \geq es_{p'}^{mk} \qquad\qquad \forall p,p', p < p', t,m,k \tag{8}$$

$$\left(2 - r_{pt}^{mk} - r_{p't}^{mk} \right) \cdot N + q_{pp't}^{mk} \cdot N +$$
$$h_{pt}^k - h_{p't}^k - e_{p'}^{mk} x_{p't} \geq es_p^{mk} \qquad\qquad \forall p,p', p < p', t,m,k \tag{9}$$

$$\left(2 - r_{pt}^{mk} - r_{pt}^{m'(k-1)} \right) \cdot N +$$
$$h_{pt}^k - h_{pt}^{(k-1)} - e_p^{m'(k-1)}x_{pt} \geq es_p^{mk} \qquad \forall p,t,k > 1, m \in M^k, m' \in M^{(k-1)} \tag{10}$$

$$h_{pt}^{|K|} + e_p^{m|K|} \cdot x_{pt} \leq w_t^{m|K|} +$$
$$\left(1 - r_{pt}^{m|K|} \right) \cdot N \qquad\qquad \forall p,t,m \in M^{|K|} \tag{11}$$

$$x_{pt}, s_{pt} \in \mathbb{N} \qquad\qquad \forall p,t \tag{12}$$

$$h_{pt}^k \geq 0 \qquad\qquad \forall p, t, k \quad (13)$$

$$r_{pt}^{mk}, q_{pp't}^{mk} \in \{0, 1\} \qquad\qquad \forall p, p', p < p', t, m, k \quad (14)$$

Objective function (1) minimizes the sum of production, inventory and setup costs. Constraints (2) deals with balance among production, inventory and demand per product p and period t. Constraints (3) define the initial inventory level for each product p, and constraints (4) impose warehousing capacity limits for all products p on a given period t. Constraints (5) limit the amount of items to be produced in a lot of product p according to the machine capacity available to produce them on stage k and period t. These constraints integrate both production planning and scheduling problems because they define that if a non-zero quantity of product is decided to be produced in the corresponding period t and stage k, this production must be allocated on some machine of the stage only if there is enough capacity available in this machine; otherwise r_{pt}^{mk} will be zero and no product p will be produced in period t. Constraints (6) ensure that the processing time needed to produce the planned quantity of all products including setup times must fit in the total capacity available on period t for the machine m of the stage k. Constraints (7) allow to allocate a product p on at most one machine per stage k and period t. Constraints (8) and (9) are disjunctive and impose the precedence relation between two products p and p' when allocated in the same machine m and stage k of period t: or product p is produced before p', or p' is produced before p. Constraints (10) ensure that a production of p in the stage k only starts after be finished in the stage $(k-1)$. Constraints (11) determine that a production of a complete lot of products p must the scheduled within the total time available for the period t. As consequence all h_{pt}^k are relative per period, i.e., must fit between 0 and the total capacity w_t^{mk}. Finally, constraints (12), (13) and (14) define non-negativity, integrality, and domain for the decision variables associates with production, assignment and scheduling.

4   MIP Heuristics for CLSP-HFS Problem

This section presents the basic structure of both relax-and-fix and fix-and-optimize methods that solve the CLSP-HFS problem by decomposing the original MIP into subproblems that are easiest to solve.

4.1   Relax-and-Fix (RF)

One of the first mentions of this method appears on [20] as a construction heuristic for production planning problems. The central idea is to partition 0–1 variables into Ω disjoint sets $Q^1, \ldots, Q^\Omega$ used to generate Ω different mixed-integer smaller subproblems that are sequentially solved to find a heuristic solution to the original problem. For the CLSP-HFS problem the RF method was adapted to handle the binary variables r and q that are related, respectively, to machine assignment and sequencing. It changes their domain along the iterations according the subset these variables belong. In the ω-th iteration RF solves the subproblem SUBMIP$_{RF}^{(\omega)}$ generated as follow: variables $r, q \in Q^{\omega}$ are restricted to be binary; variables $r, q \in \left(Q^{(\omega+1)} \cup \cdots \cup Q^{\Omega}\right)$ are relaxed; and variables $r, q \in \left(Q^1 \cup \cdots \cup Q^{(\omega-1)}\right)$ are assumed be fixed with their corresponding values computed from iteration 1 to $(\omega - 1)$. Figure 1 presents a diagram that illustrates how variables r and q behave in a partition with $\Omega = 4$ disjoint sets along iteration $\omega \in \{1, 2, 3, 4\}$ for a hypothetical example. The main steps of the RF-heuristic are illustrated in Algorithm 1. It receives as input the instance to be solved, a decomposition function that splits r and q in Ω disjoint subsets, and a time budget inspired by the *Computational Budget Allocation Scheme* (CBAS) of [12] that limits the computation of each subproblem.

Iteration	$\forall r, q \in Q^1$	$\forall r, q \in Q^2$	$\forall r, q \in Q^3$	$\forall r, q \in Q^4$
SUBMIP$_{RF}^{(1)}$ at $\omega = 1$	$r \in \{0,1\}$ $q \in \{0,1\}$	$0 \le r \le 1$ $0 \le q \le 1$	$0 \le r \le 1$ $0 \le q \le 1$	$0 \le r \le 1$ $0 \le q \le 1$
SUBMIP$_{RF}^{(2)}$ at $\omega = 2$	$r \leftarrow \hat{r}$ $q \leftarrow \hat{q}$	$r \in \{0,1\}$ $q \in \{0,1\}$	$0 \le r \le 1$ $0 \le q \le 1$	$0 \le r \le 1$ $0 \le q \le 1$
SUBMIP$_{RF}^{(3)}$ at $\omega = 3$	$r \leftarrow \hat{r}$ $q \leftarrow \hat{q}$	$r \leftarrow \hat{r}$ $q \leftarrow \hat{q}$	$r \in \{0,1\}$ $q \in \{0,1\}$	$0 \le r \le 1$ $0 \le q \le 1$
SUBMIP$_{RF}^{(4)}$ at $\omega = 4$	$r \leftarrow \hat{r}$ $q \leftarrow \hat{q}$	$r \leftarrow \hat{r}$ $q \leftarrow \hat{q}$	$r \leftarrow \hat{r}$ $q \leftarrow \hat{q}$	$r \in \{0,1\}$ $q \in \{0,1\}$

Fig. 1. Example of a RF-heuristic for the CLSP-HFS problem with $\Omega = 4$ subsets. Each line shows how the SUBMIP$_{RF}^{(\omega)}$ must be constructed on the ω-th iteration indicating which variables must be fixed (white-gray), optimized (gray), or relaxed (white). Here, $\hat{r}$ and $\hat{q}$ refer to the value assumed for the variables r and q in the imcumbent solution calculated in the past iterations that must be fixed in the current iteration.

The algorithm works as follows. In line 1 the decomposition function is applied to divide variables r and q in Ω subsets (see Sect. 5 for strategies). It is assumed that the total budget available for solving the original problem is equally distributed among the Ω subproblems (lines 2–3). Because RF-heuristic is constructive, line 4 defines the initial incumbent solution as unknown. The loop from lines 5–16 is the core of the method. It runs Ω times; at iteration ω it creates a smaller mixed-integer subproblem SUBMIP$_{RF}^{(\omega)}$ fixing the variables $r, q \in \left(Q^1 \cup \cdots \cup Q^{(\omega-1)}\right)$, keeping the variables $r, q \in Q^{\omega}$ as binary, and relaxing the integrality of variables $r, q \in \left(Q^{(\omega+1)} \cup \cdots \cup Q^{\Omega}\right)$ (line 6). SUBMIP$_{RF}^{(\omega)}$ is solved using a commercial solver upon the time budget allocated for it (line 7). Depending on the decisions made in past iterations and values assumed as fixed on the decision variables $r, q \in \left(Q^1 \cup \cdots \cup Q^{(\omega-1)}\right)$ the computed solution may be feasible or not. If feasible, the partial incumbent solution is updated (line 9), and the procedure continues; otherwise, the algorithm stops with no valid solution found and returns (line 11). The available time is recalculated using the elapsed time to solve SUBMIP$_{RF}^{(\omega)}$, distributing the remaining time uniformly

---

Algorithm 1. Relax-And-Fix(`instance, decomp-fn`, Ω, `budget`)

1: $Q^1, Q^2, \ldots, Q^\Omega \leftarrow$ `decomp-fn(instance, ` Ω `)`
2: `available-time` $\leftarrow$ `budget`
3: `rf-budget` $\leftarrow$ `available-time` / Ω
4: `incumbent` $\leftarrow \varnothing$
5: **for** $(\omega \leftarrow 1$ **to** $\Omega)$ **do**
6:     Creates SUBMIP$_{RF}^{(\omega)}$ using $Q^1, Q^2, \ldots, Q^\Omega$ for iteration ω as described
7:     `new-solution` $\leftarrow$ `solve-subproblem(`SUBMIP$_{RF}^{(\omega)}$`, rf-budget)`
8:     **if** `new-solution` is feasible **then**
9:         `incumbent` $\leftarrow$ `new-solution`
10:    **else**
11:        **return** $\varnothing$
12:    **end if**
13:    Computes the `elapsed` time to solve SUBMIP$_{RF}^{(\omega)}$
14:    Update `available-time` $\leftarrow$ (`available-time` - `elapsed`)
15:    Recompute `rf-budget` $\leftarrow$ `available-time` / $(\Omega - \omega)$
16: **end for**
17: **return** `incumbent`

---

among the subproblems SUBMIP$_{RF}^{(\omega+1)}$ to SUBMIP$_{RF}^{(\Omega)}$ (lines 13–15). After solving all subproblems, the procedure stops and returns the constructed solution (line 17).

4.2 Fix-and-Optimize (FO)

The fix-and-optimize heuristic is a procedure that starts from an initial feasible solution (e.g., computed by RF-heuristic) and tries to improve it by systematically re-optimizing smaller mixed-integer subproblems. Similar to RF-heuristic, it splits the binary variables r and q in Ω disjoint subsets $Q^1, \ldots, Q^\Omega$. However, it differs from RF-heuristic in the way the variables are fixed, using values from both initial and incumbent solutions. There is no relaxation of binary variables. Let $\dot{r}$ and $\dot{q}$ be the values of the variables r and q from the initial solution, and let $\hat{r}$ and $\hat{q}$ be the values of r and q from the incumbent solution computed by the FO-heuristic. In a given iteration ω, SUBMIP$_{FO}^{(\omega)}$ is generated setting variables $r, q \in Q^\omega$ as binaries, while $r, q \in \bigcup_{i=1}^{\Omega} Q^i \setminus Q^k$ are kept fixed using the values assumed for $\dot{r}$ and $\dot{q}$ case $r, q \in \left(Q^{(\omega+1)} \cup \cdots \cup Q^\Omega\right)$, or using the values assumed for $\hat{r}$ and $\hat{q}$ case $r, q \in \left(Q^1 \cup \cdots \cup Q^{(\omega-1)}\right)$. Figure 2 illustrate this procedure for $\Omega = 4$, highlighting the subsets that must be fixed or optimized along the Ω iterations to construct SUBMIP$_{FO}^{(\omega)}$. Because the procedure is similar to RF-heuristic, no algorithm is described, but two observations must be done about the process. Because the FO-heuristic is an improvement procedure, the initial feasible solution is assigned to the incumbent when the procedure starts. Along the iterations of FO, the incumbent solution is updated whenever a partial solution computed from SUBMIP$_{FO}^{(\omega)}$ has better objective function. When the procedure finishes, the incumbent solution is returned.

Iteration	$\forall r, q \in Q^1$	$\forall r, q \in Q^2$	$\forall r, q \in Q^3$	$\forall r, q \in Q^4$
SUBMIP$_{FO}^{(1)}$ at $\omega = 1$	$r \in \{0,1\}$ $q \in \{0,1\}$	$r \leftarrow \dot{r}$ $q \leftarrow \dot{q}$	$r \leftarrow \dot{r}$ $q \leftarrow \dot{q}$	$r \leftarrow \dot{r}$ $q \leftarrow \dot{q}$
SUBMIP$_{FO}^{(2)}$ at $\omega = 2$	$r \leftarrow \hat{r}$ $q \leftarrow \hat{q}$	$r \in \{0,1\}$ $q \in \{0,1\}$	$r \leftarrow \dot{r}$ $q \leftarrow \dot{q}$	$r \leftarrow \dot{r}$ $q \leftarrow \dot{q}$
SUBMIP$_{FO}^{(3)}$ at $\omega = 3$	$r \leftarrow \hat{r}$ $q \leftarrow \hat{q}$	$r \leftarrow \hat{r}$ $q \leftarrow \hat{q}$	$r \in \{0,1\}$ $q \in \{0,1\}$	$r \leftarrow \dot{r}$ $q \leftarrow \dot{q}$
SUBMIP$_{FO}^{(4)}$ at $\omega = 4$	$r \leftarrow \hat{r}$ $q \leftarrow \hat{q}$	$r \leftarrow \hat{r}$ $q \leftarrow \hat{q}$	$r \leftarrow \hat{r}$ $q \leftarrow \hat{q}$	$r \in \{0,1\}$ $q \in \{0,1\}$

Fig. 2. Execution of FO-heuristic for CLSP-HFS problem within $\Omega = 4$ subsets. Here, $\dot{r}$ and $\dot{q}$ denote the values assumed for the variables r and q from the initial solution, while $\hat{r}$ and $\hat{q}$ denote the values assumed for r and q from incumbent solution computed in past iterations of the FO-heuristic.

4.3   Relax-and-Fix-and-Optimize (RFO)

RFO-heuristic is the combination of RF-heuristic to construct an initial feasible solution to CLSP-HFS with FO-heuristic to improve it. Because this operation is direct, it will not be discussed here.

5   Decomposition Strategies

In this section, we present some problem-dependent strategies from literature used in this work to decompose the binary variables r and q into disjoint sets considering dimensions related to products, periods, and stages. Product-based strategies separate the binary variables r and q in $\Omega = |P|$ disjoint sets according to rules related to products and demands associates to them. Three product-based strategies were considered: Prod.P, MstProd.P, and LstProd.P. Period-based strategies uses the period as a criterion to decompose r and q variables in $\Omega = |T|$ disjoint sets. Two period-based strategies were used: Per.Fwd.T and MstPer.T. Finally, decomposition by stage separates variables r and q in $\Omega = |K|$ disjoint sets. Variables that refer to different machines that belong to each stage are placed together in the same disjoint set. Since CLSP-HFS is multistage, one strategy was proposed and implemented: Stage.K. A brief description of these strategies is given above:

- Prod.P (or lexicographical [1,9,12,13,22,24]): variables are sorted in increasing lexicographical order of the subscript $p \in P$;
- MstProd.P (or most demanded product first [1,13]): variables are sorted in decreasing order of *products demands* $d_p = \sum_{t \in T} d_{pt}$;
- LstProd.P (or less demanded product first [1]): similar to MstProd.P, but sort variables in increasing order of *products demands* $d_p = \sum_{t \in T} d_{pt}$.
- Per.Fwd.T (or chronological [1,3,4,7,10,12,13,20,24]): variables are sorted in increasing order of the subscript t of periods, with $t = \{1, \ldots, |T|\}$;

- MstPer.T (or periods with larger demand first [1]): variables are sorted in decreasing order of *overall demand* $\delta_t = \sum_{p \in P} d_{pt}$.
- Stage.K (or ascending): optimizing the more advanced stages first can outcomes unfeasible solutions depending on the availability of machines in the preceding stages, so the ascending order is followed: first consider stage one, after the second stage, and so on.

6   Numerical Experiments

In this section, some experiments were done to compare the performance of the described decomposition strategies applied to the proposed formulation. A time budget of 3600 s was available to solve each instance using the MIP model and the same budget for the combined RFO-heuristic (25% of this budget for RF and 75% for FO). The experiments were carried out in an Intel i7-8565U 1.8 GHz machine with 8 Gb of RAM. Both algorithms and formulation were implemented in Python 3 using commercial solver Gurobi 9.0.1 for Windows with default parameters set (except TimeLimit=3600, Threads=4 and MipGap=0.0000001). Statistic tests were done using R Project for Statistical Computing 3.4.4.

The tested benchmark is composed by seven datasets, named A-5, B-10, C-15, D-20, E-25, F-30 and G-40, each one containing 10 randomly generated instances that share the same structural characteristics. The fixed parameters are $|P| \in \{5, 10, 15, 20, 25, 30, 40\}$, $|T| = 8$, $|K| = 3$ and $|M^k| = 3$. The initial inventory level is $S_0^p = 0$, and the capacities available per product and per period are given, respectively, by $w_t^{mk} = 200$ and $w_t^{mk} = (|P| \cdot 200)$. The parameters that varies are presented in Table 1. They were chosen based on the works [16,23] because they have already been used in the generation of instances involving planning, scheduling or both. The demands were computed per product and period according the capacity w_{pt}^{mk} and the worst processing time among all machines of the k-th stage, denoted by $\hat{e}_p^k = \max_{m \in M^k} \{e_p^{mk}\}$, using $d_{pt} = \lfloor w_{pt}^{mk} / \sum_k \hat{e}_p^k \rfloor$.

Table 1. Parameters and ranges, based on [16,23], used to generate costs, times and inventory capacities for instances in datasets A-5, B-10, C-15, D-20, E-25, F-30,G-40.

Name	Parameter	Distribution	Value range
Production costs	cf_{pt}	Continuous Uniform	$U(10, 20)$
Inventory costs	cs_{pt}	Continuous Uniform	$U(1, 5)$
Setup costs	cu_p^m	Continuous Uniform	$U(100, 200)$
Setup times	es_p^{mk}	Continuous Uniform	$U(10, 50)$
Processing times	e_p^{mk}	Continuous Uniform	$U(1, 10)$
Inventory capacities	S_t	Discrete Uniform	$U[10, 15]$

The experiments were done as follows. First, the instances from each dataset were solved to optimality (or until the time budget expires) using the Gurobi solver and the MIP formulation. After, the combined RFO-heuristic was used to construct and improve solutions for the same instances applying the discussed strategies. The production, inventory, and setup costs for the solution computed using RFO were used to calculate a percentage deviation to the optimal solution (or to the best-known solution) $Dev = (f(RFO) - f(MIP))/f(MIP)$, a performance measure that shows how far the solution computed by the RFO is from the optimum (or the best solution found). The execution of both solver and RFO on each dataset provided a sample with ten observations used to (i) plot the confidence interval for the mean value of percentage deviations, (ii) statistically determine if there were significant differences in the outcomes generated for these strategies, and (iii) identify which pairs of strategies are different or similar in terms of performance at a 95% confidence level.

The plot presented in Fig. 3 contains the confidence interval for the mean value of percentage deviation estimated using *non-parametric Bootstrap* for each strategy and dataset. The resulting intervals suggest that some strategies differ from others in terms of performance. In special, the period-based strategies achieved total cost for planning and scheduling closest to the optimum with near-zero percentage deviation, a tighter confidence interval, and minor variation in the expected percentage deviation than those based on product and stage.

Although the plot suggests that the period-based strategies are more suitable for solving the CLSP-HFS formulation using RFO, statistical tests were used to reinforce it. Because assumptions for an ANOVA test were not met (e.g., normality of residuals), the non-parametric *Kruskal-Wallis test* [19,21] was used to check if there was a statistically significant difference in terms of percentage deviation among all strategies. It was found that, for all datasets, the corresponding p-value was on the order of 10^{-8}, concluding that the strategies are statistically different because p-value $< \alpha = 0.05$.

Therefore, a non-parametric *Pairwise Wilcoxon Rank Sum Test* with Bonferroni's correction for α was used to identify whose differs two by two at a 95% confidence level. Table 2 presents a matrix per dataset containing the Bonferroni's adjusted p-values resulting from the test for the datasets with $|P| \geq 25$ products. A p-value $< \alpha$ indicates a statistically significant difference between a pair of strategies. The cases where such a difference exists are highlighted with an asterisk character (*). These results shows that, for most cases, there is no significant difference between strategies MstPer.T and Per.Fwd.T, and among LstProf.P, MstProd.P and Prod.P, while Stage.K differs from both period-based and product-based strategies. Together with the plot of Fig. 3, these results also reinforce the suggestion that the best choices of strategies to run the RFO-heuristic considering solution quality as a criterion are MstPer.T and Per.Fwd.T because they return near-zero percentage deviation, and the corresponding sub-problems are smaller than the original, in such a way that they are solved using potentially less time that is required to run the entire MIP formulation. Although the results of the test were not reported for all datasets, the behavior observed in Table 2 holds for all.

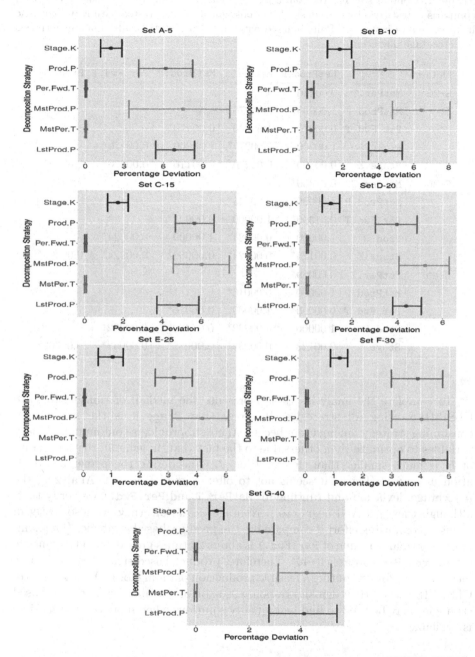

Fig. 3. Confidence intervals for the mean of the percentage deviation $\frac{f(RFO)-f(MIP)}{f(MIP)}$ resulting from execution of the RFO-heuristic using six different decomposition strategies based on period, products and stages over seven sets of instances.

Table 2. Comparison matrix containing p-values resulting from Pairwise Wilcoxon comparison test (two by two) of six decomposition strategies tested with RFO-heuristic in three sets of instances. Pairs marked with an ($*$) are statistically different in terms of percentage deviation.

Dataset		LstProd.P	MstPer.T	MstProd.P	Per.Fwd.T	Prod.P
E-25	MstPer.T	0.00270*	–	–	–	–
	MstProd.P	1.00000	0.00270*	–	–	–
	Per.Fwd.T	0.00270*	1.00000	0.00270*	–	–
	Prod.P	1.00000	0.00270*	1.00000	0.00270*	–
	Stage.K	0.00490*	0.00270*	0.00110*	0.00270*	0.0019*
F-30	MstPer.T	0.01360*	–	–	–	–
	MstProd.P	1.00000	0.00924*	–	–	–
	Per.Fwd.T	0.01360*	1.00000	0.00924*	–	–
	Prod.P	1.00000	0.00924*	1.00000	0.00924*	–
	Stage.K	0.00123*	0.00924*	0.00062*	0.00924*	0.00062*
G-40	MstPer.T	0.00069*	–	–	–	–
	MstProd.P	1.00000	0.00016*	–	–	–
	Per.Fwd.T	0.00069*	1.00000	0.00016*	–	–
	Prod.P	1.00000	0.00032*	0.04482*	0.00032*	–
	Stage.K	0.00123*	0.00032*	0.00032*	0.00032*	0.00062*

To complete the analysis, Table 3 presents the median of runtime to solve CLSP-HFS using all strategies. It shows that Stage.K is the more time-consuming strategy among them because their subproblems contain more binary variables to optimize than others due to the fact that, in general, the cardinality of K is much smaller than that of T or P. Considering the percentage deviation it returns (Fig. 3) it seems not to offer a good trade-off. Analyzing the percentage deviation and runtime for MstPer.T and Per.Fwd.T we verify that, although they achieved near-zero percentage deviation, they are also costly in terms of computing effort if compared to the product-based strategies (for example, the median runtime of Per.Fwd.T is about 22 times greater than the runtime of LstProd.P for dataset G-40). Therefore, product-based strategies seem to be more suitable for computing a feasible production planning for real-world cases of CLSP-HFS when they demand response as soon as possible, while period-based strategies may be used to find high-quality solutions when more computing time is available.

Table 3. Median of the runtime, in seconds, for different decomposition strategies used with RFO-heuristic on the tested datasets.

Dataset	LstProd.P	MstProd.P	Prod.P	Stage.K	MstPer.T	Per.Fwd.T
A-5	0.7117	0.5590	0.4997	1.7826	1.6672	1.6867
B-10	5.4010	5.4438	6.5524	85.3615	25.8521	27.5315
C-15	15.7831	11.1758	19.4317	254.7276	41.8867	46.6300
D-20	42,1319	42.4018	55.8263	310.6666	121.2430	120.1972
E-25	59.9986	47.2720	74.8139	312.0385	246.2482	263.7773
F-30	38.3576	32.2592	53.2724	310.7535	257.8930	266.3645
G-40	79.6631	82.9692	114.8287	322.0453	1766.1646	1779.9138

7 Conclusions

This paper proposed a new mixed-integer programming formulation for a multi-product, multiperiod, and multistage integrated lot sizing and scheduling problem with setup costs and solved it heuristically using a combined relax-and-fix and fix-and-optimize heuristic. Some experiments were done to evaluate the performance of the combined strategy considering six different decompositions of variables based on products, periods, and stages. They were tested in a benchmark containing seven datasets, each one with ten randomly generated instances. They were solved using a commercial solver, and the percentage deviation from the optimal solution (or to the best solution found in 3600 s) and runtime were collected for analysis. Experimental results show that period-based strategies achieved near-zero percentage deviation in most cases, while product-based strategies obtained a trade-off between solution quality and time.

For future works, we suggest to test a new decomposition criterion that merges information from stages and machines. In the current implementation, the machines that belong to the same stage are optimized together, so it is not possible to differentiate whom decision variables related to these machines are more interesting to optimize first because there is no trade-off between production and setup costs, machine capacity, and processing and setup times. Therefore, we hypothesize that separating them per machine and stage may offer a more effective and competitive decomposition. We also suggest extending the mixed-integer formulation to consider sequence-dependent setup, turning the problem more interesting in real-world cases and applications.

References

1. Araujo, K., Birgin, E., Kawamura, M., Ronconi, D.: Relax-and-fix heuristics applied to a real-world lot-sizing and scheduling problem in the personal care consumer goods industry. arXiv preprint arXiv:2107.10738 (2021)
2. Bitran, G.R., Yanasse, H.H.: Computational complexity of the capacitated lot size problem. Manag. Sci. **28**(10), 1174–1186 (1982)

3. Boas, B.E.V., Camargo, V.C., Morabito, R.: Modeling and MIP-heuristics for the general lotsizing and scheduling problem with process configuration selection. Pesquisa Operacional **41** (2021)
4. Carvalho, D.M., Nascimento, M.C.: Hybrid matheuristics to solve the integrated lot sizing and scheduling problem on parallel machines with sequence-dependent and non-triangular setup. Eur. J. Oper. Res. **296**(1), 158–173 (2022)
5. Dauzère-Péres, S., Lasserre, J.B.: An Integrated Approach in Production Planning and Scheduling, vol. 411. Springer, Heidelberg (2012)
6. Furlan, M.M., Santos, M.O.: BFO: a hybrid bees algorithm for the multi-level capacitated lot-sizing problem. J. Intell. Manuf. **28**(4), 929–944 (2017)
7. Gören, H.G., Tunalı, S.: Solving the capacitated lot sizing problem with setup carryover using a new sequential hybrid approach. Appl. Intell. **42**(4), 805–816 (2015)
8. Guo, Y., Li, W.D., Mileham, A.R., Owen, G.W.: Applications of particle swarm optimisation in integrated process planning and scheduling. Robot. Comput.-Integr. Manuf. **25**(2), 280–288 (2009)
9. Helber, S., Sahling, F.: A fix-and-optimize approach for the multi-level capacitated lot sizing problem. Int. J. Prod. Econ. **123**(2), 247–256 (2010)
10. James, R.J., Almada-Lobo, B.: Single and parallel machine capacitated lotsizing and scheduling: New iterative MIP-based neighborhood search heuristics. Comput. Oper. Res. **38**(12), 1816–1825 (2011)
11. Karimi, B., Ghomi, S.F., Wilson, J.: The capacitated lot sizing problem: a review of models and algorithms. Omega **31**(5), 365–378 (2003)
12. Lang, J.C., Shen, Z.J.M.: Fix-and-optimize heuristics for capacitated lot-sizing with sequence-dependent setups and substitutions. Eur. J. Oper. Res. **214**(3), 595–605 (2011)
13. Li, L., Song, S., Wu, C., Wang, R.: Fix-and-optimize and variable neighborhood search approaches for stochastic multi-item capacitated lot-sizing problems. Math. Probl. Eng. **2017** (2017)
14. Linn, R., Zhang, W.: Hybrid flow shop scheduling: a survey. Comput. Ind. Eng. **37**(1–2), 57–61 (1999)
15. Maravelias, C.T., Sung, C.: Integration of production planning and scheduling: overview, challenges and opportunities. Comput. Chem. Eng. **33**(12), 1919–1930 (2009)
16. Mateus, G.R., Ravetti, M.G., de Souza, M.C., Valeriano, T.M.: Capacitated lot sizing and sequence dependent setup scheduling: an iterative approach for integration. J. Sched. **13**(3), 245–259 (2010)
17. Michael, L.P.: Scheduling: Theory, Algorithms, and Systems. Springer, Heidelberg (2008)
18. Mohammadi, M., Fatemi, G.S.: Relax and fix heuristics for simultaneous lot sizing and sequencing the permutation flow shops with sequence-dependent setups. Int. J. Ind. Eng. Prod. Res. **21**(3), 147–153 (2010)
19. Montgomery, D.C., Runger, G.C.: Applied Statistics and Probability for Engineers. Wiley, Hoboken (2010)
20. Pochet, Y., Wolsey, L.A.: Production Planning by Mixed Integer Programming. Springer, Heidelberg (2006)
21. Ross, S.M.: Introduction to Probability and Statistics for Engineers and Scientists. Elsevier (2004)
22. Sahling, F., Buschkühl, L., Tempelmeier, H., Helber, S.: Solving a multi-level capacitated lot sizing problem with multi-period setup carry-over via a fix-and-optimize heuristic. Comput. Oper. Res. **36**(9), 2546–2553 (2009)

23. Saravia, K.L.L.: Modelos e algoritmos para o flowshop híbrido com tempos de preparação dependentes da sequência e da máquina. Master's thesis, Universidade Federal de Minas Gerais (2016)
24. Schimidt, T.M.P., Tadeu, S.C., Loch, G.V., Schenekemberg, C.M.: Heuristic approaches to solve a two-stage lot sizing and scheduling problem. IEEE Lat. Am. Trans. **17**(03), 434–443 (2019)
25. Toledo, C.F.M., da Silva Arantes, M., Hossomi, M.Y.B., França, P.M., Akartunalı, K.: A relax-and-fix with fix-and-optimize heuristic applied to multi-level lot-sizing problems. J. Heuristics **21**(5), 687–717 (2015)

Construct, Merge, Solve and Adapt Applied to the Maximum Disjoint Dominating Sets Problem

Roberto Maria Rosati[1,2(✉)] [ID], Salim Bouamama[3] [ID], and Christian Blum[2] [ID]

[1] DPIA, University of Udine, via delle Scienze 206, 33100 Udine, Italy
robertomaria.rosati@uniud.it
[2] Artificial Intelligence Research Institute (IIIA-CSIC) Campus of the UAB,
Bellaterra, Spain
christian.blum@iiia.csic.es
[3] Department of Computer Science, Mechatronics Laboratory (LMETR), Ferhat
Abbas University, Sétif 1, 19000 Sétif, Algeria
salim.bouamama@univ-setif.dz

Abstract. We propose a "construct, merge, solve and adapt" (CMSA) approach for the maximum disjoint dominating sets problem (MDDSP), which is a complex variant of the classical minimum dominating set problem in undirected graphs. The problem requires to find as many vertex-disjoint dominating sets of a given graph as possible. CMSA is a recent metaheuristic approach based on the idea of problem instance reduction. At each iteration of the algorithm, sub-instances of the original problem instance are solved by an exact solver. These sub-instances are obtained by merging the solution components of probabilistically generated solutions. CMSA is the first metaheuristic proposed for solving the MDDSP. The obtained results show that CMSA outperforms all existing greedy heuristics.

Keywords: Maximum disjoint dominating sets problem · Domatic partition problem · CMSA · Wireless sensor networks

1 Introduction

The identification of small dominating sets in graphs and networks is one of the classical combinatorial optimization problems in graph theory, with numerous applications ranging from biology to communication networks [14]. Given a simple, undirected graph $G = (V, E)$, a set of vertices $D \subseteq V$ is called a *dominating set* of G if every vertex $v \in V \setminus D$ is adjacent to at least one vertex $v' \in D$. The NP-hard *minimum dominating set* (MDS) problem requires to find a smallest dominating set in a given graph. The so-called *maximum disjoint dominating sets problem* (MDDSP) is a variation of the MDS problem in which a valid solution $\mathcal{S} = \{D_1, \dots, D_k\}$ consists of a collection of disjoint dominating sets D_i ($i = 1, \dots, k$) of G such that $D_i \cap D_j = \emptyset$ for all $i \neq j \in \{1, \dots, k\}$.

© The Author(s), under exclusive license to Springer Nature Switzerland AG 2023
L. Di Gaspero et al. (Eds.): MIC 2022, LNCS 13838, pp. 306–321, 2023.
https://doi.org/10.1007/978-3-031-26504-4_22

The objective function value $f(S)$ of a valid solution S is the number of disjoint dominating sets in S, that is, $f(S) := |S|$. The goal is to find a valid solution S^* that maximizes f.

Note that a highly related optimization problem is the *domatic partition problem* (DPP). Given a simple, undirected graph $G = (V, E)$ the DPP problem requires to partition the set of vertices V into the maximal number of disjoint dominating sets. In other words, a valid solution $S = \{D_1, \ldots, D_k\}$ to the DPP problem not only requires that all pairs of dominating sets $D_i \neq D_j \in S$ are disjoint $(i, j = 1, \ldots, k)$, but also that $\bigcup_{i=1}^{k} D_i = V$. Nevertheless, observe that any solution S to the MDDSP can easily be transformed into a solution S' to the DPP by adding all vertices from $V \setminus \bigcup_{i=1}^{|S|} D_i$ to any of the disjoint dominating sets of S. Note that, by adding further vertices to a dominating set D, set D does not lose its property of being a dominating set. This implies that an optimal solution to the MDDSP can easily be transformed into an optimal solution to the DPP. The value of an optimal solution to the DPP in graph G, denoted by $\gamma(G)$, is also called the *domatic number* of G. In this context, the term "domatic" was created as a composition of "dominating" and "chromatic" [8]. In the related literature, one can also find numerous references to the so-called *domatic number problem* (DNP) [8]. However, this problem only refers to finding the domatic number $\gamma(G)$ of a graph G. In other words, algorithmic approaches for solving the DNP try to identify $\gamma(G)$ without necessarily generating a corresponding solution. Therefore, by solving the DPP we simultaneously solve the DNP, but not vice versa. Finally, a special case of the DPP is the k-domatic partition problem that seeks to find a partition of a given graph into k disjoint dominating sets. Correspondingly, the k-domatic number problem asks whether a given graph can be partitioned into k dominating sets.

As mentioned above, in this work we attempt to solve the MDDSP in general graphs. Practical applications of this problem can be found especially in the context of heterogeneous multi-agent systems [18] and in wireless sensor networks (WSN), which are studied for their applications in the fields of healthcare, environmental monitoring, emergency operations and security surveillance [1]. Note that WSNs are networks composed of a rather large number of small devices called sensor nodes. These sensor nodes capture information from the environment and they are also responsible for transmitting the captured data to a base station. The power supply of sensor nodes is generally provided through batteries, which implies that their lifetime is limited. For the purpose of energy conservation, sensor nodes may change from their normal, active mode to another mode called low-energy, respectively sleep, mode. Hereby, the active mode allows capturing information and transmitting data, for example. The difference of energy consumption between the sleep and the active mode is considerable and may reach about two orders of magnitude [6,7]. When grouped into disjoint dominating sets, at any moment in time only the sensor nodes belonging to exactly one of these dominating sets are in active mode. All the others remain in sleep mode. Moreover, when the currently active sensor nodes reach a critical battery level they are put into sleep mode and the sensor nodes of the next dominating set are activated. This is repeated until all disjoint dominating sets have been used. Thus, the expected lifetime of the network is determined by the number of disjoint dominating sets times the average lifetime of a sensor in the active state. Therefore, the higher the number of dominating sets, the longer is the lifetime of the WSN.

For solving the MDDSP we propose a "construct, merge, solve and adapt" (CMSA) algorithm [3], which is a recent metaheuristic technique belonging to the field of matheuristics [12]. CMSA has shown to be suitable for other hard combinatorial optimization problems such as the minimum capacitated dominating sets problem [21], graph coloring applied to social networking [16], and the test data generation for software product lines [11]. The choice of CMSA is motivated by the fact that the algorithm relies on the availability of fast and effective greedy heuristics. In fact, most of the work on algorithms for the MDDSP and the DPP has focused on greedy heuristics so far (see Sect. 2). Another reason for selecting CMSA is that the complex structure of solutions to the problem make the application of algorithms based on local search rather hard. Finally, note that—even though the MDDSP and DPP problems have been studied since many years—to the best of our knowledge no metaheuristic or matheuristic technique for solving the problem has been proposed so far.

The remainder of the paper is organized as follows. Section 2 introduces the related work that can be found in the literature. Section 3 provides graphical examples of problem instances and solutions. Section 4 describes in detail our CMSA approach, while the experimental methodology and results are discussed in Sect. 5. Finally, Sect. 6 is dedicated to the conclusions and future work.

2   Related Work

The DPP was first introduced in [9]. In the same paper, the authors showed that the problem has an upper bound of $\delta(G) + 1$, where $\delta(G)$ is the minimum degree of all vertices in G. In other words, an optimal solution to the DPP can never have more disjoint dominating sets than $\delta(G) + 1$. Furthermore, the problem was shown to belong to the class of NP-hard problems on general graphs [13]. In fact, it remains NP-hard for co-bipartite graphs [22]. Moreover, it was shown that—unless P=NP—the MDDSP has no polynomial-time α-approximation algorithm for any constant α smaller than 1.5 [7]. Existing polynomial-time approximation schemes can be found in [10]. In addition to the above, the NP-completeness of the 3-domatic partition problem was proved for general graphs [7] and still holds when restricted to planar bipartite graphs [22] and planar unit disk graphs [19].

As mentioned before, research on the DNP[1] deals with deriving the domatic number of a given graph without the need for deriving the corresponding collection of disjoint dominating sets. A literature review yields several studies related to the DNP. Most of these studies are primarily interested in dealing with the DNP in certain families of graphs. Although not many exact approaches can be

---

[1] Remember that DNP stands for "domatic number problem".

found in the literature for DPP-like problems, being mainly due to their complexity, some exact approaches have been designed for finding optimal solutions to the DNP. In particular, the first exact deterministic exponential-time algorithm for the 3-DNP was designed in [23]. The algorithm has a running time of $\mathcal{O}(2.9416^n)$ and uses polynomial space, which is in contrast to a naive approach that runs in $\mathcal{O}(3^n)$ of time. This time complexity was later improved to $\mathcal{O}(2.695^n)$ in [24]. Combining the two main techniques typically used for the design of exact exponential-time algorithms—inclusion & exclusion, respectively measure & conquer—the work in [25] provided a fast polynomial-space algorithm that computes the domatic number in $\mathcal{O}(2.7139^n)$ time.

A natural way to build a feasible solution to the MDDSP is to greedily construct dominating sets of preferably small cardinality with the ultimate goal to maximize the number of disjoint dominating sets that can be generated.[2]

A vertex-coloring heuristic (COLOR-DDS) working in two phases and with a time complexity of $O(n^3)$, where n is the number of vertices of the graph, was proposed in [7]. First, all vertices are colored using a sequential coloring algorithm similar to the well-known Welsh-Powell algorithm in order to construct independent sets. In this context, an independent set is a subset of vertices in a graph, no two of which are connected by an edge. Vertices belonging to the same independent set then receive the same color. Subsequently, a heuristic is employed to construct disjoint dominating sets based on these independent sets.

Three other greedy heuristics, called progressive maximum degree disjoint dominating sets (P-MAX), progressive minimum degree disjoint dominating sets (PMIN) and random lowest ID disjoint dominating sets (RLID), were introduced in [19]. Their performance was evaluated and compared against COLOR-DDS both in terms of the number of disjoint dominating sets and the computation time. Note that all three greedy heuristics follow the same underlying mechanism.

An improved version of the P-MAX greedy heuristic which, in this paper, is denoted as IAM (the acronym is composed of the initials of the authors' surnames) was presented in [15]. In particular, while constructing a dominating set they consider both "white vertices" (that is, vertices not yet chosen and not yet covered) and "gray vertices" (that is, vertices not yet chosen but already covered) as possible extensions of the current dominating set. Ties concerning the employed greedy function are broken by choosing a vertex with the minimum number of neighbors (including itself) already added to one of the generated dominating sets. If the tie is still unresolved, the vertex with the smallest ID is chosen.

The currently best greedy algorithm (called MDDS-GH) for the MDDSP was more recently proposed in [2]. This algorithm can be seen as an extension of the one from [15] in a way that both enhances its performance and avoids its drawbacks. Moreover, MDDS-GH is able to solve a weighted variant of the MDDSP labeled the maximum weighted disjoint dominating sets problem, and it

---

[2] Remember that any solution to the MDDSP can be trivially transformed to a solution to the DPP by adding those vertices that do not belong to any of the disjoint dominating sets to one of the dominating sets of the MDDSP-solution.

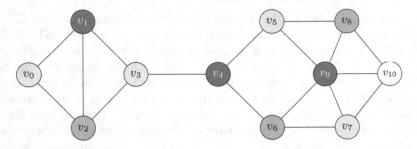

Fig. 1. A graph with 11 vertices and 17 edges. The upper bound for the domatic number is $\delta(G) + 1 = 3$. Moreover, there exists an optimal solution with 3 disjoint dominating sets (as indicated by the vertices with a background color different to white). (Color figure online)

is among the main components of a metaheuristic proposed for the same problem variant in [4].

To the best of our knowledge, no metaheuristic approach has been proposed so far to solve the MDDSP. As outlined above, existing studies focus on developing exact, approximation or greedy heuristic algorithms.

3   Graphical Problem Illustration

An example graph, together with an optimal solution, is shown for illustration purposes in Fig. 1. Let $G = (V, E)$, with $|V| = 11$ and $|E| = 17$, be the undirected graph shown in this graphic. Vertices are labeled $v_0, \ldots, v_{10}$, while edges are unlabeled. Moreover, the graphic shows an optimal solution with three dominating sets (as indicated by the background colors of the vertices). The sets are, respectively, $D_1 = \{v_0, v_3, v_5, v_7\}$, $D_2 = \{v_1, v_4, v_9\}$, and $D_3 = \{v_2, v_6, v_8\}$. From the graphic it is easy to see that all these sets are indeed dominating sets of G. Considering, for instance, the case of set D_1, the following can be observed:

1. v_0 dominates the adjacent vertices v_1 and v_2
2. v_3 dominates v_1, v_2 and v_4
3. v_5 dominates v_4, v_9 and v_0
4. v_7 dominates v_6, v_9 and v_{10}

In addition, $D_1 \cap D_2 = \emptyset$, $D_1 \cap D_3 = \emptyset$, and $D_2 \cap D_3 = \emptyset$, which means that the three dominating sets are pairwise disjoint. Hence, $S = \{D_1, D_2, D_3\}$ is a solution to the MDDSP in graph G, with objective function value $f(S) = 3$. Note that a solution to the DPP is obtained by adding v_{10}, which does not belong to any dominating set of the considered solution, to one of the three dominating sets.

A useful property of the MDDSP, that we exploit in our algorithm (details are given in Sect. 4), is the following one. As already mentioned in Sect. 2, $\delta(G) + 1$— where $\delta(G)$ is the minimum degree of all vertices in graph G—is a proven upper

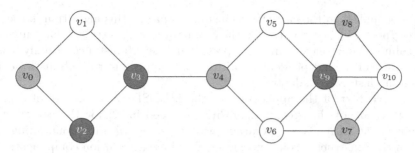

Fig. 2. A graph with 11 vertices and 16 edges. The upper bound for the domatic number is $\delta(G) + 1 = 3$, but an optimal solution has only 2 disjoint dominating sets. One of the possible optimal solutions is indicated by blue and purple vertices. Uncolored vertices do not belong to any dominating set of the displayed solution. (Color figure online)

bound for the number of disjoint dominating sets in G [9]. This upper bound is actually rather easy to be verified. Given a solution $\mathcal{S} = \{D_1, \ldots, D_k\}$ to the MDDSP in graph G, any vertex $v_k \in V$ must be dominated by a different vertex in all $D_i \in \mathcal{S}$. However, note that a vertex $v_k \in V$ with degree $deg(v_k)$ can only be dominated by itself or by any of its neighbors, that is, by at most $deg(v_k) + 1$ different vertices. Therefore, there can be at most $\delta(G) + 1$ disjoint dominating sets in G. Nevertheless, this value, which is very easy to calculate and very useful when solving the problem in practice, does not imply that a solution with $\delta(G) + 1$ disjoint dominating sets exists. That is to say, if we have a solution with value $\delta(G) + 1$ we can be sure that it is optimal, but the opposite does not hold. Compare, for example, the two graphs from Figs. 1 and 2. They are identical except for the fact that the graph from Fig. 2 does not have an edge connecting vertices v_1 and v_2. For both graphs it holds that $\delta(G) + 1 = 2 + 1 = 3$, because—in both cases—the vertex with the lowest degree has a degree of 2. However, while it is possible to find an optimal solution of value 3 for the graph in Fig. 1, an optimal solution for the graph in Fig. 2 has a value of 2.

Finally, it is worth to mention that every graph admits at least one feasible solution to the MDDSP, as V is a dominating set of $G = (V, E)$. Moreover, every graph without isolated vertices contains at least two disjoint dominating sets (see [20]). This implies that there are no infeasible problem instances.

4   The CMSA Approach to the MDDSP

In this section we describe the "construct, merge, solve and adapt" (CMSA) algorithm [3] that we designed for solving the MDDSP. CMSA is a recent matheuristic that roughly works as follows. At each iteration, the algorithm first generates a number of solutions in a probabilistic way. Second, the solution components found in these solutions are added to an initially empty sub-instance of the tackled problem instance. Third, an independent algorithm—typically an exact solver—is applied to the current sub-instance, that is, this algorithm is used to find the (possibly) best solution to the original problem instance that only contains solution components currently present in the sub-instance. Finally, based on the result of the independent algorithm when applied to the sub-instance, the current sub-instance is adapted.

In the context of the application to the MDDSP, we make use of a probabilistic version of the greedy algorithm proposed in [2] for the construction of solutions at each CMSA iteration. Moreover, the disjoint dominating sets found in the constructed solutions are regarded as the solution components. We make use of the integer linear programming (ILP) solver CPLEX (version 20.1) for solving sub-instances by means of a set packing ILP formulation. All these aspects are outlined in detail in the following. First, however, we will describe the lexicographic objective function that we designed in order to deal with plateaus in the search space.

4.1   Lexicographic Objective Function

As mentioned already before, given a solution $\mathcal{S} = \{D_1, \ldots, D_k\}$ to the MDDPS, the objective function value is simply $f(\mathcal{S}) := |\mathcal{S}|$, that is, it counts the number of disjoint dominating sets in $\mathcal{S}$. For this reason, many solutions have the same objective function value. As a consequence, the corresponding search landscape has large plateaus [26].

Consider, for example, the solution displayed in Fig. 1. $D_2 = \{v_1, v_4, v_9\}$ can be replaced by $D_2' = \{v_1, v_4, v_{10}\}$, obtained by swapping vertex v_9 with vertex v_{10}, or by $D_2'' = \{v_1, v_4, v_9, v_{10}\}$, obtained by adding vertex v_{10}. Solutions $\mathcal{S} = \{D_1, D_2, D_3\}$, $\mathcal{S}' = \{D_1, D_2', D_3\}$, and $\mathcal{S}'' = \{D_1, D_2'', D_3\}$ are all different, though, they all have the same value $f(\mathcal{S}) = f(\mathcal{S}') = f(\mathcal{S}'') = 3$.

The problem that such an objective function presents for (meta-)heuristic search processes is that—due to the lack of a suitable guidance—an algorithm may get lost on plateaus. A possible way to deal with plateaus is to use a lexicographic objective function $f^{lex}()$—as done, for example, in [5]—that uses the original objective function as a first criterion for comparing two solutions. Only when the two solutions to be compared have the same original objective function value, a second criterion is used to differentiate between them. More precisely, given two solutions $\mathcal{S}_1$ and $\mathcal{S}_2$, $\mathcal{S}_1$ is said to be lexicographically better than $\mathcal{S}_2$—that is, $f^{lex}(\mathcal{S}_1) > f^{lex}(\mathcal{S}_2)$—if and only if

1. $f(\mathcal{S}_1) > f(\mathcal{S}_2)$ **or**
2. $f(\mathcal{S}_1) = f(\mathcal{S}_2)$ **and** $r(\mathcal{S}_1) > r(\mathcal{S}_2)$

Hereby, the second criterion, $r()$, is a residual coverage function. Given a solution $\mathcal{S} = \{D_1, \ldots, D_k\}$, the idea of the residual coverage function $r(\mathcal{S})$ is to calculate the fraction of the input graph G that can be covered with the remaining vertices $V' := V \setminus \bigcup_{i=1}^{k} D_i$ which are not assigned to any of the disjoint dominating sets of $\mathcal{S}$. More specifically, $r(\mathcal{S})$ is defined as follows:

$$r(S) := \frac{|V'| + \left|\left(\bigcup_{v \in V'} N(v)\right) \setminus V'\right|}{|V|} \quad , \tag{1}$$

where $N(v)$ is the set of neighbors of v in G. In other words, $r(S)$ is the fractional size of the "residual dominating set" that can be built from vertices in V'. Note that, in general, the value of $r()$ may range between 0 and 1. The extreme case of $r(S) = 0$ is obtained if $\bigcup_{D \subset S} D = V$. The other extreme case of $r(S) = 1$ is obtained when at least one more dominating set can be constructed from the vertices in V'. In all the other cases it holds that $0 < r(S) < 1$. Thanks to this lexicographic objective function, when comparing two solutions that have the same original objective function value, in CMSA we prefer the one in which the remaining/unused vertices are closer to form an additional dominating set. The adoption of this technique eventually helps to guide the search from a solution S with objective function value $|S|$ to a solution S' with objective function value $|S| + 1$.

4.2   The CMSA Algorithm

Our CMSA algorithm for the MDDPS is based on the following main idea. Given an input graph G, let $\mathcal{C}$ be the collection of all possible dominating sets of G. If $\mathcal{C}$ could be generated in an efficient way, the MDDPS could be solved by solving the following set packing ILP formulation with an ILP solver such as CPLEX.

$$\max \sum_{D \in \mathcal{C}} x_D \tag{2}$$

$$\text{s.t.} \sum_{\{D \in \mathcal{C} | v \in D\}} x_D \leq 1 \qquad \forall v \in V \tag{3}$$

$$x_D \in \{0, 1\} \qquad \forall D \in \mathcal{C} \tag{4}$$

This ILP model is based on a binary variable x_D for each dominating set $D \in \mathcal{C}$, whereby a value of $x_D = 1$ means that D is chosen to be part of the solution. Moreover, Constraints (3) ensure that each vertex of G may be present in at most one of the chosen dominating sets. In this way, the chosen dominating sets will be pairwise disjoint.

Unfortunately, there is no efficient way for enumerating all dominating sets of a graph. And even if there was one, the size of $\mathcal{C}$ would probably be much too large to solve the above ILP model efficiently. Nevertheless, by regarding all possible dominating sets of an input graph G as the *solution components* of the problem, a CMSA algorithm can be designed—as shown in the following—that makes use of the above ILP model for solving sub-instances $\mathcal{C}' \subset \mathcal{C}$ by means of the application of an ILP solver such as CPLEX.

Algorithm 1 provides the pseudo-code of our CMSA algorithm for the MDDSP. Apart from the graph G, our algorithm takes as input the values for the following five parameters. Note that these values will be determined through a statistically-principled tuning procedure, as discussed in Sect. 5.2. The parameters are:

- n_{sols}, which fixes the number of solutions to be probabilistically generated by the construction procedure at each CMSA iteration.

Algorithm 1. CMSA for the MDDSP

1: **input:** a graph $G(V, E)$, values for n_{sols}, d_{rate}, c_{list}, age_{limit}, $Cplex_{time}$
2: $S_{bsf} \leftarrow \emptyset$; $\mathcal{C}' \leftarrow \emptyset$
3: **while** $f(S_{bsf}) < \delta(G) + 1$ **and** CPU time limit not reached **do**
4:    **for** $i \leftarrow 1,...,n_{sols}$ **do**
5:       $S_{cur} \leftarrow$ ProbabilisticGenerateSolution(G)
6:       **if** $f^{lex}(S_{cur}) > f^{lex}(S_{bsf})$ **then** $S_{bsf} \leftarrow S_{cur}$ **end if**
7:       **for all** $D \in S_{cur}, D \notin \mathcal{C}'$ **do**
8:          $age[D] \leftarrow 0$
9:          Add D to $\mathcal{C}'$
10:       **end for**
11:    **end for**
12:    $S_{exc} \leftarrow$ ApplyExactSolver$(\mathcal{C}', Cplex_{time})$
13:    **while** $r(S_{exc}) = 1$ **do**
14:       $S_{exc} \leftarrow$ ApplyRepairProcedure(S_{exc})
15:    **end while**
16:    **if** $f^{lex}(S_{exc}) > f^{lex}(S_{bsf})$ **then**
17:       $S_{bsf} \leftarrow S_{exc}$
18:    **end if**
19:    Adapt$(\mathcal{C}', S_{exc}, age_{limit})$
20: **end while**
21: **output:** S_{bsf}

- d_{rate} and c_{list}, which guide, respectively, the determinism rate in the solution construction procedure and the length of the candidate list for those steps in which non-deterministic construction is performed.
- age_{limit}, which limits the number of iterations a solution component can remain in the subinstance $\mathcal{C}'$ without being chosen by the exact solver for the best solution to the sub-instance. Note that the age of a solution component (dominating set) D is maintained in a variable $age[D]$.
- $Cplex_{time}$ (in seconds), which is the time limit for the application of CPLEX at each iteration of CMSA.

CMSA starts by the initialization of the best solution found so far, S_{bsf}, to an empty solution with objective function value equal to zero. Moreover, the sub-instance $\mathcal{C}'$ is initialized to an empty set. The main loop of CMSA starts in line 3 of Algorithm 1. At each CMSA iteration, the construct and merge steps are repeated until n_{sols} are generated; see lines 4–10. The construction procedure, specifically, is called at line 5. It performs a probabilistic version of the greedy heuristic from [2] for generating a solution S_{cur}. This greedy heuristic builds one disjoint dominating set after the other. Hereby, each disjoint dominating set is generated by adding exactly one vertex to the current partial dominating set at each construction step. This vertex is chosen from the ones that are not yet assigned to any of the dominating sets in the current partial solution. For this choice, the greedy heuristic makes use of a greedy function for evaluating all the options. For a comprehensive explanation of the procedure, we refer the interested reader to [2]. For its use in CMSA, this greedy heuristic was adapted as follows. Instead of choosing, at each construction step, the vertex with the best greedy function value, first, a random number d is drawn uniformly at random from $[0, 1)$. Then, if $d < d_{rate}$, the choice of the next vertex is done deterministically. Otherwise, a candidate list of length c_{list}, containing the maximally c_{list} best candidates according to their greedy function values, is built, and a vertex is drawn uniformly at random from this list.

After the construction of a solution (line 5), the corresponding merge step is performed in lines 7–10, that is, all those dominating sets $D \in \mathcal{S}_{cur}$ that are not yet present in sub-instance $\mathcal{C}'$ are added to $\mathcal{C}'$. Moreover, the age value $age[D]$ of each such dominating set D is initialized to zero. After generating and merging n_{sols} solutions in this way, the CMSA algorithm enters into the solve phase; see line 12. More specifically, the ILP solver CPLEX is applied in function ApplyExactSolver($\mathcal{C}', Cplex_{time}$) with a computation time limit of $Cplex_{time}$ CPU seconds for solving the sub-instance $\mathcal{C}'$. This is done by solving the ILP model stated at the beginning of this section after replacing all occurrences of $\mathcal{C}$ with $\mathcal{C}'$. The output of this function ($\mathcal{S}_{exc}$) is the best solution that can be found by CPLEX within the given CPU time.

Occasionally it may happen that $r(\mathcal{S}_{exc}) = 1$,[3] which means that from the vertices not included in any dominating set of $\mathcal{S}_{exc}$ at least one additional dominating set can be generated. This may happen for two possible reasons: (1) $\mathcal{C}'$ does not include such a dominating set, or (2) the time limit of $Cplex_{time}$ CPU seconds did not allow CPLEX to find an optimal solution to $\mathcal{C}'$. When this case is detected, a repair procedure is activated (lines 13–15) that iteratively generates additional (disjoint) dominating sets until $r(\mathcal{S}_{exc}) < 1$. Note that the reason for only applying this procedure to solutions generated by CPLEX is that the greedy procedure always builds dominating sets until no additional ones can be built.

Finally, in line 19, the sub-instance is adapted. This adaptation comprises the following steps. First, dominating sets from $\mathcal{S}_{exc}$ that may have been generated by the repair procedure and that are not already included in the sub-instance $\mathcal{C}'$ are merged with $\mathcal{C}'$ in the same way as shown in lines 7–10. Second, the age values of all dominating sets from $\mathcal{S}_{exc}$ are re-set to zero. Third, the age values of all remaining dominating sets from $\mathcal{C}'$ are incremented by one. Finally, all dominating sets $D \in \mathcal{C}'$ with $age[D] > age_{limit}$ are removed from $\mathcal{C}'$.

5 Experimental Evaluation

In this section we present the comparison of CMSA with the six greedy algorithms from the literature, namely, MDDS-GH from [2], IAM from [15], P-MAX, PMIN, RLID from [19], and COLOR_DDS from [7]. CMSA and the six greedy algorithms were implemented in C++ and compiled with GNU g++, version 9.4.0, on Ubuntu 20.04.4 LTS. All the experiments presented in this section, both for CMSA and for the greedy algorithms, were run on a machine equipped with AMD Ryzen Threadripper PRO 3975WX processor with 32 cores, with

---

[3] See Sect. 4.1 for the definition of function $r()$.

Table 1. CMSA parameters, the considered domains for parameter tuning, and the finally determined parameter values.

Parameter	RGs Domain	Value	RGGs Domain	Value
n_{sols}	$\{2, 3, ..., 200\}$	191	$\{2, 3, ..., 30\}$	20
d_{rate}	$[0.60, 1.00]$	0.93	$[0.60, 1.00]$	0.69
c_{list}	$\{2, 3, ..., 50\}$	20	$\{2, 3, ..., 30\}$	22
age_{limit}	$\{2, 3, ..., 50\}$	5	$\{2, 3, ..., 30\}$	16
$Cplex_{time}$	$\{3, 4, ..., 20\}$	7	$\{3, 4, ..., 100\}$	81

base clock frequency of $3.5\,\text{GHz}$, and 64 GB of RAM. One core was used for each experiment.

5.1   Problem Instances

We tested our algorithm on two categories of graphs, namely: random graphs (RGs) and random geometric graphs (RGGs). Hereby, the term "random" is employed because the instances were generated using a random generator. Random graphs are general graphs, characterized by two features: (1) the number of vertices, $|V|$, and (2) the average vertex degree, μ_{deg}. In these graphs, any pair of vertices may be potentially connected. RGGs, on the other hand, are characterized by $|V|$ and a radius $r_{max} < 1$. $|V|$ vertices are placed at random coordinates (x, y) on the $[0, 1]^2$ square. Then, every pair of vertices $v_i \neq v_j$ with an Euclidean distance $d(v_i, v_j) < r_{max}$ are connected by an edge. These graphs are also known as planar unit disk graphs. RGGs, furthermore, are typically used as a graph model for wireless sensor networks mentioned in Sect. 1.

In total, we tested our algorithm on 34 different combinations of $|V|$ and μ_{deg} (see Table 2), respectively 21 combinations of $|V|$ and r_{max} (see Table 3). For each of these 55 graph parameter combinations we generated 20 graphs, which makes a total of 1100 graphs. All the instances are available online for future comparison at https://bitbucket.org/maximum-disjoint-dominating-sets-problem/maximumdisjointdominatingsets-instances, together with an instance and solution validator.

5.2   Parameters Tuning

The tuning procedure was realized through *irace*, a package for automatic algorithm configuration based on iterated racing [17]. We tuned independently the parameters for random graphs and random geometric graphs. This choice is motivated by the notable difference in graph size and density between the two considered graph categories. The computation time limit given to CMSA for each application was $|V|$ CPU seconds, that is, the allowed computation time depends linearly on the graph size. Table 1 summarizes the parameters involved in the tuning, the allowed parameter value domains, and the outcome, for the two categories of graphs. All parameters have domains of natural numbers, except for d_{rate}, which takes real numbers with a precision of two digits behind the comma. We allowed quite large domains for the parameters values for the application to random graphs, while we deliberately restricted the domains of n_{sols}, c_{sols}, and age_{limit} for the application to RGG instances. This is because our RGGs are larger, which leads to the fact that CMSA generates larger ILP models that cause a high memory consumption. Accordingly, we allow a larger range for $Cplex_{time}$ for RGGs, as, in general, CPLEX may require longer running times to solve larger ILP models.

5.3 Results

Both CMSA and the six greedy algorithms were applied exactly once to each one of the 1060 graphs. As in the case of parameter tuning, the computation time of CMSA was limited to $|V|$ CPU seconds per run. The obtained results are shown in Tables 2 and 3. These tables report the values of the best solutions found by CMSA and the six greedy heuristics, averaged over the 20 graphs per graph parameter combination. The computation times indicate, in all cases, the total execution times. Note, in this context, that CMSA can only have an execution time lower than the permitted CPU time limit if solutions are found whose objective function value coincides with the upper bound $\delta(G) + 1$.

The numerical results allow making the following observations. First of all, CMSA yields better results than all six greedy approaches for all 34 RG types and for all 21 RGG types. Hereby, the advantage of CMSA over the greedy heuristics increases both with a growing graph size and with a growing graph density. In the context of the largest and densest RGs ($|V| = 250$, $\mu_{deg} = 140$) CMSA finds, on average, 5.7 disjoint dominating sets more than the best greedy heuristic (MDDS-GH), which is quite a significant improvement. The same can be observed for RGGs. In particular, CMSA finds 4.2 disjoint dominating sets more than the best greedy heuristic (P-MAX) for RGGs with $|V| = 5000$ and $r_{max} = 0.3$. Such improvements might translate into quite significant gains in the context of practical applications. Concerning computation times, clearly the advantage of the greedy heuristics in comparison to CMSA is their low computation time. Nevertheless, the before-mentioned applications of the MDDSP are not time critical, that is, there is no significant disadvantage for algorithms with a computation time of minutes (or even an hour) when compared to greedy heuristics with running times of seconds.

Concerning the comparison of the greedy heuristics among each other, the following can be observed. While MDDS-GH is clearly the best-performing greedy heuristic for the application to RGs, this is not anymore the case for RGGs. In particular, P-MAX, RLID, and COLOR_DDS generally outperform MDDS-GH especially with a growing graph density. This is especially interesting in the case of RLID and COLOR_DDS, because their execution time is about one order of magnitude lower than the one of MDDS-GH, which makes them a good

Table 2. Numerical results obtained for RGs.

Instances		CMSA		MDDS-GH		IAM		P-MAX		PMIN		RLID		COLOR_DDS	
$|V|$	μ_{deg}	obj	$t(s)$	obj	$t(s)$	obj	$t(s)$	obj	$t(s)$	obj	$t(s)$	obj	$t(s)$	obj	$t(s)$
50	15	**8.05**	28	6.75	0.000	5.95	0.001	5.70	0.000	4.80	0.000	5.00	0.000	5.40	0.000
	20	**11.30**	48	9.00	0.000	8.20	0.001	7.60	0.000	6.35	0.000	6.80	0.000	7.05	0.000
	25	**13.75**	50	11.75	0.000	10.95	0.002	9.50	0.000	8.00	0.000	8.60	0.000	9.45	0.000
	30	**16.30**	50	14.50	0.000	13.55	0.002	12.00	0.000	10.30	0.000	11.05	0.000	11.40	0.001
	35	**18.55**	50	17.25	0.001	16.50	0.002	14.40	0.000	12.30	0.000	13.25	0.000	13.50	0.001
100	20	**9.65**	91	8.45	0.001	7.55	0.008	6.60	0.000	5.45	0.001	6.10	0.000	6.25	0.001
	30	**14.70**	100	12.70	0.001	11.80	0.006	10.05	0.001	8.15	0.001	9.05	0.000	9.40	0.001
	40	**19.50**	100	17.00	0.002	16.00	0.007	13.40	0.001	10.85	0.001	11.90	0.001	12.25	0.001
	50	**24.80**	100	21.50	0.002	20.65	0.008	16.95	0.001	13.95	0.001	15.45	0.001	15.70	0.001
	60	**30.55**	100	26.30	0.002	25.35	0.007	20.65	0.001	17.20	0.001	19.25	0.001	19.20	0.002
150	30	**13.05**	143	12.00	0.003	11.00	0.012	9.55	0.001	7.75	0.001	8.70	0.001	8.70	0.001
	40	**17.35**	150	16.15	0.004	15.25	0.013	12.35	0.001	10.05	0.001	11.35	0.001	11.15	0.002
	50	**22.15**	150	20.00	0.004	19.05	0.013	15.65	0.002	12.55	0.002	13.80	0.001	14.20	0.002
	60	**26.90**	150	24.00	0.005	23.05	0.013	18.40	0.002	15.15	0.002	16.65	0.001	17.10	0.002
	70	**31.50**	150	28.10	0.006	27.15	0.015	21.50	0.002	17.75	0.002	19.30	0.001	20.00	0.003
	80	**36.95**	150	33.00	0.006	32.00	0.015	25.35	0.002	20.60	0.002	22.75	0.001	23.30	0.003
	90	**40.30**	150	36.50	0.007	35.55	0.015	28.85	0.002	24.05	0.003	26.30	0.001	26.90	0.004
200	40	**16.85**	200	15.50	0.006	14.55	0.022	12.35	0.002	9.80	0.002	10.85	0.001	11.10	0.002
	50	**20.35**	200	18.90	0.007	18.05	0.021	14.45	0.002	12.00	0.003	13.45	0.001	13.50	0.002
	60	**24.10**	200	23.30	0.008	22.35	0.022	17.60	0.003	14.10	0.003	15.65	0.002	16.00	0.003
	70	**28.05**	200	26.80	0.008	25.80	0.024	20.30	0.003	16.45	0.003	18.40	0.002	18.60	0.003
	80	**32.80**	200	30.65	0.009	29.70	0.025	23.35	0.003	18.90	0.004	21.05	0.002	21.25	0.004
	90	**38.40**	200	34.80	0.010	33.80	0.026	26.40	0.004	21.25	0.004	23.75	0.002	24.10	0.004
	100	**43.20**	200	38.50	0.010	37.60	0.026	29.55	0.004	24.15	0.004	26.75	0.002	27.20	0.005
250	50	**20.10**	250	19.00	0.006	18.00	0.035	14.30	0.003	11.35	0.004	12.85	0.002	13.00	0.003
	60	**23.75**	250	22.65	0.011	21.65	0.035	16.95	0.003	13.70	0.004	15.40	0.002	15.50	0.003
	70	**27.10**	250	26.10	0.012	25.20	0.035	19.75	0.004	15.95	0.005	17.60	0.003	18.15	0.004
	80	**30.95**	250	29.45	0.012	28.55	0.036	22.30	0.005	18.10	0.005	20.00	0.003	20.45	0.005
	90	**34.75**	250	33.50	0.013	32.50	0.038	24.80	0.005	20.45	0.005	22.60	0.003	22.80	0.005
	100	**38.85**	250	37.15	0.014	36.20	0.040	28.05	0.005	22.70	0.006	25.00	0.003	25.25	0.006
	120	**48.05**	250	45.05	0.015	44.05	0.043	33.75	0.006	27.55	0.007	30.70	0.003	30.90	0.008
	140	**59.30**	250	53.60	0.017	52.70	0.045	40.40	0.007	32.95	0.007	36.75	0.004	37.25	0.009

candidate for replacing MDDS-GH for the probabilistic solution construction in CMSA in the context of rather large and dense RGGs in future work.

Finally, the execution times of CMSA that are presented in both tables indicate if CMSA was able to obtain solutions whose objective function values coincide with the upper bound $\delta(G) + 1$. For RGs, for example, the execution times are nearly always equal to the computation time limits. Exceptions are the RGs with $|V| = 50$ and $\mu_{deg} = 15$ (9 graphs solved to proven optimality), $|V| = 50$ and $\mu_{deg} = 20$ (1 graph solved to optimality), $|V| = 100$ and $\mu_{deg} = 20$ (2 graphs solved to optimality), and $|V| = 150$ and $\mu_{deg} = 30$ (1 graph solved to optimality). Assuming that CMSA is often able to provide optimal solutions, this indicates that the upper bound $\delta(G) + 1$ is often not tight for the RG instances. This seems to be different for the RGGs. Figure 3 presents two graphics relative to the results obtained on RGGs. The first one shows—for each radius—the

Table 3. Numerical results obtained for RGGs.

Instances		CMSA		MDDS-GH		IAM		P-MAX		PMIN		RLID		COLOR_DDS	
$|V|$	r_{max}	obj	t(s)	obj	t(s)	obj	t(s)	obj	t(s)	obj	t(s)	obj	t(s)	obj	t(s)
1000	0.1	**7.90**	80	7.30	0.04	6.30	0.33	7.30	0.02	6.75	0.12	7.15	0.01	7.35	0.01
	0.2	**30.55**	50	29.30	0.12	28.30	0.47	29.95	0.05	27.25	0.12	29.40	0.02	29.35	0.04
	0.3	**71.45**	605	67.00	0.25	66.00	0.84	68.55	0.13	64.20	0.28	68.95	0.03	69.20	0.10
1500	0.1	**12.60**	77	11.65	0.21	10.65	0.94	11.75	0.17	10.60	0.28	11.40	0.18	11.40	0.12
	0.2	**48.00**	298	45.95	0.57	44.95	1.67	45.75	0.27	43.05	0.54	45.85	0.17	46.00	0.20
	0.3	**105.50**	520	101.25	1.62	100.25	2.34	102.55	0.81	95.40	1.08	102.20	0.19	102.70	0.38
2000	0.1	**16.90**	302	15.90	0.27	14.90	1.69	15.95	0.16	14.20	0.27	15.80	0.05	16.00	0.04
	0.2	**62.55**	566	60.25	4.78	59.25	10.09	60.85	1.66	56.55	4.44	60.25	0.08	60.85	0.19
	0.3	**141.75**	1047	135.45	15.21	134.45	23.40	137.35	4.53	128.65	11.19	137.80	0.15	137.90	0.69
2500	0.1	**21.00**	178	20.20	0.79	19.20	3.07	19.85	0.55	18.35	0.77	19.55	0.36	19.85	0.38
	0.2	**79.85**	930	76.70	3.26	75.70	5.63	76.75	1.50	71.40	2.48	76.40	0.42	77.00	0.60
	0.3	**177.30**	1376	169.70	6.51	168.70	9.81	173.45	2.38	160.15	4.57	171.45	0.50	172.80	1.25
3000	0.1	**23.15**	3	22.45	3.38	21.45	9.55	22.35	1.02	20.90	1.92	22.55	0.11	22.55	0.08
	0.2	**93.70**	1383	90.85	31.96	89.85	29.99	89.75	5.09	84.05	15.49	90.70	0.20	91.15	0.51
	0.3	**215.30**	1540	208.05	48.93	207.05	44.92	211.80	14.75	194.85	29.34	208.00	0.27	210.05	2.41
4000	0.1	**33.50**	731	31.35	20.63	30.35	24.59	31.55	4.83	28.90	9.30	31.35	0.22	31.55	0.17
	0.2	**129.45**	1781	125.15	45.65	124.15	48.52	125.70	16.56	117.10	31.75	125.25	0.30	126.30	1.46
	0.3	**283.65**	3282	273.45	63.11	272.45	74.15	279.90	33.42	257.75	45.87	276.65	0.59	278.75	4.80
5000	0.1	**41.30**	730	39.50	31.25	38.50	38.73	39.15	7.40	35.90	19.43	38.80	0.28	39.30	0.31
	0.2	**158.25**	2825	153.65	59.61	152.65	67.34	154.25	23.09	143.50	37.35	153.95	0.56	154.85	2.51
	0.3	**354.55**	4379	342.75	78.34	341.75	98.84	350.35	47.84	321.55	59.36	345.05	0.80	348.20	7.02

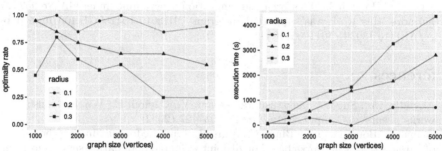

Fig. 3. Evolution of the optimality rate (left) and execution time (right) depending on the size of the RGG instance, for the three considered radii.

evolution of the optimality rate depending on the instance size. The second one displays the evolution of the average execution times. Clearly, optimal solutions to these instances coincide much more often with the upper bound. In cases ($|V| = 1500, r_{max} = 0.1$) and ($|V| = 3000, r_{max} = 0.1$), for example, all 20 graphs were solved to proven optimality by CMSA. The algorithm was able to achieve this on average in 77, respectively 3, CPU seconds. The graphics also show that with a growing graph density—a growing value of r_{max}—the values of optimal solutions seem to coincide every time less with the upper bound.

6   Conclusions

In this work we proposed a "construct, merge, solve and adapt" algorithm for the maximum disjoint dominating sets problem, which is a complex variant of the classical minimum dominating set problem in networks and graphs. The difficulty of applying standard metaheuristics to this problem is shown by the fact that, so far, only greedy heuristics, approximation algorithms, and exact techniques for special cases have been proposed in the literature. We were able to show that our algorithm clearly outperforms all existing greedy heuristics for the problem. Furthermore, it is the first matheuristic and metaheuristic approach for the maximum disjoint dominating sets problem.

Future work will deal especially with the improvement of our algorithm for even larger problem instances. As mentioned before, we noticed in the context of the application to large random geometric graphs that the greedy heuristic we use for the probabilistic construction of solutions may not be the best option. In fact, some of the other available greedy heuristics are faster and better for this case. The idea is, therefore, to adapt the choice of a suitable greedy heuristic within CMSA depending on the tackled problem instance.

Acknowledgments. This research was partially supported by TAILOR, a project funded by EU Horizon 2020 research and innovation programme under GA No 952215. Furthermore, this work was supported by grant PID2019-104156GB-I00 funded by MCIN/AEI/10.13039 /501100011033.

References

1. Akyildiz, I.F., Su, W., Sankarasubramaniam, Y., Cayirci, E.: Wireless sensor networks: a survey. Comput. Netw. **38**(4), 393–422 (2002)
2. Balbal, S., Bouamama, S., Blum, C.: A greedy heuristic for maximizing the lifetime of wireless sensor networks based on disjoint weighted dominating sets. Algorithms **14**(6), 170 (2021)
3. Blum, C., Pinacho, P., López-Ibáñez, M., Lozano, J.A.: Construct, merge, solve & adapt a new general algorithm for combinatorial optimization. Comput. Oper. Res. **68**, 75–88 (2016)
4. Bouamama, S., Blum, C., Pinacho-Davidson, P.: A population-based iterated greedy algorithm for maximizing sensor network lifetime. Sensors **22**(5), 1804 (2022)
5. Bruglieri, M., Cordone, R.: Metaheuristics for the minimum gap graph partitioning problem. Comput. Oper. Res. **132**, 105301 (2021)
6. Cardei, M., Du, D.Z.: Improving wireless sensor network lifetime through power aware organization. Wireless Netw. **11**(3), 333–340 (2005). https://doi.org/10.1007/s11276-005-6615-6
7. Cardei, M., et al.: Wireless sensor networks with energy efficient organization. J. Interconnection Netw. **3**(03n04), 213–229 (2002)
8. Chang, G.J.: The domatic number problem. Discret. Math. **125**(1–3), 115–122 (1994)
9. Cockayne, E.J., Hedetniemi, S.T.: Towards a theory of domination in graphs. Networks **7**(3), 247–261 (1977)

10. Feige, U., Halldórsson, M.M., Kortsarz, G., Srinivasan, A.: Approximating the domatic number. SIAM J. Comput. **32**(1), 172–195 (2002)
11. Ferrer, J., Chicano, F., Ortega-Toro, J.A.: CMSA algorithm for solving the prioritized pairwise test data generation problem in software product lines. J. Heuristics **27**(1), 229–249 (2021). https://doi.org/10.1007/s10732-020-09462-w
12. Fischetti, M., Fischetti, M.: Matheuristics: Handbook of Heuristics, pp. 121–153. Springer, Cham (2018)
13. Garey, M.R., Johnson, D.S.: Computers and Intractability: A Guide to the Theory of NP-Completeness. W. H. Freeman & Co., New York (1979)
14. Haynes, T.W., Hedetniemi, S., Slater, P.: Fundamentals of Domination in Graphs. CRC Press, Boca Raton (2013)
15. Islam, K., Akl, S.G., Meijer, H.: Maximizing the lifetime of wireless sensor networks through domatic partition. In: 2009 IEEE 34th Conference on Local Computer Networks, pp. 436–442. IEEE (2009)
16. Lewis, R., Thiruvady, D., Morgan, K.: Finding happiness: an analysis of the maximum happy vertices problem. Comput. Oper. Res. **103**, 265–276 (2019)
17. López-Ibáñez, M., Dubois-Lacoste, J., Cáceres, L.P., Birattari, M., Stützle, T.: The iRace package: iterated racing for automatic algorithm configuration. Oper. Res. Perspect. **3**, 43–58 (2016)
18. Mesbahi, M., Egerstedt, M.: Graph Theoretic Methods in Multiagent Networks. Princeton University Press, Princeton (2010)
19. Nguyen, T.N., Huynh, D.T.: Extending sensor networks lifetime through energy efficient organization. In: International Conference on Wireless Algorithms, Systems and Applications (WASA 2007), pp. 205–212. IEEE (2007)
20. Ore, O.: Theory of graphs (1962)
21. Pinacho-Davidson, P., Bouamama, S., Blum, C.: Application of CMSA to the minimum capacitated dominating set problem. In: Proceedings of the Genetic and Evolutionary Computation Conference, pp. 321–328 (2019)
22. Poon, S.-H., Yen, W.C.-K., Ung, C.-T.: Domatic partition on several classes of graphs. In: Lin, G. (ed.) COCOA 2012. LNCS, vol. 7402, pp. 245–256. Springer, Heidelberg (2012). https://doi.org/10.1007/978-3-642-31770-5_22
23. Riege, T., Rothe, J.: An Exact 2.9416^n algorithm for the three domatic number problem. In: Jedrzejowicz, J., Szepietowski, A. (eds.) MFCS 2005. LNCS, vol. 3618, pp. 733–744. Springer, Heidelberg (2005). https://doi.org/10.1007/11549345_63
24. Riege, T., Rothe, J., Spakowski, H., Yamamoto, M.: An improved exact algorithm for the domatic number problem. Inf. Process. Lett. **101**(3), 101–106 (2007)
25. Rooij, J.M.M.: Polynomial space algorithms for counting dominating sets and the domatic number. In: Calamoneri, T., Diaz, J. (eds.) CIAC 2010. LNCS, vol. 6078, pp. 73–84. Springer, Heidelberg (2010). https://doi.org/10.1007/978-3-642-13073-1_8
26. Watson, J.P.: An introduction to fitness landscape analysis and cost models for local search. In: Gendreau, M., Potvin, J.Y. (eds.) Handbook of Metaheuristics. International Series in Operations Research & Management Science, vol. 146, pp. 599–623. Springer, Boston (2010)

Fixed Set Search Applied to the Territory Design Problem

Tobias Cors[1], Tobias Vlćek[2], Stefan Voß[3]([✉]), and Raka Jovanovic[4]

[1] Institute of Operations Management, University of Hamburg,
Moorweidenstraße 18, 20148 Hamburg, Germany
tobias.cors@uni-hamburg.de
[2] Institute of Logistics, Transport and Production, University of Hamburg,
Moorweidenstraße 18, 20148 Hamburg, Germany
tobias.vlcek@uni-hamburg.de
[3] Institute of Information Systems, University of Hamburg, Von-Melle-Park 5,
20146 Hamburg, Germany
stefan.voss@uni-hamburg.de
[4] Qatar Environment and Energy Research Institute, Hamad bin Khalifa University,
PO Box 5825, Doha, Qatar
rjovanovic@hbku.edu.qa

Abstract. In this paper, we apply the novel fixed set search (FSS) metaheuristic in combination with mixed-integer programming to solve the Territory Design Problem (TDP). In this matheuristic approach, we select the territory centers with an extended greedy randomised adaptive search procedure (GRASP) while optimising the subproblem of the territory-center allocation with a standard mixed-integer programming solver. The FSS adds a learning procedure to GRASP and helps us to narrow down the most common territory centers in the solution population in order to fix them. This improves the speed of the optimisation and helps to find high-quality solutions on all instances of our computational study at least once within a small number of runs.

Keywords: Fixed set search · Matheuristic · Territory design problem

1 Introduction

The territory design problem (TDP) arises in the context of different domains - ranging from commercial over social to political cases; see, e.g., [26]. The problem consists of determining a partition of a set of units located in a territory meeting certain criteria.

Any problem that seeks to find optimal solutions for large instances may face demanding computations. Thus, there has been a continuous effort in the development of methods for finding near-optimal or close-to-optimal solutions in the space of greedy algorithms and metaheuristics. In this line of research, a variety of approaches have been developed, extended and hybridised in order to reduce the computational burden while increasing the solution quality. For the

L. Di Gaspero et al. (Eds.): MIC 2022, LNCS 13838, pp. 322–334, 2023.
https://doi.org/10.1007/978-3-031-26504-4_23

TDP, many approaches incorporate a local search procedure or use a variant of the greedy randomised adaptive search procedure (GRASP) [19].

Within recent research on simple but yet effective metaheuristics, the novel fixed set search (FSS) method was proposed by [15] adding a simple learning mechanism to the GRASP, directing the computational effort to promising areas of a solution space. In our study, we combine FSS with mixed integer programming in a matheuristic fashion and apply the resulting approach to the TDP.

The paper is organised as follows. Section 2 gives a brief overview over related research. In Sect. 3 we explain the TDP model our matheuristic is based on and in Sect. 4 we specify the heuristic itself. In Sect. 5 we present the structure and the results of our computational study. The implications are discussed in Sect. 6, which also concludes the paper.

2 Related Research

The problem considered in our study is a variant of the Territory Design Problem originating from the field of discrete optimisation [26]. The objective is to cluster small geographic areas, referred to as basic areas (BA), into districts by means of predefined planning criteria [31]. The TDP finds applications in a wide range of domains, such as the design of commercial, social, political or service territories [19]. Most TDP contributions are tailored to specific applications. In [18], they are consolidated and unified in a compendium as a basis for further research.

In terms of the computation time for solving the TDP, researchers face two problems. First, this concerns granularity. The choice of the size of the BAs represents a tradeoff between the accuracy of the solution and the associated computation time [22]. Second, by the nature of the problem, most models are based on the general graph-partitioning problem known to be NP-hard [7]. Thus, there has been an ongoing effort in the development of methods for finding optimal or near-optimal solutions in the space of greedy algorithms and metaheuristics. For a compilation of applications and heuristics regarding the TDP the reader is referred to [19].

Since the introduction of the GRASP algorithm by [3] as well as the earlier ideas of semi-greedy heuristics of [10], it has been advanced, hybridised and applied to many areas; see, e.g., [4–6,25]. [27] developed the first GRASP implementation for a specific TDP. They extended the GRASP to be reactive in terms of the acceptance criteria for quality solutions from the restricted candidate list. Afterwards they benchmarked the heuristic in a computational study with 500 BAs. The observed results were reported to be robust regarding high quality solutions.

In the aligning research of GRASP aiming at performance improvements, a novel Fixed Set Search (FSS) metaheuristic has been proposed by [13]. The FSS extends the GRASP algorithm by a simple learning mechanism leading the local search by elements that appear frequently in high quality solutions [15]. It was successfully applied to the traveling salesman problem [13], the minimum weighted vertex cover problem [15], the power dominating set problem [14] and the parallel machine scheduling problem [17]. The reported computational results attested the FSS a high competitiveness in the quality of found solutions compared with state-of-the-art methods, motivating our application to the TDP. An extension to a multi-objective minimum weighted vertex cover problem is proposed in [12].

The intuition for the FSS comes from the observation that often high-quality solutions, for a specific problem instance, have quite a few overlapping elements. The method relies on the idea to generate new solutions that contain such elements. That is, these elements, the fixed set, are used for the solutions that will be generated and the computational effort is dedicated to extend partial solutions towards feasibility. Using elements that frequently occur in high-quality solutions can also be based on earlier notions of adaptive memory programming and tabu search. As mentioned by [13] the idea of fixed sets is also leveraged in the construct, merge, solve & adapt (CMSA) matheuristic, which they interpreted as a specific implementation of the POPMUSIC paradigm [28]. CMSA uses an ageing mechanism for fixing solution components of repeated probabilistic solution constructions and applies an integer linear programming (ILP) solver to these generated sub-instances [1]. As the age of a component is reset to zero after each occurrence, the maximum age indirectly controls the size of the fixed set. [1] show that the hybridisation with an ILP optimisation profits from the effectiveness of the solver in the reduced sub-instances of hard problems.

From an overview for such approaches (e.g. [29]), it is concluded that it is beneficial to combine the complementary advantages of metaheuristics and mathematical programming, relating them to efficiency and effectiveness. Mostly, this is devoted towards the notion of matheuristics as they had been formulated by Maniezzo and Voß connected to model-based metaheuristics (see, e.g., [21]).

In [16], a matheuristic was developed for the 2-connected dominating set problem based on GRASP for the global search at low and a mixed-integer program (MIP) for the local search at high computational cost. Following the outlined promising directions of research, we focus on developing a matheuristic combining GRASP, MIP and FSS for the TDP. The idea of solving a simplified (or even relaxed) problem may be derived, e.g., from the use of the same mathematical model for a smaller instance obtained by fixing some of the variables; in [11] we even find the notion of an equi-model for such a proceeding.

3   Model

We define the following sets and parameters:

$\mathcal{J}$: set of BAs, indexed by j and v,

$\mathcal{I}$: set of potential locations ($\mathcal{I} \subseteq \mathcal{J}$), indexed by i,

$\mathcal{A}_j$: sets of BAs adjacent to BA j ($\mathcal{A}_j \subseteq \mathcal{J}$,)

p : number of locations,

a_{ij} : Euclidean distance between the centroid of BA i and the centroid of BA j

d_{ij} : driving time between the centroid of BA i and the centroid of BA j

w_j : weight of the BA j.

We define the following decision variables:

$x_{ij} = 1$, if BA j is assigned to district center BA i (otherwise $x_{ij} = 0$)

The design of territories can follow multiple objectives [19]. For our settings, we focus on a basic case: minimising the weighted driving time between territory centers and their assigned BAs.

$$min \quad \sum_{i \in \mathcal{I}} \sum_{j \in \mathcal{J}} d_{ij} \cdot w_j \cdot x_{ij} \tag{1}$$

subject to:

$$\sum_{i \in \mathcal{I}} x_{ij} = 1 \qquad \qquad \forall j \in \mathcal{J} \tag{2}$$

$$\sum_{i \in \mathcal{I}} x_{ii} = p \tag{3}$$

$$x_{ij} \leq x_{ii} \qquad \qquad \forall i \in \mathcal{I}, \forall j \in \mathcal{J} \tag{4}$$

$$x_{ij} \leq \sum_{v \in \mathcal{N}_{ij}} x_{iv} \qquad \qquad \forall i \in \mathcal{I}, \forall j \in \mathcal{J} \setminus \mathcal{A}_i \tag{5}$$

$$x_{ij} \in \{0, 1\} \qquad \qquad \forall i \in \mathcal{I}, \forall j \in \mathcal{J} \tag{6}$$

Objective function (1) minimises the weighted driving time over the covered area. Equation (2) secures that each BA is allocated to one district center. Constraint (3) ensures the allocation of p district centers while (4) guarantees that each BA is only allocated to realised district centers. (5) is the contiguity constraint; the idea behind it was first introduced by [31]. The constraints ensure that the resulting territories are contiguous, as each BA j allocated to i has to have at least one adjacent BA v with a closer distance $a_{iv} < a_{ij}$ allocated to the same center. This relation is expressed in

$$\mathcal{N}_{ij} = \{v \in \mathcal{A}_j | a_{iv} < a_{ij}\} \qquad \qquad \forall i \in \mathcal{I}, \forall j \in \mathcal{J}.$$

Combined with hexagonal BAs, this enforces a connected path between each BA j allocated to BA i. Constraints (6) define the variable's domains.

Unfortunately, it is often difficult to obtain optimal solutions to the problem, as our problem formulation is a variation of the NP-hard p-median problem on a general graph [20]. The number of combinatorial possibilities, after neglecting the contiguity constraint, is $\binom{|\mathcal{I}|}{p} \cdot p^{|\mathcal{J}|-p}$. Moreover, the computational effort in TDPs increases due to the contiguity constraint [24]. Thus, we propose a metaheuristic to solve the problem as indicated in the next section.

4   Heuristic

Our heuristic is based on the novel fixed set search metaheuristic. Before we present our approach, we define the following additional sets:

$\mathcal{H}$: set of potential locations ($\mathcal{H} \subseteq \mathcal{I}$),

$\mathcal{S}$: set of feasible locations ($\mathcal{S} \subseteq \mathcal{I}$),

$\mathcal{F}$: set of fixed locations ($\mathcal{F} \subseteq \mathcal{I}$),

$\mathcal{B}$: base solution ($\mathcal{B} \subseteq \mathcal{I}$),

$\mathcal{P}$: population of solutions,

$\mathcal{P}_n/\mathcal{P}_m$: population of n / m best solutions,

$Sizes$: set for FSS with $size \in Sizes$.

Our version of FSS combines a greedy approach with MIP optimisations on small problem instances, as shown in Algorithm 1 which displays one iteration of our GRASP implementation. We start with an empty fixed set $\mathcal{F}$ and generate an initial population of N solutions with GRASP as shown in lines 1 and 2 of Algorithm 2. Our greedy approach fills a set of potential locations $\mathcal{H}$ until $|\mathcal{H}| = \hat{p}$ with $\hat{p} > p$. Afterwards, we solve the relaxed MIP and check whether the solution $\mathcal{S}$ is a feasible location set without relaxation. If it is feasible, we start our local search as shown in Algorithm 3.

Algorithm 1: Pseudocode $GRASP(\mathcal{F}, p, \hat{p})$

---

1 $\mathcal{S} = \varnothing$;
2 **while** $\mathcal{S} = \varnothing$ **do**
3      $\mathcal{H} = \mathcal{F} \cup \varnothing$;
4   **while** $|\mathcal{H}| < \hat{p}$ **do**
5 | Randomly assign a location from $\mathcal{I}$ to $\mathcal{H}$;
6   **end**
7   solve the relaxed MIP with $\mathcal{H}$ as set of potential locations;
8      $\mathcal{S} = \{i \in \mathcal{H} : w_{ii} = 1\}$;
9      $z = $ value of the objective function;
10   **if** $\mathcal{S}$ *is a feasible location set for the MIP* **then**
11 | $\mathcal{S} = LS(\mathcal{S}, z, \hat{z}, p)$;
12   **else**
13 | $\mathcal{S} = \varnothing$;
14   **end**
15 **end**

In LS, we redefine $\mathcal{H}$ as $\mathcal{S}$ united with all adjacent BAs of the locations in $\mathcal{S}$. We then solve the MIP and replace $\mathcal{S}$ with the solution. We repeat this process, until we are stuck in a local minimum (the objective function is stagnant). If the solution from the greedy approach $\mathcal{S}$ in Algorithm 2, line 11 is infeasible, we start again to find a different solution. This allows to find good initial solutions, but the calculation time of each iteration depends on the size of $\mathcal{I}$, $\mathcal{J}$, p and $\hat{p}$.

After we have generated an initial population, we initialise the set $Sizes$ and start the FSS as described in the pseudocode of Algorithm 2. The procedure is

---

Algorithm 2: Pseudocode FSS

---

1 $\mathcal{F} = \varnothing$;
2 Generate initial population $\mathcal{P}$ containing N sets using $GRASP(\mathcal{F}, p, \hat{p})$;
3 Initialise $Sizes$;
4 $Size = Sizes.Next$;
5 **while** $Sizes \neq \varnothing$ **do**
6   Set $\mathcal{S}_{kn}$ to k random elements of $\mathcal{P}_n$;
7   Set $\mathcal{B}$ to a random solution in $\mathcal{P}_m$;
8      $\mathcal{F} = Fix(\mathcal{B}, \mathcal{S}_{kn}, Size)$;
9      $\mathcal{S} = GRASP(\mathcal{F})$;
10     $\mathcal{P} = \mathcal{P} \cup \{\mathcal{S}\}$;
11   **if** *Stagnant Best Solution and* $\mathcal{F} \in \mathcal{S}$ **then**
12   **if** *(Stagnant Candidates)* **then**
13   Remove $Size$ from $Sizes$;
14   **end**
15   **end**
16 **end**
17 $\mathcal{H} = \{i \in \mathcal{I} | \exists \mathcal{P}_n : i \in \mathcal{S}\}$;
18 solve the MIP with $\mathcal{H}$ as set of potential locations;
19 $\mathcal{S} = \{i \in \mathcal{H} : w_{ii} = 1\}$;
20 $\mathcal{S} = LS(\mathcal{S}, z, \hat{z}, p)$;

Algorithm 3: Pseudocode $LS(\mathcal{S}, z, \hat{z}, p)$

---

1 **while** $z < \hat{z}$ **do**
2      $\hat{z} = z$;
3      $\mathcal{H} = \mathcal{S} \cup \{i \in \mathcal{I} | \exists s \in \mathcal{S} : i \in \mathcal{A}_s\}$;
4   solve the MIP with $\mathcal{H}$ as set of potential locations;
5      $\mathcal{S} = \{i \in \mathcal{H} : w_{ii} = 1\}$;
6   z = value of the objective function;
7 **end**

---

described in more detail by [13]. Our key adjustment is the addition of a MIP optimisation based on the population of the best solutions $\mathcal{P}_n$ after the FSS terminates.

5 Computational Study

We applied the optimisation as well as the heuristic on three different problem instances with (a) $|\mathcal{J}| = 510$, (b) $|\mathcal{J}| = 1008$ and (c) $|\mathcal{J}| = 1508$ hexagonal basic units with $I = J$. The number of locations p was varied from 2 to 10 locations. To benchmark the heuristic on a realistic setting, our problem instances represented the same map section of the northern half of Hamburg, Germany, with a varying number of basic units. The weight w_j is an approximation of the population number in each BA j and represents the number of street-intersections in j, while the driving time d_{ij} was calculated on the underlying road network from OpenStreet Map [9] from the centroids of the BAs with Dijkstra's shortest path Algorithm [2].

As our GRASP approach for the initial population set $\mathcal{P}$ is combined with mixed-integer optimisations, our initial population already consists of good-quality solutions after the first iteration. We thus used a small number of solutions $N = 11$ in the initial population $\mathcal{P}$. The number of greedy-drawn locations for the relaxed MIP was $\hat{p} = 10 \cdot p$. During FSS, we always drew $k = 3$ elements from the best 11 solutions $\mathcal{P}_n$ for $\mathcal{S}_{kn}$. The set $\mathcal{P}_m$ was the subset of $\mathcal{P}_n \setminus \mathcal{S}_{kn}$. Our set $Sizes$ contained $|Sizes| = \frac{p}{2} + 1$ elements starting with 1 and the parameter $Stagnant$ was set to 3. All our calculations (the full optimisation model and our heuristic) were computed on a 6 Core AMD-4650 G CPU with 16 GB of RAM with GAMS and CPLEX. We repeated all computations five times to account for the built-in randomness of our approach.

Figure 1 displays the duration of the approach for all variations of $|\mathcal{J}|$ over the five runs of each problem instance. We clearly see that the calculation time increased approximately in a linear fashion with the number of locations p and the size of $|\mathcal{J}|$ (and thus $|\mathcal{I}|$). This was due to the additional time needed for the FSS as well as due to the larger set $\mathcal{H}$ during the MIP optimisations of the approach. Nonetheless, the problem always remained solvable with our approach within 1 h. The RAM usage never exceeded 2.5 GB, even with $|\mathcal{J}| = 1508$.

The duration of the full optimisation of our model in Sect. 3 for all five runs with a solution gap of 0.00% is displayed in Fig. 2. Note, that our computer could not solve the problem instances with $|\mathcal{J}| = 1508$ due to insufficient RAM.[1] If we compare the calculation time between the optimisation and the heuristic, we see that the instances with a low p benefit the most, as they could be solved faster up to $p = 5$ with $|\mathcal{J}| = 510$ and up to $p = 10$ (the largest p in our instances) with $|\mathcal{J}| = 1008$.

The solution gap between the optimal solution and the solution values of the five runs of our heuristic is shown in Fig. 3 for $|\mathcal{J}| = 510$ and $|\mathcal{J}| = 1008$. We see, that the gap never exceeds 0.8%. The average gap with 510 BAs is 0.05% and with 1008 BAs 0.11%. Overall, we have found the optimal solution in all five runs of the instances with known optimal solutions at least once. Moreover, the figure shows that the gap was on average smaller if p was low.

In Fig. 4 we display the value of the best heuristic solution for $|\mathcal{J}| = 1508$. The determined values are all quite close to each other. In fact, despite two exceptions, the gap between the best and worst solution of all five runs was always lower than 0.69%.

---

[1] Note added in proof: Switching to the latest version of a different solver, this picture changed but this is not reported here.

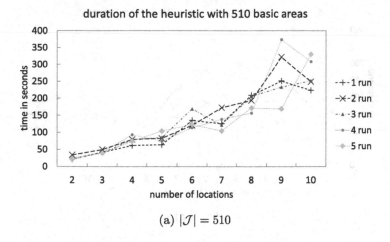

(a) $|\mathcal{J}| = 510$

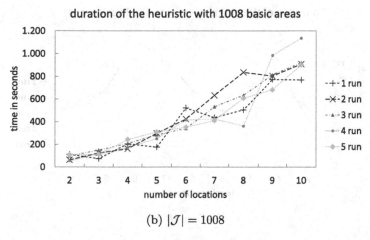

(b) $|\mathcal{J}| = 1008$

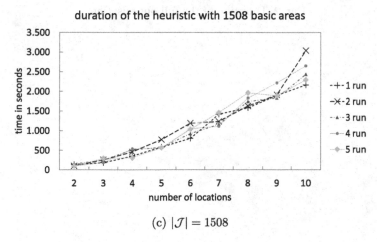

(c) $|\mathcal{J}| = 1508$

Fig. 1. Duration of the approach

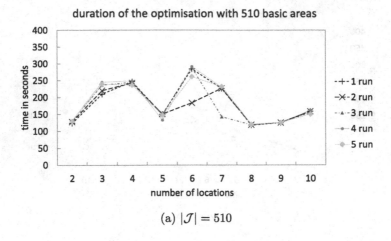

(a) $|\mathcal{J}| = 510$

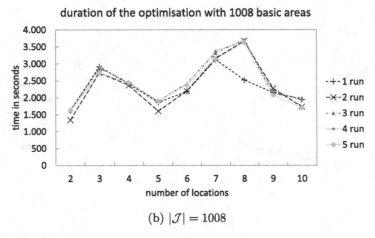

(b) $|\mathcal{J}| = 1008$

Fig. 2. Duration of the full optimisation

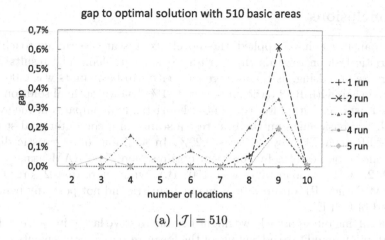

(a) $|\mathcal{J}| = 510$

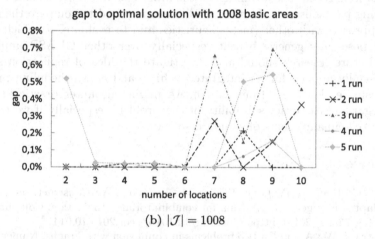

(b) $|\mathcal{J}| = 1008$

Fig. 3. Gap between optimal solution and heuristic

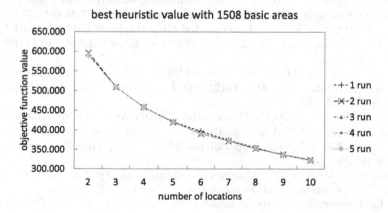

Fig. 4. Objective value of the best heuristic solution with $|\mathcal{J}| = 1508$

6   Conclusions

In this paper, we have applied the novel fixed seat search approach in a matheuristic fashion towards the territory design problem. The results of our heuristic are promising, as the average gap with 510 basic units was 0.06% and the average gap with 1008 basic units was 0.11%. Known optimal solutions were found at least once within five runs of our heuristic. For comparison, the average gap with 500 basic units on the best test instance in a computational study of the GRASP on a TDP by [27] was 2.29%. In addition, our heuristic did not require the resources of a full problem optimisation. The RAM usage topped out with 2.5 GB on the instance with 1508 BAs while more than 2 cores for the relaxed MIP and MIP optimisations in our heuristic did not pose any benefit to the speed of CPLEX.

In modifying our approach, we might be able to solve larger instances where a slight adaption could take advantage of the lower resource requirements through introducing parallelisation. Further studies are required to compare the results of our approach with other (meta-)heuristics and to evaluate, to which extent the FSS poses any generic benefits especially over other GRASP implementations. Future research should also investigate the idea of replication studies as successfully applied for an integrated vehicle and crew scheduling problem in [8]. Using earlier ideas from [23,30], we might encounter drastic improvements regarding the exact solvability of the problem, especially if we consider the expected enhancements of standard solvers.

References

1. Blum, C., Pinacho, P., López-Ibáñez, M., Lozano, J.A.: Construct, merge, solve & adapt A new general algorithm for combinatorial optimization. Comput. Oper. Res. **68**, 75–88 (2016). https://doi.org/10.1016/j.cor.2015.10.014
2. Dijkstra, E.W.: A note on two problems in connexion with graphs. Numer. Math. **1**(1), 269–271 (1959). https://doi.org/10.1007/BF01386390
3. Feo, T.A., Resende, M.G.C.: Greedy randomized adaptive search procedures. J. Global Optim. **6**(2), 109–133 (1995). https://doi.org/10.1007/bf01096763
4. Festa, P., Resende, M.G.C.: An annotated bibliography of GRASP - part I: algorithms. Int. Trans. Oper. Res. **16**(1), 1–24 (2009). https://doi.org/10.1111/j.1475-3995.2009.00663.x
5. Festa, P., Resende, M.G.C.: An annotated bibliography of GRASP - part II: applications. Int. Trans. Oper. Res. **16**(2), 131–172 (2009). https://doi.org/10.1111/j.1475-3995.2009.00664.x
6. Festa, P., Resende, M.G.C.: Hybridizations of GRASP with path-relinking. In: Talbi, E.G. (ed.) Hybrid Metaheuristics. Studies in Computational Intelligence, vol. 434, pp. 135–155. Springer, Berlin, Heidelberg (2013). https://doi.org/10.1007/978-3-642-30671-6_5
7. Garey, M.R., Johnson, D.S.: Computers and Intractability: A Guide to the Theory of NP-Completeness. W. H. Freeman & Co., New York, NY (1979)
8. Ge, L., Kliewer, N., Nourmohammadzadeh, A., Voß, S., Xie, L.: Revisiting the richness of integrated vehicle and crew scheduling. Public Transp. (2022). https://doi.org/10.1007/s12469-022-00292-6

9. Geofabrik GmbH and OpenStreetMap Contributors: OpenStreetMap Data Hamburg (2020). https://download.geofabrik.de/europe/germany/hamburg.html
10. Hart, J.P., Shogan, A.W.: Semi-greedy heuristics: an empirical study. Oper. Res. Lett. **6**(3), 107–114 (1987). https://doi.org/10.1016/0167-6377(87)90021-6
11. Hill, A., Voß, S.: An equi-model matheuristic for the multi-depot ring star problem. Networks **67**(3), 222–237 (2016). https://doi.org/10.1002/net.21674
12. Jovanovic, R., Sanfilippo, A.P., Voß, S.: Fixed set search applied to the multi-objective minimum weighted vertex cover problem. J. Heuristics **28**, 481–508 (2022). https://doi.org/10.1007/s10732-022-09499-z
13. Jovanovic, R., Tuba, M., Voß, S.: Fixed set search applied to the traveling salesman problem. Lect. Notes Comput. Sci. **11299**, 63–77 (2019). https://doi.org/10.1007/978-3-030-05983-5_5
14. Jovanovic, R., Voss, S.: The fixed set search applied to the power dominating set problem. Expert. Syst. **37**(6), e12559 (2020). https://doi.org/10.1111/exsy.12559
15. Jovanovic, R., Voß, S.: Fixed set search applied to the minimum weighted vertex cover problem. Lect. Notes Comput. Sci. **11544**, 490–504 (2019). https://doi.org/10.1007/978-3-030-34029-2_31
16. Jovanovic, R., Voß, S.: A matheuristic approach for solving the 2-connected dominating set problem. Appl. Anal. Discrete Math. **14**(3), 775–799 (2020). https://doi.org/10.2298/aadm190227052j
17. Jovanovic, R., Voß, S.: Fixed set search application for minimizing the makespan on unrelated parallel machines with sequence-dependent setup times. Appl. Soft Comput. **110**, 107521 (2021). https://doi.org/10.1016/j.asoc.2021.107521
18. Kalcsics, J., Nickel, S., Schröder, M.: Towards a unified territorial design approach – applications, algorithms and GIS integration. TOP **13**(1), 1–56 (2005). https://doi.org/10.1007/BF02578982
19. Kalcsics, J., Ríos-Mercado, R.Z.: Districting problems. In: Laporte, G., Nickel, S., Saldanha da Gama, F. (eds.) Location Sci., pp. 705–743. Springer, Cham (2019). https://doi.org/10.1007/978-3-030-32177-2_25
20. Kariv, O., Hakimi, S.L.: An algorithmic approach to network location problems. II: the p-medians. SIAM J. Appl. Math. **37**(3), 539–560 (1979). https://doi.org/10.2307/2100911
21. Maniezzo, V., Stützle, T., Voß, S. (eds.): Matheuristics: hybridizing metaheuristics and mathematical programming. Springer, Cham (2009). https://doi.org/10.1007/978-1-4419-1306-7
22. Mitchell, P.S.: Optimal selection of police patrol beats. J. Crim. Law Criminol. Police Sci. **63**(4), 577 (1972). https://doi.org/10.2307/1141814
23. Mittelmann, H.D.: Benchmarking optimization software - a (Hi)story. SN Oper. Res. Forum **1**(1), 1–6 (2020). https://doi.org/10.1007/s43069-020-0002-0
24. Önal, H., Wang, Y., Dissanayake, S.T., Westervelt, J.D.: Optimal design of compact and functionally contiguous conservation management areas. Eur. J. Oper. Res. **251**(3), 957–968 (2016). https://doi.org/10.1016/j.ejor.2015.12.005
25. Resende, M.G., Ribeiro, C.C.: Greedy randomized adaptive search procedures: advances, hybridizations, and applications. In: Gendreau, M., Potvin, J.Y. (eds.) Handbook of Metaheuristics. International Series in Operations Research & Management Science, vol. 146, pp. 283–319. Springer, Boston (2010). https://doi.org/10.1007/978-1-4419-1665-5_10
26. Ríos-Mercado, R.Z.: Research trends in optimization of districting systems. In: Ríos-Mercado, R.Z. (ed.) Optimal Districting and Territory Design. ISORMS, vol. 284, pp. 3–8. Springer, Cham (2020). https://doi.org/10.1007/978-3-030-34312-5_1

27. Ríos-Mercado, R.Z., Fernández, E.: A reactive GRASP for a commercial territory design problem with multiple balancing requirements. Comput. Oper. Res. **36**(3), 755–776 (2009). https://doi.org/10.1016/j.cor.2007.10.024

28. Taillard, E.D., Voß, S.: Popmusic—partial optimization metaheuristic under special intensification conditions. In: Ribeiro, C.C., Hansen, P. (eds.) Essays and Surveys in Metaheuristics. Operations Research/Computer Science Interfaces Series, vol. 15, pp. 613–629. Springer, Boston (2002). https://doi.org/10.1007/978-1-4615-1507-4_27

29. Talbi, E.-G.: Combining metaheuristics with mathematical programming, constraint programming and machine learning. Ann. Oper. Res. **240**(1), 171–215 (2015). https://doi.org/10.1007/s10479-015-2034-y

30. Voß, S., Lalla-Ruiz, E.: A set partitioning reformulation for the multiple-choice multidimensional knapsack problem. Eng. Optim. **48**(5), 831–850 (2016). https://doi.org/10.1080/0305215X.2015.1062094

31. Zoltners, A.A., Sinha, P.: Sales territory alignment: a review and model. Manage. Sci. **29**, 1237–1256 (1983). https://doi.org/10.1287/mnsc.29.11.1237

)

The P-Next Center Problem
with Capacity and Coverage Radius
Constraints: Model and Heuristics

Mariana A. Londe[1](✉), Luciana S. Pessoa[1], and Carlos E. Andrade[2]

[1] Department of Industrial Engineering, PUC-Rio, Rua Marquês de São Vicente,
225, Gávea, Rio de Janeiro, RJ 22453-900, Brazil
mlonde@aluno.puc-rio.br, lucianapessoa@puc-rio.br
[2] AT&T Labs Research, 200 South Laurel Avenue, Middletown, NJ 07748, USA
cea@research.att.com

Abstract. This paper introduces a novel problem of facility location, called the p-next center problem with capacity and coverage radius constraints. We formulate a mixed integer programming model for this problem, and compare the results found by CPLEX with three Biased Random-Key Genetic Algorithms variants. We also propose several instances for this problem, based on existing ones for the p-next center problem. Additionally, we analyze the effect of the radius and demand on instance difficulty. We also observe the performance gains with a relaxed capacity and demand constraint, i.e., permitting demand to be unmet by the model. Results point that the BRKGA variants had significantly better performance than CPLEX, and similar performances among themselves. Of those, `BRKGA-FI` was shown to have slightly better results than the other variants.

Keywords: Facility location · Biased random-key genetic algorithm · Metaheuristics

1 Introduction

The *p-center problem* is a classical location problem that consists of choosing p centers among n nodes in a network in order to minimize the maximum distance from any node to its closest facility [24].

The *p-next center problem* (pNCP) is an extension of the previous one that considers the possibility of a user arriving at a facility and discovering a disruption in its operation. In this scenario, the user must move to a backup point, defined as its nearest facility. The objective of the pNCP is to minimize the maximal distance a user must travel, which is composed by the distance from a point to its closest existing facility plus the distance from this facility to its backup. This problem was introduced in the context of humanitarian logistics by [1], since during an emergency there is a possibility of disruption in facilities such as shelters and hospitals. The authors also present several formulations,

L. Di Gaspero et al. (Eds.): MIC 2022, LNCS 13838, pp. 335–349, 2023.
https://doi.org/10.1007/978-3-031-26504-4_24

instances and the proof of its NP-Hardness. The first heuristic method to solve this problem was described by [19], based on the Greedy Randomized Adaptive Search Procedure (GRASP) [10] and Variable Neighborhood Search (VNS) [14]. Later, [17] presented a Biased Random Key Genetic Algorithm (BRKGA) [13] for the pNCP, alongside several new benchmarks.

The pNCP does not consider that a user may be unable to travel extensive distances during an emergency, nor the possible consequences of lack of capacity on the centers. For example, consider the case of a snake attack victim. The patient must quickly reach the closest treatment center, which may be located on the same neighborhood as the attack or on a close one. The speed of transport to the treatment facility is crucial for the patient's prospects, as the elapsed time between injury and care is a factor in severeness and lethality of the wound [6]. In fact, venom of some species of snake demand an elapsed time below a specific threshold. If there is enough anti-venom at the facility, then the patient will be efficiently treated and the emergency is resolved. However, the center may lack the medicine. There are two possible routes for treatment if this happens: (1) the patient is transferred to a supplied facility, and (2) the medicine is transported to them. On both cases the elapsed time between injury and therapy increases, and may lead to avoidable after-effects and/or death if higher then the time threshold.

Thus, the network of treatment centers must consider the maximum time elapsed between a possible attack and the closest care facility alongside the possible transference/transport time, which deal directly with the victim's perspective. The elapsed time must be below a specific threshold, otherwise the patient may have severe side-effects or death. It must also deal with a limited anti-venom capacity among all facilities, so that a center may be able to treat the highest possible number of patients without needing to bring medicine from other centers in the network. An unmet demand on a center, after all, is indicative of a probable death. Lastly, this network must deal with a demand for anti-venom that may be considered static, but distributed among many cities and/or neighborhoods.

This scenario has motivated us to present an extension of the pNCP, the *p-next center problem with capacity and coverage radius constraints*, referred as pNCPCR. We assume that the user does not know whether there is enough capacity in a center before arriving, which forces the user to go to a backup center. We also consider that there is a maximum distance between the beginning of the user's journey and the arrival at the backup center, to simulate the elapsed time between injury and treatment. Thus, this problem unifies the pNCP with a single-source capacitated facility location problem [11] and a maximum coverage location problem [22]. The pNCPCR, thus, is an attempt at increasing the effectiveness of locating and allocating facilities in emergency situations, such as the snake attack mentioned previously.

The pNCPCR focus on minimizing the maximal distance traveled by the users which exceeds the coverage radius and the amount of unmet demand on the centers. We propose a mathematical model to represent it and, due to its difficulty in finding good solutions, a Biased Random-Key Genetic Algorithm (BRKGA) was customized for this problem. Additionally, we propose several instances to evaluate the algorithms performance, and observe the effects of the coverage radius and the center capacity on the results.

The remainder of this paper is organized as follows. Section 2 brings the problem formulation together with the description of related works. The proposed method is described in Sect. 3. Experimental results are reported in Sect. 4. Section 5 closes the paper with concluding remarks.

2 Related Work and Problem Formulation

2.1 Related Work

The p-center problem has frequently been used for the assignment and location of emergency services and facilities, specially in the last decade. Huang et al. [15] studies a version of the p-center problem in which a facility cannot assume the demand of its own location, and thus needs to be allocated to another center. This situation is frequent in disaster situations, and is solved with dynamic programming. Morgan et al. [21] focus on properly allocating emergency service facilities during Islamic pilgrimages, in which excessive crowds are a concern. The authors point that, in regards to distance, coverage, and cover inequality, a genetic algorithm appears to have good balance between quality and computational time. Lastly, Yu et al. [28] considers that damages to the transport network can affect accessibility of emergency facilities, and thus introduces a multi-objective model that tries to guarantee minimum reachability of the facilities. Reachability is defined as a function of the level of damage on a given trajectory. The authors use a p-center problem to try to avoid the bi-level structure of the model, and test the approach on the Sioux Falls transportation network.

The capacitated facility location problem considers the presence of demand and capacity constraints on the facilities, something that interferes with the allocation of users to centers [11, 26]. Among the works focused on this problem, Biajoli et al. [5] uses a BRKGA to solve the two-stage capacitated facility location problem. The same problem is the focus of Souto et al. [23], who use a hybrid matheuristic, composed by clustering search, adaptive large neighborhood search, and local branching techniques. Mauri et al. [20] studies a multi-product version of the previous problem. The authors use a BRKGA to solve it, and prove that it outperforms a clustering search approach.

Meanwhile, the maximal covering location problem focus on the coverage radius, i.e., the maximal distance between user and center allocation [9, 22]. In this category, the study of Taiwo et al. [25] use a maximal covering location problem to identify potential locations for Covid-19 testing in Nigeria. The author tests several potential coverage radius, and observes their impact in the resulting network. Yang et al. [27] presents a continuous version of the maximal coverage location problem, with the aim of dynamically optimizing the organization of rescue in natural disasters. Amarilies et al. [2] uses greedy heuristics to solve the maximal coverage problem in the context of trashcan location. The results of this study were implemented on a village in Indonesia.

Several works unify the cited location problems, with the aim of increasing realism and modelling real-life-based situations appropriately. Karatas et al. [16] proposes a multi-objective facility location problem, with elements from both p-median, p-center, and maximum coverage. The authors focus on the design of a public emergency service network, and use a combination of branch-and-bound and iterative goal programming techniques to solve it. The study of Chauhan et al. [7] mixes the capacitated facility location problem with the maximal coverage problem. This was done in the context of drone launching sites. This study is extended in Chauhan et al. [8], where the uncertainty in battery consumption is also considered.

2.2   Problem Formulation

The formulation of the pNCPCR is based on the two-indexed formulation introduced by [1]. In this formulation, there are nodes named $\{1, \cdots, n\}$ inside a network. Those nodes represent locations such as neighborhoods or cities, with the related distance between them. If a facility is located on one node, then it is responsible for the demand of both that point and of the closest locations without facilities. One should note that we use indexes i, j, and k to refer either to facility or non-facility nodes, which may also be referred as users. The parameters and variables of this model are presented in Table 1.

Table 1. Parameters and decision variables definitions.

Parameters	
$C \in \mathbb{N}$	Maximal capacity of the nodes
$N_i \in \mathbb{N}$	Demand of node i
$d_{ij} \in \mathbb{N}$	Distance between nodes i and j
$R \in \mathbb{N}$	Maximal distance between user and backup center
$p \in \mathbb{N}$	Number of facilities to be assigned
Decision variables	
$y_j \in \{0,1\}$	$y_j = 1$ if a facility is opened on node j
$x_{ij} \in \{0,1\}$	$x_{ij} = 1$ if center j is the closest to node i
$w_{kj} \in \{0,1\}$	$w_{kj} = 1$ if center k is the closest to center j
$z \in \mathbb{N}$	Maximal traveled distance
$t_{kj} \in \mathbb{N}$	Capacity transported from center k to center j
$u_j \in \mathbb{N}$	Used capacity in center j
$\kappa_j \in \mathbb{N}$	Unmet demand in center j
$\delta \in \mathbb{N}$	Exceeded traveled distance

$$\min \quad \delta^2 + \sum_{i=1}^{n} \kappa_i^2 \tag{1a}$$

$$\text{s.t.} \sum_{j=1}^{n} y_j = p \tag{1b}$$

$$\sum_{\substack{j=1 \\ j \neq i}}^{n} x_{ij} = 1 \qquad \forall i \in \{1, \ldots, n\} \tag{1c}$$

$$x_{ij} \leq y_j \qquad \forall i,j \in \{1,\ldots,n\} \atop i \neq j \tag{1d}$$

$$y_j + \sum_{\substack{k=1 \\ d_{ik} > d_{ij}}}^{n} x_{ik} \leq 1 \qquad \forall i,j \in \{1,\ldots,n\} \atop i \neq j \tag{1e}$$

$$z \geq \sum_{\substack{k=1 \\ k \neq j}}^{n} d_{jk} \cdot x_{jk} \qquad \forall j \in \{1, \ldots, n\} \tag{1f}$$

$$z \geq d_{ij} \cdot (x_{ij} - y_i) + \sum_{\substack{k=1 \\ k \neq j}}^{n} d_{jk} \cdot x_{jk} \qquad \forall i,j \in \{1,\ldots,n\} \atop i \neq j \tag{1g}$$

$$z \leq R + \delta \tag{1h}$$

$$w_{jk} \leq x_{kj} \qquad \forall j,k \in \{1,\ldots,n\} \atop j \neq k \tag{1i}$$

$$w_{jk} \leq y_k \qquad \forall j,k \in \{1,\ldots,n\} \atop j \neq k \tag{1j}$$

$$t_{jk} \leq M \cdot w_{jk} \qquad \forall j,k \in \{1,\ldots,n\} \atop j \neq k \tag{1k}$$

$$C \cdot y_j \geq u_j + \sum_{\substack{k=1 \\ k \neq j}}^{n} t_{jk} \qquad \forall j \in \{1, \ldots, n\} \tag{1l}$$

$$\sum_{\substack{i=1 \\ j \neq i}}^{n} (x_{ij} - w_{ij}) \cdot N_i + \sum_{\substack{k=1 \\ k \neq j}}^{n} t_{jk} + N_j \cdot y_j - \kappa_j \leq u_j + \sum_{\substack{k=1 \\ k \neq j}}^{n} t_{kj} \quad \forall j \in \{1, \ldots, n\} \tag{1m}$$

$$\kappa_j \leq M \cdot y_j \qquad \forall j \in \{1, \ldots, n\} \tag{1n}$$

One must note that variable x_{ij} has different meanings that depend on i being a user or a facility. In the former case, the variable indicates the assignment of a facility to a user. In the latter, j is the backup center of an existing facility. Related to this, variable w_{ji} only exists if both indexes belong to facilities, and always corresponds to the assignment of j as the backup of another. In this formulation, Objective Function (1a) focuses on minimizing the total assignment cost of the network. This cost has two components. The first component is the squared excess of the maximal distance between a user and its backup center, when compared with the coverage radius. This is done to simulate the higher chance of death due to an elapsed time between injury and treatment higher than the threshold for after-effects and death. The second component is the sum of the squared value of unmet demands of each center. In a situation such as a snake attack, an unmet demand would mean an avoidable death.

Constraint (1b) guarantees that only p centers exist. Constraints (1c) and (1d) assign a reference center for each node alongside preventing self-assignment of user nodes. Constraint (1e) imposes a minimal distance when allocating a user to a reference center. Constraints (1f) and (1g) ensure the correct value of the highest distance between a user and its backup center. This value is either the distance between a reference center and its corresponding backup, or the sum of the distances between a user and its reference center, and between that center and its backup. The value of the exceeded travelling distance in relation to the coverage radius is obtained in Constraint (1h).

Constraints (1i) and (1j) guarantee that the variable w_{ji} only exists if both indexes belong to facilities, and if j is the backup center of i. The existence of this variable permits the transport of capacity between the centers, as Constraint (1k) indicates. As an example, this transferred capacity could be the transport of medicine between facilities. Constraint (1l) ensures that the used and transported capacities do not exceed the total available capacity of the center. Finally, Constraints (1m) and (1n) regulate the flow of demand and capacity in a given center, alongside the existence of unmet demands. The sum of the demands of the users allocated to a center, of the capacity transported to other centers, and of the demand in the center should not exceed the sum of the capacity used in it and of the capacity transported to it. If this equilibrium is violated there is an amount of unmet demand in the center, which is then penalized in the objective function.

3   Customizing the BRKGA for the pNCPCR

We developed an algorithm based on the Multi-Parent Biased Random-Key Genetic Algorithm with Implicit Path-Relinking (BRKGA-MP-IPR [4]), which is a multi-parent variant of the standard BRKGA [13]. This algorithm was chosen due to its good performance in capacitated location problems [5, 20]. In addition, BRKGA is the state-of-the-art algorithm for the pNCP [17].

3.1   Evolutionary Process

The Biased Random-Key Genetic Algorithm (BRKGA) begins by creating p populations composed by $|\mathcal{P}|$ individuals, which are called chromosomes. Each gene of a chromosome is a real-value number in the interval $[0, 1]$.

The *decoder* procedure associates a solution and the corresponding fitness value with a chromosome. The individuals are then ranked by their fitness values. The solutions with highest quality belong to the elite set $\mathcal{P}_e$, while the remaining are in the non-elite set.

On each generation three procedures are performed to obtain new populations. The *Reproduction* procedure copies all chromosomes in the elite set. *Mutants generation* deletes $|\mathcal{P}_m|$ individuals from the non-elite set and randomly creates the same amount of individuals. The remaining $|\mathcal{P}| - |\mathcal{P}_e| - |\mathcal{P}_m|$ chromosomes are generated with *Crossover*.

For the Multi-Parent BRKGA, π_t parents are selected for the *Crossover* procedure. From those, π_e belong to the elite set. The parents' fitness values are ranked and associated with probabilities by the bias function $\Phi(r)$. Then, each gene is taken from a parent according to its rank, defined by the comparison of its fitness value among all parents. The steps of reproduction, mutants generation, and crossover procedures are repeated until a stopping criterion is met. If there is an improvement in the best solution on a given generation, the local search procedure detailed in Sect. 3.2 is performed on the new best solution.

An intensification strategy for BRKGA used in this study is the Implicit Path Relinking (IPR) procedure. Path-relinking explores the neighborhood obtained in the path between two distinct solutions [12]. IPR is considered implicit due to being performed on the chromosomes of the BRKGA solution space, not on the decoded solution [4]. After path-relinking, the algorithm may migrate some elite solutions between different populations, if g iterations have passed without improvement in the best solution. Likewise, if I_s generations have passed without improvement, then the shaking procedure presented in Sect. 3.2 may be called. Finally, if $I_s \cdot R_m$ iterations passed without improvement, then a full reset is performed.

3.2   Chromosome Representation and Decoder

For the pNCPCR, the chromosome is a vector with n genes, with n being the number of nodes in the network. The decoder procedure may be divided in four phases. The first phase is the selection of centers. In it, the chromosome is sorted and the first p nodes are chosen as reference centers. The second phase corresponds to the allocation of user nodes to the closest centers, and the computation of used capacity in each center.

In the third and fourth phases, we allocate backup centers to the reference centers. However, in the third phase the backup center is only chosen if it has enough extra capacity to comport the excess demand of the reference center. If there are no candidate backup centers that obey this constraint, the center is allocated its closest center as backup in the fourth phase, and the unmet demand of the center is recorded.

The fitness value of the solutions is obtained as such: the highest value from the distance to backup center among all nodes is compared with the coverage radius. If higher, the excess distance is penalized with its squared value. The squared value of the unmet demand is then added to the squared excess distance.

Warm-Start Solution. The introduction of good solutions in the initial population is noted to increase performance of the algorithm. For the pNCPCR, the constructive heuristic starts by selecting the first node as a center. The remaining are selected from the $p - 1$ closest nodes to the first node. The decoding procedure is then performed to the resulting set of centers and non-centers.

Exploitation Strategies. We consider three exploitation strategies apart from IPR on our approach. The local search procedure, the shaking procedure, and the reset procedure are noted to lead to better algorithm performance. The local search observes the solutions found by swapping a user and a center node. It may use the first improvement or the best improvement strategies. The shaking procedure randomly exchanges an amount of user and center nodes on the elite chromosomes, and the random restart of the non-elite solutions. The reset procedure is the random restart of all individuals.

4 Experimental Results

4.1 Instances

We generated 1,652 instances derived from pNCP instances. There are four instance groups, whose differences lie in individual center capacity and coverage radius. The instances were based on the 132 proposed by [1] and the 281 proposed by [17].

The process of generating instances for the pNCPCR has three phases. The first is the definition of demand for each node, which was randomly obtained from the interval $[10, 50]$. The second phase is the calculus of the individual center capacity $C = (\sum_{i=1}^{n} D_i)/(p \cdot DC)$. The parameter "DC" in this equation refers to the demand/capacity equilibrium of a given instance. If the instance is in category "high demand", then the sum of demands is 85% of the total capacity. If it belongs to "low demand", then this percentage is equal to 40%.

The third phase of the instance generation process is the selection of the coverage radius. As the original instance solution values have a median of approximately 95, and a minimum of about 45, the "high radius" category has its distance limit randomly chosen in the interval $[45, 95]$. It is expected that, due to the capacity and demand of the nodes, the maximal distances will increase in relation to the original instances. The "low radius" category has values randomly chosen in the interval $[15, 30]$, i.e. at one-third of the previous category.

Each instance name refers to its number of nodes, number of centers, demand category, and radius category, in that order. As an example, instance pmed1_10_5_h_1 has 10 nodes and 5 centers, and belongs to the "high demand" and "low radius" categories.

4.2 Computational Environment and Parameter Settings

The computational experiments were performed in a cluster of identical machines with an Intel Xeon E5530 CPU at 2.40 GHz and 120 GB of RAM running CentOS Linux. The formulation proposed in Sect. 2 was solved with IBM ILOG CPLEX 20.1 solver. Heuristics proposed in this paper were implemented in C++ language using the BRKGA-MP-IPR framework [4]. All BRKGA variants use four threads, and all runs are limited to 30 wall-clock minutes or 1,000 generations without improvement on the best solution.

We run CPLEX with three different setups. In the first, CPLEX uses four threads and stops either when it finds an optimal integer solution, or it reaches the maximum time of 30 min (CPLEX-30min). This setup is meant to achieve direct comparison with the other methods proposed in this paper. In the second setup, we run CPLEX for one day, using 24 threads (CPLEX-1d). Such configuration looks to find optimal solutions or, at least, compute the best possible bounds. Since this configuration uses far more time and computer power per run than the other configuration, we use the results only for reporting. The third setup removes the parameter κ_i from the formulation, which turns constraint (1m) into a non-relaxed version (CPLEX-ST). In practice, this means that it is forbidden to have unmet demands on the centers. This configuration was meant to observe the effect of the relaxation of the constraint, and its results portrait a more realistic approach.

We named the BRKGA variants as follows: BRKGA-NLS for the variant without local search (pure BRKGA evolution); BRKGA-FI for the variant with first-improvement local search; and BRKGA-BI for the variant with best-improvement local search. We performed 30 independent runs of each BRKGA variation for each instance. The parameters used for the BRKGA variations were suggested by the irace package [18], and may be seen in Table 2.

4.3   Mathematical Model Results

Our first task is to find optimal solutions or as-best-as-possible solutions for the 1,652 instances.

Note that Model (1) admits at least one feasible solution for each instance. However, this is not true for CPLEX-ST due to the use of the hard demand constraint, for which there are instances without feasible solutions.

Table 3 shows the performance of the three CPLEX variants. CPLEX-1d had the least amount of instances without any solution found, at 722 (44%), with a significantly smaller average Gap%. This is expected, as it has considerably more computational power and available time to explore the different solutions. Following it is CPLEX-30min, with 1,169 (70%) infeasible instances, and, lastly, CPLEX-ST, with 1,258 (76%). Again, this scenario is not unexpected, as the use of a hard constraint would diminish the pool of feasible solutions.

Now the effect of the radius and demand category is explored in regards to instance feasibility and non-optimum runs. One should note that several instances considered infeasible by CPLEX-30min, with the higher computational time from CPLEX-1d, were considered feasible and, on some cases, an optimum solution was found. This means that, in this case, we may consider an infeasible instance as a non-optimal one, and need to combine the results of those two categories in the analysis. Table 4 presents the percentage of instances without an optimal solution for the three CPLEX variants regarding their demand and

Table 2. Best parameter configurations suggested by irace for BRKGA variations.

	BRKGA					IPR					Shaking		LS%
	$|\mathcal{P}|$	$\mathcal{P}_e\%$	$\mathcal{P}_m\%$	π_e,π_t	Φ	p	md	sel	$ps\%$	I_{ipr}	I_s	R_m	
BRKGA-BI	4085	0.22	0.22	5,10	r^{-2}	2	0.27	RE	0.23	226	60	1.86	0.96
BRKGA-FI	3912	0.28	0.46	5,10	r^{-2}	3	0.05	RE	0.63	471	100	1.77	–
BRKGA-NLS	4075	0.20	0.12	5,10	r^{-2}	1	0.08	RE	0.83	206	96	1.71	–

Table 3. Model performance on all instances. Columns "Dem." and "Rad." detail the demand and radius categories, respectively. Column "# Opt" presents the number of instances with optimum found. Column "# NoSol" shows the amount of instances considered infeasible by the respective algorithm, i.e. the configuration found no feasible solutions for the instance. Column "# Fea" has the number of instances in which feasible and non-optimal solutions were found. "Avg. Gap%" is the average percentage of gap for non-optimal and non-infeasible instances. The fifth line in each section presents the summary of results for all instances, independently of demand and radius categories.

	Dem.	Rad.	# Opt	# NoSol	# Fea	Avg. gap%
CPLEX-30min	High	High	43	78	292	84
	High	Low	45	77	291	63
	Low	High	87	290	36	93
	Low	Low	81	296	36	80
	Both	Both	256	1,169	227	78
CPLEX-1d	High	High	120	164	120	89
	High	Low	108	193	108	64
	Low	High	36	173	36	75
	Low	Low	35	192	35	63
	Both	Both	299	722	299	75
CPLEX-ST	High	High	41	337	41	94
	High	Low	43	333	43	82
	Low	High	31	295	31	84
	Low	Low	38	293	38	74
	Both	Both	153	1,258	153	84

radius categories. Note that for all cases, CPLEX has more difficulty in finding feasible solutions on the high demand category. In fact, the effect of the demand is more severe than that of the radius – something reasonable considering Model (1). The effect of the radius category is more subtle, and becomes more apparent with the longer running time and computational power of CPLEX-1d.

4.4   BRKGA Results

To compare the algorithms, we analyze the results regarding the solution quality and computational effort. For solution quality, we compute the classical Relative Percentage Deviation (RPD) and associated averages as defined in [3], with a small modification to prevent division by zero. Let $\mathcal{I}$ be a set of instances. Let $\mathcal{A}$ be the set of algorithms, and assume that set $\mathcal{R}_A$ enumerates the independent runs for algorithm $A \in \mathcal{A}$ (as defined in Sect. 4.2, 30 runs for the heuristics and one run for CPLEX-30min). We defined C_{ir}^A as the total cost obtained by algorithm A in instance i on run r, and C_i^{best} as the best total cost found across all algorithms for instance i. In order to deal with the possibility of having a

Table 4. CPLEX results in regards to categories and percentage of infeasible and non-optimal instances. The percentages presented refer to the amount of non-optimal and infeasible instances in relation to the total.

		High rad. (%)	Low rad. (%)
CPLEX-30min	High Dem.	90	89
	Low Dem.	79	80
CPLEX-1d.	High Dem	69	73
	Low Dem.	50	55
CPLEX-ST.	High Dem	91	91
	Low Dem.	79	80

best known solution equal to zero, which happens on 11 instances, we introduce the constant $c^* > 0$ on the computation. This is done to prevent the division by zero that would happen on those cases. The RPD from the best solution i is, thus, defined as

$$RPD_{ir}^A = \frac{C_{ir}^A - C_i^{best} + c^*}{C_i^{best} + c^*} \times 100, \quad \forall A \in \mathcal{A}, \ i \in \mathcal{I}, \ \text{and} \ r \in \mathcal{R}_A. \quad (2)$$

The BRKGA variants did not have the same difficulties as CPLEX-30min with infeasible solutions, finding at least one feasible solutions for all instances. Figure 1 presents a boxplot with RDP distributions for each algorithm. Note that the y-axis is plotted on a log scale to enhance the visualization. The reason is that most of the algorithms found optimal solutions frequently, skewing the display on a linear scale. In fact, the median of the RPD distributions of the three BRKGA variants was zero, which indicates that at least half of the instances reached the best or optimal solution.

However, all three algorithms had runs with considerably high RPDs, something that skews the results. In fact, the mean of all distributions was higher than the value of the 0.75 percentile, and the maximal RPD for BRKGA-BI and BRKGA-FI variants was 15,132. BRKGA-NLS had a maximal RPD of 15,902. BRKGA-NLS presented slight better results than its counterparts, with $1,486 \pm 2,952\%$; BRKGA-FI produced $1,475 \pm 2,937\%$; and BRKGA-BI had an average of $1,480 \pm 2,943\%$. Since those results are too close to call, we applied the pairwise Wilcoxon rank-sum test with Bonferroni p-value adjustment method among all algorithms. With a confidence interval of 95%, we cannot affirm there is a significant difference on the results of the three BRKGA variations.

Table 5 presents the results in regards to the instances with optimal and non-optimal found by CPLEX-30min. One may note the BRKGA variants had a poor performance on the instances with known optima, finding the optimal solution on only 5% of instances and on circa 5% of runs. The performance of all three BRKGA variants was considerably similar for those instances. For instances with unknown optima, BRKGA performance was considerably better. In fact, all three variants found the best known solution for those instances in, in average, 80% of

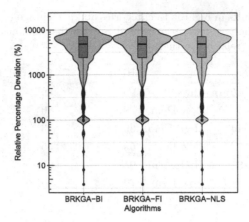

Fig. 1. Distribution of relative percentage deviations for each algorithm. Note that, since the data is plotted in log scale, zero deviations are not shown, although the algorithms have reached them.

Table 5. Algorithm performance on all instances.

Algorithm	Known optima (256 instances)			Unknown optima (1340 instances)		
	# Opt	% Opt	% Run	# Best	% Best	% Run
BRKGA-BI	13	5.08	5.01	1106	79.23	71.91
BRKGA-FI	13	5.08	5.08	1146	82.09	74.35
BRKGA-NLS	13	5.08	5.07	1143	81.88	73.03

instances. This was done on at least 70% of runs. One may note that BRKGA-FI had the best performance among the algorithms, with higher percentages on all criteria. BRKGA-BI had the worst performance of the algorithms.

Finally, Fig. 2 presents the cumulative probability of finding the best or optimal solution in relation to running time of the algorithms. Note that all BRKGA variants had better performances than CPLEX-30min on 1,800 s. In addition, BRKGA-FI is shown to be better than the other variants, with higher chance of finding the best solution on lower times. Note also that all three BRKGA variants tended to finish before the maximum permitted time. This means the BRKGA is reaching the upper limit on the number of iterations without improving the best solution, which indicates quick convergence.

When considering all results presented in this section, there are many conclusions that may be observed. The first is that BRKGA was shown to have a better performance than that of CPLEX-30min, when considering feasible solutions and cumulative probability of finding the best solution. The second consideration is that BRKGA-FI performed better than the other BRKGA variants, even if we cannot affirm the presence of a difference between the RPD distributions of the three algorithms. The last conclusion is that the BRKGA variants converged

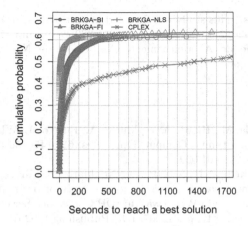

Fig. 2. Running time empirical distributions to the best solution values for all instances. The identification marks correspond to 2% of the points plotted for each algorithm.

quickly, and tends to find the optimal solutions, as shown by the medians of the RPD distributions being zero.

5   Conclusions

In this work, we presented the p-next center problem with capacity and coverage radius constraints, referred as pNCPCR. This is an extension of the p-next center problem introduced by [1], itself an extension of the classical p-center problem. The novel pNCPCR was inspired on a situation of a snake attack, in which the victim cannot be treated on the closest facility due to a lack of medicine. We formulate a mathematical model for the pNCPCR, and develop a Biased Random-Key Genetic Algorithm (BRKGA) to solve this problem.

In order to observe the effectiveness of our proposed approach, we generated 1,652 instances based on the ones used in [17], and divided in four categories. To find optimal solutions to those instances, we used three CPLEX configurations. Of those, CPLEX-30min is used in the comparison with BRKGA. We also observed the effects of the flow constraint in the proposed model, by running CPLEX-ST with a harder version of this constraint.

CPLEX-30min results were compared with BRKGA variants, which differ in local search approaches. One may node that all three BRKGA variants had better performances than CPLEX-30min, and that the performances of those algorithms was very similar, with BRKGA-FI being slightly better than the other variants.

There are several possible future works based on this research. One may consider non-uniform facility capacities, something that the model and instances proposed in this paper do not. In addition, the problem could also be formulated as a two-stage problem with stochastic demand.

Acknowledgements. This study was financed in part by CNPq, PUC–Rio, FAPERJ (Project Numbers E-26/211.086/2019, E-26/211.389/2019, and E-26/211.588/2021), and the Coordenação de Aperfeiçoamento de Pessoal de Nível Superior – Brasil (CAPES) – Finance Code 001.

References

1. Albareda-Sambola, M., Hinojosa, Y., Marín, A., Puerto, J.: When centers can fail: a close second opportunity. Comput. Oper. Res. **62**, 145–156 (2015). https://doi.org/10.1016/j.cor.2015.01.002
2. Amarilies, H.S., Redi, A.P., Mufidah, I., Nadlifatin, R.: Greedy heuristics for the maximum covering location problem: a case study of optimal trashcan location in kampung Cipare-Tenjo-West java. In: IOP Conference Series: Materials Science and Engineering, vol. 847, p. 012007. IOP Publishing (2020)
3. Andrade, C.E., Silva, T., Pessoa, L.S.: Minimizing flowtime in a flowshop scheduling problem with a biased random-key genetic algorithm. Expert Syst. Appl. **128**, 67–80 (2019). https://doi.org/10.1016/j.eswa.2019.03.007
4. Andrade, C.E., Toso, R.F., Gonçalves, J.F., Resende, M.G.C.: The multi-parent biased random-key genetic algorithm with implicit path-relinking and its real-world applications. Eur. J. Oper. Res. **289**(1), 17–30 (2021). https://doi.org/10.1016/j.ejor.2019.11.037
5. Biajoli, F.L., Chaves, A.A., Lorena, L.A.N.: A biased random-key genetic algorithm for the two-stage capacitated facility location problem. Expert Syst. Appl. **115**, 418–426 (2019). https://doi.org/10.1016/j.eswa.2018.08.024
6. Bochner, R., Fiszon, J.T., Machado, C., et al.: A profile of snake bites in brazil, 2001 to 2012. Clinical Toxicology (2014). https://doi.org/10.4172/2161-0495.1000194
7. Chauhan, D., Unnikrishnan, A., Figliozzi, M.: Maximum coverage capacitated facility location problem with range constrained drones. Transp. Res. Part C: Emerg. Technol. **99**, 1–18 (2019). https://doi.org/10.1016/j.trc.2018.12.001
8. Chauhan, D.R., Unnikrishnan, A., Figliozzi, M., Boyles, S.D.: Robust maximum coverage facility location problem with drones considering uncertainties in battery availability and consumption. Transp. Res. Rec. **2675**(2), 25–39 (2021)
9. Drezner, Z., Hamacher, H.W.: Facility Location: Applications and Theory. Springer Science & Business Media, Cham (2004)
10. Feo, T.A., Resende, M.G.: Greedy randomized adaptive search procedures. J. Global Optim. **6**(2), 109–133 (1995). https://doi.org/10.1007/BF01096763
11. Filippi, C., Guastaroba, G., Speranza, M.G.: On single-source capacitated facility location with cost and fairness objectives. Eur. J. Oper. Res. **289**(3), 959–974 (2021). https://doi.org/10.1016/j.ejor.2019.07.045
12. Glover, F.: Tabu search and adaptive memory programming—advances, applications and challenges. In: Barr, R.S., Helgason, R.V., Kennington, J.L. (eds.) Interfaces in Computer Science and Operations Research. Operations Research/Computer Science Interfaces Series, vol. 7, pp. 1–75. Springer, Boston (1997). https://doi.org/10.1007/978-1-4615-4102-8_1
13. Gonçalves, J.F., Resende, M.G.C.: Biased random-key genetic algorithms for combinatorial optimization. J. Heuristics **17**(5), 487–525 (2011). https://doi.org/10.1007/s10732-010-9143-1
14. Hansen, P., Mladenović, N.: Variable neighborhood search. In: Burke, E.K., Kendall, G. (eds.) Search Methodologies, pp. 211–238. Springer, Boston (2005). https://doi.org/10.1007/0-387-28356-0_8

15. Huang, R., Kim, S., Menezes, M.B.: Facility location for large-scale emergencies. Ann. Oper. Res. **181**(1), 271–286 (2010). https://doi.org/10.1007/s10479-010-0736-8
16. Karatas, M., Yakıcı, E.: An iterative solution approach to a multi-objective facility location problem. Appl. Soft Comput. **62**, 272–287 (2018)
17. Londe, M.A., Andrade, C.E., Pessoa, L.S.: An evolutionary approach for the p-next center problem. Expert Syst. Appl. **175**, 114728 (2021). https://doi.org/10.1016/j.eswa.2021.114728
18. López-Ibáñez, M., Dubois-Lacoste, J., Cáceres, L.P., Birattari, M., Stützle, T.: The irace package: iterated racing for automatic algorithm configuration. Oper. Res. Perspect. **3**, 43–58 (2016). https://doi.org/10.1016/j.orp.2016.09.002
19. López-Sánchez, A.D., Sánchez-Oro, J., Hernández-Díaz, A.G.: GRASP and VNS for solving the p-next center problem. Comput. Oper. Res. **104**, 295–303 (2019). https://doi.org/10.1016/j.cor.2018.12.017
20. Mauri, G.R., Biajoli, F.L., Rabello, R.L., Chaves, A.A., Ribeiro, G.M., Lorena, L.A.N.: Hybrid metaheuristics to solve a multiproduct two-stage capacitated facility location problem. Int. Trans. Oper. Res. **28**(6), 3069–3093 (2021). https://doi.org/10.1111/itor.12930
21. Morgan, A.A., Khayyat, K.M.J.: Improving emergency services efficiency during illamic pilgrimage through optimal allocation of facilities. Int. Trans. Oper. Res. **29**(1), 259–300 (2022). https://doi.org/10.1111/itor.13026
22. Murray, A.T.: Maximal coverage location problem: impacts, significance, and evolution. Int. Reg. Sci. Rev. **39**(1), 5–27 (2016). https://doi.org/10.1177/0160017615600222
23. Souto, G., Morais, I., Mauri, G.R., Ribeiro, G.M., González, P.H.: A hybrid matheuristic for the two-stage capacitated facility location problem. Expert Syst. Appl. **185**, 115501 (2021)
24. Suzuki, A., Drezner, Z.: The p-center location problem in an area. Locat. Sci. **4**(1–2), 69–82 (1996). https://doi.org/10.1016/S0966-8349(96)00012-5
25. Taiwo, O.J.: Maximal covering location problem (MCLP) for the identification of potential optimal COVID-19 testing facility sites in nigeria. Afr. Geogr. Rev. **40**(4), 395–411 (2021)
26. Wu, L.Y., Zhang, X.S., Zhang, J.L.: Capacitated facility location problem with general setup cost. Comput. Oper. Res. **33**(5), 1226–1241 (2006). https://doi.org/10.1016/j.cor.2004.09.012
27. Yang, P., Xiao, Y., Zhang, Y., Zhou, S., Yang, J., Xu, Y.: The continuous maximal covering location problem in large-scale natural disaster rescue scenes. Comput. Ind. Eng. **146**, 106608 (2020)
28. Yu, W.: Reachability guarantee based model for pre-positioning of emergency facilities under uncertain disaster damages. Int. J. Disaster Risk Reduction **42**, 101335 (2020). https://doi.org/10.1016/j.ijdrr.2019.101335

Automatic Configuration of Metaheuristics for Solving the Quadratic Three-Dimensional Assignment Problem Using Irace

Imène Ait Abderrahim[1]([✉]) [iD] and Thomas Stützle[2] [iD]

[1] Khemis Miliana University, Khemis Miliana, Algeria
i.aitabderrahim@univ-dbkm.dz
[2] Université libre de Bruxelles (ULB), Bruxelles, Belgium
Thomas.Stuetzle@ulb.be

Abstract. Metaheuristic algorithms are traditionally designed follow-ing a manual and iterative algorithm development process. The perfor-mance of these algorithms is, however, strongly dependent on their cor-rect tuning, including their configuration and parametrization. This is labour-intensive, error-prone, difficult to reproduce and explores only a limited number of design alternatives. To overcome manual tuning, the automatic configuration of algorithms is a technique that has shown its efficiency in finding performance-optimizing settings of parameters. This paper contributes to overcoming the challenge of automatically config-ured metaheuristics using the iterated racing for automatic algorithm configuration irace applied to the quadratic three-dimensional assign-ment problem. In particular, we use particle swarm optimization (PSO), a tabu search (TS), an iterated local search (ILS) and two hybrid algo-rithms PSO-TS and PSO-ILS. Of these algorithms, the tabu search algo-rithm and the PSO-ILS worked the best. The results show that the algo-rithm automatic configuration enables identifying an ideal tuning of the parameters and reaching better results when compared to a manual con-figuration, in similar execution time.

1 Introduction

Permutations are used in many combinatorial optimization problems to represent a candidate solution. Such problems are known as permutation-based problems in the literature and can be found in many theoretical and real domains such as scheduling, routing and assignment. One is the quadratic three-dimensional assignment problem (Q3AP), which is known to be $\mathcal{NP}$-hard [12]. It is an exten-sion of the quadratic assignment problem (QAP) and as such known to be dif-ficult for the exact solution. In fact, we can not solve large-size instances using exact methods, because the size of the search space for an instance of size n from Q3AP is $n!^2$, much more than the already quite difficult QAP [12,25].

Metaheuristics have proved their efficiency to solve large and difficult opti-mization problems and they get optimal solutions or very good approximate

L. Di Gaspero et al. (Eds.): MIC 2022, LNCS 13838, pp. 350–364, 2023.
https://doi.org/10.1007/978-3-031-26504-4_25

solutions with respect to the optimal. Yet, the performance of optimization algorithms requires a considerable amount of computational effort and expert knowledge [28]. Several elements can influence on algorithm's performance like the characteristics of the algorithm, the problem being solved and the environment in which the execution will be performed. To guide the configuration process, there are several automatic algorithm configuration tools, also called configurators, proposed in the literature. Some of the methods of these are ParamILS [16], SMAC [15], GGA [3] and irace [19]. These configuration tools have obtained very good settings for different types of algorithms and improved their performance in many cases [14, 27].

In this paper, we propose to solve the problem of the Q3AP using automatically configured metaheuristic algorithms, where we configure some parameters of the metaheuristics and try to find the best settings for these parameters that improve the performance of these algorithms. We believe that a well-performed configuration procedure can significantly improve the performance of the metaheuristics for solving a complex problem and that by automatically setting the parameters of the metaheuristics we show how this kind of approach of algorithm configurations is used.

The remainder of the paper is organized as follows. In Sect. 2 we define the Q3AP and review methods to solve the Q3AP. Then, in Sect. 3, we describe the different algorithms used in this paper to solve the Q3AP. In Sect. 4, we define the automatic configuration of algorithms. Section 5 reports the experimental results; Sect. 6 concludes the work.

2   The Q3AP

The Q3AP can be seen as an assignment problem where the objective is to optimize a quadratic function over a three-dimensional assignment polytope. Therefore, one formulation of the Q3AP is representing the objective function as a quadratic expression [12]. Mostly, this quadratic formulation is used for the case of exact algorithms, using the reformulation linearization technique (RLT-1). However, for the case of metaheuristics which rely on underlying iterative improvement algorithms, another permutation-based formulation of the Q3AP can be represented by two permutations π and ψ of the numbers in the set $\{1, 2, \cdots n\}$ [12]. This formulation is derived from that of the QAP [13], where

$$\min \left\{ f(\pi, \psi) = \sum_{i=1}^{n} \sum_{j=1}^{n} C_{i\pi(i)\pi(j)j\psi(i)\psi(j)} + \sum_{i=1}^{n} b_{i\pi(i)\psi(i)} \right\} \tag{1}$$

where: π and ψ are permutations over the set $\{1, \cdots, n\}$. With this formulation, one can see that that the size of the search space is equal to $n! \times n!$.

Iterative improvement algorithms for the Q3AP typically make use of a double 2-exchange neighbourhood relation, in which a double two solutions are neighbored if their permutations differ in exactly two positions. That are, the neighborhoods $N(\pi)$ and $N(\psi)$ of permutations π and ψ, respectively.

As such, the Q3AP was first introduced by William P. Pierskalla in 1967 [26] and then revisited by Peter Hahn et al. [12] as the first work for solving the Q3AP problem. They implemented a sequential branch-and-bound algorithm based on Lagrangian relaxation to extend the development and implementation of effective lower bounds. From another side they also proposed metaheuristics approaches, where they adapted four approximate methods, known for being very successful on solving the QAP, for solving the Q3AP. These methods are the simulated annealing (SA) algorithm of Conolly [8], a tabu search (TS) algorithm, the fast ant (FANT) algorithm [30] and the iterated local search (ILS) [29]. The ILS was considered the best-performing one of the four metaheuristics.

Other approaches were proposed in the literature for solving the Q3AP problem. In the exact approaches, we could find the work of Galea et al. [9], Mezmaz et al. [24]. The best performing exact algorithm is due to Mittelmann and Salvagnin [25] where they addressed a challenging Q3AP instance REAL-16 and reached the optimum of this instance for the first time using exact methods.

Metaheuristics are another class of approaches that have been explored to solve the Q3AP problem. Next to the four approaches proposed by Hahn [12], we find also the work presented by Luong et al. [22], where they proposed a GPU-based parallel iterative tabu search algorithm where the steps of TS are executed on the CPU, but the generation and evaluation of the large size neighborhood are executed on GPU. Wu et al. [31] were investigating the modulation diversity(MoDiv) design problem for HARQ in CoMP-MIMO system aiming to minimize the bit error rate (BER) upper bound where they formulated the MoDiv design into a Q3AP. They presented an iterated local search (ILS) algorithm. The rest of the contributions are from the hybrid population-based algorithms family. In the work of Loukil et al. [20], they proposed a two-level parallel hybrid evolutionary algorithm (EA) with a simulated annealing algorithm. Lipinski [18] suggests a hybrid algorithm that combines an EA with a local search method. In Mehdi et al. [23], they proposed a new cooperative hybrid scheme that combines an evolutionary algorithm and a branch-and-bound method to solve large benchmarks of permutation-based problems. Another work is undertaken by Abderrahim et al. [1] where they propose a particle swarm optimization (PSO) algorithm hybridized with an iterated local search embedding a tabu search. Gmys et al. [11] propose a comparative study of high productivity and high-performance programming languages for parallel metaheuristics where they tested their algorithms by solving the Q3AP. They implemented an ILS algorithm and a genetic algorithm (GA) hybridized with a local search (LS).

3   Metaheuristics for the Q3AP

We perform experiments to compare the performance of the parameters when using the metaheuristics: PSO, TS and ILS for solving the Q3AP [2]. Some modifications have been considered on the parameters of the metaheuristics, where we have included new parameters for the configuration design process. This includes in TS and in ILS, the number and the type of perturbations and the acceptance criteria, as well as we have fixed a few parameters such as the constants c_1 and c_2 for PSO and the max run-time for the algorithms. As a result, we have four algorithms to tune namely TS and ILS, PSO-TS, and PSO-ILS. Since we also included the PSO algorithm, we start with that one.

3.1   Particle Swarm Optimization for Q3AP

Particle swarm optimization (PSO) was inspired by nature behaviours [17]. It is a method for optimizing continuous functions that exploits the interaction of the simple behaviours of individuals in a swarm to form an organized and coherent, collective behaviour. In this section, we describe the adapted discrete PSO for Q3AP of Abderrahim et al. [2].

In PSO, each particle is represented by its position vector x_i and its velocity vector v_i and moves through the search space by updating the values of the x_i and v_i vectors at every iteration. As the solution of the Q3AP is the double permutation of size n, for each particle, the position of a particle is represented as a double vector $x_i = (x_i^1, x_i^2)$ and the velocity is represented by one vector v_i only. For a given particle i, its velocity $v_i(t+1)$ and its position $x_i(t+1)$ at iteration $(t+1)$, for that, some parameters are needed, where the parameters r_1 and r_2 are random numbers in $[0, 1]$. c_1 and c_2 are positive constants called coefficient of the self-recognition and coefficient of the social component respectively. The variable ω is the inertia factor in which value is typically set up to vary linearly from 1 to near 0 during the iterated process and it is a parameter to control the impact of the previous velocities on the current velocity. The adapted PSO method for Q3AP is illustrated in Algorithm 1. Since the PSO is a continuous method, we have to adapt it to discrete problems by a supplementary procedure that can allow this kind of configuration. There are many methods that meet the above-mentioned mapping condition such as the great value priority (GVP) rule [7]. The GPV was used because it showed that it does not break the structure of the permutation. This rule of GVP notes the search space of n dimension as Ω^n, and the space of the Q3AP, whose scale is n, it is labelled Φ^n. This g corresponds to the rule of mapping of $\Omega \rightarrow \Phi$, namely: $g : \Omega^n \rightarrow \Phi^n$ and g must meet the following terms. (1) For each vector X in Ω^n you can find a unique corresponding permutation p, and it is noted $g(X) = p$. (2) A certain permutation order from X must be found by g so that it can reflect the priority order relation in the permutation. Namely if $i > j$ then by using g it means that $x_i > x_j$, where ">" expresses the array relation between the elements. For details on the GVP see Algorithm 2.

It can be given a concrete example of Algorithm 2, suppose the vector (0.3, 0.4, 0.1, 0.7) then after sort operation, we get the ordered set {0.7, 0.4, 0.3, 0.1}, finally, we get the permutation {4, 2, 1, 3}.

Algorithm 1. PSO for Q3AP [2]

1: initialise parameters of PSO: NP, $c_1 = 2$, $c_2 = 2$, r_1, r_2, ω;
2: initialise the position and velocity vectors of particles randomly;
3: **for** each particle $i = 1$ to NP **do**
4:     $p_{bestid} = x_{id}$; // initialise the best local positions with the initial positions
5: **end for**
6: Evaluate $f(x_{id})$ of each particle;
7: Evaluate g_{bestd}; // best value among all p_{bestid};
8: **repeat**
9:     **for** i to 1 to NP **do**
10:         Update velocity
11:         Update position
12:         find sequence using GVP rule
13:         evaluate the fitness of each particle
14:         **if** $f(x_{id}) < f(p_{bestid})$ **then**; // Evaluation
15:             $p_{bestid} = x_{id}$;// Update p_{bestid}
16:         **end if**
17:         **if** $f(p_{bestid}) < f(g_{bestd})$ **then**; // Evaluation
18:             $g_{bestd} = p_{bestid}$; // Update g_{bestd}
19:         **end if**
20:     **end for**
21: **until** ($iteration < Max\_PSO\_iteration$ and $elapsed\_time < Max_{Time}$)
22: Print the best solution found

3.2 TS for Q3AP

The tabu search (TS) technique was developed by Glover [10]. TS is based on the neighbourhood search. TS uses memory to direct the search and to escape local optima, typically by forbidding certain solutions or solution components to avoid reversing moves and revisiting solutions. In our TS algorithm, we start by setting the parameters and initialising the initial solution. Once the initialisation step is finished, we apply the main iterations of the TS algorithm. We first find the best non-tabu neighbour and after that, we update the tabu list. A new feature that is introduced by our TS is that a perturbation is done either to a single permutation or to the double permutation of the solution. At this level, we call a parameter $Swap \in (Single, Both)$ that decides whether the perturbation will be applied to a single permutation or to both of them. The number of perturbations is taken based on a random number, where the number of perturbations is between 2 and n and is decided by a parameter $NbrPrt$. After the perturbation step, we get a new solution s. The stopping criterion is based on the run-time (Max_{Time}) where the algorithm stops when the time is out. The value of Max_{Time} is computed from the size of the instance and the value of the parameter $timeLL$ where $Max_{Time} = timeLL \times n$. The algorithm can be seen in Algorithm 3.

Algorithm 2. GVP Rule [7]

```
 1: Input: X = (x₁, x₂, ⋯ , xₙ); xᵢ ∈ R
 2: Output: p = (p₁, p₂, ⋯ , pₙ); pᵢ ∈ {1, 2, ⋯ , n}
 3: k = 0
 4: for i = 1 to n do
 5:     k=1
 6:     for j=1 to n and j ≠ i do
 7:         if (xⱼ > xᵢ) then
 8:             k = k +1
 9:         end if
10:     end for
11:     p[k] = i
12: end for
13: Return p
```

3.3   ILS for Q3AP

The iterated local search (ILS) is a simple and powerful stochastic local search method [21]. The ILS algorithm explores the search space of local optima. It starts from an initial solution s and returns a local optimum. It then applies a perturbation to go from the space of local optima to the intermediate space of all solutions. Then, a local search is applied again and we reach hopefully different local optima. If this passes an acceptance test, it becomes the next element of the walk in the local optima space [21]. As a local search, we used the TS method.

In our ILS algorithm, we start by setting the parameters and do the initialisation of the initial solution. We are considering in this algorithm the following parameters: $alpha$, $Swap$, $nSwap$, $AccpCrt$, $timeLL$ and $Tconst$ when needed. Once the initialisation step is finished, we apply one iteration of the TS algorithm as mentioned in the previous section. At the level of acceptance criteria, for the ILS, three types of acceptance criteria have been used in our algorithm, namely 0, 1 and 2. Acceptance criteria 0 takes the last solution, type 1 takes the best of the two solutions and acceptance criteria 2 means that we are using the Metropolis criteria from the Simulated Annealing style for acceptance. The ILS method process consists in a loop over the TS method, where the stopping criterion of ILS is based on the run-time (Max_{Time}). The value of Max_{Time} is computed from the size of the instance and the value of the parameter $timeLL$ where $Max_{Time} = timeLL \times n$ and the TS stopping criteria is based on the number of iterations defined as $MaxIters = kk \times n$. The ILS can be seen in the Algorithm 4.

3.4   PSO-TS and PSO-ILS for Q3AP

Two hybrid algorithms have been implemented to solve the Q3AP problem, they are the PSO-TS and PSO-ILS. In our hybrid algorithm, we start by setting all the necessary parameters. Then we begin the PSO algorithm until the PSO stopping criteria is reached, as described in Algorithm 1. After that, the mapping

Algorithm 3. TS for Q3AP

1: Set parameters: $alpha, Swap, nSwap, timeLL = 10s$
2: $elapsed\_time = 0$
3: $s \leftarrow s_0$ // Create a random solution
4: $s_{best} = s$
5: $Eval(s)$ // Evaluate the solution ;
6: Initialise Tabu list
7: **repeat**
8:     Find the best non-tabu neighbour of s
9:     Update Tabu list
10:    **if** $(Swap ='' Both'')$ **then**
11:        $s \leftarrow PermuteBoth(s')$
12:    **else**
13:        $s \leftarrow PermuteSingle(s')$
14:    **end if**
15:    **if** $f(s) < f(s_{best})$ **then**
16:        $s_{best} \leftarrow s$
17:    **end if**
18: **until** $(elapsed\_time >= Max_{Time})$
19: Print the best solution found

---

rule GVP is applied to the best particle obtained so far by the PSO algorithm in order to map it to a permutation with discrete values. This permutation is given as starting solution to the TS algorithm, which is then executed. When the TS algorithm reaches its stopping criteria, which is the number of iterations in this case ($MaxIters$), it gives back the best solution found so far. At this time, the hybrid algorithm keeps always the best solution found so far in memory before repeating the process. The process stops when the time of execution is exhausted, which is the global stopping criteria Max_{Time}. The hybrid algorithm is given in Algorithm 5. Note that the same execution is applied for the second hybrid algorithm, PSO-ILS.

4   Automated Algorithm Configuration

4.1   The Automatic Configuration of Algorithms

Algorithm configuration is the process of finding a set of values for the algorithm parameters that ideally expose high empirical performance for instances of a particular class of problem [4,16]. For algorithm configuration we need a configuration scenario that contains a *parameter search space*, a set of *training* and *test* instances, and a *budget* of configuration [14,19]. Roughly, the parameters can be classified into two types. Firstly, these are the parameters that define the categorical side of the algorithmic components (e.g. the crossover operator for an evolutionary algorithm, or the branching strategy for an exact algorithm). Secondly, there are the parameters that are in charge of the control of the selection of the components of an algorithm (e.g. the length of a tabu list, or the size

Algorithm 4. ILS for Q3AP

1: Get parameters: $alpha, Swap, nSwap, AccpCrt, kk, timeLL = 10$
2: $elapsed\_time = 0$
3: $s \leftarrow s_0$; // Create a random solution;
4: $s = TabuSearch()$
5: $s_{best} \leftarrow s$
6: $Eval(s)$;// Evaluate the solution ;
7: **repeat**
8:     **if** $(Swap =" Both")$ **then**
9:         $s \leftarrow PermuteBoth(s')$;
10:     **else**
11:         $s \leftarrow PermuteSingle(s')$;
12:     **end if**
13:     Initialise Tabu list;
14:     $s' = TabuSearch()$
15:     $s \leftarrow AcceptanceCriterion(s, s', Tconst)$
16: **until** ($elapsed\_time >= Max_{Time}$)
17: Print the best solution found

Algorithm 5. Hybrid PSO with TS to solve Q3AP

1: Set parameters: $alpha, Swap, nSwap, AccpCrt, timeLL = 10, c_1 = c_2 = 2$
2: $f^* = +\infty$ //initialisation of best cost
3: **repeat**
4:     $Gbest \leftarrow PSO()$
5:     $s \leftarrow GVP(Gbest)$
6:     $s' \leftarrow TS(s)$
7:     **if** $(Eval(s') < f^*)$ **then**
8:         $s^* = s'; f^* = Eval(s')$
9:     **end if**
10: **until** $elapsed\_time >= Max_{Time}$
11: Print the best solution found

of a perturbation). Their domain commonly corresponds to numbers of integer or continuous parameters [6]. Generally, it is represented by *categorical* or *ordinal* variables in the first case, while in the second case as numerical ones, that is, integer or real-valued ones. Further, a parameter can have dependencies relations with the values of other parameters. The configuration of algorithms can be used for optimizing different measures of performance, which are generally the average solution quality or the time of execution [14, 19].

The irace package [19] is a configurator of algorithms that implements configuration procedures based on iterated racing and we use it here as the tool for automatic algorithm configuration [6]. This tool is a general-purpose configurator that requires defining a scenario of the configuration as described above. The supported parameter types are categorical, ordinal, integer and real parameters. irace applies a racing procedure. In such a procedure several configurations are evaluated on bigger subsets of the training instance set and statistical tests are established to identify the algorithm configurations that get worse performance than the others. If the statistical tests speak against configurations these will be eliminated from the race and the further execution will continue with those that survived. This holds since either only one or a set of configurations survive or the termination criterion is accomplished. After one iteration new configurations are then generated from a probabilistic model. It is updated to be centred around the best configurations and simultaneously this model is decreased in a variance of the previous one. In this way, the irace convergence gets a high-performing areas of the parameter search space, while increasing the performance estimation by increasing the number of instances in which elite configurations are evaluated.

4.2   Parameters Configuration

In the automatic configuration, for each target parameter, an interval or set of values must be defined according to the type of the parameter. There is no limit to the size of the set or the length of the interval, but keep in mind that larger ranges could increase the difficulty of the tuning task. For simplification, the description for the parameter space is given as a table. Each line of the table defines a configurable parameter $< name > < label > < type > < domain >$ [|*condition*] [19] where each field is defined as follows.

$< name >$: The name of the parameter as an unquoted alpha-numeric string.
$< label >$: A label for this parameter. This is a string that will be passed together with the parameter.
$< type >$: The type of the parameter, either integer, real, ordinal or categorical given as a single letter: 'i', 'r', 'o' or 'c'.
$< range >$: The range or set of values of the parameter delimited by parentheses; for example, (0,1) for a real parameter in the range $[0, 1]$ or a categorical parameter with item a, b, c, and d.
[|*condition*] : An optional condition that determines whether the parameter is enabled or disabled, thus making the parameter the condition evaluates to false, then no value is assigned to this parameter, and neither the parameter value nor the corresponding label are passed to the *TargetRunner*.

The experiments presented in this paper configure different parameters of the algorithms to a set of different algorithm benchmarks for the Q3AP problem. The experiments presented in this paper configure different parameters of the algorithms to a set of different algorithm benchmarks for the Q3AP problem. Here, please note that the algorithms could only be improved further, but we expect that the algorithms would in their improved version be rather similar. The total number of the tuned parameters is six plus three fixed parameters for the completion. The goal of the configuration process is to find the best parameter settings that improve the quality of the solution. The optimization algorithm parameters are the following:

alpha: It is a parameter used to compute the size of the tabu list, which takes real values from the interval $]0.0, 0.5]$. The formula used in this case is $tabu.size = \lfloor \alpha \times neighborhood\_size \rfloor$.

Swap: This is a categorical parameter, which defines the kind of perturbations to be applied to the solution (*Single, Both*). *Single* means that the perturbation is applied in one of the permutations, whereas *Both* means that the perturbation is applied in both permutations of the current solution.

nSwap: Abbreviation of "number of Swaps". This parameter is used in the formula $NbrPrt = 2 + \lfloor nSwap \times n - 1 \rfloor$, where the value of $NbrPrt$ defines the number of perturbations to be applied to the permutation. The values of $nSwap$ are real values from the interval $[0.0, 1.0]$.

kk: This parameter is an integer and is used to set the number of the iterations for the local search algorithms (the tabu search in ILS) depending on the instance size. It holds that $maxIters = kk \times n$ where $kk \in [1, 200]$.

AccpCrt: This parameter is categorical. It is used to set an acceptance criteria on the current solution compared to the previous one. This parameter takes one of the three following values $(0, 1, 2)$ where

- 0: is using a selection of the last solution.
- 1: it is using the method better. It consists on selecting the best solution between the current solution and the previous one.
- 2: it consists of using the system as in the method of simulated annealing Metropolis by accepting some different solution with a condition of a small error. In this case, the value of the temperature T is computed using the next parameter $Tconst$. The probability error p of acceptance of the new solution is computed by $p = -(Eval(s') - Eval(s))/T$, where $T = Tconst \times Eval(s')$ and $Eval(s')$ and $Eval(s)$ are the fitness of s and s', the best solution so far and the current solution, respectively.

Tconst: It is a conditional parameter of type real. It is related to the acceptance criteria (2), where we need to compute the temperature T as in simulated annealing Metropolis. Thus, the formula is like following $T = Tconst \times Eval(s')$, where $Tconst \in [0.0, 0.1]$ and $Eval(s')$ is the fitness of the current solution.

The last parameters $timeLL$, c_1 and c_2 are all fixed through the whole configuration. The parameter $timeLL$ is used to compute the run-time for the algorithms according to $Max_{Time} = timeLL \times n$, where $timeLL = 10$, in our case. The parameters c_1 and c_2 represents the acceleration coefficients for the particles in PSO algorithm, where $c_1 = c_2 = 2$.

The configuration of the parameters is declared in a parameter file for tuning the algorithms, as shown in Table 1. The first column is the name of the parameter; the second column is the label, typically the command-line switch that controls this parameter, which irace will concatenate to the parameter value when invoking the target algorithm; the third column gives the parameter type (either *integer, real, ordinal* or *categorical*); the fourth column gives the domain.

Table 1. Structure of parameters file

Name	Switch	Type	Values
swap	"–swap"	c	(single, both)
alpha	"–alpha"	r	(0.00, 0.50)
nSwap	"–nSwap"	r	(0.0, 1.0)
kk	"–kk"	i	(1, 200)
AccpCrt	"–AccpCrt"	c	(0, 1, 2)
Tconst	"–Tconst"	r	(0.0, 0.1)

5   Experimental Evaluation

5.1   Experimental Setup

Our experiments consist on applying irace, a tool for automatic algorithm configuration, to the different proposed algorithms in order to find the best algorithm configuration and parameter setting, given a set of training problem instances and a tuning budget. The tuning budget is defined as the maximum number of runs of the algorithm. The best configuration found by irace is then applied to a set of test instances, different from the training set in order to assess its performance.

For the training process, we have generated 30 instances of size n, where $n = 10, 11, 12, 13, 14, 15, 16, 17, 18, 19, 20$. The instances are divided into 3 groups a, b and c. Instances from category 'a' are matrices where values are in $(0,5)$, instances from category 'b' take the values from the interval $(0,10)$ and 'c' category values are in the interval $(0,20)$, where we generated them in such a way that they would look like the Nugent instances from QAPLIB [5]. In the test process, we used 6 instances from the Nugent family on QAPLIB [5]: nug8, nug12, nug13, nug15, nug20 and nug30. We have chosen these specific instances for the test because they are the most cited in the literature. The algorithms are run with different random seeds on the test instances. The tuning budget considered here for one run of irace is 5000 runs of an algorithm. At the end of the run, irace prints the best configurations as a table and as command-line parameters.

The goal was to test if the performance of the candidate configurations corresponds to the execution of the optimization algorithms parameters. The tests were run under Cluster Rocks 6.2, which is based on CentOS 6.2. The machine used was 2 AMD Opteron (2 GHz), 8 cores, 12 MB cache and 16 GB RAM.

5.2   Experimental Results

The experiments were run for the same execution time for the same instance in all the different algorithms, where $Max_{Time} = timeLL \times n$.

The Table 2 shows the values of the best configurations given by irace for each algorithm. From the experiments, we saw that in most configurations, the parameter "$swap$" is set to "$both$", which seems to increase the quality of the

Table 2. Results of the parameters

Algorithm	swap	alpha	nSwap	kk	AccptCrt	Tconst
TS	Single	0.0051	0.9904	–	0	–
ILS	Both	0.1386	0.0495	196	0	–
PSO-TS	Both	0.2576	0.5054	–	0	–
PSO-ILS	Both	0.1946	0.9727	194	0	–

Table 3. Solution costs for the PSO and TS algorithms based on the same computation time.

Inst.	PSO		TS	
	Best	Avg	Best	Avg
nug8	504	809.2	134	134
nug12	2072	2786.6	580	682.6
nug13	3656	5066.8	1974	2057.6
nug15	4804	7501.2	2230	2605.6
nug20	34832	38293.8	25792	26279.6
nug30	109698	114590.8	72450	73581.8

solution. Anyway, the TS prefers to use a swap in one permutation. For the calculation of tabu list size, the value of the parameter "*alpha*" is average, this means that the tabu list influences more on the algorithm performance; the only exception is again the TS where the values of this parameter are small close to zero, which is translated that the tabu list is small. The parameter "*nSwap*" has large values in most configurations, which means that the more you can permute the better are our results. What is maybe surprising is that the acceptance criteria "*AccpCrt*" is always a random selection, that is 0. This means it doesn't focus only on the best solution neighbors but tries to find other better solutions in the search space. Finally, we have always the kk at a high value. This parameter influences on the number of iterations of the local searches in ILS algorithms.

We first use the PSO algorithm alone and compare it to the TS algorithm on the same smaller computation time. This is the observation of whether a population algorithm alone is able to be similar to a local search algorithm. The results in Table 3 do say it is not the case. In fact, without needing a null hypothesis test, it is clear that the PSO algorithm alone can not get close to the performance of, for example, the TS.

We go now to compare the fitness of the four algorithms based on the best values and averages given by irace. As it is clear in Table 4, the fitness averages in TS, ILS, PSO-TS and PSO-ILS algorithms are significantly promising. While comparing the results given by these four algorithms we realised that the TS algorithm beats the others in terms of quality of solution where the success in the tested instances is 100% for the instances nug8, nug12 and nug15 and very close to the best known values for the rest of the instances.

Table 4. Solution costs obtained by the different algorithms on the test instances

Inst.	TS		PSO-TS		ILS		PSO-ILS	
	Best	Avg	Best	Avg	Best	Avg	Best	Avg
nug8	134	134.0	134	134.0	134	134.0	134	134.6
nug12	580	580.0	580	588.0	580	588.5	580	594.4
nug13	1912	2022.2	1912	1962.0	1912	1965.3	1912	1962.7
nug15	2230	2230.0	2230	2256.7	2230	2449.7	2230	2324.4
nug20	25590	25889.0	26024	26349.8	25590	26120.8	26084	26443.2
nug30	71680	72210.2	72652	73769.2	72830	74268.8	72714	73513.0

From the experimental results given by the test process, the TS is outperforming the other algorithms on instances nug20 and nug30 based on a Wilcoxon-rank sum test on the instances. There can be different reasons for this. One, can be that the run-time is not high enough. This may be also the reason that for the instance that was found to be difficult, the nug13 instance, also the TS was somehow worse performing than the other algorithms. Yet, another reason may be that the generated training instances are not good for highlighting the algorithms. This was the case in our test instances, where the best-so-far on the training instance (algorithm PSO-ILS) was on the test instances outperformed by the TS algorithm.

Finally, we report about the run time of experiments in irace, where irace takes $timeLL \times n = 200$s for the biggest instances. This allows us also to compare roughly the times on the nug12 and nug13 instances, which in our case are 120 and 130 s on a single core of the 2 GHz AMD Opteron, to the average time it took for a single core 440 MHz of an HPJ5000 machine, which were for the ILS 2162.2 and 7519.2 s [12].

6   Conclusion

In this paper, we propose to exploit the available knowledge on algorithmic components and parameter setting strategies for metaheuristics in the form of automatically configurable algorithms for solving the problem of Q3AP. Overall, we had four metaheuristics: the tabu search (TS), iterated local search (ILS) and two hybrid ones, where, the PSO with the tabu search algorithm (PSO-TS) and the iterated local search variant (PSO-ILS). irace could find settings that significantly improve over the default settings of the algorithms. However, what is striking is that the PSO-ILS was the best on the training instances, but on the test instances it was the TS algorithm which was best performing. Hence, this is something we have to test in the future.

Other directions for future work can be considered for the extension. First, it would be interesting to extend the experimental part by allowing the PSO parameters to undergo the tuning. We could re-consider the great value priority rule to let by the tuning decide whether it still is chosen. Anyway, the parameters of the PSO and also some additional parameters of the TS and the ILS would be interesting for the tuning. A second direction is to extend our work by investigating other single-based and population-based SLS algorithms for the quadratic three-dimensional assignment problem. Ideally, these extensions would allow us to compare the automatically designed algorithms in this paper against other automatically designed metaheuristics, to study the impact of the automatic configuration on the different metaheuristics and to understand how and which algorithms can be combined.

References

1. Abderrahim, I.A., Loukil, L.: Hybrid PSO-TS approach for solving the quadratic three-dimensional assignment problem. In: 2017 First International Conference on Embedded and Distributed Systems (EDiS), pp. 1–5 (2017)
2. Abderrahim, I.A., Loukil, L.: Hybrid approach for solving the Q3AP. Int. J. Swarm Intell. Res. (IJSIR) **12**, 98–114 (2021)
3. Ansótegui, C., Sellmann, M., Tierney, K.: A gender-based genetic algorithm for the automatic configuration of algorithms. In: Gent, I.P. (ed.) CP 2009. LNCS, vol. 5732, pp. 142–157. Springer, Heidelberg (2009). https://doi.org/10.1007/978-3-642-04244-7_14
4. Birattari, M., Stützle, T., Paquete, L., Varrentrapp, K.: A racing algorithm for configuring metaheuristics. In: Proceedings of GECCO 2002, pp. 11–18. Morgan Kaufmann Publishers (2002)
5. Burkard, R.E., Karisch, S.E., Rendl, F.: QAPLIB-a quadratic assignment problem library. J. Glob. Optim. **10**(4), 391–403 (1997)
6. Pérez Cáceres, L., Pagnozzi, F., Franzin, A., Stützle, T.: Automatic configuration of GCC using irace. In: Lutton, E., Legrand, P., Parrend, P., Monmarché, N., Schoenauer, M. (eds.) EA 2017. LNCS, vol. 10764, pp. 202–216. Springer, Cham (2018). https://doi.org/10.1007/978-3-319-78133-4_15
7. Congying, L., Huanping, Z., Xinfeng, Y.: Particle swarm optimization algorithm for quadratic assignment problem. In: Proceedings of 2011 International Conference on Computer Science and Network Technology, vol. 3, pp. 1728–1731 (2011)
8. Connolly, D.T.: An improved annealing scheme for the QAP. Eur. J. Oper. Res. **46**(1), 93–100 (1990)
9. Galea, F., Hahn, P.M., LeCun., B.: A parallel implementation of the quadratic three-dimensional assignment problem using the bob++ framework. In: 21st Conference of the European Chapter on Combinatorial Optimization (ECCO XXI) (2008)
10. Glover, F.: Tabu search - part I. ORSA J. Comput. **1**(3), 190–206 (1989)
11. Gmys, J., Carneiro, T., Melab, N., Talbi, E.G., Tuyttens, D.: A comparative study of high-productivity high-performance programming languages for parallel metaheuristics. Swarm Evol. Comput. **57**, 100720 (2020)
12. Hahn, P.M., et al.: The quadratic three-dimensional assignment problem: exact and approximate solution methods. Eur. J. Oper. Res. **184**(2), 416–428 (2008)
13. Hillier, F.S., Connors, M.M.: Quadratic assignment problem algorithms and the location of indivisible facilities. Manag. Sci. **13**(1), 42–57 (1966)
14. Hoos, H.H.: Automated algorithm configuration and parameter tuning. In: Hamadi, Y., Monfroy, E., Saubion, F. (eds.) Autonomous Search, pp. 37–71. Springer, Heidelberg (2011). https://doi.org/10.1007/978-3-642-21434-9_3

15. Hutter, F., Hoos, H.H., Leyton-Brown, K.: Sequential model-based optimization for general algorithm configuration. In: Coello, C.A.C. (ed.) LION 2011. LNCS, vol. 6683, pp. 507–523. Springer, Heidelberg (2011). https://doi.org/10.1007/978-3-642-25566-3_40
16. Hutter, F., Hoos, H.H., Leyton-Brown, K., Stützle, T.: ParamILS: an automatic algorithm configuration framework. JAIR **36**, 267–306 (2009)
17. Kennedy, J., Eberhart, R.: Particle swarm optimization. In: Proceedings of ICNN'95 - International Conference on Neural Networks, vol. 4, pp. 1942–1948 (1995)
18. Lipinski, P.: A hybrid evolutionary algorithm to quadratic three-dimensional assignment problem with local search for many-core graphics processors. In: Fyfe, C., Tino, P., Charles, D., Garcia-Osorio, C., Yin, H. (eds.) IDEAL 2010. LNCS, vol. 6283, pp. 344–351. Springer, Heidelberg (2010). https://doi.org/10.1007/978-3-642-15381-5_42
19. López-Ibáñez, M., Dubois-Lacoste, J., Pérez Cáceres, L., Stützle, T., Birattari, M.: The irace package: iterated racing for automatic algorithm configuration. Oper. Res. Perspect. **3**, 43–58 (2016)
20. Loukil, L., Mehdi, M., Melab, N., Talbi, E.G., Bouvry, P.: Parallel hybrid genetic algorithms for solving Q3AP on computational grid. Int. J. Found. Comput. Sci. **23**(02), 483–500 (2012)
21. Lourenço, H.R., Martin, O.C., Stützle, T.: Iterated Local Search, pp. 320–353. Springer, Boston (2003)
22. Luong, T.V., Loukil, L., Melab, N., Talbi, E.G.: A GPU-based iterated tabu search for solving the quadratic 3-dimensional assignment problem. In: ACS/IEEE International Conference on Computer Systems and Applications - AICCSA 2010, pp. 1–8 (2010)
23. Mehdi, M., Charr, J.C., Melab, N., Talbi, E.G., Bouvry, P.: A cooperative tree-based hybrid GA-B&B approach for solving challenging permutation-based problems. In: Proceedings of GECCO 2011, pp. 513–520. Association for Computing Machinery, New York (2011)
24. Mezmaz, M., Mehdi, M., Bouvry, P., Melab, N., Talbi, E.G., Tuyttens, D.: Solving the three dimensional quadratic assignment problem on a computational grid. Cluster Comput. **17**, 205–217 (2014)
25. Mittelmann, H.D., Salvagnin, D.: On solving a hard quadratic 3-dimensional assignment problem. Math. Program. Comput. **7**(2), 219–234 (2015). https://doi.org/10.1007/s12532-015-0077-3
26. Pierskalla, W.P.: The multidimensional assignment problem. Oper. Res. **16**(2), 422–431 (1968)
27. Stützle, T., López-Ibáñez, M.: Automatic (offline) configuration of algorithms. In: GECCO (Companion), pp. 681–702. ACM Press, New York (2015)
28. Stützle, T., López-Ibáñez, M.: Automated design of metaheuristic algorithms. In: Gendreau, M., Potvin, J.-Y. (eds.) Handbook of Metaheuristics. ISORMS, vol. 272, pp. 541–579. Springer, Cham (2019). https://doi.org/10.1007/978-3-319-91086-4_17
29. Stützle, T.: Iterated local search for the quadratic assignment problem. Eur. J. Oper. Res. **174**(3), 1519–1539 (2006)
30. Taillard, E.D.: Fant: Fast ant system. Technical report, Istituto Dalle Molle Di Studi Sull Intelligenza Artificiale (1998)
31. Wu, W., Mittelmann, H., Ding, Z.: Modulation design for mimo-comp harq. IEEE Commun. Lett. **21**(2), 290–293 (2017)

Hyper-parameter Optimization Using Continuation Algorithms

Jairo Rojas-Delgado[1](✉)[iD], J. A. Jiménez[2][iD], Rafael Bello[3][iD],
and J. A. Lozano[1,4][iD]

[1] Basque Center for Applied Mathematics, Bilbao, Spain
{jrojasdelgado,jlozano}@bcamath.org
[2] Universidad de las Ciencias Informáticas, Havana, Cuba
ja.jimenez@uci.cu
[3] Universidad Central de Las Villas, Santa Clara, Cuba
rbellop@uclv.edu.cu
[4] Intelligent Systems Group, University of the Basque Country
UPV/EHU, Donostia, Spain
ja.lozano@ehu.eus

Abstract. Hyper-parameter optimization is a common task in many application areas and a challenging optimization problem. In this paper, we introduce an approach to search for hyper-parameters based on continuation algorithms that can be coupled with existing hyper-parameter optimization methods. Our continuation approach can be seen as a heuristic to obtain lower fidelity surrogates of the fitness function. In our experiments, we conduct hyper-parameter optimization of neural networks trained using a benchmark set of forecasting regression problems, where generalization from unseen data is required. Our results show a small but statistically significant improvement in accuracy with respect to the state-of-the-art without negatively affecting the execution time.

Keywords: Hyper-parameter · Optimization · Continuation · Machine learning

1 Introduction

Hyper-parameters appear in many application areas such as scientific simulation studies, material design, drug discovery and especially in machine learning applications. Hyper-parameters affect the execution time, memory consumption and, even more importantly, the ability to generalize from unseen data [7]. A hyper-parameter can be of categorical, discrete or continuous nature and influences the performance of a given algorithm.

Hyper-parameters can represent many concepts and behaviours. For example, when considering the k nearest neighbourhood algorithm, the value of the nearest points k can be considered a discrete hyper-parameter. Similarly, the number of centroids of the k-means algorithm, the maximum depth of a decision tree and the number of layers of a multi-layer neural network are examples of discrete hyper-parameters. Hyper-parameters can be of categorical nature,

© The Author(s), under exclusive license to Springer Nature Switzerland AG 2023
L. Di Gaspero et al. (Eds.): MIC 2022, LNCS 13838, pp. 365–377, 2023.
https://doi.org/10.1007/978-3-031-26504-4_26

for example: when we select one among several types of activation functions or training algorithms for a neural network, or the type of kernel for a support vector machine. In addition, there are also many examples of continuous hyper-parameters such as the learning rate of Stochastic Gradient Descent (SGD). Further references to these algorithms and their hyper-parameters can be found in any of the available textbooks [2,7,13].

There are two main approaches to hyper-parameter optimization: manual and automatic methods. The manual approach assumes that there is an understanding of how the hyper-parameters affect the algorithm, hence, by selecting such hyper-parameters directly[1], no optimization is conducted. On the other hand, the automatic approaches use an optimization method to find hyper-parameters. The use of automatic hyper-parameter optimization greatly reduces the need for this understanding, but its use comes at the expense of costlier computation.

In the last few years there has been an increase in the efforts to address automatic hyper-parameter optimization [4,11,14,18,20]. In the literature, several important algorithms have been introduced for this task, mostly based on Gaussian Processes and Bayesian Optimization, for example, Sequential Model-based Algorithm Configuration (SMAC) [8] and Tree Parzen Estimators (TPE) [4].

In the optimization field, the term continuation method refers to the general approach of starting to solve a surrogate of the true fitness function and transforming it progressively during the course of iterations to the true fitness function [19]. The way the surrogate fitness function is transformed to the true fitness function is usually done via homotopy [1]. Informally, two continuous functions are called homotopic if one can be continuously deformed into the other.

We introduce a simple approach to perform automatic hyper-parameter optimization based on continuation algorithms. We assume that the fitness of a hyper-parameter can be calculated using an iterative approach. The main contribution of our work is that we show that, using a simple increase of the budget for hyper-parameter optimization and not a given fixed budget, the accuracy of optimization can be improved without affecting the execution time. Our continuation approach can be seen as a heuristic to obtain lower fidelity surrogates of the fitness function. Further experimental work suggests that our approach outperforms similar state-of-the-art hyper-parameter optimization algorithms. In addition, we link our approach and other similar algorithms to continuation methods, which could become an area of future theoretical analysis.

The organization of this paper is as follows. Section 2 examines related work, Sect. 3 presents continuation algorithms and Sect. 4 describes our approach to perform hyper-parameter optimization via continuation. In Sect. 5 we provide details on the hyper-parameter optimization problem of neural networks. We present the results of comparing our continuation approach for hyper-parameter optimization with several state-of-the-art hyper-parameter optimization algorithms. Finally, some conclusions and recommendations are given.

---

[1] For example, based on the researcher experience or based on some heuristics collected from results of previous works.

2   Related Work

Continuation is a general approach for building several lower fidelity surrogates of the fitness function of an optimization problem. The key ideas of continuation can be summarized in two points: first the surrogates range from lower fidelity to higher fidelity during the course of the optimization process and secondly, information from the exploration/exploitation process of the lower fidelity surrogates is used to enhance the search when the higher fidelity surrogates are used. In the next section, we provide a more formal definition of continuation.

Using surrogates for optimization can be traced back to Tovey [21] in which the author introduces an approach called surrogate function swindle. Surrogates have evolved to consider statistical models that learn the fitness function during optimization in a form of a general meta-model for the surrogate-based optimization framework [17]. Meta-model surrogates are similar to continuation as their accuracy increases during the course of iterations, therefore, the surrogates increase their fidelity during optimization. In the area of hyper-parameter optimization, the use of surrogates have inspired the development of methods such as the Sequential Model-Based Optimization (SMBO), Tree Parzen Estimators (TPE) introduced by Bergstra et al. [5] and Sequential Model-Based Optimization for General Algorithm Configuration (SMAC) [8].

The adaptive resource allocation framework is another line of work related to building surrogates. The general idea is to allocate an increasing amount of resources to hyper-parameter values that are promising and prune under-performing hyper-parameter values. Resources can be iterations, examples of the learning data set or the execution time. Two recent examples of such algorithms are the Successive Halving (SH) algorithm [10] and its successor the Hyperband (HB) algorithm [15]. However, the information from the previous evaluations of the surrogates is not used to explore the hyper-parameter search space. Instead, hyper-parameters are all sampled from a random search in the first step and then under-performing hyper-parameters are discarded using SH. More recently, Falkner et al. [6] introduced an approach that hybridizes Bayesian Optimization with HB algorithm (BOHB) using previous surrogate evaluations to guide the search as devised by the original authors of HB.

Basically, the work by Falkner et al. [6] relies on HB to determine how many configurations to evaluate with which budget, but, it replaces the random selection of configurations at the beginning of each HB iteration with a model-based search. The relationship of the resource allocation methods with continuation comes from the assumption that as more resources are allocated to promising hyper-parameter sets, the fitness function converges to an optimum [10]. A similar area of interest is multi-fidelity optimization. Wu et al. [22] suggest the use of a machine learning model to learn which hyper-parameters evaluate with which budget. Generally speaking, methods that can choose their fidelity are very appealing and more powerful than the conceptually simpler resource allocation methods previously discussed. However, given a small budget, it is reasonable to raise concerns about the convergence of such multi-fidelity approaches, especially when the model has a small number of fitness evaluations to learn [9].

Our continuation based approach aims to use lower-to-higher fidelity surrogates similar to the work by Falkner *et al.* [6]. However, our proposal uses an approach different to performing successive halving on a set of predefined hyper-parameter values, whether or not they are sampled using a model. Our approach can be combined with a model-search algorithm such as TPE. Moreover, we agree on the hypothesis discussed by [9] that a simpler heuristic to achieve lower-to-higher fidelity surrogate evaluations can make better use of fixed schedules than using a model, such as in the multi-fidelity approach.

3   Hyper-parameter Optimization via Continuation

In supervised machine learning, we train models to fit data. Usually, when we say data we are referring to a set of examples to learn from, that is $E = \{e_1, ..., e_p\}$ where $e_i = (x_1, ..., x_d, y_1, ..., y_t)$, $x = (x_1, ..., x_d) \in X$ and $y = (y_1, ..., y_t) \in Y$. Often, x_i is referred as the input feature and y_j is known as the output feature or ground-truth. We denote $\hat{X} = \{x_1, ..., x_p\}$ as the set that contains only the input features of each example and $\hat{Y} = \{y_1, ..., y_p\}$ as the set that contains only the output features of each example. Here, we will not distinguish whether x_i or y_j are continuous, discrete, ordinal, nominal or a combination.

Let $\hat{f} : X \times \Omega \times W \to Y$ be a model estimated from a set of examples E using hyper-parameters $\omega = (\omega_1, ..., \omega_h) \in \Omega$ and learning parameters $w \in W$. Again, we will not distinguish whether ω_i are continuous, discrete, ordinal, nominal or a combination of the previous. Considering a loss function $L(.,.)$ that measures the error between the ground-truth and the model estimation, we can formally define the hyper-parameter optimization problem as follows:

$$\omega = \underset{\omega \in \Omega}{\operatorname{argmin}} \, L(\hat{Y}, \{\hat{f}(x, \omega, w(\omega)) : x \in \hat{X}\}) \tag{1}$$

where $w : \Omega \to W$ is the result of minimizing the given loss considering a specific set of hyper-parameters as defined in the following lines. Usually, hyper-parameter optimization involves two nested cycles of optimization when searching hyper-parameters in machine learning settings. This is because, in order to calculate the loss function, first, we must perform an elementary-cycle of optimization to find the model learning parameters $w(\omega)$. That is, with a given set of hyper-parameters, solve:

$$w(\omega) = \underset{w \in W}{\operatorname{argmin}} \, L(\hat{Y}, \{\hat{f}(x, \omega, w) : x \in \hat{X}\}) \tag{2}$$

In practice, we use two different identically distributed and independent sets of examples to solve Eq. (1) and to solve Eq. (2). This is to reduce the bias induced during hyper-parameter optimization. These are the so-called validation set and training set respectively.

Hyper-parameter optimization is a class of challenging optimization problems, whose objective function tends to be non-smooth, discontinuous, unpredictably varying in computational cost and includes continuous, nominal and/or discrete variables [12]. We call parent-algorithm to the optimization algorithm that deals with choosing the hyper-parameters related to the model estimation by solving Eq. (1). Moreover, we refer to the optimization algorithm that solves Eq. (2) as the base-algorithm.

Continuation, embedding or homotopy methods have long served as useful tools in modern mathematics. For a complete and formal definition of continuation, please refer to [1]. For the sake of our discussion, considering the function $w : \Omega \rightarrow W$, we call homotopy or continuation of w to the deformation $H : \Omega \times [0,1] \rightarrow W$ such that:

$$H(\boldsymbol{\omega}, 0) = g(\boldsymbol{\omega}), H(\boldsymbol{\omega}, 1) = w(\boldsymbol{\omega}) \tag{3}$$

where $\boldsymbol{\omega} \in \Omega$ and $g : \Omega \rightarrow W$ is a surrogate function of $w(.)$ but less costly to evaluate, ideally but not necessarily, smooth and convex. Actually, all $H(\boldsymbol{\omega}, j)$ with $j \in [0,1)$ are surrogates of $w(.)$ with an increased degree of fidelity.

We do not consider the exact solution of Eq. (2), instead, we consider the solution of such an optimization problem after being solved using an iterative approach. Let $w(\boldsymbol{\omega}, i)$ denote the solution of solving Eq. (2) for a number of i iterations. The somehow vague term iteration may refer to the number of iterations of SGD, a total wall-clock time or a total number of training examples.

We assume there is a maximum budget M for the number of iterations that can be allocated to a single evaluation of a given set of hyper-parameters, usually given by some practical limitation such as a maximum wall-clock time or a maximum number of examples in a training set. Therefore, we set $H(\boldsymbol{\omega}, i/M) = w(\boldsymbol{\omega}, i)$ for $0 \leq i \leq M$ as our continuation transformation. The main assumption of our work is that $w(\boldsymbol{\omega}, i)$ is a consistent estimator of $w(\boldsymbol{\omega})$, that is: $\Pr[||w(\boldsymbol{\omega}, i) - w(\boldsymbol{\omega})|| > \epsilon] \rightarrow 0$ when $i \rightarrow \infty$, which is referred to as weak consistency [7].

4   Our Continuation Approach for Hyper-parameter Optimization

Our proposal for hyper-parameter optimization uses $w(\boldsymbol{\omega}, i)$ as a surrogate of the fitness function in Eq. (2). Considering the practical limitations, practitioners usually have a limited number of iterations M used to evaluate each hyper-parameter set. Therefore, even without using explicitly continuation, they do not use the true fitness function, but, a surrogate determined by $w(\boldsymbol{\omega}, M)$. Considering a sequence of t surrogates given by $w(\boldsymbol{\omega}, m_1), ..., w(\boldsymbol{\omega}, m_t)$, we define a continuation transformation by making $m_1 < ... < m_t = M$ on the basis of the consistency assumption.

We denote the number of iterations of the parent-algorithm as $n_1, ..., n_t$ such that each surrogate $w(\boldsymbol{\omega}, m_i)$ is used for a number n_i iterations of the parent algorithm. The general idea is to accumulate the search experience of the surrogates at the beginning of the exploration of the hyper-parameter space and after some iterations of the parent-algorithm become more aggressive using higher fidelity surrogates. Figure 1 shows a graphical illustration of our continuation

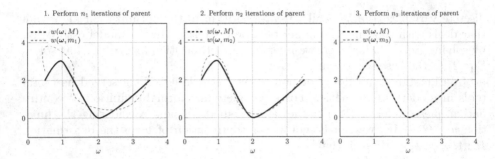

Fig. 1. Representation of optimization by continuation over the course of iterations of the parent-algorithm. Here, we use $t = 3$ surrogates and, for simplicity, only one hyper-parameter.

approach applied to the hyper-parameter optimization problem with $t = 3$ surrogates and a single hyper-parameter denoted as ω. In the horizontal axis, we represent the hyper-parameter values and in the vertical axis, we represent the loss value obtained after solving Eq. (2).

Considering Fig. 1, in the first n_1 iterations of the parent-algorithm, we use the surrogate $w(\omega, m_1)$ which is correlated to $w(\omega, M)$ but is not exactly the same. This allows the parent-algorithm to recognize promising regions of the search space and focus on such regions in future iterations. This way, the parent-algorithm uses the information gathered up to this point to continue the search in the following n_2 iterations. Notice that when using the last surrogate, $m_3 = M$, however, in the continuation case $w(\omega, m_3)$ it is only used for n_3 iterations of the parent-algorithm.

Algorithm 1 describes our continuation approach to hyper-parameter optimization. The input of our algorithm is a total budget B, the maximum number of iterations of the base-algorithm given by M and a number of surrogates t. The total budget B is the number of base-algorithm iterations used during the entire hyper-parameter optimization process and M is the maximum number of iterations of the base-algorithm that can be allocated to a single surrogate evaluation. Our algorithm takes this maximum budget and sets the number of iterations assigned to each surrogate (m_i) and the number of iterations of the parent-algorithm used for each surrogate (n_i) as follows:

- Each surrogate is associated with a predefined number of base-algorithm iterations, such as, while using the i-th surrogate, the base-algorithm runs with a budget of $m_i = \lfloor M \cdot i/t \rfloor$ optimization steps. As $m_i \leq M$, the i-th surrogate does not use the maximum budget, obtaining a surplus of $q_i = M - m_i$ iterations each time the parent algorithm uses such surrogate to evaluate a hyper-parameter set. The surplus can be used to perform an additional number of iterations of the parent-algorithm.
- In our algorithm, we use the surplus of the i-th surrogate to perform additional parent-algorithm iterations of the $(t - i)$-th surrogate. This way, the

---

Algorithm 1: Continuation hyper-parameter optimization.

input : Total budget: B.
input : Maximum number of base-algorithm iterations per surrogate: M.
input : Number of surrogates: t
output: Best set of hyper-parameters.

1   Initialize: L := [] ;   // List of hyper-parameters
2   **for** $i := 1$ **to** t **do**
3   Set: $m_i := \lfloor M \cdot i/t \rfloor$;
4   Set: n_i using Equation(4);
5   **for** $j := 1$ **to** n_i **do**
6   H:= EvaluateNext(L, m_i);
7   L := L ∪ H;
8   **end**
9   **end**
10   **return** GetBest(L);

---

maximum surplus gets assigned to additional parent-algorithm iterations of the surrogate with the largest base-algorithm budget, the second maximum surplus gets assigned to additional parent-algorithm iterations of the surrogate with the second largest base-algorithm budget and so on. Therefore, the number of iterations of the parent-algorithm when using the i-th surrogate is given by the following expression:

$$
n_i = \begin{cases} \lfloor n \rfloor, & \text{if } i < t - i \\ \lfloor n + nq_i/m_i \rfloor, & \text{if } i = t - i \\ \lfloor n + nq_i/m_i + nq_{t-i}/m_i \rfloor, & \text{otherwise} \end{cases} \tag{4}
$$

where $n = B/tM$ is an initial predefined number of parent-algorithm iterations. In total, our algorithm distributes approximately B optimization steps of the base-algorithm consuming the specified total budget.

In line 1 of Algorithm 1, we initialize the list of hyper-parameters found so far as an empty list. We set the number of iterations for the base-algorithm and for the parent-algorithm in line 3–4. In lines 5–7, the parent-algorithm performs n_i iterations and this is where the actual hyper-parameter space exploration occurs. In our case, we consider that the parent-algorithm observes a previous history of hyper-parameter evaluations L and generates a new hyper-parameter set evaluated using a budget of m_i iterations. The previous behaviour is encoded in the procedure EvaluateNext(., .) in line 6. This is similar to Bayesian Optimization methods widely used for hyper-parameter optimization, and in our work, we use the TPE algorithm to perform this step. The best hyper-parameter set is returned in line 10 using the procedure GetBest(.) which simply returns the hyper-parameter set in L with the lowest loss value.

5   Results and Discussion

In this section, we present the empirical results of comparing our continuation hyper-parameter optimization approach with state-of-the-art methods in the literature. First, we present a benchmark of similar methods and describe additional experimental settings used in this work. Latter, we present a comparison of the accuracy of the different studied approaches and compare the meta-parameters of our continuation approach, i.e., the number of surrogates compared to the successive halving meta-parameter of HB and BOHB.

Similar Methods. In our study, we compare several hyper-parameter optimization approaches. We consider the following algorithms:

- TPE: Tree Parzen Estimators algorithm [4].
- HB: Hyperband algorithm with random search [15].
- BOHB: HB with Bayesian Optimization [6].
- CTPE: TPE algorithm where the evaluation budget is given by our continuation approach. This is the approach introduced in this work.

Code and Data Sets. The source code and data used in the experimental study is available at https://github.com/ml-opt/continuation-hpo. Furthermore, we use a benchmark set of forecasting regression problems from the UCI Machine Learning Repository [16]. We include a detailed specification of the transformations performed to such data sets in the publicly available code repository. Table 1 presents a list of the data sets considered in this work.

Table 1. List of benchmark forecasting regression problems used in the optimization of neural networks hyper-parameters.

	Type	Examples	Features
Facebook Metrics (FM)	Regression	495	17
Forest Fires (FF)	Regression	517	10
Aquatic Toxicity (AT)	Regression	545	8
Fish Toxicity (FT)	Regression	907	6
Airfoil Noise (AN)	Regression	1502	5
Concrete Strength (CS)	Regression	1029	8

Hyper-parameter Settings. Through our experiments, we use a fixed artificial neural network architecture consisting of a two-layer perceptron and ReLU activation function. Furthermore, we use a fixed maximum budget of $M = 100$ epochs of SGD per artificial neural network. This is a maximum budget but either our continuation approach CTPE, HB or BOHB can allocate smaller budgets during optimization.

Hyper-parameters such as the SGD learning-rate (Lr) and momentum (M), the size of the mini-batches (Sb) and the number of neurons in each layer of

Table 2. Detailed description of artificial neural network hyper-parameters optimized by the different algorithms.

	Type	Min	Max	Description
Lr	Real	0.001	0.1	SGD learning rate
M	Real	0.8	0.999	SGD momentum
Sb	Int	40	200	Number of examples in each mini-batch
H1	Int	1	40	Number of neurons in the first hidden layer
H2	Int	1	40	Number of neurons in the second hidden layer

the two-layer perceptron (H1, H2) are fine-tuned using the automatic hyper-parameter optimization approaches studied in this work. Table 2 shows a detailed specification of such hyper-parameters.

Additional Settings. To ensure a fair comparison, we allocate an equal total budget to the different methods, i.e., $B = 10000$ epochs. For example, given that we set a maximum budget for each surrogate of $M = 100$ epochs, we perform 100 iterations of the TPE algorithm accounting for a total budget of $B = 10000$ epochs. Similarly, we set an initial predefined number of iterations $n = 100$ of CTPE with a maximum budget per surrogate of $M = 100$ epochs, which ensures a distribution of a total budget of $B = 10000$ epochs. Moreover, for HB and BOHB we perform $100/(\lfloor \log_\eta(M) \rfloor + 1)$ iterations, where η is the successive halving hyper-parameter, accounting for the specified total budget.

Statistical Validation. In our experiments, we split the examples into three disjoint subsets for training (60%), validation (20%) and test (20%). For each hyperparameter optimization method, we repeat the experiments ten times in order to perform meaningful statistical comparisons. We further conduct statistical pairwise comparisons of our continuation approach for hyper-parameter optimization using a Bayesian hierarchical correlated t-test [3].

5.1   Accuracy Analysis

We study the performance of the different hyperparameter optimization methods. The split parameter of HB and BOHB and the number of surrogates of the CTPE algorithm are chosen after conducting a grid search. The next section provides further details and a careful empirical analysis of such meta-parameters. Table 3 shows the mean squared error of the neural networks using unseen examples during training. In the first column, we show the name of the forecasting problem. We report the mean of ten measurements truncated to two decimal places while highlighting in bold the lowest value found.

We further conduct statistical pairwise comparisons between the different hyper-parameter optimization methods using a Bayesian hierarchical correlated t-test. Figure 2 plots 50 thousand Monte Carlo samples of the probabilities of one method fitness value being larger than the other in barycentric coordinates.

Table 3. Mean Squared Error of hyper-parameter optimization algorithms while configuring neural networks hyper-parameters in a benchmark set of regression problems.

	TPE	HB	BOHB	CTPE
FM	2.50E–03	1.16E–03	**8.62E–04**	1.69E–03
FF	5.28E–02	4.66E–02	4.65E–02	**4.27E–02**
AT	2.48E–02	1.88E–02	1.88E–02	**1.63E–02**
FT	1.22E–02	1.16E–02	1.10E–02	**1.01E–02**
AN	1.51E–02	1.53E–02	1.33E–02	**1.19E–02**
CS	1.50E–02	1.74E–02	**1.47E–02**	1.51E–02

The three vertices are denoted as M1, R, M2 and represent the probability of method M1 finding worse solutions than method M2, practical equivalence and vice-versa respectively. Here, Pr[R] is the probability of $|M1 - M2| < 1.0E\text{-}03$ which stands for practical equivalence.

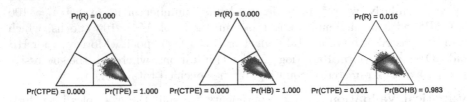

Fig. 2. Pairwise comparisons between hyper-parameter optimization methods configuring neural networks hyper-parameters in a benchmark set of forecasting problems.

In this scenario, the probability of the method CTPE being worse than any other method is small according to our statistical validation framework. Here, it is important to consider that we define practical equivalence when the difference between the two methods is smaller than 1.0E-03, which may be different for other application domains. Nevertheless, the results show that our continuation approach for hyper-parameter optimization does not introduce overfitting in the hyper-parameter optimization process and is able to find hyper-parameters that lead to better generalization.

In addition, HB and BOHB methods require the optimization of the base-algorithm to be paused when performing successive halving. After successive halving, HB and BOHB resume the optimization of the base-algorithm. This step introduces a difficulty for applications with numerous parameters, such as complex simulations or training large neural networks, that need to save and restore their current state. Conversely, our approach delivers improved accuracy without having to save and restore the base-optimization state.

5.2   Additional Meta-parameters

We qualitatively analyze the difficulty of configuring the additional meta-parameters of our continuation approach and HB. Our continuation approach requires specifying the number of surrogates (t) used to assign different budgets during optimization. Conversely, HB requires specifying the split parameter (η) used during successive halving. Such discrete meta-parameters affect the accuracy and/or the execution time of the algorithms and in practice require allocating further training budget in order to find their values.

Figure 3 shows the accuracy of our continuation approach (in blue violin plots) and HB (in yellow violin plots) using different values of t and η (in the horizontal axis). The plot shows the mean squared error of ten trials.

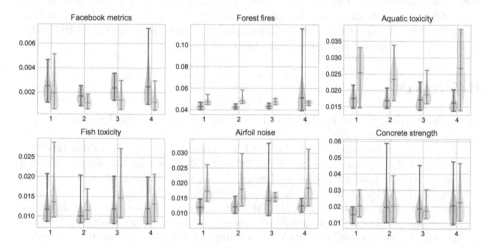

Fig. 3. Comparison between our continuation approach for hyper-parameter optimization (in blue) and HB (in yellow). The horizontal axis represents different values for t and η. The vertical axis represents the Mean Squared Error using unseen examples. (Color figure online)

Considering Fig. 3, the selection of the values of t and η is closely related to the forecasting regression problem at hand. With the exception of the problem Facebook metrics, in general, the selection of the hyper-parameter η in the case of HB is more sensitive than the selection of t used by our continuation approach. This can be seen in the higher variance and higher errors obtained when considering different values of η if compared with the different values of t. This suggests that, in the scenario of selecting an arbitrary default value for t or η, chances are that our continuation approach will outperform the accuracy of HB. A similar comparison can be conducted between CTPE and BOHB. However, BOHB has even more meta-parameters than HB, making the comparison more challenging. In our work, we used the default meta-parameters of BOHB[2].

---

[2] Available at: https://automl.github.io/HpBandSter.

6   Conclusions

In this paper, we presented a continuation approach for hyper-parameter optimization that can be coupled with existing algorithms designed for this task. We observe an improvement in the accuracy compared with regular Bayesian algorithms such as TPE and similar resources allocation methods, such as HB and BOHB, without affecting the execution time. We expect that continuation algorithms can become a suitable theoretical framework to explain why many of the newly introduced hyper-parameter optimization methods based on transformations of the fitness function or transformation of the search operator are expected to work properly.

Acknowledgements. This research has been partially supported by the Spanish Ministry of Sciences, Innovation and Universities through BCAM Severo Ochoa accreditation SEV-2017–0718; by the Basque Government through the program BERC 2022–2025; and by Elkartek Project KK.2021/00091.

References

1. Allgower, E.L., Georg, K.: Numerical Continuation Methods: An Introduction, vol. 13. Springer, Cham (2012)
2. Alpaydin, E.: Introduction to Machine Learning. MIT press, Cambridge (2020)
3. Benavoli, A., Corani, G., Demšar, J., Zaffalon, M.: Time for a change: a tutorial for comparing multiple classifiers through Bayesian analysis. J. Mach. Learn. Res. **18**(77), 1–36 (2017). http://jmlr.org/papers/v18/16-305.html
4. Bergstra, J., Yamins, D., Cox, D.: Making a science of model search: hyperparameter optimization in hundreds of dimensions for vision architectures. In: Dasgupta, S., McAllester, D. (eds.) Proceedings of the 30th International Conference on Machine Learning. Proceedings of Machine Learning Research, vol. 28, pp. 115–123. PMLR, Atlanta, Georgia (2013). http://proceedings.mlr.press/v28/bergstra13.html
5. Bergstra, J.S., Bardenet, R., Bengio, Y., Kégl, B.: Algorithms for hyper-parameter optimization. In: Shawe-Taylor, J., Zemel, R.S., Bartlett, P.L., Pereira, F., Weinberger, K.Q. (eds.) Advances in Neural Information Processing Systems 24, pp. 2546–2554. Curran Associates, Inc. (2011). http://papers.nips.cc/paper/4443-algorithms-for-hyper-parameter-optimization.pdf
6. Falkner, S., Klein, A., Hutter, F.: BOHB: robust and efficient hyperparameter optimization at scale. In: Dy, J., Krause, A. (eds.) Proceedings of the 35th International Conference on Machine Learning. Proceedings of Machine Learning Research, vol. 80, pp. 1437–1446. PMLR, Stockholmsmässan, Stockholm Sweden (2018). http://proceedings.mlr.press/v80/falkner18a.html
7. Goodfellow, I., Bengio, Y., Courville, A.: Deep Learning. MIT press, Cambridge (2016)
8. Hutter, F., Hoos, H.H., Leyton-Brown, K.: Sequential model-based optimization for general algorithm configuration. LION **5**, 507–523 (2011)
9. Hutter, F., Kotthoff, L., Vanschoren, J.: Automated Machine Learning. Springer (2019). https://doi.org/10.1007/978-3-030-05318-5

10. Jamieson, K., Talwalkar, A.: Non-stochastic best arm identification and hyperparameter optimization. In: Gretton, A., Robert, C.C. (eds.) Proceedings of the 19th International Conference on Artificial Intelligence and Statistics. Proceedings of Machine Learning Research, 09–11 May 2016, vol. 51, pp. 240–248. PMLR, Spain (2016). http://proceedings.mlr.press/v51/jamieson16.html

11. Klein, A., Falkner, S., Springenberg, J.T., Hutter, F.: Learning curve prediction with Bayesian neural networks. In: International Conference On Learning Representation (ICLR), vol. 51, pp. 240–248 (2017). https://openreview.net/forum?id=S11KBYclx&noteId=r15rc0-Eg

12. Koch, P., Golovidov, O., Gardner, S., Wujek, B., Griffin, J., Xu, Y.: Autotune: a derivative-free optimization framework for hyperparameter tuning. In: Proceedings of the 24th ACM SIGKDD International Conference on Knowledge Discovery & Data Mining, KDD 18, Association for Computing Machinery, New York, pp. 443–452 (2018). https://doi.org/10.1145/3219819.3219837, https://doi.org/10.1145/3219819.3219837

13. Kubat, M.: An introduction to machine learning. Springer (2017). https://doi.org/10.1007/978-3-319-63913-0

14. Law, H.C., Zhao, P., Chan, L.S., Huang, J., Sejdinovic, D.: Hyperparameter learning via distributional transfer. In: Wallach, H., Larochelle, H., Beygelzimer, A., d Alché-Buc, F., Fox, E., Garnett, R. (eds.) Advances in Neural Information Processing Systems 32, pp. 6804–6815. Curran Associates, Inc. (2019). http://papers.nips.cc/paper/8905-hyperparameter-learning-via-distributional-transfer.pdf

15. Li, L., Jamieson, K., DeSalvo, G., Rostamizadeh, A., Talwalkar, A.: Hyperband: a novel bandit-based approach to hyperparameter optimization. J. Mach. Learn. Res. **18**(1), 6765–6816 (2017)

16. Lichman, M.: UCI machine learning repository (2013). http://archive.ics.uci.edu/ml

17. Lukšič, Ž, Tanevski, J., Džeroski, S., Todorovski, L.: General meta-model framework for surrogate-based numerical optimization. In: Yamamoto, A., Kida, T., Uno, T., Kuboyama, T. (eds.) DS 2017. LNCS (LNAI), vol. 10558, pp. 51–66. Springer, Cham (2017). https://doi.org/10.1007/978-3-319-67786-6_4

18. Maclaurin, D., Duvenaud, D., Adams, R.P.: Gradient-based hyperparameter optimization through reversible learning. In: Proceedings of the 32Nd International Conference on International Conference on Machine Learning, ICML 2015, JMLR.org, vol. 37, pp. 2113–2122 (2015). http://dl.acm.org/citation.cfm?id=3045118.3045343

19. Mobahi, H., Fisher, J.W.: A theoretical analysis of optimization by gaussian continuation. In: AAAI, pp. 1205–1211 (2015)

20. Probst, P., Boulesteix, A.L., Bischl, B.: Tunability: importance of hyperparameters of machine learning algorithms. J. Mach. Learn. Res. **20**(53), 1–32 (2019). http://jmlr.org/papers/v20/18-444.html

21. Tovey, C.A.: Simulated simulated annealing. Am. J. Math. Manag. Sci. **8**(3–4), 389–407 (1988). https://doi.org/10.1080/01966324.1988.10737246

22. Wu, J., Toscano-Palmerin, S., Frazier, P.I., Wilson, A.G.: Practical multi-fidelity Bayesian optimization for hyperparameter tuning. In: Adams, R.P., Gogate, V. (eds.) Proceedings of The 35th Uncertainty in Artificial Intelligence Conference. Proceedings of Machine Learning Research, vol. 115, pp. 788–798. PMLR (2020). http://proceedings.mlr.press/v115/wu20a.html

Selecting the Parameters of an Evolutionary Algorithm for the Generation of Phenotypically Accurate Fractal Patterns

Habiba Akter[1]([✉])(iD), Rupert Young[1](iD), Phil Birch[1](iD), Chris Chatwin[1](iD),
and John Woodward[2](iD)

[1] Department of Engineering and Design, School of Engineering and Informatics,
University of Sussex, Brighton, UK
{h.akter,r.c.d.young,p.m.birch,c.r.chatwin}@sussex.ac.uk
[2] School of Electronic Engineering and Computer Science,
Queen Mary University of London, London, UK
j.woodward@qmul.ac.uk

Abstract. This paper describes the selection of parameters of an Evolutionary Algorithm (EA) suitable for optimising the genotype of a fractal model of phenotypically realistic structures. To achieve the proposed goal an EA is implemented as a metaheuristic search tool to find the coefficients of the transformation matrices of an Iterated Function System (IFS) which then generates regular fractal patterns. Fractal patterns occur throughout nature, a striking example being the fern patterns modelled by Barnsley. Thus the algorithm is evaluated using the IFS for the fern fractal using the EA-evolved parameters.

Keywords: Metaheuristic search algorithm · Evolutionary algorithm · Fractal structure · Iterated function system · Optimisation

1 Introduction

Classic geometry is not adequate to explain the more complex patterns often observed in nature [14]. Some good examples of these are the roughness of a coastline, the leaf structures of ferns and maple leaves, the silhouette of a tree in winter, and the branching of mammalian lungs. Benoît Mandelbrot first introduced the idea of a "fractal" to categorise complex chaotic patterns [7,24]. Mainly, fractals can be of two types [4]: (i) geometric or regular fractals (self-similar objects i.e., if the object is zoomed in, it will still look similar to the original shape) and (ii) non-geometric or irregular (unlike the geometric fractals, they do not have the self-similarity property).

L. Di Gaspero et al. (Eds.): MIC 2022, LNCS 13838, pp. 378–390, 2023.
https://doi.org/10.1007/978-3-031-26504-4_27

The continued improvement of computers has facilitated the ability of the researchers, who are interested in studying fractals to compute the structures and plot them as images [6,25]. One of the initial examples of a computer-generated fractal is the Mandelbrot's set [23]. Later, the generation of complex phenotypic structures, common in biological organisms, particularly plants, has been explored in the pioneering work of Barnsley [1–3]. He has shown how Iterated Functions Systems (IFS) can generate computer images that have remarkable likeness with biological phenotypes. In the IFS, for each fractal structure, a limited number of parameters control the final output image which could be determined by conducting a thorough analysis of the images. In this paper the use of an EA as a metaheuristic search algorithm is explained to effectively determine the near-optimal parameters.

Evolutionary computation has a wide range of applications as a metaheuristic search technique in different fields. In the area of engineering, researchers have the liberty to select the values for the EA parameters to drive the search successfully. In the task of generating a phenotype that mimics biological organisms, the parameters need to be selected carefully. The parameters of an EA are selected that can efficiently find an optimal solution in the search space containing the parameters to generate a complex phenotype observed in the real world.

1.1 Overview of the Paper

Section 2 specifies the optimisation problem to generate the selected fractal of the Barnsley Fern taken as a example for the paper. It also includes a brief description of the parameters of the GA that need to be selected.

Section 3 evaluates the algorithm. For different sets of GA-parameters, the algorithm evolves the coefficients of the IFS to generate Barnsley Fern.

Section 4 concludes the paper and states the future scope.

2 Problem Specification

2.1 Selected Fractal: The Barnsley Fern

As mentioned in Sect. 1, the computer-generated image of the fractal structure of the Barnsley fern is the target phenotype. The generation of a fern structure using an IFS was first described by Barnsley [3]. This fractal mimics the structure of a natural Black Spleenwort fern which demonstrates a self-similar fractal. Depending on the target image, an IFS needs tens of thousands of iterations [2,5].

The IFS for the Barnsley fern is stochastic as it has four transformation functions each of which is selected with a certain probability. Equation 1 represents four affine transformations f_1 to f_4 of an IFS generating any self-similar fractal:

$$\begin{bmatrix} x_{n+1} \\ y_{n+1} \end{bmatrix} = \begin{bmatrix} a & b \\ c & d \end{bmatrix} \begin{bmatrix} x_n \\ y_n \end{bmatrix} + \begin{bmatrix} k \\ l \end{bmatrix} \tag{1}$$

These four functions are responsible for generating different parts of the fern. Each of them yields a new attractor and is selected with a certain probability, p. a through d, k and l are the coefficients. The values of a to d control the generation of the final patterns. x_i and y_j are the locations of the points plotted in the image of the target fractal.

60000 iterations of an IFS using Eqs. 2, 3, 4 and 5 generate the fern shown in Fig. 1 [3,5]. Table 1 includes the probability of selecting each of the transformation functions and the portion it generates.

$$x_{n+1} = 0$$
$$y_{n+1} = 0.16 y_n \tag{2}$$

$$x_{n+1} = 0.85 \times x_n + 0.04 \times y_n$$
$$y_{n+1} = -0.04 \times x_n + 0.85 \times y_n + 1.6 \tag{3}$$

$$x_{n+1} = 0.2 \times x_n - 0.26 \times y_n$$
$$y_{n+1} = 0.23 \times x_n + 0.22 \times y_n + 1.6 \tag{4}$$

$$x_{n+1} = -0.15 \times x_n + 0.28 \times y_n$$
$$y_{n+1} = 0.26 \times x_n + 0.24 \times y_n + 0.44 \tag{5}$$

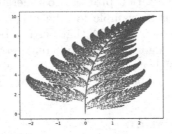

Fig. 1. A fern generated from IFS with parameter values given by Barnsley

Table 1. Probability of selecting each transformation function to generate the image of fern shown in Fig. 1

Equations	Functions	Probability	Portion generated
Eqn. 2	f_1	$p_1 = 1\%$	Stem
Eqn. 3	f_2	$p_2 = 85\%$	Smaller leaflet
Eqn. 4	f_3	$p_3 = 7\%$	Largest left-hand leaflet
Eqn. 5	f_4	$p_4 = 7\%$	Largest right-hand leaflet

The coefficients of Eqs. 2 to 5 above generate the attractors which control the fractal. An attractor is a set of numerical values the IFS evolves towards. Instead of using the pre-selected values, this is the point at which EA is implemented. After thorough research, a Genetic Algorithm (GA) is used to evolve the values automatically to successfully generate the phenotype of a Barnsley fern. It is one of the oldest and most frequently used EAs [20, 28].

The GA searches and evolves 3 different sets of the coefficient values of the matrices a to d for f_2, f_3 and f_4. f_1 simply represents the fern stem [2].

2.2   GA Parameters

- **Initialisation:** An initial population set, P is randomly generated after evaluating with different sizes, N and paying attention so that it does not cost too much memory and time [16, 17, 22]. Each chromosome in P is a set of 12 variables, w_0 to w_{11}. The first four, w_0 to w_3 are used as the coefficients of the 2nd transformation, w_4 to w_7 for the 3rd affine transformation and w_8 to w_{11} for the 4th affine transformation.

  The upper range and lower range of w_n, denoted with w_{max} and w_{min}, also need to be selected carefully which is shown later in the results of Sect. 3.
- **Evaluation:** To evaluate each set of the weight matrix (a to d), fractal dimension is proposed as the fitness function. Fractal dimension is one of the main concepts for studying fractals as it provides information related to the complexity of the structures [11–13, 19]. Amongst the different types of fractal dimensions, the box-counting dimension (BCD) or grid method has been selected to be used as the fitness function for the evaluation of the population.

  As explained earlier, the set of values of x and y are the points to be plotted for generating a fractal image. The graph containing the x and y points is divided into a number of foursquare boxes of a pre-selected size. Assuming n as the number of boxes covering the points in the graph and ϵ as the size or side length of boxes, Eq. 6 calculates the dimension D, which equates to fitness, F:

$$F = \lim_{\epsilon \to 0} \frac{\log n}{\log \frac{1}{\epsilon}} \tag{6}$$

If n versus ϵ is plotted, a straight line with least square method is obtained and the absolute value of the slope is the final value of BCD i.e., the fitness value, optimised in the GA. To calculate the BCD of any particular set of parameters, different values of ϵ, are used within a scale of ϵ_{min} and ϵ_{max}.

A higher fractal dimension means a better and more complex fractal pattern [13, 26]. Hence, a higher value of F results in a fitter chromosome i.e., the GA is solving a "maximisation" problem. It is important to note that the value of the BCD depends on the size of the box.

It should be emphasised that the objective function is designed in a way that it does not rely on the final output image to calculate the box-counting dimension. Rather, from the GA-output (i.e. the values of the coefficients w), the locations of the points (x and y) are obtained which are then used in the box-counting calculation.

- **Mating Selection:** The chromosomes of P are sorted in the descending order of F and the size of the mating pool, N_{mp} is selected as 80% of the population size, N i.e.,

$$N_{mp} = \frac{80\% \times N}{100} \tag{7}$$

This high value is chosen to keep some of the low fitness-valued chromosomes as well to ensure the diversity [9,21,27].

- **Reproduction:** At te stage of reproduction, the Single-point crossover is implemented on the selected parents. Here the parameter values to be determined are crossover probability, ρ_c and crossover rate, r_c.

  Similarly, for mutation, the values of the parameters, mutation probability, ρ_m and mutation rate, r_c are to be selected. This is the most crucial part requiring hundreds of trials to determine a near-optimal solution. However, research shows that in biology, the crossover probability varies from 50% to 80% and the mutation probability is relatively low [8,15,18]. [10,29] suggest that a higher probability of crossover and lower probability of mutation makes the GA more efficient.

- **Environmental Selection:** Similar to the mating selection, at this stage, Eq. 6 is used to select the coefficients based on their fractal dimension. Here, the offspring are merged with the best chromosome from the current iteration. It makes sure that the best set of genes is never lost.

- **Terminating condition:** The GA terminates when it reaches the maximum number of iterations. After some test runs, it is set as 40 since after that the fitness does not improve.

Table 2 summarises the parameters of the GA which need to be selected.

Table 2. Parameters for the algorithm to generate Fern

Parameters	Notations
Population size	N
Upper limit of the variables	w_{max}
Lower limit of the variables	w_{min}
Crossover probability	ρ_c
Mutation probability	ρ_m
Crossover rate	r_c
Mutation rate	r_m
Terminating condition	i_{max}

3   Results and Evaluation

This section includes the results of the algorithm using different set of parameters. For a clear overview, the parameters chosen for different scenarios are shown in Table 3. Equation 6 is used to calculate the fitness value of the population setting the scale of the boxes within the range $\epsilon_{min} = -0.4$ and $\epsilon_{max} = 0.9$.

Table 3. Parameter values for the algorithm to generate the Black Spleenwort fern

Parameters	Set 1	Set 2	Set 3	Set 4	Set 5
N	80	80	80	80	80
F	BCD	BCD	BCD	BCD	BCD
ϵ_{min}	−0.4	−0.4	−0.4	−0.4	−0.4
ϵ_{max}	0.9	0.9	0.9	0.9	0.9
w_{max}	1	0.75	0.75	0.50	0.40
w_{min}	−1	−0.75	−0.75	0.50	−0.40
ρ_c	60%	60%	50%	60%	60%
ρ_m	10%	10%	40%	10%	10%
r_c	0.06	0.06	0.05	0.06	0.06
r_m	0.002	0.002	0.02	0.002	0.002
i_{max}	40	40	40	40	40

To begin with, an initial population of size 80 is randomly generated keeping the variables within the range of $w_{min} - -1$ to $w_{max} - 1$. For a crossover probability of 60% and a mutation probability of 20%, the fitter chromosomes undergo the reproduction process, where the values after crossover and mutation are also within the range stated. The algorithm is run for 40 iterations. Figure 2 shows the two images after the 10th and the 20th iterations.

Figure 3 shows the images generated using the GA-output after the iterations 30 and 40.

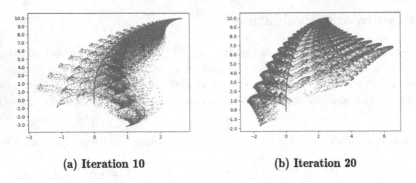

(a) Iteration 10   (b) Iteration 20

Fig. 2. Computer images generated using Set 1 parameters in Table 3

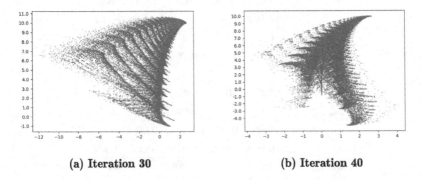

(a) Iteration 30   (b) Iteration 40

Fig. 3. Computer images generated using Set 1 parameters in Table 3

The BCD for 2a, 2b, 3a and 3b are calculated as $0.875, 0.8934, 0.931$ and 0.941 (using the method explained in Sect. 2.2). It is observed from Fig. 2 and 3 that the images do not resemble the Black Spleenwort Fern.

For the next test, the range of the variables is changed and replace with $w_{min} = -0.75$ and $w_{max} = 0.75$, keeping the other parameters the same. Note that, the scale of the box size, ϵ is the same for all tests to make a proper comparison of the fitness values.

From the images generated, a slight improvement can be seen after the 30th iteration (Fig. 4c) as compared to Fig. 3a. It is also clear that after the 40th iteration, the fern shown in Fig. 4d starts taking on a more realistic shape. The fitness value of this image is 1.235.

Since the range of w, $w_{min} = -0.75$ and $w_{max} = 0.75$ is giving better results with a better BCD value, the next set of results are generated keeping them the same. But the crossover probability is changed from 60% to 50% and mutation probability is altered to 40%. The crossover and mutation rate, r_c and r_m are set as 0.05 and 0.02 for this test.

Figure 5 includes the results for these values.

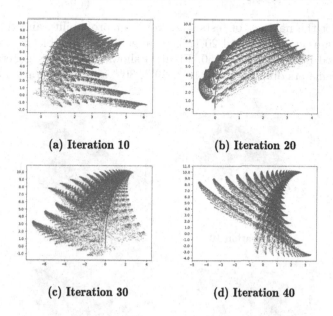

(a) Iteration 10   (b) Iteration 20

(c) Iteration 30   (d) Iteration 40

Fig. 4. Computer images generated using Set 2 parameters in Table 3

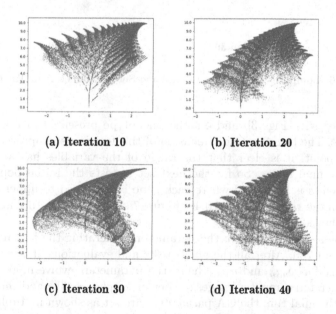

(a) Iteration 10   (b) Iteration 20

(c) Iteration 30   (d) Iteration 40

Fig. 5. Computer images generated using Set 3 parameters in Table 3

The fitness value of the image shown in Fig. 5d is 1.221. However, it is clear from Fig. 5 and the fitness values that lowering the values of the parameters rho_c and increasing rho_m do not improve the results.

Hence, for the next set of tests, the crossover probability and the mutation probability are set as 60% and 20%. The range of the variables is also changed and restricted from −0.50 to 0.50. Figure 6 shows the images generated using the IFS-parameters obtained from 10th, 20th, 30th and 40th iteration of the GA.

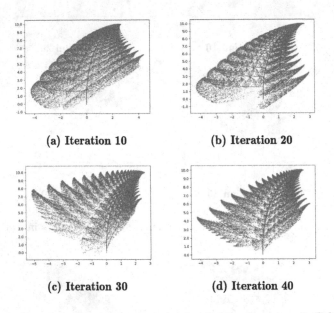

(a) Iteration 10   (b) Iteration 20

(c) Iteration 30   (d) Iteration 40

Fig. 6. Computer images generated using Set 4 parameters in Table 3

Compared with Figs. 3b and 4d, the phenotype presented in Fig. 6d shows better result. The box-counting dimension of this fern is also improved to 1.621.

At this point, it is clear that the range of the variables has a significant effect on the final image. So for the next set of tests the initial population of 80 chromosomes is generated where each gene is restricted to upper and lower limits within the range −0.40 to 0.40. Figure 7 shows the resulting images from each 10th iteration.

The fitness measurement of the parameters generating the fern in Fig. 7d is calculated as 1.755. After all the above tests and evaluations, it is evident that a lower range for w_{min} and w_{max} helps the parameters evolve more efficiently. Also the generated computer images are better for $rho_c = 60\%$ and $rho_m = 20\%$. Hence, for the final run, the GA parameters are set as shown in Table 4.

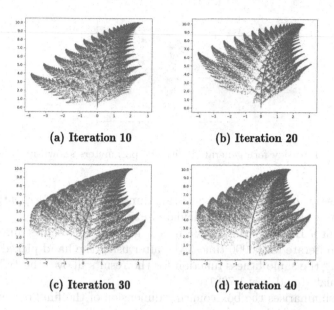

(a) Iteration 10   (b) Iteration 20

(c) Iteration 30   (d) Iteration 40

Fig. 7. Computer images generated using Set 5 parameters in Table 3

Table 4. Final set of GA parameters

GA-Parameters	Values
N	80
w_{max}	0.35
w_{min}	−0.35
F	1.87
ρ_c	70%
ρ_m	20%
r_c	0.06
r_m	0.002
i_{max}	40

Figure 8 shows the final phenotype generated by using the values which are evolved after 40 iterations of the GA.

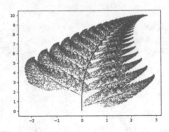

Fig. 8. Barnsley fern generated with the parameters shown in Table 4

The fitness value, i.e. the box-counting dimension of the fern pattern shown in Fig. 8, has a value of 1.86991, calculated using Eq. 6.

To obtain a Black Spleenwort fern of a good box-counting dimension, the IFS needs to iterate for 60000 times even after using the hand-picked parameter values. Using the same fitness function for the results above, the BCD of Fig. 1 is also calculated and it is 1.7381.

Table 5 summarises the box-counting dimension of the final outputs for GA-tuned and hand-picked parameters.

Table 5. Box-counting dimensions of the fractals

Images	BCD values
Figure 3b	0.941
Figure 4d	1.221
Figure 5d	1.235
Figure 6d	1.621
Figure 7d	1.755
Figure 8	1.869
Figure 1	1.7381

This is evidence that the evolved variable sets of GA can generate more complex fractal patterns.

4   Conclusions and Future Work

This paper has demonstrated that a Genetic Algorithm (GA) can successfully explore the search space to find a near-optimal set of parameters of an Iterated Function System (IFS). Thus, the GA can be used to generate natural fractal-based biological organisms. This work has used the Barnsley fern as the target phenotype. The contribution of this work is in using a GA to select the parameters of an IFS, rather than tuning these parameters by hand. Using the box-counting dimension as a fitness measure it is concluded that the parameters can successfully generate complex realistic fractals. Another contribution of this work is the calculation of box-counting dimension from the set of the parameter values generated and subsequently evolved by the GA. This is in contrast with the traditional method in which the box counting takes the image directly as input, making the GA faster and more accurate.

The algorithm is expected to work for other self-similar fractals. Future work will involve searching parameters for other self-similar fractal structures observed in nature, e.g. the Maple leaf and Romanesco Broccoli. A further investigation will also be conducted to implement a GA suitable for generating irregular fractals.

Acknowledgement. This work was funded by the Leverhulme Trust Research Project Grant RPG- 2019-269 which the authors gratefully acknowledge.

References

1. Barnsley, M., Hutchinson, J.E., Stenflo, Ö.: V-variable fractals and superfractals. arXiv: preprint math/0312314 (2003)
2. Barnsley, M.F.: Fractals Everywhere. Academic press, Boston (2014)
3. Barnsley, M.F., Demko, S.: Iterated function systems and the global construction of fractals. Proc. Roy. Soc. London. A. Math. Phys. Sci. **399**(1817), 243–275 (1985)
4. Bayırlı, M., Selvi, S., Çakılcıoğlu, U.: Determining different plant leaves' fractal dimensions: a new approach to taxonomical study of plants (2014)
5. Bourke, P.: Macintosh IFS manual. Retrieved from Paul Bourke: http:// paulbourke.net/fractals/ifs (1990)
6. Bunde, A., Havlin, S.: Fractals in Science. Springer, Cham (2013)
7. Campbell, P., Abhyankar, S.: Fractals, Form, Chance and Dimension (1978)
8. Chiu, C.S.: A genetic algorithm for multiobjective path optimisation problem. In: 2010 Sixth International Conference on Natural Computation, vol. 5, pp. 2217–2222. IEEE (2010)
9. Collet, P., Lutton, E., Raynal, F., Schoenauer, M.: Polar ifs+ parisian genetic programming= efficient ifs inverse problem solving. Genet. Program Evolvable Mach. **1**(4), 339–361 (2000)
10. Supervised by Dr Chris Phillips, H.A.: PhD Thesis: AS Domain Tunnelling for User-Selectable Loose Source Routing. PhD thesis, Queen Mary Univesity of London (2020)
11. Falconer, K.: Fractal Geometry: Mathematical Foundations and Applications. John Wiley, Hoboken (2004)
12. Fernández-Martínez, M., Sánchez-Granero, M.: Fractal dimension for fractal structures: a hausdorff approach revisited. J. Math. Anal. Appl. **409**(1), 321–330 (2014)
13. Fernández-Martínez, M., Sánchez-Granero, M.: Fractal dimension for fractal structures. Topology Appl. **163**, 93–111 (2014)
14. Frame, M., Urry, A.: Fractal Worlds: Grown, built, and Imagined. Yale University Press, New Heaven and London (2016)
15. Goldberg, D.E., Deb, K.: A comparative analysis of selection schemes used in genetic algorithms. In: Foundations of Genetic Algorithms, vol. 1, pp. 69–93. Elsevier (1991)

16. Goldberg, D.E., Deb, K., Clark, J.H., et al.: Genetic algorithms, noise, and the sizing of populations. Complex Syst. **6**, 333–333 (1992)
17. Harik, G., Cantú-Paz, E., Goldberg, D.E., Miller, B.L.: The gambler's ruin problem, genetic algorithms, and the sizing of populations. Evol. Comput. **7**(3), 231–253 (1999)
18. Hassanat, A., Almohammadi, K., Alkafaween, E., Abunawas, E., Hammouri, A., Prasath, V.: Choosing mutation and crossover ratios for genetic algorithms-a review with a new dynamic approach. Information **10**(12), 390 (2019)
19. Husain, A., Reddy, J., Bisht, D., Sajid, M.: Fractal dimension of coastline of Australia. Sci. Rep. **11**(1), 1–10 (2021)
20. Katoch, S., Chauhan, S.S., Kumar, V.: A review on genetic algorithm: past, present, and future. Multimed. Tools Appl. **80**(5), 8091–8126 (2021)
21. Liu, F., Tang, X., Yang, Z.: An encoding algorithm based on the shortest path problem. In: 2018 14th International Conference on Computational Intelligence and Security (CIS), pp. 35–39. IEEE (2018)
22. Macready, W.G., Wolpert, D.H.: Bandit problems and the exploration/ exploitation tradeoff. IEEE Trans. Evol. Comput. **2**(1), 2–22 (1998)
23. Mandelbrot, B.B., Evertsz, C.J., Gutzwiller, M.C.: Fractals and Chaos: The Mandelbrot Set and Beyond, vol. 3. Springer, Cham (2004)
24. Mandelbrot, B.B., Mandelbrot, B.B.: The Fractal Geometry of Nature, vol. 1. WH freeman and Co., New York (1982)
25. Mandelbrot, B.B., Passoja, D.E., Paullay, A.J.: Fractal character of fracture surfaces of metals. Nature **308**(5961), 721–722 (1984)
26. Pedro, S.S.: Fractal dimensions of leaf shapes. https://www.math.tamu.edu/ mpilant/math614/StudentFinalProjects/SanPedro_Final.pdf (2009)
27. Schrijver, A.: Combinatorial optimization: polyhedra and efficiency (algorithms and combinatorics). J.-Oper. Res. Soc. **55**(9), 1018–1018 (2004)
28. Slowik, A., Kwasnicka, H.: Evolutionary algorithms and their applications to engineering problems. Neural Comput. Appl. **32**(16), 12363–12379 (2020). https://doi.org/10.1007/s00521-020-04832-8
29. Véhel, J.L., Lutton, E.: Optimization of fractal: function using genetic algorithms. Ph.D. thesis, INRIA (1993)

Addressing Sustainability in Precision Agriculture via Multi-Objective Factored Evolutionary Algorithms

Amy Peerlinck[✉] and John Sheppard

Montana State University, Bozeman, MT, USA
{amy.peerlinck,john.sheppard}@montana.edu

Abstract. Precision agriculture is a research area that uses technology from engineering and computer science to improve all aspects of agriculture, including but not limited to crop health, irrigation, and fertilizer application. In agriculture, questions of sustainability often arise: How do we minimize environmental impact while simultaneously helping farmers maximize their net return? In this paper, we present a method to optimize crop yield production in winter wheat, with the goal of seeking to increase farmers' production. However, only focusing on optimizing production can lead to poor sustainability if an unnecessary amount of fertilizer is applied or farming equipment is put under undo stress. We therefore seek to address these impacts on sustainability by including objectives that directly address these concerns. Our method utilizes a new approach to solve multi-objective optimization that uses overlapping subpopulations, known as a Multi-Objective Factored Evolutionary Algorithm. Our results indicate that including overlapping subpopulations in the multi-objective optimization context is beneficial for exploration of the objective space. Our results also indicate that including these sustainability-driven objectives does not significantly impact net return or yield.

1 Introduction

Sustainable Agriculture has been defined by the U.S. Congress in the 1990 Farm Bill as "an integrated system of plant and animal production practices having a site-specific application that will, over the long term: Satisfy human food and fiber needs; Enhance environmental quality and the natural resource base upon which the agricultural economy depends; Make the most efficient use of non-renewable resources and on-farm resources and integrate, where appropriate, natural biological cycles and controls; Sustain the economic viability of farm operations; Enhance the quality of life for farmers and society as a whole [30]." Based on this definition, we investigate the effect of adding two sustainability-focused objectives to the problem of optimizing net return for winter wheat production, creating a multi-objective optimization (MOO) problem.

Supported by NSF grant 1658971 and USDA Grant NR213A750013G021.

L. Di Gaspero et al. (Eds.): MIC 2022, LNCS 13838, pp. 391–405, 2023.
https://doi.org/10.1007/978-3-031-26504-4_28

In our work, we focus on defining "prescription maps" that dictate levels of nitrogen fertilizer to be applied on a field planted with winter wheat. We address two problems: first is generating the experimental prescription maps, which are used to gather data on the field to determine the nitrogen response of the crop, and second is generating prescription maps based on estimating crop response to determine how much fertilizer to apply to maximize net return. Each problem has a unique objective: maximizing stratification and maximizing net return. However, they both have potential sustainability issues, which we address by considering two additional objectives: minimizing overall fertilizer applied to mitigate environmental impact, and minimizing jumps in variable rate application to reduce the economic impacts that can result from undue strain on equipment. Due to the resulting multi-objective nature of these problems, meta-heuristic approaches are often employed to explore possible solutions along the different objective axes [15]. This results in searching for a set of "non-dominated" solutions, i.e., solutions where no other options exist that improves the result for one objective without deteriorating another objective. [26].

In this paper, we consider three different MOO algorithms. First we apply the Non-Dominated Sorting Genetic Algorithm II (NSGA-II) [8] and the Co-operative Co-evolutionary NSGA (CC-NSGA-II) [19] to validate our approach. Then we extend the Factored Evolutionary Algorithm (FEA) [27] to create a multi-objective implementation which we call MO-FEA [21], where the specific implementation using NSGA-II is denoted F-NSGA-II. Then we apply all three approaches to the two problems in Precision Agriculture (PA): experimental and optimal prescription map generation.

2   Background

2.1   Prescription Maps

As discussed previously, our interesting in this paper is in developing prescription maps (experimental and optimal—Fig. 1) that optimize objectives balancing net return and sustainability. Experimental prescription maps aim to spread pre-defined fertilizer rates evenly across a field to determine crop response based on different input (e.g., nitrogen) levels [22]. Possible fertilizer rates are pre-defined to ensure that the different levels are represented evenly across the field. In our experiments, we focus on stratifying the rates based on previous years' yield information. In other words, we look at which parts of the field had high, medium, and low yield the year before and distribute the fertilizer across these 3 levels, where the specific objective function is based on the work in [22].

Optimal prescription maps specify fertilizer rates to apply based on crop response and economic models to maximize expected net return. These maps depend upon the ability to predict based on the prescribed inputs, general field information, and satellite data such as the normalized difference vegetation index (NDVI). The classic way to approach yield prediction is to use linear regression; however, this approach is limited in its ability to represent the yield response curves. As a result, machine learning approaches such as Random Forests [16]

Fig. 1. Example of an actual experimental prescription map (left) and an actual optimal prescription map (right) for field Sec35Mid. Colors indicate fertilizer rates where red-to-green indicates increasing rates. (Color figure online)

and Deep Learning [31] have become more popular. We use these models to predict the yield and use the result to determine the expected net return (NR):

$$NR = Y \times P - AA \times CA - FC, \tag{1}$$

where Y is the expected crop yield, P is the crop selling price, AA is the "as-applied" fertilizer rate, CA is the fertilizer cost, and FC reflects any fixed costs associated with production.

2.2 Multi-Objective Optimization

Multi-Objective Optimization is the process of optimizing multiple objective functions simultaneously [15]. Formally and without loss of generality, assume we wish to minimize k objectives. Then MOO consists of solving

$$\min_{\mathbf{x} \in \mathcal{X}} \mathbf{f}(\mathbf{x}) = \{f_1(\mathbf{x}), f_2(\mathbf{x}), \ldots, f_k(\mathbf{x})\},$$

with $k \geq 2$ objective functions $f_i : \mathbb{R}^n \to \mathbb{R}$ that have conflicting goals and $\mathbf{x} = [x_1, x_2, \ldots, x_n]^\top$ denotes the decision variables. Then for $f_i \in F^k$, F^k represents the objective space, and $\mathcal{X} \in \mathbb{R}^n$ represents the solution space.

One of the most approachable ways to perform MOO is to transform the set of objectives into a single objective [6]. The weighted sum approach does this by assigning weights to each objective function, where the weights sum to 1. The ϵ-constraint method transforms all objectives except one into constraints which are bound by a value $R\epsilon$. Then single-objective methods can be applied directly.

Such transformation approaches are limited by the fact that a user needs to decide which objective is more important. Alternatively, if a set of solutions is desired, the problem needs to be solved multiple times with different weights. To avoid such pitfalls, meta-heuristic approaches are often employed.

There are two different general classes of meta-heuristic approaches: local search and population-based search [9]. These meta-heuristics use the concept of Pareto optimality to return a set of potential solutions spread across the objective space.

When exploring algorithms that are generalized for any MOO problem, Pareto-based solutions are often favored [33]. More specifically, metaheuristic methods track the Pareto front of possible solutions along the different dimensions in the solution space [26]. Pareto dominance is used to determine solution quality. Formally, a point $\mathbf{x}^* \in \mathcal{X}$ is Pareto-dominant if $\forall f_i \in \mathbf{f}, \forall \mathbf{x} \in \mathcal{X}, \mathbf{x} \neq \mathbf{x}^*, f_i(\mathbf{x}^*) \leq f_i(\mathbf{x})$, and $\exists f_j \in \mathbf{f}, f_j(\mathbf{x}^*) < f_j(\mathbf{x})$. where $\mathbf{x} \prec \mathbf{y}$ denotes that $\mathbf{x}$ dominates $\mathbf{y}$. Then the Pareto optimal front (PF^*) is the set of points mapped from the set of Pareto optimal solutions onto the objective space F^k to form the boundary of the set of non-dominated solutions.

The Non-Dominated Sorting Genetic Algorithm (NSGA) was introduced by Srinivas et al. in 1994 [26], and improved in 2002 by Deb et al. to create NSGA-II [8]. It was then adjusted further to address MOO problems with more than three objectives to create NSGA-III [7]. However, we chose NSGA-II since it has been shown to be the better choice on 3-objective problems [13]. NSGA-II is an elitist GA that finds Pareto non-dominated solutions and uses a crowding distance measure to maintain diversity in subsequent generations. The parent population P_t and the offspring population Q_t are combined into one population $R_t = P_t \cup Q_t$. R_t is sorted based on the non-domination principle, and individuals are assigned to different non-domination sets based on how good the solution is. If an set of non-dominated solutions is larger than the remaining slots for the next population, a second elimination is performed based on crowding distance.

Cooperative co-evolutionary algorithms (CCEAs) were introduced by Potter and De Jong [23]. CCEAs are based on symbiotic relationships found in nature, where different species live together and improve each others' standard of life. To mimic this, a problem is divided into smaller components, each represented by a different population. In the first version of CCEA, single-objective problems with n dimensions are decomposed into n 1-dimensional subproblems, each with their own subpopulation. These subpopulations are then evolved separately and periodically recombined to form a complete solution. An individual's fitness is based, not only on how well it solves its own part of the problem, but also on its ability to cooperate with other partial solutions. CCEA is applied frequently to MOO, and the resulting algorithms are called co-operative co-evolutionary multi-objective optimisation algorithms (CCMOEA) [19]. Several studies found that a co-operative approach can be beneficial for finding a well spread out Pareto front when compared to the single population MOEA [12,19].

3   Related Work

The question of sustainability in agriculture has existed since the start of precision agriculture as an area of study [2]. But addressing these sustainability issues in practice has proven more difficult. Variable Rate Application (VRA) involves technology that allows farmers to apply different input rates to different parts of the field to control their production and reduce cost more precisely

[25]. By using VRA, farmers are able to apply less fertilizer overall than if they were to apply a uniform rate across an entire field, thus also improving their sustainability practice [2].

Several studies have shown that VRA can help with sustainability [5,32]; however, this is not always the case. Some studies found that using VRA increases the cost for the farmer [28]. To the best of our knowledge, three studies have applied MOO algorithms to VRA prescription maps to provide a set of potential prescriptions. In [22], a simple weighted-sum approach is applied to maximize stratification for experimental prescriptions while trying to minimize jumps in input levels between consecutive cells. Zheng *et al.* applied a multi-objective fireworks optimization algorithm for variable-rate fertilization [34] with the goal of finding the optimal fertilization for oil crops based on three objectives: yield, energy consumption, and spatial effects. An additional study used an economic optimization model to determine the fitness of fertilizer prescription maps and irrigation strategies for optimal crop yield in western Switzerland [18]. The authors integrated climate change into a genetic algorithm model and found that climate change increases income risk for farmers.

4   Optimizing Prescriptions with MO-FEA and NSGA-II

Classic CCEA only creates subpopulations that have distinct variables, i.e., there is no overlap between subpopulations. Strasser *et al.* proposed including overlap in subpopulations to create the Factored Evolutionary Algorithm (FEA). FEA defines a factor architecture (FA) to decompose the set of variables into groups, in a way that permits overlap. As such, FEA includes CCEA as a special case [27]. Due to the possibility of overlap, FEA combines principles from both cooperative and competitive co-evolution, and FEA with overlapping FA's has been shown to perform well on combinatorial optimization problems such as NK-landscapes [27] and Bayesian network abductive inference [11]. We extended FEA to handle MOO based problems in a way that still supports existing single-population MOO algorithms to work within the framework [21].

In MO-FEA, in addition to maintaining an archive of non-dominated solutions $\mathcal{N}$, we keep a set of global solutions $\mathbf{G}$ that are non-dominated for that generation, i.e., $\mathbf{G}$ is replaced each generation based on the non-dominated solutions found by the current subpopulations. Initially, each subpopulation is assigned the same global solution $g \in \mathbf{G}$ for evaluating the fitness of its individuals. As the algorithm progresses, a random global solution $g_r \in \mathbf{G}$ is chosen for each subpopulation without replacement to ensure that as many of the found non-dominated solutions are represented across the subpopulations.

In this work, we assume a field is divided into m cells, each of which will receive a prescription of the amount of nitrogen to apply. To create an experimental prescription we use a set of fertilizer levels to apply to the field (e.g., $F = \{0, 20, 40, 60, 80, 120\}$). This means we are solving a combinatorial optimization problem. On the other hand, the optimized prescription specify fertilizer rate on a continuum with a lower and upper bound. Thus we are solving a continuous

---

Algorithm 1: Factored NSGA-II

    input   : number of cells m, fertilizer values F, FEA generations it_{FEA},
                    population size n, NSGA-II iterations it_{NSGA}
    initialize: individual $X \leftarrow \{x_0, \ldots, x_m\}; x_i \in F$, global solution set $\mathbf{X} \leftarrow \{X\}$,
                  subpopulations $\leftarrow \{s_j \subset X\}$, non-dominated archive $\mathcal{N} \leftarrow \{\}$

1 **while** *FEA generations* $< it_{FEA}$ **do**
2   **for** *each $s_j \in$ subpopulations* **do**
3 | $N''_{s_j} \leftarrow$ NSGA-II(s, n, it_{NSGA}, X)
4   **end**
5 $\mathcal{N}' \leftarrow \{\}$
6   **for** *each variable x_i* **do**
7 $X \leftarrow$ random$(\mathbf{X})$
8   **for** *each s_j where $x_i \in s_j$* **do**
9 $n'' \leftarrow \arg\min_{\text{crowding-distance}}(N''_{s_j})$
10 $\mathcal{N}' \leftarrow \mathcal{N}' \cup n''$
11 $X(i) \leftarrow n''_i$
12 $\mathcal{N}' \leftarrow \mathcal{N}' \cup X$
13   **end**
14   **end**
15 $\mathcal{N}' \leftarrow$ non-dominated$(\mathcal{N}')$
16 $\mathbf{X} \leftarrow \mathcal{N}'$
17 $\mathcal{N} \leftarrow \mathcal{N} \cup \mathcal{N}'$
18   **for** *each $s_j \in$ subpopulations* **do**
19 $X \leftarrow$ random$(\mathbf{X})$
20 $s_j(X) \leftarrow X$
21 $\arg\max_{\text{crowding-distance}}(\{p_0, \ldots, p_m\} \in s_j) \leftarrow X$
22   **end**
23 $\mathcal{N} \leftarrow$ non-dominated$(\mathcal{N})$
24 **end**
25 **return** $\mathcal{N}$

---

optimization problem. In our experiments, we set the lower and upper limit of the fertilizer rate to $F = \{0, \ldots, 150\}$, since the farmers we work with impose an upper limit of 150 lbs nitrogen/acre.

Algorithm 1 shows the basic operation of F-NSGA-II, which is our implementation of MO-FEA where each subpopulation is optimized using NSGA-II [8]. During the "Compete" step (lines 5–17), overlapping subpopulations use the non-dominated solution in $\mathcal{N}''$ with the best crowding distance to represent the current decision variable. Each potential solution for every decision variable is then saved in a temporary solution set $\mathcal{N}'$. In addition, a randomly chosen non-dominated solution from the subpopulation is added to $\mathcal{N}'$ to increase exploration. $\mathcal{N}'$ is evaluated for non-dominance, only keeping Pareto optimal solutions, and the resulting $\mathcal{N}'$ forms the new set of global solutions $\mathbf{G}$. Then $\mathcal{N}'$ is added to $\mathcal{N}$, which is re-evaluated for non-dominance. Then $\mathbf{G}$ is shared across the subpopulations, where the worst solution in the subpopulation is replaced by the chosen g. The fitness scores are then updated, completing the "Share" step (lines 18–22). This constitutes one iteration of MO-FEA.

5   General Experimental Approach

We pose the following hypothesis: Including sustainability-focused objectives to minimize overall fertilizer rate and reduce impact on farm equipment does not significantly degrade Montana winter wheat harvest profit. To evaluate our hypothesis we examine the cropping of winter wheat using both experimental and optimal prescription maps as applied to Montana fields using three MOO-algorithms, denoted NSGA-II (single population), CC-NSGA-II (cooperative co-evolutionary), and F-NSGA-II (factored).

5.1   Factor Architecture

The factor architecture for MO-FEA and CCMOEA is based on the length of a single strip, i.e., one group includes the cells from one end of the field to the opposite end where the applicator has to turn around. This gives us the distinct groupings as used in CCMOEA. To introduce overlap, we calculate a number p by multiplying the number of cells in each strip by 0.2 and rounding the resulting number upward such that $p \leq 1$. A new group is then created by combining the last p cells of a strip and the first p cells of the subsequent strip.

5.2   Objective Functions

To generate nitrogen prescriptions, we optimize the following objectives:

1. Maximize net return or stratification (see Sect. 2.1)
2. Minimize application level changes between adjacent cells
3. Minimize overall fertilizer rate

For the second objective, large jumps in fertilizer rate between consecutive cells puts strain on the farming equipment, increasing wear and tear. In turn, this leads to the farmer having to repair or replace equipment more frequently, increasing cost and waste, which has negative ecological impacts. To address this, we adjust the jump score presented in [22] to handle continuous fertilizer rates as follows: $Fitn_{jumps} = \sum_{i=1}^{c-1} |F(map_i) - F(map_{i+1})|$, where $F(map_i)$ is the fertilizer rate for the ith cell on the field. The jump score now sums over the absolute difference in applied fertilizer between adjacent cells determined by an "as-applied" map.

Lastly, we want to mitigate the effect fertilizer, such as synthetic nitrogen, has on the environment by reducing the overall amount of fertilizer applied to a field. This reduces pollution of the atmosphere by limiting greenhouse gas emission and can help avoid polluting waterways [17], which can result in a loss of drinkable water and the death of aquatic life. The overall fertilizer application is calculated by summing the fertilizer prescribed in each cell: $Fitn_{fert} = \sum_{i=1}^{c} F(map_i)$.

5.3 Evaluation Metrics

The hypervolume indicator (HV) is one of the most commonly used evaluation metrics in MOO [3]. Its popularity is partially because the only information needed to calculate the HV of a Pareto Front approximation is a reference point. This is in contrast to measures such as Generational Distance, which requires the true Pareto Front to be known. Since we do not know the true Pareto Front for our problems, the HV is a natural choice to gain insight in the size of the covered objective space [35]. In these experiments, we min-max normalize the objective scores and use a negative net return to support minimization. Given this, we know the worst possible solution in the objective space is $\{1, 1, 1\}$, which we use for computing HV. To assess the diversity of the Pareto Front approximations, we use the spread indicator SI [14], corresponding to the sum of the width for each objective, indicating how wide the solutions are spread across the objective space. For each algorithm, the average HV and SI were calculated across 5 runs. An ANOVA test with $\alpha = 0.05$ was performed to assess statistical significance across the algorithms, followed by a pairwise t-test with $p = 0.005$.

Finally, to compare two Pareto fronts, denoted as $\mathbf{X}'$ and $\mathbf{X}''$, we use the coverage C of the fronts as presented in [36]. We define the total non-dominated set, or union front, $\mathbf{X}^*$ to be the result of combining the fronts from the three algorithms' representative runs g as

$$\mathbf{X}^* = \text{nondom}\left(\bigcup_i^g \mathbf{X}'_i\right).$$

$\mathbf{X}^*$ can then be used to calculate what percentage of the original non-dominated set is included in $\mathbf{X}^*$: $AC(\mathbf{X}', \mathbf{X}^*)$. We then find the relative coverage of the non-dominated sets when compared to the union front, calculated as

$$AC(\mathbf{X}', \mathbf{X}^*) = \frac{|\{\mathbf{x}' \in \mathbf{X}' : \exists \mathbf{x}^* \in \mathbf{X}^* : \mathbf{x}^* \preceq \mathbf{x}'\}|}{|\mathbf{X}^*|}.$$

The union front's HV and SI are also calculated and compared.

5.4 Prescription Evaluation

For both experiments, we choose four different non-dominated solutions from the approximate Pareto Front created by each algorithm for each of the five runs of the algorithms. These solutions are based on the three extreme points in the Pareto Optimal set: minimum jump score, maximum net return, and minimum fertilizer rate. The centroid for these three solutions, $\mathbf{x}_c$, is found as follows,

$$\mathbf{x}_c^j = \frac{1}{3}\sum_{i=1}^3 \mathbf{x}_i^j, \ \forall j \in k,$$

where k represents the objectives. The non-dominated solution closest to this centroid (based on the Lebesgue measure [4]), is used as the fourth solution. We compare the four different types of prescription maps using an ANOVA test ($\alpha = 0.05$) to evaluate the impact of different objectives on net return.

6   Experimental Prescription Results

6.1   Parameters and Data

For the experimental prescriptions, each of the algorithms was set to terminate after the non-dominated archive did not change for 5 iterations. We chose this approach since farmers are able to generate their own experimental trials, and they often create these trials shortly before they need to fertilize. Thus, we wanted to mimic the shortened runtime desired by the farmers. Mutation rate and crossover rate were set to 0.1 and 0.9 respectively, based on results from [22]. Note that we used swap mutation, and the parents for crossover are selected using tournament selection with tournament size 5. The remaining parameters are the population sizes for all three algorithms and the number of iterations NSGA-II needs to be run on the subpopulations for F-NSGA-II and CC-NSGA-II. To determine these parameters, a grid search was performed. Four different population sizes were considered, $\{100, 200, 500, 800\}$, and three different iteration limits, $\{50, 100, 200\}$. Based on the results of the grid search, a population size of 500 was chosen for all algorithms, and an iteration limit of 100 was chosen for both CC-NSGA-II and F-NSGA-II.

For our experimental maps, we collected data on three fields from two farms. We use the farmer designations for these fields (i.e., Henrys, Sec35Mid, and Sec35West). Previously, we trained a convolutional neural network (CNN) based on prior experiments to predict yield from the wheat harvested on these fields [1]. We use this CNN to predict yield based on the fertilizer applied. We create an initial, random prescription based on the field boundary. The farmer provides information on the width of their fertilizer application equipment and which fertilizer rates to apply across the field. For our experiments, the cell sizes for the fields are 120 ft by 300 ft. For all prescriptions, six different fertilizer rates were specified in pounds per acre: $F = \{20, 40, 60, 80, 100, 120\}$. Once the initial grid was created, the cells were ordered based on the "as-applied" route the farmer takes across the field to apply fertilizer.

6.2   Results and Discussion

We found significant differences ($\alpha = 0.05$) between yield predicted for each of the fields, as well as the different algorithms for each field. However, no significant difference was found between the results for different objectives, confirming our hypothesis that ethical objectives do not impact yield. For each field, the yield predictions for a specific prescription are averaged to create Fig. 2. A summary for each field of the hypervolume, spread indicator, and adjust coverage for each algorithm's non-dominated sets averaged over 5 runs, as well as the unionized non-dominated set, are given in Table 1. Adjusted coverage uses a randomly selected run for each algorithm to avoid bias. This is repeated five times and the five separate calculations are averaged to get the final adjusted coverage score.

The prescription maps across all three algorithms, as well as the union front, produce consistent yield predictions with small fluctuations between the different

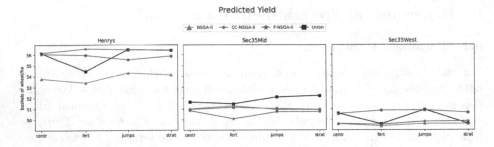

Fig. 2. Yield prediction results averaged across the entire field.

Table 1. Experimental prescriptions: Hypervolume (HV), spread (SI), and adjusted coverage (AC) results for each algorithm and the union front.

		NSGA-II	CC-NSGA-II	F-NSGA-II	Union Front
Henry's	HV	0.463	0.387	0.498	0.589
	SI	0.467	0.329	0.856	1.112
	AC	38.9%	11.3%	49.8%	100%
Sec35Mid	HV	0.465	0.397	0.504	0.593
	SI	0.440	0.289	0.824	1.011
	AC	30.2%	9.5%	60.3%	100%
Sec35West	HV	0.469	0.396	0.474	0.578
	SI	0.615	0.468	1.184	1.619
	AC	27.7%	26.2%	46.1%	100%

objectives, as can be seen in Fig. 2. The statistical results confirm what can be assessed visually: there is no significant difference between the predicted yield values across the different objectives, including those for the union front.

When looking at coverage between algorithms (Table 1), we can see that F-NSGA-II seems to cover more non-dominated solutions in the objective space than the other algorithms since it contributes the largest percentage of solutions to the union front. NSGA-II also makes large contributions to the union front, while CC-NSGA-II has the smallest contribution for all results. This could potentially be explained by the use of disjoint subpopulations in CCEA, since its disjoint nature means that a part of the solution space may be left unexplored. On the other hand, FEA uses the overlap to find more diverse solutions across the subpopulations, not only by saving a non-dominated solution from each subpopulation, but through the replacement of single variables in the global solution as well. The hypervolume and spread indicator results further confirm our hypothesis that F-NSGA-II explores more of the objective space. Across all fields, F-NSGA-II has the largest spread and HV for its approximate Pareto front. According to a pairwise t-test, the HV results were found to be significantly different for all three fields; however, no significant difference was found for the spread indicator results for Henrys and Sec35Mid between F-NSGA-II and NSGA-II. When comparing the algorithms' results to the union front, we see that the union front HV and spread are not much larger than those found by F-NSGA-II, which is in line with the aforementioned coverage results.

7   Optimal Prescription Results

7.1   Parameters and Data

As in our previous experiments, mutation rate and crossover rate were set to 0.1 and 0.9 respectively. Swap mutation was used, and the parents for crossover were selected using tournament selection with tournament size 5. The remaining parameters are the population sizes for all three algorithms and the number of iterations NSGA-II needs to be run on the subpopulations for F-NSGA-II and CC-NSGA-II. For these experiments, our stopping criterion is the number of fitness evaluations. This approach leads to an increase in runtime; however, since farmers are not yet able to create their own optimal prescription maps, we can generate these maps beforehand, negating the need for a reduced runtime. To achieve this, we set the number of generations and population size such that each algorithm has approximately the same number of function evaluations. For NSGA-II, this resulted in setting a population size of 500 and running the algorithm for 500 iterations, yielding $500 \times 500 = 250,000$ FEs. Our instance of CC-NSGA-II has 24 groups, rounded up to 25 for ease of calculation, resulting in CC-NSGA-II being run 10 times where each subpopulation is of size 50, and NSGA-II is run for 20 generations on each subpopulation: $10 \times 20 \times 50 \times 25 = 250,000$. We used the same logic for F-NSGA-II, where population size is decreased to 25 to accommodate the increase in the number of subpopulations.

For this experiment, we look at field Sec35Mid since it is a more complex field and time constraints prevented us from finishing our experiments on the other two fields. The cell sizes are adjusted to be smaller (120 by 240 feet) to provide more refined optimal fertilizer rates. Based on the provided field data, we trained a regression model for yield prediction, where yield is reported in number of bushels per acre. We used the trained model to predict yield for the prescribed fertilizer; the predicted yield can then be substituted into Eq. 1. We used economic data provided by the US Department of Agriculture to determine crop price and fertilizer cost [29].

We used a Random Forest (RF) as a regression model to predict yield. The RF model [16] was implemented via Scikit Learn [20] and evaluated via 10-fold cross-validation on the available data set. The RF yielded a root mean squared error of 2.97. Note that we were not able to use the neural network applied with the experimental maps since those maps could be evaluated offline while these needed direct interfacing between the optimizer and the network. Currently, that interface is still under construction.

7.2   Results and Discussion

Applying an ANOVA test to the net return results for the different types of prescription maps confirms that there is no significant difference in net return.

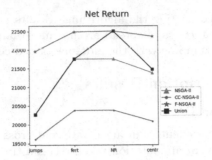

Fig. 3. Net return for the four different prescription maps.

Table 2. Experimental prescriptions: Hypervolume (HV), spread (SI), and adjusted coverage (AC) results for each algorithm and the union front on Sec35Mid.

	NSGA-II	CC-NSGA-II	F-NSGA-II	Union Front
HV	0.52	0.32	0.42	0.53
SI	1.67	1.65	1.68	1.68
AC	75%	0%	25%	100%

When we visually inspect the net return values in Fig. 3, we can see that the difference in net return when focusing on different objectives is minimal for each algorithm. Interestingly, the lowest net return is found consistently when focusing on minimizing jumps. Currently, the net return calculation does not include the cost of wear on equipment. If farmers could gather data on how large jump rates impact them economically, we could refine our net return calculation.

When evaluating the algorithms' performance, we find there is no significant difference between the SI, but there is a significant difference for the HV ($p = 0.05$). Overall, NSGA-II performed better than F-NSGA-II and CC-NSGA-II for HV and coverage. F-NSGA-II, however, did consistently find solutions having a higher net return, indicating it explored a different part of the search space than the other two algorithms, which is reflected in the higher SI score (Table 2).

Overall, single population NSGA-II performs better on the optimal prescription map problem; however, this could be due to the use of FE's as a stopping criterion. In [10], results indicate that using FE's may result in an unfair stopping condition. This idea is supported by our experimental map results, where we use convergence of the non-dominated archive as the stopping criterion. In this scenario, F-NSGA-II performs better; we believe this is because NSGA-II may be converging early on, before moving out of initial optima.

8 Conclusion

Multi-Objective Optimization provides a way for sustainability issues to be addressed when optimizing fertilizer prescriptions in precision agriculture. In this paper, we investigated adding two sustainability-focused objectives to existing precision agriculture problems, that of creating experimental and optimal fertilizer prescription maps. For both problems three competing objectives were optimized: the base objective, stratification or net return maximization respectively, and two sustainability objectives, fertilizer rate jump minimization, and overall fertilizer rate minimization. We applied three different MOO algorithms, NSGA-II, CC-NSGA-II, and F-NSGA-II, of which the latter is an adaptation of the Factored Evolutionary Algorithm in which overlapping subpopulations are used to find an approximate Pareto front. We found that all three MOO algorithms could find optimized prescription maps successfully, and that including these sustainability objectives had minimal impact on yield and net return. Based on these results, we confirmed our hypothesis that focusing on sustainability need not significantly influence net return, thus indicating a strong justification for modifying farming practices to incorporate such objectives, thereby reducing environmental impact. Furthermore, our results indicate that using overlapping subpopulations increases exploration of the objective space when compared to the single population and disjoint subpopulation alternatives.

9   Future Work

As next steps, we plan to investigate adding temporal objectives, such as minimizing variation in net return across several years, and including the impact climate change might have on crop response [18]. Another goal is to investigate the effect of different yield prediction approaches when creating optimized prescription maps. In other words, how much influence does accurate yield prediction have on prescribing the correct fertilizer rate? Or is it more important to use a model that accurately describes the shape of the yield response curve?

Results found by Strasser et al. indicate that the factor architecture could impact optimization [27]. However, work by Pryor et al. shows that the specific factor architecture may not matter as long as the overlap ensures that all variables are connected [24]. Based on these differing results, we plan to explore different factor architectures and their influence, not only regarding the problem of optimizing prescription maps, but by looking at different multi-objective benchmark problems such as DTLZ and the multi-objective knapsack problem.

Lastly, we would like to note that using function evaluations as a stopping criterion could impact results in an unfair way [10]. To this end, we intend to explore different ways to evaluate how long an algorithm should run, for example, based on the amount of change in the non-dominated archive as was done in the experimental design, or a lack of change in hypervolume. We would then evaluate the influence of different stopping criteria on runtime to enable a transition to farmers creating their own optimal prescription maps.

References

1. Anonymous, A.: Reduced-cost hyperspectral convolutional neural networks. J. Appl. Remote Sens. 14(3), 036519–036519 (2020)

2. Bongiovanni, R., Lowenberg-Deboer, J.: Precision agriculture and sustainability. Precis. Agric. **5**(4), 359–387 (2004)
3. Bringmann, K., Friedrich, T.: Approximation quality of the hypervolume indicator. Artif. Intell. **195**, 265–290 (2013)
4. Burk, F.: Lebesgue Measure and Integration: An Introduction. John Wiley, Hoboken (2011)
5. De Koeijer, T., Wossink, G., Verhees, F.: Environmental and economic effects of spatial variability in cropping: nitrogen fertilization and site-specific management. In: The Economics of Agro-Chemicals, pp. 187–200 (2018)
6. Deb, K.: Multi-objective optimization. In: Burke, E., Kendall, G. (eds.) Search Methodologies, pp. 403–449. Springer, Boston (2014). https://doi.org/10.1007/978-1-4614-6940-7_15
7. Deb, K., Jain, H.: An evolutionary many-objective optimization algorithm using reference-point-based nondominated sorting approach, part i: solving problems with box constraints. IEEE Trans. Evol. Comp. **18**(4), 577–601 (2013)
8. Deb, K., Pratap, A., Agarwal, S., Meyarivan, T.: A fast and elitist multi-objective genetic algorithm: NSGA-II. IEEE Trans. Evol. Comp. **6**(2), 182–197 (2002)
9. Ehrgott, M., Gandibleux, X.: A survey and annotated bibliography of multiobjective combinatorial optimization. OR-Spektrum **22**(4), 425–460 (2000)
10. Engelbrecht, A.P.: Fitness function evaluations: a fair stopping condition? In: 2014 IEEE Symposium on Swarm Intelligence, pp. 1–8 (2014)
11. Fortier, N., Sheppard, J., Strasser, S.: Abductive inference in Bayesian networks using distributed overlapping swarm intelligence. Soft. Comput. **19**(4), 981–1001 (2015)
12. Goh, C., Tan, K., Liu, D., Chiam, S.: A competitive and cooperative co-evolutionary approach to multi-objective particle swarm optimization algorithm design. Eur. J. Oper. Res. **202**, 42–54 (2010)
13. Ishibuchi, H., Imada, R., Setoguchi, Y., Nojima, Y.: Performance comparison of NSGA-II and NSGA-III on various many-objective test problems. In: 2016 IEEE Congress on Evolutionary Computation (CEC), pp. 3045–3052. IEEE (2016)
14. Ishibuchi, H., Shibata, Y.: Mating scheme for controlling the diversity-convergence balance for multiobjective optimization. In: GECCO, pp. 1259–1271 (2004)
15. Ishibuchi, H., Tsukamoto, N., Nojima, Y.: Evolutionary many-objective optimization: a short review. In: IEEE CEC, pp. 2419–2426 (2008)
16. Jeong, J.H., et al.: Random forests for global and regional crop yield predictions. Public Libr. Sci. **11**(6), e0156571 (2016)
17. Kim, S., Dale, B.E.: Effects of nitrogen fertilizer application on greenhouse gas emissions and economics of corn production. Environ. Sci. Technol. **42**(16), 6028–6033 (2008)
18. Lehmann, N., Finger, R.: Optimizing whole-farm management considering price and climate risks. In: 123rd European Association of Agricultural Economists Seminar (2012)
19. Maneeratana, K., Boonlong, K., Chaiyaratana, N.: Multi-objective optimisation by co-operative co-evolution. In: Yao, X., et al. (eds.) PPSN 2004. LNCS, vol. 3242, pp. 772–781. Springer, Heidelberg (2004). https://doi.org/10.1007/978-3-540-30217-9_78
20. Pedregosa, F., et al.: Scikit-learn: machine learning in Python. J. Mach. Learn. Res. **12**, 2825–2830 (2011)
21. Peerlinck, A., Sheppard, J., Maxwell, B.: Using deep learning in yield and protein prediction of winter wheat based on fertilization prescriptions in precision agriculture. In: International Conference on Precision Agriculture (2018)

22. Peerlinck, A., Sheppard, J., Pastorino, J., Maxwell, B.: Optimal design of experiments for precision agriculture using a genetic algorithm. In: IEEE CEC, pp. 1838–1845 (2019)
23. Potter, M.A., Jong, K.A.D.: Cooperative coevolution: an architecture for evolving coadapted subcomponents. Evol. Comput. **8**(1), 1–29 (2000)
24. Pryor, E., Peerlinck, A., Sheppard, J.: A study in overlapping factor decomposition for cooperative co-evolution. In: 2021 IEEE Symposium Series on Computational Intelligence (SSCI), pp. 01–08. IEEE (2021)
25. Raun, W.R., et al.: Improving nitrogen use efficiency in cereal grain production with optical sensing and variable rate application. Agron. J. **94**(4), 815–820 (2002)
26. Srinivas, N., Deb, K.: Multiobjective optimization using nondominated sorting in genetic algorithms. Evol. Comput. **2**(3), 221–248 (1994)
27. Strasser, S., Sheppard, J., Fortier, N., Goodman, R.: Factored evolutionary algorithms. IEEE Trans. Evol. Comp. **21**(2), 281–293 (2017)
28. Thrikawala, S., Weersink, A., Fox, G., Kachanoski, G.: Economic feasibility of variable-rate technology for nitrogen on corn. Am. J. Agric. Econ. **81**(4), 914–927 (1999)
29. United States Department of Agriculture: Agricultural prices (2022). https://usda. library.cornell.edu/concern/publications/c821gj76b?locale=en
30. U.S. Congress: Agricultural research, extension, and teaching. In: U.S. Code Title 7, chap. 64. No. 3103, U.S. Government Publishing Office, Washington, DC, USA (2011). https://www.gpo.gov/
31. Van Klompenburg, T., Kassahun, A., Catal, C.: Crop yield prediction using machine learning: a systematic literature review. Comput. Electron. Agric. **177**, 105709 (2020)
32. Whitley, K.M., Davenport, J.R., Manley, S.R.: Differences in nitrate leaching under variable and conventional nitrogen fertilizer management in irrigated potato systems. In: Robert, P.C., Rust, R.H., Larson, W.E. (eds.) Proceedings of the 5th International Conference on Precision Agriculture, pp. 1–9. American Society of Agronomy, Madison (2000)
33. Yu, X., et al.: Set-based discrete particle swarm optimization based on decomposition for permutation-based multiobjective combinatorial optimization problems. IEEE Trans. Cybern. **48**(7), 2139–2153 (2017)
34. Zheng, Y.J., Song, Q., Chen, S.Y.: Multiobjective fireworks optimization for variable-rate fertilization in oil crop production. Appl. Soft Comput. **13**(11), 4253–4263 (2013)
35. Zitzler, E.: Evolutionary Algorithms for Multiobjective Optimization: Methods and Applications. Ph.D. thesis, Swiss Federal Institute of Technology (1999)
36. Zitzler, E., Deb, K., Thiele, L.: Comparison of multiobjective evolutionary algorithms: empirical results. Evol. Comp. **8**(2), 173–195 (2000)

Modeling and Solving the K-Track Assignment Problem

Jakob Preininger, Felix Winter[✉], and Nysret Musliu

Christian Doppler Laboratory for Artificial Intelligence and Optimization for
Planning and Scheduling, DBAI, TU Wien, Karlsplatz 13, 1040 Vienna, Austria
{preininger,winter,musliu}@dbai.tuwien.ac.at

Abstract. In the industrial production of cleaning supplies, larger production quantities are stored in storage boilers and from there they are filled into household-sized bottles. An interesting problem arises in the planning of this process in which production orders have to be assigned to these storage boilers at predetermined times. It turns out that this problem corresponds to a variant of the problem known in the literature as the k-track assignment problem or operational fixed job scheduling problem (OFJSP), which is a classical NP-hard optimization problem. In this paper we investigate and compare different modeling approaches including a CP model, a direct ILP model, a network flow based reformulation as well as a simulated annealing approach. We evaluate these methods on a large set of instances for this problem and on benchmark instances for a related problem. We show that the simulated annealing approach provides very good solutions and outperforms other known solution approaches for larger instances. Our methods have been applied in real-life scenarios, where they have been able to obtain optimal solutions in a short time.

Keywords: K-track assignment · Fixed job scheduling · Exact methods · Simulated annealing · Real-life application

1 Introduction

In the production planning of industrial manufacturing of cleaning supplies after producing the following problem arises during bottling. After scheduling jobs for filling at certain times considering the prior process of production, one has to make an assignment of these jobs to storage boilers taking into account that these boilers are only suitable for certain jobs and are only available at certain times. This assignment should be done in such a way that a maximum number of jobs can be assigned to the storage boilers such that a minimum number of adjustments are necessary for the production plan.

Reframing these boilers as machines at which the jobs are processed this becomes the following problem, which is known as the generalized k-track assignment problem or operational fixed job scheduling problem (OFJSP). We are given a set $M = \{1, \ldots, k\}$ of machines and a set $J = \{1, \ldots, n\}$ of jobs. Every

L. Di Gaspero et al. (Eds.): MIC 2022, LNCS 13838, pp. 406–420, 2023.
https://doi.org/10.1007/978-3-031-26504-4_29

job j has given start time $start_j$ and end time end_j and a subset of eligible machines $M_j \subseteq M$. The goal is to maximize the number of jobs scheduled such that on every machine no two jobs are running simultaneously. In other words to find a biggest subset of jobs J_0 with assignments $x_j \in M_j$ for $j \in J_0$ such that if the jobs $j, k \in J_0$ overlap then $x_j \neq x_k$.

Arkin and Silverberg [2] showed that this problem is NP-complete w.r.t. the number of machines k, which makes the problem interesting from a computational point of view. Therefore this and similar problems have been extensively studied in the past. A survey of methods and results can be found in [10]. Contributions in recent years include e.g. [1,7,13,14].

An exact algorithm was proposed by Brucker and Nordmann [4]. They solved the problem using a direct dynamic programming algorithm, which works well for small instances with a small number of machines. For a larger number of machines though the algorithm has no reasonable runtime since it is of order $O(n^k k! k^k)$ for n jobs and k machines and additionally runs into memory storage problems since $O(k^k)$ states have to be stored in memory during the computations. Reformulating the problem as an integer linear program (ILP) turns out to be a better approach for exactly solving the problem, when the problem instances are bigger (with $k \geq 5$ machines). However, since the problem remains to be of exponential runtime w.r.t. the number of machines exactly solving the k-track assignment problem is still not feasible for large instances.

In this paper using MiniZinc [12], a solver-independent constraint modeling language, we compare various modelings including a direct ILP formulation, a CP formulation as well as a new network flow reformulation that is an improved variation of the approach presented in [3], which turns out to give good upper bounds for the solution when exactly solving the problem takes too much time. We introduce a simulated-annealing scheme which achieves good solutions for larger instances, where exact methods cannot provide solutions in a reasonable time. Finally, to show the robustness of our method, we compare with a state-of-the-art metaheuristic approach [13] on a slightly more general problem (the OFJSP with spread time constraints) that was investigated in that paper. Our simulated annealing approach gives competitive results and outperforms this approach for large instances. Further, the metaheuristic approach has been successfully deployed for solving real life boiler assignment problems and is currently used in practice.

2 Solver-Independent Modeling Approaches for the K-Track Assignment Problem

Using the MiniZinc constraint modeling language we investigated three modeling approaches for our problem. The advantage of solver-independent MiniZinc formulations is that they can be used by different solvers including MIP and CP solvers.

2.1   Constraint Programming Model

The constraint programming model for the generalized k-track assignment problem uses the following input parameters:

Input Parameters

- A set of k machines M.
- A set of n Jobs J.
- A set $M_j \subseteq M$ of eligible machines for each job $j \in J$.
- A set $O \subseteq J \times J$ of all pairs (j, k) where $j \neq k$ are overlapping jobs.

We handle the requirement that two jobs with overlapping time intervals cannot be run on the same machine by introducing constraints for each pair of overlapping jobs on their joint eligible machines. By not including start and end times in our model but instead computing all pairs of overlapping jobs in a preprocessing phase we can save valuable model compilation time. Therefore, start and end times are not explicitly mentioned in the input parameters of our model.

Decision Variables. The following decision variables are used to model the machine assignments for each job j (a value *null* indicates that the job is not assigned to any machine):

- $x_j \in M_j \cup \{null\}$ $\forall j \in J$

Note that we implicitly model the requirement that each job can only be assigned to its set of eligible machines by restricting the variable domains accordingly.

Constraints. The set of constraints that forbids the assignment of overlapping jobs to the same machine is specified as follows:

$$(x_j = x_k = null) \vee (x_j \neq x_k)   \forall (j, k) \in O \tag{1}$$

Cost Function. The cost function counts the number of jobs assigned to any machine:[1]

$$maximize \sum_{j \in J} [x_j \neq null] \tag{2}$$

2.2   Integer Linear Programming Model

In addition to the constraint programming model we provide an integer linear programming model for the k-track assignment problem as follows:

---

[1] Here [·] denotes the Iverson bracket (i.e. $[A] = 1$ if A is true and $[A] = 0$ otherwise).

Input Parameters

- A set of k machines M
- A set of n Jobs J
- A set $M_j \subseteq M$ of eligible machines for each job $j \in J$
- A set T of triples $(j, k, m) \in J \times J \times M$ where j, k is a pair of jobs that overlap and are both eligible on machine m

For the input parameters of the linear model we calculate in a preprocessing step a set of triples $(j, k, m) \in J \times J \times M$ where j, k is a pair of jobs that overlap and are both eligible on machine m. Based on the resulting set T we then define a set of linear constraints for each triple that implements the conditions for non-overlapping job assignments on a single machine.

Decision Variables. The set of decision variables defines a Boolean variable for each pair of job and machine, where a value of 1 indicates corresponding job assignment:

- $x_{j,m} \in \{0, 1\} \quad \forall j \in J, m \in M$

Constraints. The first set of constraints in the linear model ensures that each job is assigned to at most one machine:

$$\sum_{m \in M} x_{j,m} \leq 1 \quad \forall j \in J \tag{3}$$

The second set of constraints models the requirement that jobs can only be assigned to eligible machines:

$$x_{j,m} = 0 \quad \forall j \in J, m \in M \setminus M_j \tag{4}$$

The third set of constraints forbids overlapping job assignments on the same machine:

$$x_{j,m} + x_{k,m} \leq 1 \quad \forall (j, k, m) \in T \tag{5}$$

Cost Function. The cost function of the linear model counts all job assignments:

$$maximize \sum_{(j,m) \in J \times M} x_{j,m} \tag{6}$$

2.3   Network Flow Reformulation

In this section we further propose a network flow reformulation of the k-assignment problem which is related to the model proposed in [3], but has a variation regarding assignments of jobs to different machines. The main idea behind this formulation is to model the job assignments by creating an individual flow network for each machine. Vertices in a graph are modeling the jobs that

Algorithm 1: Generating a network for a given machine of the k-track assignment problem.

Function $CreateNetworkForMachine(m)$

    $J_m = \{j \in J : m \in M_j\}$
    $V_m = \{s_m, t_m\} \cup J_m$
    $A_m = \{\}$
    $L = []$
    $L.append([s_m])$
    ▷ sort jobs by earliest start (break ties by earliest end)
    $L.append(sort([j \in J_m]))$
    $L.append([t_m])$
    for $i \in \{1, \ldots, |L| - 1\}$ **do**
        $v = L[i]$
        $w = t_m$
        for $j \in \{i+1, \ldots, |L| - 1\}$ **do**
            if $start_{(L[j])} \geq end_v$ **then**
                $w = L[j]$
                break

        $A_v = (v, w)$
        if $w \neq t_m$ **then**
            $A_v = A_v \cup \{(v, x) : x \in J_m \wedge start_x \in [start_w, end_w]\}$

        $A_m = A_m \cup A_v$

    return $G_m = (V_m, A_m)$

---

are eligible on the respective machine, and the decisions about the selected flow path through a network determine the jobs assigned to the machine. In contrast to the model in [3] our model uses additional Boolean variables for each job and machine pair that are set to 1 if and only if the job is assigned to the respective machine. Thus, the number of variables needed for the maximization objective is highly reduced. Furthermore, we significantly reduce the number of arcs in the network by omitting redundant paths and hence reduce the number of variables in the model. This can be helpful for the evaluation of larger instances and to find good upper bounds fast.

Computing the Machine Networks. Algorithm 1 describes how a network $G_m = (V_m, A_m)$ for each machine $m \in M$ consisting of a set of vertices V_m and a set of directed arcs A_m is created for an instance of the k-track assignment problem.

The procedure shown in Algorithm 1 first selects the set of relevant jobs (i.e. the jobs that are eligible for the given machine) and then creates the set of vertices V_m by creating a vertex for each relevant job in addition to a single source- s_m and sink vertex t_m. To create the set of directed arcs, the algorithm then creates an ordered list L of all vertices as follows: The list starts with the source vertex and is followed by all job vertices ordered increasingly by the job's start time (if multiple jobs have identical start times, jobs with earlier end times are selected first). Finally, the list includes the sink vertex at the last position.

Then the procedure creates outgoing arcs A_v for each vertex $v \in V_m$ by first determining the next non-overlapping successor vertex w from list L, which is determined by iterating through the remaining list and selecting the first job

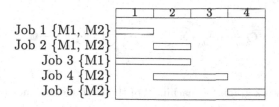

Fig. 1. A simple example k-track assignment problem instance consisting of five jobs and two machines.

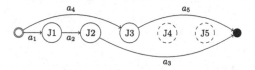

Fig. 2. Network graph for machine 1 of the example instance from Fig. 1

vertex which has a start time that is larger or equal to the end time of the job related to v (i.e. the first non-overlapping job). In case no such job vertex exists, the sink vertex is selected as successor w. If w is not the sink, the set of A_v further includes an arc v, w plus additional arcs to each job vertex where the job times are overlapping with the job related to w. Finally, the set of all network arcs A_m unifies the arc sets created for all vertices.

Machine Networks Example. We now illustrate networks created for a simple instance using two machines as an example. Figure 1 visualizes the jobs of an example instance consisting of 5 jobs and 2 machines using a gantt chart.

In this example job 1 and job 2 are eligible on machine 1 as well as machine 2 and their scheduled processing times range from 0–1 and and 1–2 respectively. Job 3 is only eligible on machine 1 and processed in the time from 0–2, whereas jobs 4 and 5 are only eligible on machine 2 and scheduled from 1–3 and 3–4 respectively.

Figures 2 and 3 illustrate the networks created with Algorithm 1 for both machines in this example.

In Fig. 2 we can see that the network for machine 1 only includes vertices representing jobs 1, 2, and 3 but does not include jobs 4 and 5 as they are not eligible on these machines (jobs 4 and 5 are drawn with a dashed line to indiciate their absence). Furthermore, the figure shows two outgoing arcs a_1 and a_4 from the source vertex, creating paths from the source to either job 1 or job 3. Vertices appearing on separated paths in the network indicate that their related jobs cannot be assigned at the same time to this machine as they are overlapping, whereas jobs on the same path are non overlapping (like e.g. job 1 and job 2 in this example). Figure 3 illustrates the network for machine 2 in a similar way.

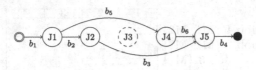

Fig. 3. Network graph for machine 2 of the example instance from Fig. 1

Network Flow Model. Using the machine networks we introduced in the previous section, we can formally define the model of the network flow reformulation as follows:

Input Parameters

- Set of k machines M
- Set of n jobs J
- G_m: The network graph for each machine $m \in M$. The set of nodes is $V_m = \{s_m, t_m\} \cup \{n_{m,j} : j \in J\}$ where s_m, t_m are the source and sink nodes, and $n_{m,j}$ are nodes representing each of the jobs. The set of directed arcs is given as A_m. Let further $\delta^+(i)$ and $\delta^-(i)$ denote the set of ingoing and outgoing arcs for a node i.

Decision Variables

- $j_i \in \{0,1\} \quad \forall i \in J$: Boolean variables which are set to 1 iff the associated job is assigned to any machine.
- $f_{i,j} \in \{0,1\} \quad \forall i \in M, j \in A_i$: Boolean variables, that are set to 1 iff arc j is part of the selected path in the network of machine i .

Constraints

- Flow conservation constraints for job nodes:

$$\sum_{a \in \delta^-(n_{m,j})} f_{i,a} - \sum_{b \in \delta^+(n_{m,j})} f_{i,b} = 0 \quad \forall i \in M, j \in J \tag{7}$$

- Flow conservation constraints for source nodes:

$$\sum_{a \in \delta^-(s_m)} f_{i,a} - 0 = 1 \quad \forall i \in M \tag{8}$$

- Flow conservation constraints for sink nodes:

$$0 - \sum_{b \in \delta^+(t_m)} f_{i,b} = -1 \quad \forall i \in M \tag{9}$$

- Constraints to channel the values of the job selection variables to the arc selection variables:

$$j_i \leq \sum_{m \in M, a \in \delta^+(n_{m,i})} f_{m,a} \quad \forall i \in J \tag{10}$$

Algorithm 2: Simulated Annealing

Function *Simulated Annealing*($T_{init}, T_{final}, t_{limit}$)
   ▷ Initialize temperature and cooling rate
   $T \leftarrow T_{init}$
   $c \leftarrow 1$
   ▷ Initialize empty solution
   $x_j \leftarrow null \quad \forall j \in J$
   $z \leftarrow x$
   while $t_{elapsed} < t_{limit} \wedge T > T_{final}$ **do**
      ▷ Generate random neighbor
      $y_j \leftarrow x_j \quad \forall j \in J$
      $j \leftarrow Random(J)$
      $m \leftarrow Random(M_j \cup \{null\})$
      $y_j \leftarrow m$
      if $Accept(y, T)$ **then**
         $x \leftarrow y$
         if $Cost(x) < Cost(z)$ **then**
            $z \leftarrow x$

      ▷ Update Cooling rate based on elapsed time
      $c \leftarrow UpdateCoolRate(T, T_{final}, t_{limit}, t_{elapsed})$
      ▷ Update temperature
      $T \leftarrow cT$
   return (x)

Objective Function

$$maximize \sum_{i \in J} j_i \tag{11}$$

3 A Simulated Annealing Approach

Additionally to the exact modeling approaches above we propose a local search approach based on simulated annealing [9] for solving the k-track assignment problem. The pseudo code of the simulated annealing approach that we apply for our problem is presented in Algorithm 2.

We use an empty solution where no jobs are assigned as an initial solution and randomly generate a single local search move in each iteration that assigns a random job to a random machine (including the *null* machine). We accept a move based on a geometrical cooling scheme, where moves are always accepted if they improve the solution and accepted if they do not improve the solution with probability $e^{-\delta/T}$, where δ is the delta cost of the move and T is the current temperature. After every move the temperature is adjusted by a cooling rate c which is recalculated at every step based on the current and final temperature and on the time remaining in the algorithm, assuming that future moves are applied at approximately the same speed as previous ones. So the temperature starts at a given starting temperature T_{init} and is approximately geometrically cooled down to the final temperature T_{final} at the end of the given time limit of the algorithm.

Note that this version of the simulated annealing approach only needs two parameter inputs apart from the instance itself, namely the initial temperature

T_{init} and T_{final} which makes this approach rather flexible for different runtimes and sizes of instances as long as the average delta costs of moves remains the same.

Furthermore, for the evaluation of candidate solutions we extend the objective function as follows:

$$minimze \sum_{j \in J}[x_j = null] + M \sum_{j,k \in O} [x_j = x_k \neq null] \qquad (12)$$

Here M is a number bigger than the total number of jobs, i.e. we penalize violations of the overlapping jobs requirement in the objective function so that a single violation is more costly than any other cost.

Additionally we included the possibility to adjust the objective to include weights for each job and a spread time constraint. I.e.

$$min \sum_{j \in J}[x_j = null] * w_j + M \sum_{j,k \in O \cup S} [x_j = x_k \neq null] \qquad (13)$$

where w_j is a given weight for every job and $S \subseteq J \times J$ is the set of all pairs (j, k) where $j \neq k$ are jobs which violate the following spread time constraint (cf. [13])

$$end_j - start_k > St \vee end_k - start_j > St \qquad (14)$$

where St is a given spread time parameter.

We note that for the k-track assignment problem we consider in this paper, the weight of the jobs are normalized to 1, while in the spread time problems these weights can vary. This adjustment enables us to compare our approach with the state-of the art approach for the spread time variant of the k-track assignment problem [13].

4   Computational Results

We conducted all of our experiments on a computing cluster with 10 identical nodes, each having 24 cores, an Intel(R) Xeon(R) CPU E5–2650 v4 @ 2.20 GHz and 252 GB RAM.

4.1   Generation of Instances

Although the k-track assignment problem has been studied for a long time, to the best of our knowledge large instances for this problem are not available. The instances used in [4] include at most 5 machines and 100 jobs and are very small. For the related problem called operational fixed job scheduling with spread-time constraints [13] there exists a larger set of instances and we compare on these instances in the next section. However, we also generated additional large instances for the problem we investigate in this paper to compare the performance of our approaches.

We developed a random instance generator that generates a given number of machines and jobs such that the jobs have a uniformly distributed random length up to 1/10 of the total length of the scheduling horizon which we set to 10000. We further select uniformly random distributed start times for each job so that all jobs fit into the time horizon. To randomly select the eligible machines we put each machine to the set of eligible machines for each job with a probability of 1/2. Using our random instance generator, we generated a set of 10 groups of instances with 10 samples with different configurations regarding the number of jobs and machines. Out of these 100 instances we selected the two first instances of each group to compare the approaches. The different size parameters of the instances are shown in Columns 2–3 of Table 1, where each row represents a single instance. All instances will be made publicly available.

4.2 Comparison of Results

We used the MIP-solvers Gurobi [8] and Cplex [6] as well as CP solver Chuffed [5] to solve the instances using our models presented in Sect. 2 within a runtime limit of 1 h. Our initial experiments showed that the ILP and the network flow models obtained best results and therefore in this paper we present only the results of these models. As our models use a large number of variables and constraints for both the linear model and the network flow model we configured Cplex and Gurobi to use the barrier algorithm to solve the root linear relaxation instead of the standard simplex algorithm.

For the evaluation of our simulated annealing approach we conducted ten repeated experiments (as it is a stochastic approach) for each instance under a time limit of 1 h.

Table 1 summarizes the best solution costs achieved by the evaluated solution approaches. Columns 2–3 display the size parameters for each instance, whereas Columns 4–11 show from left to right: The best solution cost achieved by Cplex with the linear model (cplex lin), Cplex with the network flow model (cplex net), Cplex with the network flow model from [3] (cplex bc), Gurobi with the linear model (gurobi lin), Gurobi with the network flow model (gurobi net), Gurobi with the network flow model from [3], the average solution costs achieved over the 10 repeated runs with simulated annealing (SA avg), and the overall best solution cost achieved with simulated annealing (SA best).

We can see in the results shown in Table 1 that Cplex is able to solve larger instances than Gurobi and the linear model produces better results for some of the larger problems than the network flow model. However, for the largest instances (Instances 11–12, Instances 17–20) both Cplex and Gurobi fail to find a solution in the given time or only find solutions far away from the best bounds produced with SA. In these cases the simulated annealing approach produced the best results. However, simulated annealing could not produce competitive results for smaller instances compared to exact results. Furthermore, we see that

Table 1. Summary of the best solution costs for evaluated approaches

Instance	Jobs	Machines	cplex lin	cplex net	cplex bc	gurobi lin	gurobi net	gurobi bc	SA avg	SA best
Instance 1	100	20	**100**	**100**	**100**	**100**	**100**	**100**	100	100
Instance 2	100	20	**100**	**100**	**100**	**100**	**100**	**100**	100	100
Instance 3	500	20	**379**	**379**	**379**	**379**	**379**	**379**	370.5	372
Instance 4	500	20	**382**	**382**	**382**	**382**	**382**	**382**	374.0	375
Instance 5	1000	20	**410**	**410**	297	**410**	**410**	**410**	392.9	395
Instance 6	1000	20	**408**	**408**	**408**	**408**	**408**	**408**	393.4	397
Instance 7	2000	20	**802**	688		**802**	735		768.0	772
Instance 8	2000	20	**799**	302		**799**	738		772.8	778
Instance 9	5000	20	**1242**	358			296		1179.5	1185
Instance 10	5000	20	**1272**	346			278		1200.7	1212
Instance 11	10000	20		384			305		1673.5	**1679**
Instance 12	10000	20		350			301		1662.2	**1670**
Instance 13	1000	50	**909**	519	846	**909**	818	781	885.5	888
Instance 14	1000	50	**926**	511	857	**926**	817		900.4	905
Instance 15	2000	50	**1279**	650			430		1235.9	1239
Instance 16	2000	50	**1210**	675			460		1272.2	**1277**
Instance 17	5000	50		786			570		1991.8	**1996**
Instance 18	5000	50		814			541		1986.8	**1997**
Instance 19	10000	50							2735.6	**2743**
Instance 20	10000	50					578		2758.3	**2771**

results produced by the network flow formulation from [3] can be improved by the network flow reformulation in this paper for the large majority of the larger instances. Actually, for instances 11–12, 17–18, and instance 20 the new network flow model was the only exact technique that was able to produce valid solutions within the runtime.

Table 2 further provides a summary of the best upper bounds achieved by the evaluated MIP solvers. Columns 4–9 show from left to right: The best bound achieved by Cplex with the linear model (cplex lin), Cplex with the network flow model (cplex net), Cplex with the network flow model from [3] (cplex bc), Gurobi with the linear model (gurobi lin), Gurobi with the network flow model (gurobi net), and Gurobi with the network flow model from [3] (gurobi bc). Additionally, Columns 10–12 in Table 2 display the bound achieved by solving the initial linear programming relaxation using the linear model (relax lp), the network flow model (relax net lp), and the network flow model from [3] (relax bc lp).

The results displayed in Table 2 show that although solving the LP-relaxation of the direct linear model only gives a trivial bound for all instances, the final bound produced with this model within the runtime limit of 1 h can provide good results for many instances and sometimes even better bounds than the network flow model. The LP-relaxation bound produced with the network flow model on the other hand seems to be providing an optimal bound for instances 1–8. For instances 9–10 and 13–16 the LP-relaxation of the network flow model also provided the best bound results although we could not verify that the LP-bound is optimal. These results indicate that our network flow model can be very useful to provide strong bounds for this problem, especially as solving the

Table 2. Summary of the best upper bounds achieved by the evaluated integer linear programming approaches.

Instance	Jobs	Machines	cplex lin	cplex net	cplex bc	gurobi lin	gurobi net	gurobi bc	relax lp	relax net lp	relax bc lp
Instance 1	100	20	100	100	100	100	100	100	100	100	100
Instance 2	100	20	100	100	100	100	100	100	100	100	100
Instance 3	500	20	379	379	379	379	379	379	500	379	379
Instance 4	500	20	382	382	382	382	382	382	500	382	382
Instance 5	1000	20	410	410	410	410	410	410	1000	410	410
Instance 6	1000	20	408	408	408	408	408	408	1000	408	408
Instance 7	2000	20	802	802	802	802	802	802	2000	802	802
Instance 8	2000	20	799	799		799	799		2000	799	799
Instance 9	5000	20	1242	5000		2004	1242		5000	1242	
Instance 10	5000	20	1272	5000		2053	1272		5000	1272	
Instance 11	10000	20	10000	10000		10000	10000		10000		
Instance 12	10000	20	10000	10000		10000	10000		10000		
Instance 13	1000	50	909	909	909	909	909	909	1000	909	909
Instance 14	1000	50	926	926	926	926	926	926	1000	926	926
Instance 15	2000	50	1279	2000		1974	1279		2000	1279	
Instance 16	2000	50	1320	2000		1975	1320		2000	1320	
Instance 17	5000	50	5000	5000		5000	2108		5000		
Instance 18	5000	50	5000	5000		5000	5000		5000		
Instance 19	10000	50	10000	10000		10000	10000		10000		
Instance 20	10000	50	10000	10000		10000	10000		10000		

Table 3. Summary of the best solutions and upper bounds found.

Instance	$|J|$	$|M|$	best	ub	Instance	$|J|$	$|M|$	best	ub
Instance 1	100	20	100	100	Instance 11	10000	20	1679	1784
Instance 2	100	20	100	100	Instance 12	10000	20	1670	1780
Instance 3	500	20	379	379	Instance 13	1000	50	909	909
Instance 4	500	20	382	382	Instance 14	1000	50	926	926
Instance 5	1000	20	410	410	Instance 15	2000	50	1279	1279
Instance 6	1000	20	408	408	Instance 16	2000	50	1277	1320
Instance 7	2000	20	802	802	Instance 17	5000	50	1996	2108
Instance 8	2000	20	799	799	Instance 18	5000	50	1997	2086
Instance 9	5000	20	1242	1242	Instance 19	10000	50	2743	2893
Instance 10	5000	20	1272	1272	Instance 20	10000	50	2771	2948

LP-relaxation is a tractable problem. We further see that the new network flow reformulation is able to provide good LP-relaxation bounds within 1 h for four additional instances compared to the existing network flow model from [3].

In Table 3 we summarize our overall best solutions found for each instance. To find upper bounds for the large instances (Instances 11–12 and 17–20) we gave the solver for the LP-relaxation more time. Upper bounds for these solutions could be found within 10 h.

4.3   Comparison to the Literature

To compare our simulated annealing approach with the state of the art in the literature, we used the benchmark instances provided by [13] and compared with their results. To the best of our knowledge this paper reports best existing results for this problem. Rossi et al. [13] consider the spread time variant of the k-track assignment problem (operational fixed job scheduling problem) which is a more general problem. As described in Sect. 3 we adapted our simulated annealing model to incorporate violations of the spread time constraint with a M penalty similar to the violation of the overlapping constraint. Therefore, our simulated annealing approach can be used both for the k-track assignment problem and the operational fixed job scheduling problem [13].

The approach of [13] features a greedy algorithm as well as a hybrid grouping genetic algorithm (GGA) as a metaheuristic which starts with the solution of the greedy algorithm.

Table 4 illustrates the results for the different large sized instances from [13]. In Table 4 n refers to the number of jobs, m to the number of machines and r, p, w are parameters that change the distributions of job start- and endtimes as well as job weights and are given in detail in [13]. The results in "GGA average" are taken directly from [13] while the results in "SA average" were computed via our simulated annealing approach with a time limit of 30 s. In contrast the results of GGA reported in [13] are obtained using no time limit, but the algorithm is stopped when no further improvements can be made, which for their larger instances with 500 jobs is reached after approximately 2 min. To evaluate the metaheuristic approach for these new instances we further have to configure the initial and final temperature as the parameters of our algorithm. Based on some manual tuning attempts we selected a T_{init} value of 10^3 and a T_{final} value of 10^{-3}. Starting from these default values, we further used the state-of-the-art parameter tuning software SMAC [11] to automatically tune all of the parameters (Parameter value ranges were restricted to $T_{init} \in [1, 10^6]$ and $T_{final} \in [10^{-5}, 1]$. The tuning process was then started with the metaheuristic and all large instances from [13] as the training set. We set the runtime limit for each individual run to 2 min and set the overall wallclock time limit to 2 days. The resulting parameter configuration which we used for our final experiments is as follows: $T_{init} = 10^{2.4477457159156693}$ and $T_{final} = 10^{-0.16850206022148662}$.

While for smaller instances with only 250 jobs and 20 machines, the GGA gives slightly better results on average than the simulated annealing approach, SA starts to give better results for larger instances (500 jobs and/or 50 machines) and as we can see from Table 4 SA gives better results than GGA for 23 large instances, whereas GGA gives better results for 13 large instances. In general we can conclude our SA gives very good results for large instances compared to the existing state-of-the-art approach.

4.4   The Deployment of Our Method on Real-Life Scenarios

In the industrial manufacturing of cleaning supplies the production planning process includes scheduling jobs for producing the material as well as jobs for filling this material into household-sized bottles. During the time the latter jobs are running the material has to be stored in storage boilers. At any given time there are only a certain number of these boilers available and certain boilers are

Table 4. Comparing GGA with SA on instances with spread time constraint

n	m	r	p	w	GGA average	SA average	n	m	r	p	w	GGA average	SA average
250	10	1	1	1	**3901.65**	3634.7	250	20	1	1	1	**5922.1**	2523.15
250	10	1	1	2	**6069.25**	5859.05	250	20	1	1	2	7856.45	**13827.05**
250	10	1	1	3	**30889.9**	29872.75	250	20	1	1	3	43476.3	**77100.65**
250	10	1	2	1	**5251.2**	4748.95	250	20	1	2	1	**9224.45**	4724.25
250	10	1	2	2	**4665.9**	4452.3	250	20	1	2	2	6972.55	**11987.5**
250	10	1	2	3	**26557.15**	25348.8	250	20	1	2	3	38962.55	**61233.15**
250	10	2	1	1	**4467.85**	4173.45	250	20	2	1	1	**6236.7**	2533.7
250	10	2	1	2	**6066.9**	5865.3	250	20	2	1	2	7224.25	**13989.6**
250	10	2	1	3	**33414.65**	32282.3	250	20	2	1	3	39236.35	**77821.35**
250	10	2	2	1	**5217.5**	4743.75	250	20	2	2	1	**7833.75**	3833.1
250	10	2	2	2	**5015.5**	4771.05	250	20	2	2	2	6244.25	**10143.75**
250	10	2	2	3	**27082.8**	25927.5	250	20	2	2	3	34657.9	**56183.85**
500	10	1	1	1	5185.8	**5624.95**	500	20	1	1	1	**10013.95**	3588.05
500	10	1	1	2	**9962.85**	7706.65	500	20	1	1	2	10663.75	**16672.15**
500	10	1	1	3	**52538.7**	42943.85	500	20	1	1	3	68481.3	**88965**
500	10	1	2	1	5556.85	**8606**	500	20	1	2	1	**11006.3**	8579.35
500	10	1	2	2	6366.85	**6752.1**	500	20	1	2	2	10170.95	**13134.45**
500	10	1	2	3	35916.55	**37915.9**	500	20	1	2	3	60385.65	**76302.8**
500	10	2	1	1	4596.2	**5994.8**	500	20	2	1	1	**9040.75**	3440.35
500	10	2	1	2	**8432.65**	7043.65	500	20	2	1	2	13153.05	**15856.7**
500	10	2	1	3	**41272.75**	38324	500	20	2	1	3	66183.85	**84577.05**
500	10	2	2	1	5576.35	**7203.4**	500	20	2	2	1	**10548.75**	7798.7
500	10	2	2	2	5847.8	**6006.85**	500	20	2	2	2	8741.1	**13110.25**
500	10	2	2	3	**33418.4**	33222.15	500	20	2	2	3	51625.95	**68031.75**

only suitable for certain materials. Hence in a subsequent planning step these scheduled jobs with given start and end times then have to be assigned to these storage boilers. This assignment problem can then be reformulated as the k-track assignment problem we studied in this paper.

Our simulated annealing approach has been adopted by our industry partners and is now used in their production planning process. SA has been already used very successfully in several real-life scenarios that include 24–66 jobs and 22 machines (these instances will also be made publicly available), where it could provide optimal solutions in less than a second. Moreover, the simulated annealing approach is very flexible in practice as it guarantees to give a solution after any given running time and adapts easily to new instances of different sizes.

5   Conclusion

We investigated the real-life problem of assigning already scheduled jobs to fill bottles with cleaning supplies to storage boilers and identified it as a variant of the well studied problem of k-track assignment or operational fixed job scheduling (OFJSP). We compared results of state-of-the-art solvers for the ILP formulation as well as a new network flow formulation of the problem for generated instances and illustrated that the proposed network flow model is useful to calculate upper bounds for larger instances. In addition we proposed a simulated annealing approach which gives good results when the instances cannot be solved in reasonable time by exact methods. Our simulated annealing was also compared with the state-of-the-art approach from the literature on the existing instances for the operational fixed job scheduling problem. It turns out that the simulated annealing approach improves this state-of-the-art approach on bigger instances and is very flexible and therefore suitable to be used in practice.

A further investigation of instances with various further parameter configurations and investigation of hybrid methods remains as an interesting research topic for the future.

Acknowledgments. The financial support by the Austrian Federal Ministry for Digital and Economic Affairs, the National Foundation for Research, Technology and Development and the Christian Doppler Research Association is gratefully acknowledged.

References

1. Angelelli, E., Bianchessi, N., Filippi, C.: Optimal interval scheduling with a resource constraint. Comput. Oper. Res. **51**, 268–281 (2014)
2. Arkin, E.M., Silverberg, E.B.: Scheduling jobs with fixed start and end times. Discret. Appl. Math. **18**, 1–8 (1987)
3. Barcia, P., Cerdeira, J.O.: The k-track assignment problem on partial orders. J. Sched. **8**, 135–143 (2005)
4. Brucker, P., Nordmann, L.: The k-track assignment problem. Computing **52**, 97–122 (1993)
5. Chu, G.: Improving combinatorial optimization. Ph.D. thesis, University of Melbourne, Australia (2011). http://hdl.handle.net/11343/36679
6. IBM Corporation: IBM ILOG CPLEX 12.10 User's Manual (2019)
7. Eliiyi, D., Azizoglu, M.: Heuristics for operational fixed job scheduling problems with working and spread time constraints. Int. J. Prod. Econ. **132**, 107–121 (2011)
8. Gurobi Optimization: Gurobi Optimizer Reference Manual (2020). http://www.gurobi.com
9. Kirkpatrick, S., Gelatt, C.D., Vecchi, M.P.: Optimization by simulated annealing. Science **220**(4598), 671–680 (1983)
10. Kovalyov, M., Ng, C., Cheng, T.: Fixed interval scheduling: models, applications, computational complexity and algorithms. Eur. J. Oper. Res. **178**, 331–342 (2007)
11. Lindauer, M., Eggensperger, K., Feurer, M., Falkner, S., Biedenkapp, A., Hutter, F.: SMAC v3: Algorithm Configuration in Python. GitHub (2017)
12. Nethercote, N., Stuckey, P.J., Becket, R., Brand, S., Duck, G.J., Tack, G.: MiniZinc: towards a standard CP modelling language. In: Bessière, C. (ed.) CP 2007. LNCS, vol. 4741, pp. 529–543. Springer, Heidelberg (2007). https://doi.org/10.1007/978-3-540-74970-7_38
13. Rossi, A., Singh, A., Sevaux, M.: A metaheuristic for the fixed job scheduling problem under spread time constraints. Comput. Oper. Res. **37**(6), 1045–1054 (2010)
14. Zhou, H., Bai, G., Deng, S.: Optimal interval scheduling with nonidentical given machines. Clust. Comput. **22**(3), 1007–1015 (2019). https://doi.org/10.1007/s10586-018-02892-z

Instance Space Analysis
for the Generalized Assignment Problem

Tobias Geibinger[2]($\boxtimes$) (iD), Lucas Kletzander[1,2] (iD), and Nysret Musliu[1,2] (iD)

[1] Christian Doppler Laboratory for Artificial Intelligence and Optimization
for Planning and Scheduling, TU Wien, Vienna, Austria
[2] Databases and Artificial Intelligence Group, Institute for Logic and Computation,
TU Wien, Vienna, Austria
{tgeibing,lkletzan,musliu}@dbai.tuwien.ac.at

Abstract. In this work, we consider the well-studied Generalized
Assignment Problem and investigate the performance of several metaheuristic methods. To obtain insights on strengths and weaknesses of
these solution approaches, we perform Instance Space Analysis on the
existing instance types and propose a set of features describing the hardness of an instance. This is of interest since the existing benchmark set is
dated and rather limited and the known instance generators might not
be fully representative. Our analysis for metaheuristic methods reveals
that this is indeed the case and finds several gaps, which we fill with
newly generated instances thus adding diversity and providing a new
benchmark instance set. Furthermore, we analyze the impact of problem
features on the performance of the methods used and identify the most
important ones.

1 Introduction

The goal of the *Generalized Assignment Problem* (GAP) [24] is to find an assignment of n tasks to m agents such that each task is assigned, no agent exceeds
their capacity, and the overall cost is minimal. The GAP is an important problem
that arises in several domains including scheduling, telecommunication, facility
location and transportation.

This problem has been extensively investigated in the literature and different exact and heuristic approaches have been proposed (see reviews of existing
methods [20, 34]). The existing approaches have been evaluated mainly on the
small set of existing benchmark instances. Although various methods exist for
this problem, optimal solutions are still not known for all instances and there is
no extensive study that analyses the strengths and weaknesses of methods on a
larger set of instances. Therefore, there is a need for a deep investigation of the
performance of various state-of-the-art solving paradigms and a critical study
that would enable the generation of a large set of instances with a sufficient
diversity. This is very important to get insights on strengths and weaknesses of

L. Di Gaspero et al. (Eds.): MIC 2022, LNCS 13838, pp. 421–435, 2023.
https://doi.org/10.1007/978-3-031-26504-4_30

various solution approaches. Moreover, the identification of features that characterize this problem well and have an impact on the performance of solution methods has not been investigated before.

The main contributions of this paper are:

- We implement and compare various metaheuristic approaches based on Simulated Annealing, Tabu Search and Min-Conflicts heuristics.
- We propose for the first time an extensive set of features to characterize GAP instances.
- Based on initial Instance Space Analysis, we generate more than 1700 new instances to better cover the instance space for this problem.
- We perform a deep Instance Space Analysis, and based on new instances and features, we investigate the performance of solvers in different regions of the instance space. Our analysis gives interesting insights regarding the importance of features, and strengths and weaknesses of solvers in the instance space.

2   Problem Definition

Formally, the GAP is given by the following MIP formulation [31, 33]:

$$min \sum_{i=1}^{m} \sum_{j=1}^{n} c_{ij} x_{ij} \tag{1}$$

$$\sum_{j=1}^{n} a_{ij} x_{ij} \leq b_i \qquad\qquad 1 \leq i \leq m \tag{2}$$

$$\sum_{i=1}^{m} x_{ij} = 1 \qquad\qquad 1 \leq j \leq n \tag{3}$$

$$x_{ij} \in \{0, 1\} \qquad\qquad 1 \leq i \leq m, 1 \leq j \leq n \tag{4}$$

Here c_{ij} is the cost of assigning task j to agent i, a_{ij} is the capacity used when agent i performs task j, and b_i is the capacity of agent i. For each combination of task and agent, we have a Boolean decision variable x_{ij} indicating that task j is assigned to agent i. Furthermore, $B = \sum_{1 \leq i \leq m} b_i$ denotes the total capacity of an instance. Constraint (2) enforces that for each agent, the sum of the usages of all assigned tasks is less or equal than their capacity. Furthermore, it is ensured by constraint (3) that each task is assigned to exactly one agent. The objective (1) states that we want to minimise the sum of all costs, where the cost of the task is determined by the agent it is assigned to. The GAP is known to be NP-hard [7].

3   Related Work

The Generalized Assignment Problem (GAP) has been studied extensively since it was first proposed [31]. The survey paper [20] and book chapter [34] are excellent introductions to algorithms for the GAP. Besides MIP formulations, the most commonly found solution approaches for the GAP are either branch and bound [2,12], column generation [25], MIP relaxation methods [6], or metaheuristics and hybrid techniques [3,21,33,35].

The best known results were achieved with exact methods, for the hardest instances most of the best bounds where found by Avella et al. [1] using a cutting plane method. For the remaining instances, Posta et al. [22] achieved the best known results using a reformulation of the problem into a sequence of decision problems. Furthermore, this leaves only 12 instances from the standard benchmark set where the optimal solution is not known. However, these upper bounds are reached using very long computation times of up to 24 h per instance. One upper bound is also improved by a hybrid heuristic [30] where the GPU is used for parallelism, but still computing for more than one hour.

Instance Space Analysis (ISA) [28] is a recent methodology proposed by Smith-Miles and co-workers [26] that offers a more objective assessment of the relative power of algorithms by projecting the instances of a problem into a 2D plane, the *instance space*, revealing relationships between properties of the instances (features) and the performance of the algorithms, and identifying regions of good algorithm performance, called *footprints* [29]. The method is an extension to Rice's algorithm selection framework [23]. Applications include blackbox optimization [15,16], machine learning [5,17], personnel scheduling [11], course timetabling [4], job shop scheduling [32], multi-objective optimization [37], and methods to generate new instances based on the results of the instance space analysis [14,27].

4   Algorithms

In this section, we present the solution approaches used in our evaluation. All implementations, instances, logs, and random seeds are available online.[1,2]

We implemented Simulated Annealing (SA), Tabu Search (TS) and Min Conflicts (MC) for the GAP. SA [10], TS [8] and MC [13] are well-known metaheuristics, which work by repeatedly applying small changes (moves) to a candidate solution.

Our implementations are written in Rust 1.54.0[3] and employ two types of moves. One is called *shift* and operates by selecting a task in the current solution and moves it to another agent. The other type of move is called *swap* and exchanges the assignment of two tasks in the current solution which are not assigned to the same agent. Whereas SA always selects a move at random, TS explores both neighborhoods exhaustively and applies the best non-tabu move. MC in turn selects a task at random and then performs the best move involving this task out of the two neighborhoods.

In SA, improvements are always accepted, while worsening moves are accepted with a certain probability that depends on the difference in the objective function and a parameter called *temperature*. Higher temperatures lead to higher acceptance probabilities. Over the course of the search, the temperature

---

[1] https://dbai.tuwien.ac.at/user/tgeibing/gap/mic22.zip.
[2] https://dbai.tuwien.ac.at/user/tgeibing/gap/instances/.
[3] https://www.rust-lang.org/.

is slowly decreased. In order to penalize an infeasible assignment, we calculate its *overspill*, which is the total amount of exceeded agent capacity. The weighted overspill is then added to the objective value of the candidate solution. Further parameters of SA are

- the *initial* and *minimum temperature*,
- the *cooling rate*, i.e., the factor with which the temperature is multiplied after each iteration,
- *shift/swap probability*, i.e., the probability that shift is selected instead of swap at each iteration,
- the *overspill weight*, and
- the *cooling mode*, which is either
  - *standard*: From the initial temperature the cooling rate is applied until the minimum temperature is reached,
  - *reheating*: the temperature is reset whenever the minimum is hit, or
  - *dynamic*: the cooling rate is determined dynamically such that the minimum is reached at the end of the runtime.

As stated TS includes the concept of tabu moves, which denote moves that are not allowed to be performed. To faciliate this the metaheuristic keeps a FIFO list of a given length of those moves and as long as they are in the list they are tabu. In our case, whenever a move gets performed, we include the reversing move in the tabu list. The length of the tabu list is set as with *tabu list length*.

In order to escape local optima in MC, it is often combined with *random walks*. Our implementation includes a parameter *rw prob*, which indicates the probability that instead of the current MC move, a random walk, i.e., a completely random move, is performed.

The concrete values used for each parameter are given as follows. For SA we set the *initial temperature* to $1,000,000$, the *initial temperature* to 0.5, the *shift/swap probability* to 20%, *overspill weight* was $850,000$ and *dynamic* was chosen as the *cooling mode*. The *tabu list length* in TS was set to $49,974$ and *rw prob* for MC was 1%. Note that the chosen *tabu list length* is quite large. This is most likely the case because TS performs best on small to medium instances, where millions of moves are performed in each run of the algorithm. Furthermore, as stated above, we only add the directly reversing moves to the tabu list.

To configure the parameters, we considered the automated algorithm tuner SMAC[4] [9] and additional experiments with several parameter settings. SMAC was run for several days ($1,000$ target algorithm runs) on the same hardware as was used in the evaluation and on a tuning set of 680 instances of types A–F (c.f. Sect. 5). The runtime for each configuration was 15 min. The parameters for TS and MC are the result of this tuning. However, it turned out that the parameter configuration suggested by SMAC for SA performed worse than the manually set values given above.

Our variant of SA for the GAP differs from the one given in [21] in several ways. The first difference is that we allow moves which make the current

---

[4] https://github.com/automl/SMAC3.

incumbent solution infeasible. Furthermore, we do not start from a feasible initial solution, but rather from a random assignment and we apply simple moves. Similarly, in difference to the generalized swap moves described by Osman et al. [21], our TS variant only utilizes simple shifts and swaps. We also do not utilize *ejection chains* like the ones used by Yagiura et al. [35].

5   Instances

A widely used [33, 35, 36] set of benchmark instances for the GAP can be found online[5]. Those instances have five types depending on how they were generated. The first four types were introduced in [3] and are defined as follows:

Type A: The usages a_{ij} and costs c_{ij} are integers taken uniformly from $[5, 25]$ and $[10, 50]$ respectively, and the capacity of any agent is $0.6\frac{n}{m}15 + 0.4r$, where $r = max_{1 \leq i \leq m} \sum_{1 \leq j \leq n, \sigma(j)=i} a_{ij}$ and $\sigma(j) = min(\{1 \leq i \leq m \mid c_{ij} = min(\{c_{lj} \mid 1 \leq l \leq m\})\})$ for any task j.

Type B: The usages a_{ij} and costs c_{ij} are the same as for type A and the capacity of each agent is 70% of the value given for type A.

Type C: The usages a_{ij} and costs c_{ij} are the same as for type A and the capacity of an agent i is $b_i = 0.8 \sum_{1 \leq j \leq n} a_{ij}/m$.

Type D: The usages a_{ij} are taken uniformly from $[1, 100]$, $c_{ij} = 111 - a_{ij} + e$ where e is an integer taken uniformly from $[-10, 10]$, and the agent capacities are defined as in type C.

In [36] the authors introduced a new instance type E.

Type E: $a_{ij} = 1 - 10 \cdot \ln(e_1)$ where e_1 is taken uniformly from $(0, 1]$, $c_{ij} = 1000/a_{ij} - 10e_2$ where e_2 is taken uniformly from $[0, 1]$, and the agent capacities are defined as in type C.

5.1   Results on Existing Instances

All our algorithms were evaluated on a cluster with 13 nodes, each having 2 Intel Xeon CPUs E5-2650 v4 (max. 2.90 GHz, 12 physical cores, no hyperthreading), with a memory limit 20 GB. Since the metaheuristics are not deterministic, we ran them on each instance 3 times and chose the average of those results.

The original instance set is comprised by 57 instances in five different types as described above. First, this is a very low number of instances, which might lead to conclusions that depend on some specifics of the given instances. Second, our initial tests showed that all given instances were rather easy for MIP methods to obtain high quality solutions in a short time, which might not be the case for instances outside of ranges imposed by the given types. To the best of our knowledge, the best bounds for the open instances from the literature are given by Avella et al. [1], Posta et al. [22], and Souza et al [30]. In difference to our focus, those works aim to obtain new bounds and thus consider much longer timeouts than we do (up to 24 h). Furthermore, the technique by Posta et al. [22]

---

[5] http://www.al.cm.is.nagoya-u.ac.jp/~yagiura/gap/.

is not designed to obtain good solutions fast, but rather to prove new optimal solutions. Furthermore, to the best of our knowledge these implementations are also not publicly available.

Based on our initial experiments, we concluded that for a deep analysis of the performance of our methods a much larger and more diverse set of instances is needed, as the comparison of methods only on the existing set can give misleading conclusions about the performance. Therefore, we initially extended the original set with new instances and in the later phase we generated additional instances based on ISA.

5.2   Extended Instance Set

At this stage, two extensions were initially made:

Type F: The usages a_{ij} and costs c_{ij} are integers taken uniformly from $[50, 500]$ and $[10, 1000]$ respectively, and the capacity of an agent i is $b_i = e \sum_{1 \leq j \leq n} a_{ij}/m$ where e is taken uniformly from $[0.15, 0.3]$. The intention behind this type F is to make the capacities of the agents as tight as possible such that finding a feasible solution is already hard.

In order to increase the number of instances, we used the 57 existing instances of types A, B, C, D and E. Furthermore, we created 570 new instances with the same types and sizes, as well as 154 instances of type F with between 80 and 200 agents and 900 and 5000 tasks. In total, our first extended benchmark set contains 781 instances.

6   Instance Space Analysis

To gain more insights on the distribution of the instances and the detailed behaviour of the algorithms we use *Instance Space Analysis (ISA)* [28]. This analysis was used to generate additional instances to ensure that we cover different regions of the instance space.

6.1   Concept and Methodology

Figure 1 shows the underlying framework of ISA. The *problem space* $\mathcal{P}$ contains all instances of the GAP, from which we have a subset of instances **I**.

For these instances, we obtain meta-data in two forms. First, each instance $x \in \mathbf{I}$ is represented by the feature vector $f_x \in \mathcal{F}$ in the *feature space* $\mathcal{F}$. Second, a performance measure $y_{\alpha,x} \in \mathcal{Y}$ is obtained for each combination of an instance x and an algorithm $\alpha \in \mathcal{A}$, where $\mathcal{A}$ is the *algorithm space*, representing the set of algorithms that we use to solve the GAP. $\mathcal{Y}$ is the *performance space* measuring the performance of algorithms.

While this could be used to directly learn a selection mapping S, ISA uses the meta-data to compute a mapping from the feature space to a 2D *instance space*, representing each instance by a point $z \in \mathbb{R}^2$ in a way to maximize linearly observable trends across the instance space. This step provides several benefits,

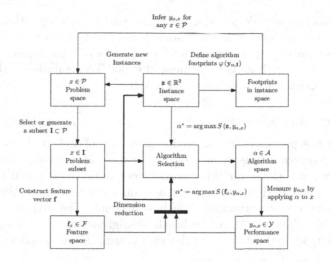

Fig. 1. Summary of the ISA framework [28].

including the potential to assess the adequacy of the given instances and features, allowing insights regarding which parts of the problem space $\mathcal{P}$ should be explored with new problem instances. A selection mapping S can be computed from the instance space with the additional benefit of a clear visualisation. Algorithm performance can be evaluated in detail by computing *footprints* $\varphi(\mathbf{y}_{\alpha,\mathbf{I}})$, which are regions of strong algorithm performance. The selected features and their distribution can be investigated to gain insights into instance hardness and diversity. The footprints can further be used to infer algorithm performance even for unseen parts of the problem space.

The general methodology of ISA consists of six steps:

1. Collect feature values $\mathbf{F}$ and performance metrics $\mathbf{Y}$ for the set of instances $\mathbf{I}$ run on the algorithm portfolio $\mathcal{A}$.
2. Construct an instance space using a feature selection process on $\mathbf{F}$ and $\mathbf{Y}$.
3. Generate machine learning predictions for automated algorithm selection.
4. Generate algorithm footprints and metrics.
5. Analyze the instance space.
6. If needed, generate additional meta-data and go to step 2.

The steps 2 to 4 are done using the *Instance Space Analysis* toolkit [18] available on Github[6]. Step 2 is performed using four main tasks: A binary measure of good performance, bounding and scaling are applied first. Then a relevant subset of features that explain algorithm difficulty and eliminate redundancies is chosen. Next a projection to the 2D instance space that maximizes linear trends in features and algorithms is computed. Finally, bounds for the instance space can be computed.

---

[6] https://github.com/andremun/InstanceSpace.

Step 3 uses Support Vector Machines trained for each algorithm that take the coordinates of an instance in the instance space and map them to either good or bad performance. Step 4 aims at finding clusters of high density with good instances for each algorithm using a method called Triangulation with Removal of Areas with Contradicting Evidence (TRACE).

6.2 Performance Measure

As stated in the previous section, an integral part of ISA is the performance space which requires a measurement of performance for each algorithm on each instance. A natural choice for this measurement in our case would be the cost of the found solution. However, the metric is does not take the evolution of the solution quality over the running time into account. The *area score* which is used in the MiniZinc Challenge[7] to break ties, gives exactly this information. Our (slightly modified) definition of the area score is given as follows

$$
\begin{aligned}
area(a, i) = {}& 0.25 \cdot solTime(a, i, 1) \\
& + 0.5 \cdot \frac{\sum_{1 \leq j \leq n+1} solCost(a, i, j-1) \cdot (solTime(a, i, j) - solTime(a, i, j-1))}{UB(i) - LB(i) + 1} \\
& + 0.25 \cdot totalTime(a, i),
\end{aligned}
$$

where

- n is the number of solutions found by algorithm a on instance i,
- $solTime(a, i, j)$ denotes the time when a found the j-th solution on i ($solTime(a, i, n+1) = totalTime$),
- $solCost(a, i, j)$ is the cost of the j-th solution ($solCost(a, i, 0) = UB(i)$),
- $totalTime(a, i)$ is total time needed by algorithm a on i,
- $UB(i)$ and $LB(i)$ denote the maximum and minimum cost of any solution found by any algorithm for i.

Since a minimal area score does not always imply minimal cost, we need to combine both notions for our measurement. To achieve this we first introduce some additional notation. By $minCost(i)$ and $maxArea(i)$, we denote the best cost and respectively worst area score of any algorithm for instance i. The cost of algorithm a on instance i is in turn given by $cost(a, i)$. Now, combing both cost and area score we obtain $absScore(a, i) = cost(a, i) - minCost(i) + \frac{area(a,i)}{maxArea(i)}$, where any improvement in cost weights higher than the area score. In order to ease comparison of different instances with costs of varying magnitude, we additionally scale the score from zero to one using the following exponential function

$$
relScore(a, i) = \begin{cases} 1.001^{-absScore(a,i)} & \text{if } a \text{ found at least one solution} \\ 0 & \text{otherwise.} \end{cases}
$$

Note that this also takes care of the cases where an algorithm did not produce any solution. In our analysis, we consider the performance of an algorithm a on instance i "good" if $relScore(a, i) > 0.999$.

---

[7] https://www.minizinc.org/challenge.html.

6.3 Features

To apply Instance Space Analysis, we need to define features which in turn need to be able to explain the hardness of instances and the different performances of algorithms to obtain useful results from the analysis. While a subset of six features is eventually used for the instance space, we define a larger set of 78 potentially useful features, using three categories.

Direct Instance Features. The following 15 features can be directly obtained from the parameters of an instance.

- Number of tasks n.
- Number of agents m.
- The number of possible assignments $n \times m$.
- The minimum, maximum, median and average task cost.
- The minimum, maximum, median and average task usage.
- The minimum, maximum, median and average agent capacity.

Advanced Instance Features. Based on a closer observation of potential hard instances, we identified 56 further features.

- The ratio of agents to tasks given by $\frac{n}{m}$.
- The variance and interquartile range of all agent capacities, task costs and task usages.
- The minimum, maximum, median, average, variance and interquartile range of the relative capacities, i.e., of the set $\{b_i / \sum_{1 \leq j \leq m} b_j \mid 1 \leq i \leq m\}$.
- The minimum, maximum, median, average, variance and interquartile range of the relative task usages, i.e., of the set $\{a_{ij}/b_i \mid 1 \leq i \leq m, 1 \leq j \leq n\}$.
- The min, max, median, average, variance and interquartile range of $minTaskCostUsageRatio$, $maxTaskCostUsageRatio$, $medTaskCostUsageRatio$ and $avgTaskCostUsageRatio$, which for a task j are given by the minimum, maximum, median and average of $\{c_{ij}/a_{ij} \mid 1 \leq i \leq m\}$ respectively.

For the remaining advanced features, we are going to make use of the *lower bound assignment* of an instance. This assignment is obtained by assigning each task an agent where it has minimal cost ignoring capacities (ties are broken by taking the smaller agent) and its cost is a lower bound for the optimal cost for the instance. Let y_{ij} represent such a lower bound assignment. Then for an agent i, we define its overusage as $o_i = \max\{\sum_{1 \leq j \leq n} a_{ij}y_{ij} - b_i, 0\}$ and its slack as $s_i = \max\{\sum_{1 \leq j \leq n} b_i - a_{ij}y_{ij}, 0\}$. Based on those notions, we can define the following additional features.

- The relative amount of overusage in the lower bound assignment
  $relLbOverusage = \sum_{1 \leq i \leq m} o_i \ / \ B$.
- The relative amount of slack in the lower bound assignment
  $relLbSlack = \sum_{1 \leq i \leq m} s_i \ / \ B$.
- The min, max, median, average, variance and interquartile range of $relLbAgentOverusage$ which for an agent i is given by o_i/b_i.
- The min, max, median, average, variance and interquartile range of v $relLbAgentSlack$ which for an agent i is given by s_i/b_i.

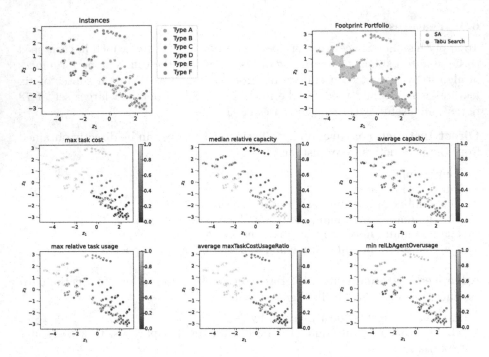

Fig. 2. Type and feature distribution for the extended instance set on metaheuristics

Model Features. We use MIP and CP formulations of GAP in MiniZinc [19] to obtain further features from the FlatZinc conversion:

- Number of Boolean and integer CP variables,
- Number of Boolean CP constraints,
- Number of evaluated reified CP constraints,
- Number of MIP integer variables and constraints.

6.4 Initial Instance Space Analysis

The initial Instance Space Analysis was performed on the extended set of 781 instances belonging to the types A to F, using fifteen minutes per run. Parameters were mostly set to the defaults, except that we do not bound outliers in the data and utilize MATLAB's own SVM implementation. Furthermore, we restrict the feature selection to six features.

The ISA gives the following projection (the prime indicates the normalized values of the features):

$$\begin{pmatrix} z_1 \\ z_2 \end{pmatrix} = \begin{pmatrix} -0.635 & 1.5615 \\ -0.6347 & 0.0499 \\ 0.606 & -0.18 \\ -0.0216 & -0.6918 \\ -0.6509 & -0.3889 \\ 0.1744 & 0.2453 \end{pmatrix}^{\mathsf{T}} \cdot \begin{pmatrix} maxTaskCost' \\ maxRelTaskUsage' \\ avgCapacity' \\ medRelCapacity' \\ avgMaxTaskCostUsage' \\ minRelLbAgentOverusage' \end{pmatrix} \tag{5}$$

Figure 2 shows the results of the initial analysis regarding the selection of features and the resulting distribution of the instances in the instance space. Types A, B, C again form a combined cluster that is however somewhat interwoven with D. E and F are again rather clearly separated from the rest. The effect of the new instances clustering together can again be observed as well as the features mostly focusing on separating the types.

The portfolio footprint is again very segmented even though SA is chosen for most types.

6.5   New Instance Generator

According to the problems revealed by the initial ISA, we propose a new instance generator that increases diversity by filling the gaps in the instance space and filling the boundary regions between the types to get a more precise picture of the algorithm behaviour. This is done by mixing the different construction approaches and unifying different selection boundaries:

- Usage a_{ij}: Random choice of either integers taken uniformly from $[u_{min}, u_{max}]$ where u_{min} is randomly chosen in $[1, 50]$ and u_{max} randomly in $[\max\{25, u_{min} + 10\}, 500]$ (subtype A covering A, B, C, D, F), or the distribution from type E (subtype E)
- Cost c_{ij}: Random choice of either integers taken uniformly from $[10, c_{max}]$ where c_{max} is randomly chosen in $[50, 1000]$ (subtype A covering A, B, C, F), or $c_{ij} = \max\{1, u_{max} - a_{ij} + e\}$ where e is an integer taken uniformly from $[-10, 10]$ (subtype D), or the distribution from type E (subtype E)
- Capacity b_i: Random choice of either the capacity function from type A multiplied with a factor uniformly chosen from $[0.7, 1]$ (subtype A covering A, B), or the capacity function from type F where e is uniformly chosen from $[c_{min}, c_{max}]$ with c_{min} uniformly chosen from $[0.15, 0.8]$ and c_{max} from $[\max\{0.3, c_{min}\}, 0.8]$ (subtype C covering C, D, E, F)

An instance is now generated as follows:

- Randomly choose a subtype for usage, cost, and capacity, leading to one of the 12 subtypes AAA, AAC, ADA, ADC, AEA, AEC, EAA, EAC, EDA, EDC, EEA, EEC.
- Randomly choose boundaries for the whole instance based on the subtypes.
- Randomly choose the individual values for agents and tasks within the previously selected boundaries.

Both the extended instance set using the types A to F and the set of 1011 new instances as well as the new generator are publicly available online (See footnote 2).

6.6   Instance Space Analysis for the Full Instance Set

This section describes the analysis on the total set of instances, combining the new instance set with 1011 instances and the extended old set with 781 instances.

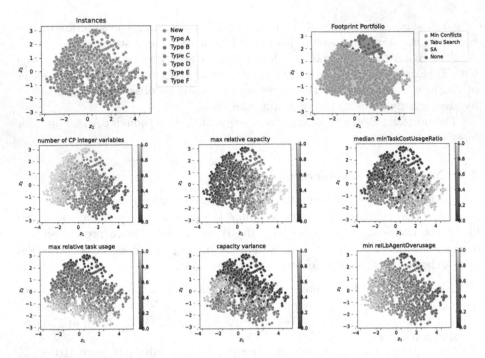

Fig. 3. Type and feature distribution and portfolio footprint of the metaheuristics for the full instance set

The ISA produced the following projection and set of features for our metaheuristic approaches:

$$
\begin{pmatrix} z_1 \\ z_2 \end{pmatrix} = \begin{pmatrix} -0.2883 & -0.7266 \\ -0.2031 & -0.3703 \\ -0.0443 & -0.3273 \\ 0.378 & -0.2846 \\ -0.652 & -0.1612 \\ -1.0177 & 0.185 \end{pmatrix}^{\mathsf{T}} \cdot \begin{pmatrix} maxRelTaskUsage' \\ capacityVariance' \\ maxRelCapacity' \\ medMinTaskCostUsageRatio' \\ minAgentLbRelOverusage' \\ nrCpIntVars' \end{pmatrix} \tag{6}
$$

Figure 3 shows the distribution of instances and features as well as the footprints of the metaheuristic portfolio in the new instance space. Note that the projection is now again different, but the instance space is still nicely covered. SA also still performs well for most of the instances, but its footprint is now completely connected. The footprint of TS now also covers a large part of the top of the instance space, which consists completely of instances created by the new generator. However, between the TS and SA footprints there is an area that is either covered by no footprint or by MC but in a very segmented way.

Regarding the selected features, again there is one main axis using a size measure, this time the number of CP integer variables, and the max relative capacity. This axis corresponds to the one using the number of tasks and the median relative capacity for the exact methods. Present again, but more closely aligned with the first main axis this time, is the median $minTaskCostUsageRatio$.

The median relative task usage is replaced by the max, which forms an axis together with the capacity variance (instead of the IQR) with lower values on the top to top right, and higher values on the bottom to bottom left (even though the highest values for capacity variance are more distributed). This axis also corresponds the most to the differences in the algorithm portfolio. Finally, min $relLbAgentOverusage$ is distributed along the z_1 axis.

The analysis shows that despite the different set of methods, five out of six chosen features are very similar between the two analyses, strengthening the impression that versions of these selected features are important to explain the performance of different methods and the hardness of certain instances.

7 Conclusion

In conclusion, we presented a comparison of various metaheuristics for the General Assignment Problem, showing that we can provide high quality solutions for the existing instance set in short time. However, the small size and low diversity make it difficult to draw insightful conclusions regarding the performance of various approaches. Therefore, we extended the dated benchmark instance set in various ways, using Instance Space Analysis to guide the development of a diverse new set of instances that is not clustered into distinct individual classes with very similar properties. For this purpose, we proposed several new instance features with the aim to describe instance hardness. We identified several features that are important for algorithm performance, In particular, our experiments have shown that features describing the distribution of agent capacities and task costs, as well the ratio between cost and usage, correlate well with the hardness of an instance. While our experiments have shown that most methods have their regions of strengths and weaknesses and that Simulated Annealing gives best results among metaheuristic approaches.

As the main aim of this paper was to generate new instances and identify important features, we focused on meta-heuristic solution methods that obtain good results in a short time. For future work, it would be interesting to investigate and compare more advanced exact approaches on the new set of instances.

Acknowledgments. This work has been funded by the Austrian security research programme KIRAS of the Federal Ministry of Agriculture, Regions and Tourism (BMLRT). Furthermore, the financial support by the Austrian Federal Ministry for Digital and Economic Affairs, the National Foundation for Research, Technology and Development and the Christian Doppler Research Association is gratefully acknowledged.

References

1. Avella, P., Boccia, M., Vasilyev, I.: A computational study of exact knapsack separation for the generalized assignment problem. Comput. Optim. Appl. **45**(3), 543–555 (2010)

2. Cattrysse, D., Degraeve, Z., Tistaert, J.: Solving the generalised assignment problem using polyhedral results. Eur. J. Oper. Res. **108**(3), 618–628 (1998)
3. Chu, P.C., Beasley, J.E.: A genetic algorithm for the generalised assignment problem. Comput. Oper. Res. **24**(1), 17–23 (1997)
4. De Coster, A., Musliu, N., Schaerf, A., Schoisswohl, J., Smith-Miles, K.: Algorithm selection and instance space analysis for curriculum-based course timetabling. J. Sched. 1–24 (2021)
5. Fernandes, L.H.d.S., Lorena, A.C., Smith-Miles, K.: Towards understanding clustering problems and algorithms: an instance space analysis. Algorithms **14**(3), 95 (2021)
6. Fisher, M.L.: The lagrangian relaxation method for solving integer programming problems. Manag. Sci. **50**(12_supplement), 1861–1871 (2004)
7. Garey, M.R., Johnson, D.S.: Computers and Intractability: A Guide to the Theory of NP-Completeness. W. H. Freeman, New York (1979)
8. Glover, F.: Future paths for integer programming and links to artificial intelligence. Comput. Oper. Res. **13**(5), 533–549 (1986)
9. Hutter, F., Hoos, H.H., Leyton-Brown, K.: Sequential model-based optimization for general algorithm configuration. In: Coello, C.A.C. (ed.) LION 2011. LNCS, vol. 6683, pp. 507–523. Springer, Heidelberg (2011). https://doi.org/10.1007/978-3-642-25566-3_40
10. Kirkpatrick, S., Gelatt, C.D., Vecchi, M.P.: Optimization by simulated annealing. Science **220**(4598), 671–680 (1983)
11. Kletzander, L., Musliu, N., Smith-Miles, K.: Instance space analysis for a personnel scheduling problem. Ann. Math. Artif. Intell. **89**(7), 617–637 (2021)
12. Martello, S., Toth, P.: Knapsack Problems: Algorithms and Computer Implementations. Wiley, Hoboken (1990)
13. Minton, S., Johnston, M.D., Philips, A.B., Laird, P.: Solving large-scale constraint-satisfaction and scheduling problems using a heuristic repair method. In: Proceedings of the 8th National Conference on Artificial Intelligence (AAAI 1990), pp. 17–24. AAAI Press/The MIT Press (1990)
14. Muñoz, M.A., Smith-Miles, K.: Generating custom classification datasets by targeting the instance space. In: Proceedings of the Genetic and Evolutionary Computation Conference Companion, pp. 1582–1588 (2017)
15. Muñoz, M.A., Smith-Miles, K.: Generating new space-filling test instances for continuous black-box optimization. Evol. Comput. **28**(3), 379–404 (2020)
16. Muñoz, M.A., Smith-Miles, K.A.: Performance analysis of continuous black-box optimization algorithms via footprints in instance space. Evol. Comput. **25**(4), 529–554 (2017)
17. Muñoz, M.A., Villanova, L., Baatar, D., Smith-Miles, K.: Instance spaces for machine learning classification. Mach. Learn. **107**(1), 109–147 (2018)
18. Muñoz, M.A., Smith-Miles, K.: Instance space analysis: a toolkit for the assessment of algorithmic power (2020)
19. Nethercote, N., Stuckey, P.J., Becket, R., Brand, S., Duck, G.J., Tack, G.: MiniZinc: towards a standard CP modelling language. In: Bessière, C. (ed.) CP 2007. LNCS, vol. 4741, pp. 529–543. Springer, Heidelberg (2007). https://doi.org/10.1007/978-3-540-74970-7_38
20. Öncan, T.: A survey of the generalized assignment problem and its applications. INFOR: Inf. Syst. Oper. Res. **45**(3), 123–141 (2007)
21. Osman, I.H.: Heuristics for the generalized assignment problem: simulated annealing and tabu search approaches. OR Spektrum **17**, 211–255 (1995)

22. Posta, M., Ferland, J.A., Michelon, P.: An exact method with variable fixing for solving the generalized assignment problem. Comput. Optim. Appl. **52**(3), 629–644 (2012)
23. Rice, J.: The algorithm selection problem. In: Advances in Computers, vol. 15, pp. 65–118. Elsevier (1976)
24. Ross, G.T., Soland, R.M.: A branch and bound algorithm for the generalized assignment problem. Math. Program. **8**(1), 91–103 (1975)
25. Savelsbergh, M.: A branch-and-price algorithm for the generalized assignment problem. Oper. Res. **45**(6), 831–841 (1997)
26. Smith-Miles, K., Baatar, D., Wreford, B., Lewis, R.: Towards objective measures of algorithm performance across instance space. Comput. Oper. Res. **45**, 12–24 (2014)
27. Smith-Miles, K., Bowly, S.: Generating new test instances by evolving in instance space. Comput. Oper. Res. **63**, 102–113 (2015)
28. Smith-Miles, K., Muñoz, M.A.: Instance space analysis for algorithm testing: methodology and software tools (2021)
29. Smith-Miles, K., Tan, T.T.: Measuring algorithm footprints in instance space. In: 2012 IEEE Congress on Evolutionary Computation, pp. 1–8. IEEE (2012)
30. Souza, D.S., Santos, H.G., Coelho, I.M.: A hybrid heuristic in GPU-CPU based on scatter search for the generalized assignment problem. Procedia Comput. Sci. **108**, 1404–1413 (2017)
31. Srinivasan, V., Thompson, G.L.: An algorithm for assigning uses to sources in a special class of transportation problems. Oper. Res. **21**(1), 284–295 (1973)
32. Strassl, S., Musliu, N.: Instance space analysis and algorithm selection for the job shop scheduling problem. Comput. Oper. Res. **141**, 105661 (2022)
33. Woodcock, A.J., Wilson, J.M.: A hybrid tabu search/branch & bound approach to solving the generalized assignment problem. Eur. J. Oper. Res. **207**(2), 566–578 (2010)
34. Wu, W., Yagiura, M., Ibaraki, T.: Generalized assignment problem. In: Handbook of Approximation Algorithms and Metaheuristics, Second Edition, Volume 1: Methologies and Traditional Applications, pp. 713–736. Chapman and Hall/CRC (2018)
35. Yagiura, M., Ibaraki, T.: Recent metaheuristic algorithms for the generalized assignment problem. In: Proceedings of the International Conference on Informatics Research for Development of Knowledge Society Infrastructure (ICKS 2004), pp. 229–237 (2004)
36. Yagiura, M., Yamaguchi, T., Ibaraki, T.: A variable depth search algorithm with branching search for the generalized assignment problem. Optim. Methods Softw. **10**(2), 419–441 (1998)
37. Yap, E., Munoz, M.A., Smith-Miles, K., Liefooghe, A.: Instance space analysis of combinatorial multi-objective optimization problems. In: 2020 IEEE Congress on Evolutionary Computation (CEC), pp. 1–8. IEEE (2020)

Decision Support for Agri-Food Supply Chains in the E-Commerce Era: The Inbound Inventory Routing Problem with Perishable Products

D. Cuellar-Usaquén[1], C. Gomez[1], M. Ulmer[2], and D. Álvarez-Martínez[1(✉)]

[1] Universidad de Los Andes, Bogotá D.C., Colombia
{dh.cuellar,gomez.ch,d.alvarezm}@uniandes.edu.co
[2] Otto-von-guericke universität magdeburg, Magdeburg, Germany
marlin.ulmer@ovgu.de

Abstract. We consider an integrated planning problem that combines purchasing, inventory, and inbound transportation decisions in an agri-food supply chain where several suppliers (farmers) offer a subset of products with different selling prices and available quantities. We provide a mixed-integer programming formulation of the problem and a matheuristic decomposition that divides the problem into two stages. First, the purchasing and inventory problem is solved. Second, the capacitated vehicle routing problem is solved using a split CVRP procedure. Computational experiments on a set of generated test instances show that the matheuristic can solve instances of large size within reasonably short computational times, providing better solutions than its MIP counterpart. In future work, it is proposed to develop heuristic approaches to validate the performance of the presented matheuristic and to try other routing cost approximations.

Keywords: Agri-food supply chain · Inbound transportation · Inventory routing problem · Perishable products

1 Introduction

The spread of online shopping through e-commerce platforms has disrupted not only traditional business models but also the supply chains that support them, with a growth of 7 to 10% in European countries in recent years [1]. Consumers now can access a global supply of products that can be delivered to any location within short times. In turn, producers (even small ones) can access competitive markets that were previously attainable only for large corporations with expensive infrastructure for distribution and marketing. These technology-based trade relationships have increased democratization in access to markets and provided efficiencies and convenience for both consumers and producers.

The agriculture sector has especially benefited from these technology-based business models, as small farmers can move away from intermediaries which

L. Di Gaspero et al. (Eds.): MIC 2022, LNCS 13838, pp. 436–448, 2023.
https://doi.org/10.1007/978-3-031-26504-4_31

traditionally provided distribution channels but take a large share of the revenue of the end-markets [2]. E-commerce platforms, although intermediaries as well, provide more transparent relationships with final customers, such as restaurants or hotels demanding unique characteristics from specific producers (e.g., organic, fair-trade) at competitive prices. These advantages contribute to sustainability in global commerce by closing historical gaps in competitiveness between small and large players [3].

The shift towards a market based on several small producers implies coordinating a two-echelon supply chain with a network of participants (rather than a single provider). First, there is an echelon in which products are collected from suppliers and taken to a distribution center where inventories are managed (first mile). Then, there is an echelon in which products are distributed to end customers (last mile). The design and operation of efficient supply chains is crucial to enabling more competitive markets, in contrast to traditional structures characterized by a concentration of large agricultural companies [4].

The supply chains induced by e-commerce in the agricultural sector have special features whose treatment is incipient in the literature. The joint treatment of procurement logistics and inventory management (i.e., first mile logistics) has been little studied. Often, problems assume there is a supplier that guarantees the provision of products under a direct delivery, instead of addressing the logistics of picking up products from distributed suppliers with changing prices and availability. Moreover, the fact that such procurement strategy must respond to a dynamic demand of perishable products is a challenging realistic feature that has not been considered in the literature. In practice, companies struggle to coordinate procurement strategies with inventory management of fresh agricultural products. For example, perishable food waste in 2017 reached losses of 47 billion USD per year in China and 218 billion USD in the United States [5]. Therefore, solving such integrated problem efficiently is paramount to achieve the benefits associated to e-commerce.

The objective of this research is to develop and test algorithms that can efficiently support first mile logistics decisions as part of a decision support system for agri-food supply chains in the context of e-commerce. At this stage, we present a two-stage matheuristic scheme that integrates decisions about the quantities to purchase of each products, the inventory levels to satisfy the demand of perishable products, as well as the selection of suppliers and routing of vehicles to replenish products at a warehouse. Section 2 provides an overview of the literature on this problem, while Sects. 3 and 4 detail the characteristics of the problem and the solution approach, respectively. Section 5 provides a set of computational experiments and analysis, and Sect. 6 concludes.

2   Literature Review

The majority of the research that considers integrated inventory management and routing decisions focuses on outbound routing problem, which is most commonly referred to as the Inventory Routing Problem (IRP). The most studied variant of the IRP is known as the Vendor Managed Inventory (VMI) problem, in which customers transfer the responsibility of inventory management to a vendor. The vendor knows the stock levels of their customers and must plan a distribution scheme to maintain adequate levels for all products of all customers. A general review of the IRP is presented in [6] and [7].

Few authors address the first mile problem with inbound transportation and inventory decisions, as most problems assume that ordered products simply arrive at the warehouse, disregarding the selection of suppliers and the logistics of collecting products from them. [8] and [9] consider a multiperiod, multi-supplier, many-to-one supply chain structure problem with a single assembly plant in which each supplier provides a distinct part type. In both cases, the problem is deterministic, and the solution approach is approximate optimization. In [10], a decomposition matheuristic is developed to solve an assembly, production, inventory routing problem with inbound transportation. The problem consists in selecting the suppliers to visit, their order, and the inventory level at the supplier and the plant, considering only one type of product. Later, in [11], the authors solve the same problem considering different products available by means of a branch and cut (B&C) algorithm.

In this work, suppliers are allowed to have different prices and availability for each product in each period, responding to the nature of distributed markets in the context of e-commerce. The supplier's inventory is considered an exogenous factor that cannot be managed. The work that most closely resembles such setting is presented in [12]. The authors propose a non-linear model, test its performance on a single test instance, and consider price discounts in the suppliers. There is work that considers product perishability, inventory management, and routing decisions together [13] and [14], but the authors assume direct shipment from suppliers to the warehouse and do not consider varying selling prices.

In the first mile problem proposed in this research, the company must plan the procurement logistics (i.e., which suppliers to visit, in which order, and how much to buy of each product from each supplier) based on the estimated demands from customers, the current inventory levels, and the supplier characteristics (location, as well as product prices and availability). The Multi-Vehicle Traveling Purchaser Problem (MV-TPP) addresses the portion of the stated problem regarding purchase and routing decisions (See [15]), but does not consider inventory management. MV-TPPs, can be classified according to the following four categories referring to the available supply, demand, vehicle capacity, and purchasing policy, as discussed in [16]. Table 1 presents a comparison between the different MV-TPP variants relative to this work.

The focus of this paper is on the inbound transportation corresponding a restricted, capacitated, general Multi Vehicle Traveling Purchaser Problem with non-split purchases plus inventory management of perishable products at the warehouse.

Table 1. MVTPP variants comparison

	Non-split	Non-split	Split	Split	
Unrestricted	[16, 17, 19–21, 24, 25]		Invalid		Capacited
Unrestricted	[18]		Invalid		Uncapacited
Restrcited	Invalid	**Our work**	Invalid	[19, 22]	Capacited
Restricted	Invalid	[23]	Invalid	[18]	Uncapacited
	Unitary	General	Unitary	General	

3   Problem Definition and Mathematical Formulation

The inbound inventory routing problem with perishable products (IB-IRP-PP) addressed in this work consists of a many-to-one system composed of a set of M suppliers and a single warehouse. Over the discrete periods t a planning horizon T, the warehouse satisfies a deterministic demand, d_{kt}, of the k products in set K. The products are purchased and collected from the geographically dispersed suppliers using a homogeneous fleet F of vehicles v located at the warehouse, each with capacity Q. The suppliers must be visited by only one vehicle, and the total quantities purchased in any supplier must not exceed the vehicle capacity (i.e., non-split constraints are considered). At period t, product k can be purchased from a subset of suppliers $M_{kt} \subseteq M$; each supplier i has their own selling price p_{ikt} and available quantity q_{ikt} of each product. At each period, the warehouse can purchase more than demanded of any product and store the remaining units in inventory to supply future demand. This encourages a holding cost h_{kt}. The warehouse has unlimited storage capacity. Each product has a perishable nature represented by the subset O_k that contains the periods that the product can remain in inventory before perishing. We define the problem on a complete undirected graph with nodes set $N = M \cup \{0\}$, where 0 represent the warehouse, and a set of edges $E = \{(i,j) : i,j \in N, i < j\}$.

The decisions to make are: the quantity to be purchased of each product at each supplier and each period; the quantity to maintain in inventory of each product at the end of each period; the selection of suppliers to be visited; and the order in which each vehicle visits suppliers in each period (i.e., the routes). The warehouse needs to simultaneously minimize the purchasing, inventory, and transportation costs for the entire planning horizon. It is easy to show that the (IB-IRP-PP) is NP-hard since the Multi-Vehicle Traveling Purchaser Problem (TPP) is a special case of it for each period. The problem can be formulated as the following mixed-integer program:

Variables

- I_{kto}: inventory level of product k of age o at the end of period t ($o = 0$ indicates the product is fresh, whereas $o = |O_k|$ is the latest age acceptable for product k)
- r_{kt}: quantity of product k to be replenished at period t

- y_{kto}: quantity of product k of age o to be shipped at period t
- $x_{ijtv} = \begin{cases} 1 \text{ if arc (i,j) is traversed by vehicle v at period t} \\ 0 \text{ otherwise} \end{cases}$
- $w_{itv} = \begin{cases} 1 \text{ if supplier i is visited by vehicle v at period t} \\ 0 \text{ otherwise} \end{cases}$
- z_{iktv}: quantity of product k purchased at the supplier i at period t by vehicle v.

Objective function

$$\min \sum_{t \in T} \left(\sum_{v \in F} \left(\sum_{(i,j) \in E} c_{ij} x_{ijtv} + \sum_{k \in K} \sum_{i \in M_k} p_{ikt} z_{iktv} \right) + \sum_{k \in K} \sum_{o \in O_k} h_{kt} I_{kto} \right) \quad (1)$$

Subject to

$$I_{k1o} = I_{k0o} - y_{k1o}, \qquad\qquad \forall k \in K, \forall o \in O_k | o > 0 \qquad (2)$$

$$I_{kt0} = r_{kt} - y_{kt0}, \qquad\qquad \forall k \in K, \forall t \in T \qquad (3)$$

$$I_{kto} = I_{kt-1o-1} - y_{kto}, \qquad\qquad \forall k \in K, \forall t \in T | t > 1, \forall o \in O_k | o > 0 \qquad (4)$$

$$\sum_{o \in O_k} y_{kto} = d_{kt}, \qquad\qquad \forall k \in K, \forall t \in T \qquad (5)$$

$$\sum_{v \in F} \sum_{i \in M_{kt}} z_{iktv} = r_{kt}, \qquad\qquad \forall k \in K, \forall t \in T \qquad (6)$$

$$\sum_{v \in F} z_{iktv} \leq q_{ikt}, \qquad\qquad \forall k \in K, \forall t \in T, \forall i \in M_{kt} \qquad (7)$$

$$z_{iktv} \leq q_{ikt} w_{itv}, \qquad\qquad \forall k \in K, \forall t \in T, \forall i \in M_{kt}, \forall v \in F \qquad (8)$$

$$\sum_{k \in K} \sum_{i \in M_{kt}} z_{iktv} \leq Q, \qquad\qquad \forall t \in T, \forall v \in F \qquad (9)$$

$$\sum_{v \in F} w_{itv} \leq 1, \qquad\qquad \forall t \in T, \forall i \in M \qquad (10)$$

$$\sum_{(i,j) \in \delta^+(\{m\})} x_{ijtv} = \sum_{(i,j) \in \delta^-(\{m\})} x_{ijtv} = w_{mtv}, \qquad \forall v \in F, \forall t \in T, \forall m \in M \quad (11)$$

$$u_{itv} - u_{jtv} + |N| x_{ijtv} \leq |N| - 1, \qquad \forall t \in T, \forall v \in F, \forall i \in M, \forall j \in M \qquad (12)$$

$$I_{kto} \geq 0, \qquad\qquad \forall k \in K, \forall t \in T, \forall o \in O_k \qquad (13)$$

$$r_{kt} \geq 0, \qquad\qquad \forall k \in K, \forall t \in T \qquad (14)$$

$$y_{kto} \geq 0, \qquad\qquad \forall k \in K, \forall t \in T, \forall o \in O_k \qquad (15)$$

$$x_{ijtv} \in \{0, 1\}, \qquad\qquad \forall (i, j) \in E, \forall t \in T, \forall v \in F \qquad (16)$$

$$w_{itv} \in \{0, 1\}, \qquad\qquad \forall i \in M, \forall t \in T, \forall v \in F \qquad (17)$$

$$z_{iktv} \geq 0, \qquad\qquad \forall i \in M_{k,t}, \forall k \in K, \forall t \in T, \forall v \in F \qquad (18)$$

The objective function (1) minimizes the total purchasing, inventory, and transportation costs. The holding cost is only considered in the warehouse. Initialization and inventory flow balance for the products of different ages is imposed through constraints (2)–(4). Constraint (5) guarantees demand satisfaction. Constraints (6) and (7) ensure to buy the quantity to be replenished and respect the quantities available from each supplier. Constraint (8) limits the quantity to be purchased at a supplier depending on the capacity of the vehicle that visits them. Constraints (9) and (10) limit not purchase more than vehicle capacity and the supplier just to be visited only by one vehicle. These are the non-split constraints. Constraints (11) and (12) rule the visiting tour feasibility. Equations (11) impose that, for each visited supplier, exactly one arc must enter and leave the relative node, where, for any subset N' of nodes, $\delta^+(N') := \{(i, j) \in E : i \in V', j \notin V'\}$ and $\delta^-(N') := \{(i, j) \in E : i \notin V', j \in V'\}$. Inequalities (12) are connectivity constraints that prevent the creation of sub-tours by controlling the order of visits of the suppliers. Miller-Tucker-Zemlin (MTZ) formulation is used [26]. The constraints (13)–(18) correspond to the domain of the variables.

4 A Two-Stage Matheuristic Decomposition

Algorithm 1. Two-stage matheuristic decomposition

Output: Solution *incumbent*;

1: *Initialize* $\leftarrow \hat{c}_{it}$
2: **while** termination condition not satisfied **do**
3:    $z_{ikt}, I_{kto}, w_{it} \leftarrow SolvePurchaseAndInventory(T, M, K, O_k, p_{ikt}, q_{ikt}, h_{kt}, \hat{c}_{it}; Q)$
4:    $R \leftarrow GenerateRoutes(T, M, c_{ij}, z_{ikt}, w_{it}; Q)$
5:    $CurrentSolution \leftarrow assembleSolution(R, z_{ikt}, w_{it}, I_{kto})$
6:    $\hat{c}_{it} \leftarrow updatedRoutingEstimation(Routes, c_{ij})$
7:    **Update** *Incumbent* **if** *CurrentSolution* **is better**
8: **end while**
9: **return** *Incumbent*

In this section, we present a two-stage matheuristic decomposition for the IB-IRP-PP. Our algorithm decomposes the problem into two separate subproblems. The first sub-problem determines for each period the inventory levels and the quantity of each product to be purchased from each supplier, taking into account the perishability of the products. An approximate transportation cost $(\hat{c}_{it})$ is used to estimate the actual cost of visiting supplier i at period t. This approximation is made as routing decisions are not considered at this stage. The second subproblem solves, for each period t, a separate Capacitated Vehicle Routing Problem (CVRP) using the purchasing decisions found in the first stage. The solutions to the routing subproblems are then used to update the approximate transportation cost $(\hat{c}_{it})$ of the first stage to obtain different purchase and inventory levels in the next iteration. This procedure is repeated for a number of iterations to reach a local optimum or until a stopping condition is met. Algorithm 1 presents an overview of the matheuristic.

4.1   Stage 1 - Solving Purchasing and Inventory Decisions

The first subproblem aims to generate a good replenishment and inventory plan by solving a simplified problem in which we use an approximate transportation cost based on the estimation of the actual cost of visiting supplier i at period t. The objective function presented in (1) is reformulated as follows:

$$\min \sum_{t \in T} \left(\sum_{i \in M} \hat{c}_{it} w_{it} + \sum_{k \in K} \sum_{i \in M_k} p_{ikt} z_{ikt} + \sum_{k \in K} \sum_{o \in O_k} h_{kt} I_{kto} \right) \quad (19)$$

We define the first stage model with the objective function (19) subject to constraints (2)–(9), omitting the index $v \in F$ corresponding to the fleet of vehicles. Solving this model (line 3 - Algorithm 1) results in a (sub-optimal) purchasing (z_{ikt}) and inventory (I_{kto}) plan that respects perishability.

4.2   Stage 2 - Routing Decisions

The second stage solves, for each period t, a Capacitated Vehicle Routing Problem using the purchasing decisions found in the first stage (line 4 - Algorithm 1). The routing procedure is presented in Algorithm 2. We fix the values of variables $\bar{w}_{it}$, $\bar{z}_{ikt}$. First, with the values of $\bar{w}_{it}$, a Nearest Neighbour Algorithm is run to obtain the order in which selected suppliers will be visited (line 2 - Algorithm 2) as presented in [27]. Then, with this general tour and the quantities to be purchased at each supplier, $\bar{z}_{ikt}$, a split C-VRP procedure is developed to obtain the vehicle routes that respect vehicle capacities [28]. The augmented graph is built (line 3 - Algorithm 2) and the shortest path problem is solved using the Bellman-Ford algorithm [29] (line 4 - Algorithm 2). The solution is assembled (line 6 - Algorithm 1) with the routes and the values of z_{ikt}, w_{it} and I_{kto}, and if the solution is better than the incumbent, it is updated (line 7 - Algorithm 2).

4.3   Connection Between Stages

The information flow between the two stages is through parameter $\hat{c}_{it}$, which must be updated at each iteration. At iteration 0 ($iter = 0$), in line 1 (Algorithm 1) this parameter is initialized with the direct shipping cost (i.e., $\hat{c}_{it}^{\,iter} = c_{0i} + c_{i0}, \forall i \in M, \forall t \in T$). At the end of each iteration, the cost $\hat{c}_{it}^{\,iter}$ is updated after vehicles' routes have been obtained for each period t (line 7). There are two ways of updating this parameter. First, if supplier i is part of a route at period t, the cost of visiting them in the next iteration ($iter = iter + 1$) will be $\hat{c}_{it}^{\,iter} = (\hat{c}_{it}^{\,iter-1} + c_{i_p i} + c_{i i_s} - c_{i_p i_s})/2$, where i_p and i_s are the predecessor and successor nodes of supplier i in their current route in that period, respectively. Second, if node i is not visited in any of the routes, then we set $\hat{c}_{it}^{\,iter} = (\hat{c}_{it}^{\,iter-1} + c_{insertion})/2$, where $c_{insertion}$ is equal to the cost of the cheapest insertion into an existing route in that period. This is based on the assumption that when a supplier i is eliminated from their route, an acceptable route can be obtained by connecting the predecessor and successor suppliers. Similarly, when inserting supplier i, an acceptable route can be obtained with the best insertion among all the routes in a specific period. The two stages are executed until the stopping criterion is reached (line 2).

Algorithm 2. GenerateRoutes

Input: List T, M, **List** c_{ij} : transportation cost between nodes i and j, $z_{ikt}, w_{it}; Q$
Output: Routes R : Routes of period t;
  1: **for** $t = 1$ **To** $|T|$ **do**
  2:     $Tour \leftarrow NearestNeighbourAlgorithm(M, w_{it}, c_{ij})$
  3:     $Graph \leftarrow GenerateAugmentedGraphCVRP(M, w_{it}, c_{ij}, Tour, Q)$
  4:     $R_t \leftarrow BellmanFordShortestPathAlgorithm(Graph)$
  5: **end for**
  6: **return** R

---

5   Computational Experiments

The MIP and the decomposition matheuristic were implemented in Python 3.7, with Gurobi 9.1.1 as a solver for exact models. All computational experiments were performed on a 2.11 GHz processor with 16 GB of RAM.

5.1   Data Sets

We built a data set of 240 instances, taking into account the inventory characteristics of [8] and supplier characteristics of [15]. The number of suppliers, products, and periods were set as $M \in \{10, 25, 50, 100, 150\}$, $K \in \{10, 25, 50, 100\}$ and $T \in \{5, 10, 21\}$. The supplier locations were generated in a $[0, 1000] \times [0, 1000]$ square according to a uniform distribution and routing costs c_{ij} as truncated Euclidean distances. Each product k at period t is associated with $|M_{kt}|$ randomly selected suppliers, where $|M_{kt}|$ is uniformly generated number in $[1, |N| - 1]$.

Parameter q_{ikt} of offered quantities is randomly taken in $[1, 15]$. Parameter λ is used to control the number of suppliers in a feasible solution through the product demand $d_{kt} := [\lambda \max_{i \in M_{kt}} q_{ikt} + (1 - \lambda) \sum_{i \in M_{kt}} q_{ikt}], \forall k \in K, \forall t \in T,$ with $0 < \lambda < 1$. The lower the value of λ, the higher the number of suppliers in a solution. This parameter was set as $\lambda \in \{0.5, 0.9\}$. The selling price p_{ikt}, and the holding cost h_{kt} were uniformly generated in $[1, 500]$. The latest age acceptable for product k was uniformly generated in $[1, |T|]$. To find a feasible vehicle capacity Q, we solve a model with objective function min $z = Q$ subject to (2)–(10), omitting the index $v \in F$ corresponding to the fleet of vehicles. The result of this model is a feasible capacity value, which is multiplied by 1.2 and rounded up to avoid a hard constraint. The number of vehicles $v \in F$ in the fleet is obtained by $|F| = \lceil \sum_{t \in T} \sum_{k \in K} d_{k,t}/Q \rceil$. Finally, two replicates were generated for each combination of M, K, T y λ.

5.2   Stopping Condition

In order to measure the impact of the stopping criterion on the quality of solutions and computational time, we execute 500 iterations of the matheuristic procedure (line 2 - Algorithm 1). In 83% of the instances, the incumbent is updated no more than 50 consecutive iterations. We proceed to evaluate the stopping criterion as the maximum number of consecutive iterations without the incumbent solution being improved. We evaluated three values for the stopping criterion (10, 25 and 50 iterations without improvement). Figure 1 shows the computational time for the matheuristic for the proposed values. Larger values of the stopping criterion lead to larger times on average and higher variability.

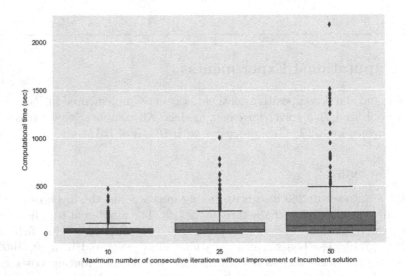

Fig. 1. Computational time for different stopping criteria

As expected, the quality of solutions increases as the stopping criterion increases, with the value of 50 maximum iterations without improvement achieving the best results for all instances. However, a value of 25 in the stopping criterion achieves the best solution for an 88% of the instances, while using only a 49% of the computational time (on average). Similarly, the value of 10 for the stopping criterion achieves the best solutions for 77% of the instances, while taking 23% of the computational time (on average). We adopt the stopping criterion of 25 iterations without incumbent improvement in order to balance computational time and quality of solutions.

5.3 Results

A time-limit of 3600 s was set to test the performance of the MIP with respect to the matheuristic with the selected stopping criteria. Table 2 presents the results obtained by the MIP and the proposed matheuristic decomposition. Instances that do not appear in the table cannot be compared because the model did not find an integer solution within 3600 s or the computer mem-

Table 2. Comparison MIP and Matheuristic

M	K	T	λ	ID	MIP				Matheuristic (MH)		
					OF	Time (s)	BestBound	$Gap_{B\&B}$	OF	Time (s)	Δ_{MIP-MH}
10	10	5	0.5	1	179999	3601	174089.73	3.28	180835	0.89	0.46
10	10	5	0.5	2	173967	3603	167315.63	3.82	174396	0.83	0.25
10	10	5	0.9	1	152601	3603	148486.29	2.7	153044	0.91	0.29
10	10	5	0.9	2	168418	3602	164659.47	2.23	170265	0.95	1.1
10	10	10	0.5	1	351791	3606	310593.85	11.71	350653	2.17	−0.32
10	10	10	0.5	2	334332	3605	301934.26	9.69	336872	2.33	0.76
10	10	10	0.9	1	287696	3605	264497.63	8.06	286580	2.95	−0.39
10	10	10	0.9	2	255647	3604	248834.19	2.66	257094	1.39	0.57
10	25	5	0.5	1	336682	3603	328169.4	2.53	341168	1.13	1.33
10	25	5	0.5	2	338057	3602	321125.79	5.01	341580	1.22	1.04
10	25	5	0.9	1	290924	3602	282531.74	2.88	293026	1.2	0.72
10	25	5	0.9	2	281152	3602	266002.45	5.39	284913	0.95	1.34
10	25	10	0.5	1	651117	3606	598765.96	8.04	655208	3.11	0.63
10	25	10	0.5	2	699029	3607	654945.61	6.31	707306	2.97	1.18
10	25	10	0.9	1	613872	3608	575706.69	6.22	615916	2.77	0.33
10	25	10	0.9	2	646987	3606	612270.47	5.37	650546	3.16	0.55
25	10	5	0.5	1	313066	3617	167515.35	46.49	318562	3.25	1.55
25	10	5	0.5	2	285319	3617	215350.49	24.52	286671	6.76	0.47
25	10	5	0.9	1	155489	3611	125146.56	19.51	159081	4.33	2.31
25	10	5	0.9	2	215384	3617	103543.67	51.93	214455	7	−0.43
25	10	10	0.5	1	616183	3669	375955.31	38.99	569111	12.54	−7.64
25	10	10	0.9	1	333216	3701	189534.13	43.12	329694	11.91	−1.06
25	10	10	0.9	2	320842	3649	219608.05	31.55	321824	10.12	0.31
25	25	5	0.5	1	548035	3620	463297.97	15.46	553392	4.05	0.98
25	25	5	0.5	2	555895	3619	429518	22.73	564814	3.95	1.6
25	25	5	0.9	1	373487	3622	242635.23	35.04	370630	5.36	−0.76
25	25	5	0.9	2	343751	3621	236583.57	31.18	338638	10.01	−1.49

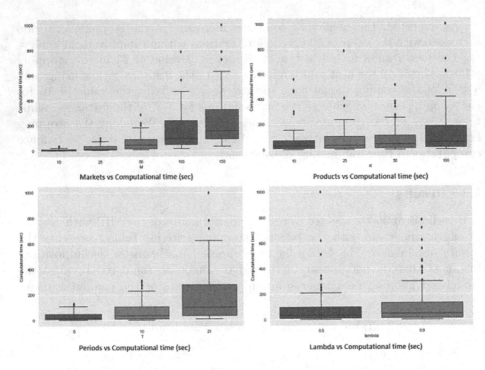

Fig. 2. Computational time depending on the increase in problem size

ory was not sufficient. As the MIP model could not find an optimum solution in 3600 s for any instances, the results presented in the MIP columns are the best integer solution found. Also, the *Bestbound* and the gap calculated by Gurobi ($GAP_{B\&B}$) are reported. The column Δ_{MIP-MH} is the percentage difference between the solutions obtained for both approaches, and is calculated as $\Delta_{MIP-MH} = ((BKS_{MIP} - BKS_{MH})/BKS_{MIP}) * 100$.

Table 2 shows that when the size of the instance increases, the $GAP_{B\&B}$ of the model also increases due to the limited capacity of the exact model to solve large instances. On the other hand, when the size of the instance increases, the proposed matheuristic finds better solutions than the MIP with significantly lower CPU time required. These results provide a preliminary confirmation about the potential of the proposed methodology. Currently, adjustments on the proposed strategies and more comprehensive experiments are being developed in order to determine, with better statistical significance, which variations of instances and strategies lead to better results.

Figure 2 shows the increase in computational time depending on the increase in problem size. The largest increase in time occurs as the number of periods increases, followed by the increase in the number of suppliers.

6   Conclusions

We presented a matheuristic decomposition approach for the inbound inventory routing problem with perishable products, responding to challenges of the agri-food supply chain in the context of e-commerce. The key contribution is the integration of inventory decisions to satisfy a demand of perishable products with procurement decisions, including a selection of supplier for each product (with varying locations and product availability and price) and routing decisions. As a first step, the proposed matheuristic approach obtains good quality solutions within reasonable computation times, given the limitations of the exact model. Current results provide an initial confirmation on the potential of the proposed approach. However, ongoing work is devoted to the execution of a more comprehensive set of computational experiments that allow to reach more solid conclusions regarding the variants of the approach that are better suited for each type of instance. The next step of the research is the incorporation of stochasticity in the demand and product prices and availability at the supplier, as well as the possibility to update decisions dynamically, as usually allowed in e-commerce platforms.

References

1. Viu-Roig, M., Alvarez-Palau, E.J.: The impact of E-Commerce-related last-mile logistics on cities: a systematic literature review. Sustainability **12**(16), 6492 (2020)
2. Gu, W., Archetti, C., Cattaruzza, D., Ogier, M., Semet, F., Speranza, M.G.: A sequential approach for a multi-commodity two-echelon distribution problem. Comput. Industr. Eng. **163**, 107793 (2022)
3. Prajapati, D., Chan, F.T., Daultani, Y., Pratap, S.: Sustainable vehicle routing of agro-food grains in the e-commerce industry. Int. J. Prod. Res. **60**, 7319–7344 (2022)
4. Majluf-Manzur, Á.M., González-Ramirez, R.G., Velasco-Paredes, R.A., Villalobos, J.R.: An operational planning model to support first mile logistics for small fresh-produce growers. In: Rossit, D.A., Tohmé, F., Mejía Delgadillo, G. (eds.) ICPR-Americas 2020. CCIS, vol. 1408, pp. 205–219. Springer, Cham (2021). https://doi.org/10.1007/978-3-030-76310-7_17
5. Ji, Y., Du, J., Han, X., Wu, X., Huang, R., Wang, S., Liu, Z.: A mixed integer robust programming model for two-echelon inventory routing problem of perishable products. Phys. A **548**, 124481 (2020)
6. Andersson, H., Hoff, A., Christiansen, M., Hasle, G., Løkketangen, A.: Industrial aspects and literature survey: combined inventory management and routing. Comput. Oper. Res. **37**(9), 1515–1536 (2010)
7. Coelho, L.C., Cordeau, J.F., Laporte, G.: Thirty years of inventory routing. Transp. Sci. **48**(1), 1–19 (2014)
8. Moin, N.H., Salhi, S., Aziz, N.A.B.: An efficient hybrid genetic algorithm for the multi-product multi-period inventory routing problem. Int. J. Prod. Econ. **133**(1), 334–343 (2011)
9. Mjirda, A., Jarboui, B., Macedo, R., Hanafi, S., Mladenović, N.: A two phase variable neighborhood search for the multi-product inventory routing problem. Comput. Oper. Res. **52**, 291–299 (2014)

10. Chitsaz, M., Cordeau, J.F., Jans, R.: A unified decomposition matheuristic for assembly, production, and inventory routing. INFORMS J. Comput. **31**(1), 134–152 (2019)
11. Chitsaz, M., Cordeau, J.F., Jans, R.: A branch-and-cut algorithm for an assembly routing problem. Eur. J. Oper. Res. **282**(3), 896–910 (2020)
12. Çabuk, S., Erol, R.: Modeling and analysis of multiple-supplier selection problem with price discounts and routing decisions. Appl. Sci. **9**(17), 3480 (2019)
13. Rohmer, S.U.K., Claassen, G.D.H., Laporte, G.: A two-echelon inventory routing problem for perishable products. Comput. Oper. Res. **107**, 156–172 (2019)
14. Wei, M., Guan, H., Liu, Y., Gao, B., Zhang, C.: Production, replenishment and inventory policies for perishable products in a two-echelon distribution network. Sustainability **12**(11), 4735 (2020)
15. Manerba, D., Mansini, R., Riera-Ledesma, J.: The traveling purchaser problem and its variants. Eur. J. Oper. Res. **259**(1), 1–18 (2017)
16. Bianchessi, N., Irnich, S., Tilk, C.: A branch-price-and-cut algorithm for the capacitated multiple vehicle traveling purchaser problem with unitary demand. Discret. Appl. Math. **288**, 152–170 (2021)
17. Baldacci, R., Dell'Amico, M., González, J.S.: The capacitated m-ring-star problem. Oper. Res. **55**(6), 1147–1162 (2007)
18. Bianchessi, N., Mansini, R., Speranza, M.G.: The distance constrained multiple vehicle traveling purchaser problem. Eur. J. Oper. Res. **235**(1), 73–87 (2014)
19. Choi, M.J., Lee, S.H.: The multiple traveling purchaser problem. In: Proceedings of the Fortieth International Conference on Computers and Industrial Engineering (CIE), pp. 1–5 (2010)
20. Gendreau, M., Manerba, D., Mansini, R.: The multi-vehicle traveling purchaser problem with pairwise incompatibility constraints and unitary demands: a branch-and-price approach. Eur. J. Oper. Res. **248**(1), 59–71 (2016)
21. Hoshino, E.A., De Souza, C.C.: A branch-and-cut-and-price approach for the capacitated m-ring-star problem. Discret. Appl. Math. **160**(18), 2728–2741 (2012)
22. Manerba, D., Mansini, R.: A branch-and-cut algorithm for the multi-vehicle traveling purchaser problem with pairwise incompatibility constraints. Networks **65**(2), 139–154 (2015)
23. Manerba, D., Mansini, R.: The nurse routing problem with workload constraints and incompatible services. IFAC-PapersOnLine **49**(12), 1192–1197 (2016)
24. Riera-Ledesma, J., Salazar-González, J.J.: Solving school bus routing using the multiple vehicle traveling purchaser problem: a branch-and-cut approach. Comput. Oper. Res. **39**(2), 391–404 (2012)
25. Riera-Ledesma, J., Salazar-González, J.J.: A column generation approach for a school bus routing problem with resource constraints. Comput. Oper. Res. **40**(2), 566–583 (2013)
26. Miller, C.E., Tucker, A.W., Zemlin, R.A.: Integer programming formulation of traveling salesman problems. J. ACM (JACM) **7**(4), 326–329 (1960)
27. Cuellar-Usaquén, D., Gomez, C., Álvarez-Martínez, D.: A GRASP/Path-Relinking algorithm for the traveling purchaser problem. Int. Trans. Oper. Res. **30**(2), 831–857 (2021)
28. Prins, C.: A simple and effective evolutionary algorithm for the vehicle routing problem. Comput. Oper. Res. **31**(12), 1985–2002 (2004)
29. Magzhan, K., Jani, H.M.: A review and evaluations of shortest path algorithms. Int. J. Sci. Technol. Res. **2**(6), 99–104 (2013)

A Multi-objective Biased Random-Key Genetic Algorithm for the Siting of Emergency Vehicles

Francesca Da Ros[1,2]([✉]), Luca Di Gaspero[1], David La Barbera[3],
Vincenzo Della Mea[3], Kevin Roitero[3], Laura Deroma[4], Sabrina Licata[2],
and Francesca Valent[5]

[1] Dipartimento Politecnico di Ingegneria e Architettura, Università di Udine, via delle Scienze 208, 33100 Udine, Italy
daros.francesca001@spes.uniud.it, luca.digaspero@uniud.it

[2] SOC Istituto di Igiene ed Epidemiologia Clinica, Azienda Sanitaria Universitaria Friuli Centrale, p.zzale SM della Misericordia 15, 33100 Udine, Italy
sabrina.licata@uniud.it

[3] Dipartimento di Scienze Matematiche, Informatiche e Fisiche, Università di Udine, via delle Scienze 208, 33100 Udine, Italy
labarbera.david@spes.uniud.it, {vincenzo.dellamea,kevin.roitero}@uniud.it

[4] SOC Igiene e Sanità Pubblica, Azienda Sanitaria Universitaria Friuli Centrale, p.zzale SM della Misericordia 15, 33100 Udine, Italy
laura.deroma@asufc.sanita.fvg.it

[5] Servizio Epidemiologia Clinica Valutativa, Azienda Provinciale per i Servizi Sanitari di Trento, via Alcide Degasperi 79, 38123 Trento, Italy
francesca.valent@apss.tn.it

Abstract. We propose the development and application of a multi-objective biased random-key genetic algorithm to identify sets of ambulance locations in a rural-mountainous area. The algorithm involves a discrete event simulator to estimate the objective functions, thus we want to minimize the response time while maximizing the area served within the standard time. It is applied to the case of the mountainous area of the Italian region of Friuli Venezia Giulia. Preliminary results are encouraging, as the best case for each objective shows that the average response time decreases of 28.9%, the 90th percentile for the response time decreases of 43.0%, the number of municipalities served within the standard time increases of 8 units during the day, and of 26 units during the night.

Keywords: BRKGA · Multi-objective optimization · Emergency medical service

1 Introduction

In this paper, we address the location of emergency vehicles (i.e., ambulances) in a sparsely populated rural-mountainous area. Emergency medical service (EMS)

Supported by EASY-NET project (NET 2016-02364191).

includes ambulance, paramedic, and pre-hospital services. Its primary goal is to provide timely first medical aid to patients involved in emergency situations. The response time indicates the time incurred from when the emergency call is received to the moment in which an ambulance arrives at the emergency location [8]. It depends on several factors including ambulance location, dispatching policies, busy fractions (i.e., the closest ambulance is already serving another emergency), etc. To achieve high-performing services, accurate EMS planning is strategically, tactically, and operationally crucial: improving EMS enables a higher probability of patient survival [20], therefore these systems should be designed to minimize response times and maximize the number of emergency calls served within a given time limit [21]. Several authors discuss the association between EMS and patient priority: as the response time should always be the lowest possible, priority queues should be in place as real case scenarios are characterized by resource limitations, payoff of the treatments, and duration of the service [10]. Ball et al. [2] measured the association between ambulance dispatch priorities and patient conditions, highlighting different time standards based on different priority levels. Fargetta et al. [5] considered patient selection for treatment examining their priorities on three levels.

The ambulance location problem has been tackled by many researchers during the last four decades, therefore several models have been proposed. Extensive reviews on the evolution of the modeling approaches are offered in [1,12,14,16]. However, the majority of the solutions focus on highly populated urban areas and are unsuitable for rural zones as they tend to consider demographic inputs (e.g., inhabitants, density, etc.). In fact, maximizing the covered demand in this sense may result in low-density areas remaining uncovered [16], leading to an unfair EMS, as part of the population will systematically receive inefficient treatment (i.e., the standard service time is not respected).

This work deals with the location of ambulance stations in a sparsely populated rural-mountainous area. The goal of our study is to understand if a new geographical distribution of the existing ambulances, that considers both the covered area and the past emergency calls, may result in reducing inequities in accessing emergency services (i.e., emergencies in the entire area should be efficiently served). We propose the development and application of a multi-objective biased random-key genetic algorithm (BRKGA) that identifies sets of ambulance locations. The algorithm involves a discrete event simulator (DES) to estimate the objective functions, thus it tries to minimize the response time while maximizing the area served within the standard time (i.e., 18 min in the current practice). To achieve the objective, we aim to answer the following research questions: *(i)* does a more geographically sparse distribution of ambulance sites account for a more efficient service in a rural-mountainous area? If so, to what extent? *(ii)* To what extent does an optimization-simulation loop improve the results of the ambulance locations? The current study contributes to the literature on the ambulance location problem as follows: *(i)* to the best of our knowledge, BRKGA has never been applied to this topic; *(ii)* it extends the literature on the topic related to rural areas. The rest of this paper is organized as follows. Section 2 characterizes the case study providing a solid description of the study area; it further argues on the formulation of the BRKGA, mainly examining the encoding and decoding system. Section 3 exposes and discusses the preliminary results, whereas Sect. 4 reports some conclusions.

2 Methods

2.1 Study Area

The area of study is the mountainous area (i.e., municipalities with an altitude equal to or greater than 450 m above mean sea level) of the Italian region Friuli Venezia Giulia (FVG). This area extends over 4,322 km^2 (54.49% of FVG area), counts 83 municipalities (39% of FVG municipalities), and has approximately 167,500 inhabitants (14% of FVG population) [9]. Thus, despite being geographically extended, the area is scarcely populated. We focus on data collected in the context of the Emergency Health Service and Medical Emergency of the FVG region in the time interval from January 2018 to December 2020. Emergencies in the mountainous area occurred as follows: in 2018 11.5% of the FVG emergencies occurred in the interested zone, in 2019 11.69%, and in 2020 12.39%. A preliminary analysis of this data (not reported for space constraints) has highlighted that emergencies occur sparsely during the week (i.e., it is likely to have long periods with no emergencies). Furthermore, an increase in the number of emergencies *per capita* is registered in the months from December to March, which are characterized by a high presence of tourists [18]. The FVG EMS consists of: *(i)* a regional dispatch center, that receives and evaluates emergency calls, and consequently dispatches the most appropriate vehicle; *(ii)* the rescue units that provide first medical aid to patients involved in emergencies. Particularly, the Italian system works in a Franco-German style, meaning that the rescue unit is allowed to treat the patient directly on the emergency site. FVG emergency vehicle fleet includes advanced life support (ALS) and basic life support (BLS) ambulances, medical cars, and medical helicopters. Additionally, each vehicle is categorized by its shift, namely a vehicle can operate during a daily shift (8:00–20:00), a night shift (20:00–8:00), or a 24 h shift that includes both. We do not consider medical helicopters, therefore, the emergencies served by them are not taken into account in the investigation. Providing a location for the medical cars is out of the scope of this paper since they need to operate in presence of a doctor and are located at hospital sites [6]. However, their usage is accounted for in the simulations. Particularly, they are considered as per their current location and shift.

2.2 Biased Random-Key Genetic Algorithm

The biased random-key genetic algorithm (BRKGA) was firstly proposed in [7] and is based on the random-key genetic algorithm (RKGA) introduced in [3]. In such algorithms, chromosomes are presented as vectors of random real numbers within the interval $[0, 1)$. Eventually, thanks to a decoder, chromosomes are transformed into solutions for the optimization problem at hand. The initial population consists of p chromosomes of random value within the interval $[0, 1)$. The population of random keys evolves a certain number of times (generations). At each generation, the decoder calculates the fitness of all individuals. Based on this value, individuals are divided into two groups: those who have the best fitness (i.e., elite population) (p_e) and the remaining ones. The following generation is made up as follows: *(i)* all the elite population is copied into the new generation, *(ii)* a given number of chromosomes is randomly generated (i.e., mutants) (p_m), *(iii)* a given number of chromosomes is produced as the result of the mating of the previous generation (Fig. 1).

BRKGA is said to be biased because of the way parents are selected for mating: one parent is selected at random from the elite population and one from the remaining population. Furthermore, the offspring inherits the vector component of the elite parent with a probability $\rho_e \geq 0.5$. Consequently, the evolution between generations highlights that the only problem-dependent part of the BRKGA is how the decoding of the chromosomes into a proper solution is made (Fig. 2).

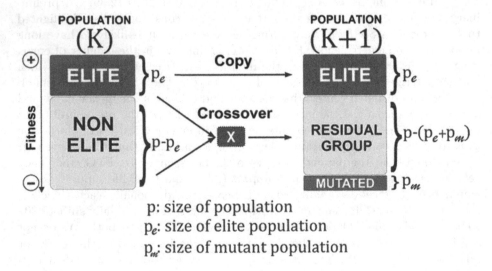

p: size of population
p_e: size of elite population
p_m: size of mutant population

Fig. 1. Generation of a new population in BRKGA [11]

Encoding to a Vector of Random Keys. A solution is encoded as a vector $V = (v_0, ..., v_{n-1})$ of size $n = \sum_{i \in \{0,1\}} \sum_{j \in \{0,1,2\}} N_{ij}$ where v_x is a random number in the interval $[0, 1)$ and N_{ij} indicates the number of ambulances of type i working toward the shift j. In particular, $i = 0$ indicates an ALS ambulance, and $i = 1$ indicates a BLS ambulance, while $j = 0$ indicates the daily shift, $j = 1$ indicates the night shift, and $j = 2$ indicates the 24 h shift.

Decoding from a Vector of Random Keys. The decoding process of a chromosome is based on the fact that each gene represents the municipality where an ambulance of a specific type operating in a specific shift is located. Specifically, given m municipalities ranging from 0 to $m-1$, municipality s is decoded when the random key r respects Eq. (1):

$$\frac{s}{m} \leq r < \frac{s+1}{m} \tag{1}$$

A set of constraints checks if a decoded solution is feasible or not; in particular, a municipality cannot host more than one ambulance during the 8:00–20:00 time slot (this includes ambulances working toward the daily shift and the 24 h shift) and during the 20:00–8:00 time slot (this includes ambulances working toward the night shift and the 24 h shift).

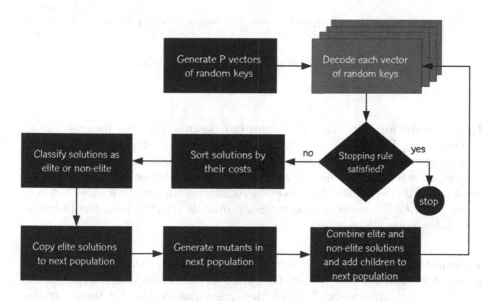

Fig. 2. Flowchart of a BRKGA [7]

Figure 3 provides a complete encoding and decoding example of a feasible solution that accounts for seven municipalities ($m = 7$), and considers a total of 9 ambulances, organized as $N_{00} = 3$, $N_{10} = 1$, $N_{01} = 2$, $N_{11} = 2$, $N_{02} = 1$, and $N_{12} = 0$. For instance, $v_0 = 0.100$ respects Eq. 1 when municipality 0 ($s = 0$) is picked. Eventually, no municipality hosts more than one ambulance per time slot; in fact, during the 8:00–20:00 time slot, ambulances are located in municipalities 0, 2, 3, 4, and 6, whereas during the 20:00–8:00 time slot, ambulances are located in municipalities 1, 2, 3, 4, and 6.

To calculate the fitness of a solution, a simulation is run with the ambulance locations proposed by the decoder and considering 10 weeks randomly selected from the historical data. The objective functions consider the following aspects:

(i) minimizing the average response time, (ii) minimizing the 90th percentile for response time, (iii) maximizing the number of municipalities covered within the standard time during the day, (iv) maximizing the number of municipalities covered within the standard time during the night. Considering the last two objectives, the BRKGA acts as a multi-objective Maximal Covering Location Problem with unitary demands [21].

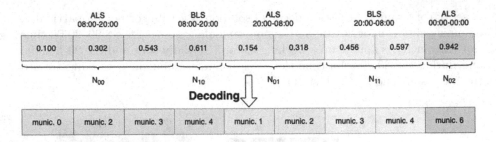

Fig. 3. Example of decoding of a feasible solution.

Implementation Details. The algorithm described in this paper is implemented using Pymoo [4], a Python-based optimization framework. The DES is implemented in C++ language and makes use of osrm-backend [15], a routing engine written in C++ designed to run on OpenStreetMap data [17], and SimCpp20 [19], which is a discrete-event simulation framework for C++. A simulation run for 10 weeks of emergencies randomly selected from the historical data takes approximately 10 s.

Parameter Setting. The problem-independent part of the BRKGA involves parameters that need to be set. Gonçalves et al. [7] suggested experimental ranges to determine them. To tune the appropriate values for these parameters, we used SMAC3 [13], a tool for algorithm configuration to optimize the parameters of arbitrary algorithms. The best configuration found is as follows: 257 generations, a population (p) of 107 individuals, an elite population (p_e) of $0.13p$, a mutant population (p_m) of $0.11p$, and a probability for the offspring to inherit the allele of the elite parent (ρ_e) of 0.74.

3 Preliminary Results

Comparisons are made with a simulation run on the current stations. The BRKGA considers the same number of ambulances of the current situation and preserves their types and working shifts. Therefore, it locates 16 ambulances considering 83 municipalities. A municipality is covered if it can be served within 18 min. Preliminary results (cf. Table 1) show that the solutions provided by the

BRKGA outperform the current one. The sparse distribution of the ambulances is guaranteed by the constraints, as a solution is feasible only if during the same shift a maximum of one ambulance is located in a municipality. As the current distribution includes more than one ambulance per municipality, the results suggest that a sparse location helps in reducing the response times. Furthermore, the 90th percentile for the response time can decrease of 43.0%, which suggests a substantial improvement for the worst cases.

Table 1. Comparison between a set of solutions provided by the BRKGA and the current distribution.

Indicator	Current situation	BRKGA (1)	BRKGA (2)	BRKGA (3)
Municipalities covered during the day (#)	71	79	76	73
Municipalities covered during the night (#)	43	69	68	64
Average response time (s)	788.64	650.83	619.81	560.85
90th percentile for response time (s)	1,905.00	1,175.00	1,085.90	1,117.1

4   Conclusion

This work addresses the location of ambulances in a sparsely populated rural-mountainous area using a multi-objective BRKGA alongside a DES. The case study is performed on the Italian region of FVG. Preliminary results show that using a meta-heuristic approach together with simulation decreases the response time indicators as well as maximizes the covered zone. Further research involves the following areas: *(i)* BRKGA further specification: additional objectives that consider an estimated survival rate, and different patient priorities, the consideration of a more precise decoder, and the use of a local search to enhance the BRKGA results. *(ii)* DES further specification: the consideration of different emergency scenarios (e.g., disasters), the addition of details so that it better adheres to reality. *(iii)* Additional testing: consideration for the entire region setting (i.e., urban, rural, and maritime areas).

References

1. Aringhieri, R., Bruni, M.E., Khodaparasti, S., van Essen, J.T.: Emergency medical services and beyond: addressing new challenges through a wide literature review. Comput. Oper. Res. **78**, 349–368 (2017)
2. Ball, S.J., et al.: Association between ambulance dispatch priority and patient condition. Emerg. Med. Australas. **28**, 716–724 (2016)
3. Bean, J.C.: Genetic algorithms and random keys for sequencing and optimization. ORSA J. Comput. **6**(2), 154–160 (1994). https://doi.org/10.1287/ijoc.6.2.154
4. Blank, J., Deb, K.: Pymoo: Multi-objective optimization in Python. IEEE Access **8**, 89497–89509 (2020)

5. Fargetta, G., Scrimali, L.: A multi-stage integer linear programming problem for personnel and patient scheduling for a therapy centre. In: Proceedings of the 11th International Conference on Operations Research and Enterprise Systems - ICORES, pp. 354–361 (2022). https://doi.org/10.5220/0010902500003117

6. Giunta Regionale Regione Autonoma Friuli Venezia Giulia: Allegato a DGR FVG n. 2039 del 16 ottobre 2015 'LR 17/2014, Art. 37 - Piano dell'emergenza urgenza della regione Friuli Venezia Giulia: approvazione definitiva' (2015)

7. Gonçalves, J., Resende, M.: Biased random-key genetic algorithms for combinatorial optimization. J. Heuristics **17**, 487–525 (2011). https://doi.org/10.1007/s10732-010-9143-1

8. Hammami, S., Jebali, A.: Designing modular capacitated emergency medical service using information on ambulance trip. Oper. Res. Int. J. **21**(3), 1723–1742 (2019). https://doi.org/10.1007/s12351-019-00458-4

9. Istituto Nazionale di Statistica (ISTAT): Confini delle unità amministrative a fini statistici al 1° gennaio 2022 (2022). https://www.istat.it/it/archivio/222527

10. Jacobson, E.U., Argon, N.T., Ziya, S.: Priority assignment in emergency response. Oper. Res. **60**(4), 813–832 (2012)

11. Júnior, B., Costa, R., Pinheiro, P., Luiz, J., Araújo, L., Grichshenko, A.: A biased random-key genetic algorithm using dotted board model for solving two-dimensional irregular strip packing problems. In: 2020 IEEE Congress on Evolutionary Computation (CEC) (2020). https://doi.org/10.1109/CEC48606.2020.9185794

12. Li, X., Zhao, Z., Zhu, X., Wyatt, T.: Covering models and optimization techniques for emergency response facility location and planning: a review. Math. Methods Oper. Res. **74**, 281–310 (2011)

13. Lindauer, M., et al.: SMAC3: A versatile Bayesian optimization package for hyperparameter optimization (2021)

14. Liu, Y., Yuan, Y., Shen, J., Gao, W.: Emergency response facility location in transportation networks: a literature review. J. Traffic Transp. Eng. (English Edition) **8**, 153–169 (2021)

15. Luxen, D., Vetter, C.: Real-time routing with OpenStreetMap data. In: Proceedings of the 19th ACM SIGSPATIAL International Conference on Advances in Geographic Information Systems, GIS 2011, pp. 513–516. ACM, New York (2011). https://doi.org/10.1145/2093973.2094062

16. Neira-Rodado, D., Escobar-Velasquez, J., McClean, S.: Ambulances deployment problems: categorization, evolution and dynamic problems review. ISPRS Int. J. Geoinf. **11**, 1–37 (2022). https://doi.org/10.3390/ijgi11020109

17. OpenStreetMap: OpenStreetMap (2022). https://www.openstreetmap.org

18. Regione Autonoma Friuli Venezia Giulia: Regione in cifre 2021 (2021). https://www.regione.fvg.it/rafvg/cms/RAFVG/GEN/statistica/FOGLIA3/FOGLIA74/

19. Schütz, F.: SimCpp20 (2021). https://github.com/fschuetz04/simcpp20.git

20. Sudtachat, K.: Strategies to improve the efficiency of Emergency Medical Service (EMS) systems under more realistic conditions (2014). https://tigerprints.clemson.edu/all_dissertations/1359

21. Yin, P., Mu, L.: Modular capacited maximal covering location problem for the optimal siting of emergency vehicles. J. Appl. Geogr. **34**, 247–254 (2012)

Simulated Annealing for a Complex Industrial Scheduling Problem

Quentin Perrachon[1,2(✉)], Alexandru-Liviu Olteanu[1], and Marc Sevaux[1]

[1] Lab-STICC, UMR 6285, CNRS, Université Bretagne Sud, Lorient, France
quentin.perrachon@univ-ubs.fr
[2] HERAKLES, Vannes, France

1 Introduction

We focus on a complex industrial problem extending the job shop scheduling problem (JSSP). We consider multiple jobs consisting of operations to be scheduled and assigned to a set of available resources. Each operation may require multiples necessary resources to be processed and each of its necessary resources must be selected from a corresponding subset of resources. Our goal is to take into account constraints often seen in an industrial context to produce solutions that are easier to exploit directly. For this purpose, the extensions of the classical job shop scheduling problem that we tackle are as follows:

Flexibility: resources must be selected from a subset of compatible resources;
Multi-resource: each operation may require more than one resource;
Unavailability: each resource may be unavailable during certain time periods. Operations interrupted by unavailabilities must be restarted;
Partially necessary resources: some resources may not be required for the entire duration of an operation;
Non-linear routing: precedence constraints between operations of a same job are not necessarily linear.

Finding a solution to our problem corresponds to finding a valid assignment for all necessary resources of all operations and finding sequences of operations on each resource while satisfying all constraints of our problem. In this paper, we will focus on jobs with due dates and our main optimization criterion will be the minimisation of the total lateness ($\sum L_j$ in the classical $\alpha|\beta|\gamma$ notation [9]).

Job shop scheduling problems have seen extensive research due to their many real-life applications. The flexibility extensions of the problem (Flexible Job Shop Scheduling Problem) has also been a popular topic [2]. The other extensions studied here are less popular and, most often, only deal with the minimization of the makespan criterion, which generally is not the most pertinent criterion in an industrial context. Dauzère-Pérès and Pavageau [4] proposed an approach to a similar problem, extending previous works [3,5] on an integrated neighborhood structure for job shop scheduling problems using disjunctive graph representation. Mauguière et al. [8] presented a job shop scheduling problem with

L. Di Gaspero et al. (Eds.): MIC 2022, LNCS 13838, pp. 457–463, 2023.
https://doi.org/10.1007/978-3-031-26504-4_33

unavailabilities also using the disjunctive graph representation. To our knowledge, no work on scheduling problems involving multi-resource, flexibility and unavailabilities has so far been published.

2 Description of the Problem

We consider a set of n operations $O = \{o_i \mid \forall i \in 1..n\}$ that need to be scheduled on m resources $M = \{m_r \mid \forall r = 1..m\}$. Each operation o_i is associated to one of g jobs $J = \{J_j \mid \forall j \in 1..g\}$, with $job(o_i) \in J$ indicating this association. Each job has a due date denoted with d_j.

We divide each operation o_i into $q(i)$ consecutive stages $o_i = \{o_{is} \mid \forall s = 1..q(i)\}$. Each stage corresponds to a different phase of the operation where a set of resources different from neighboring stages is required. Each operation o_i requires $m(i)$ resources to be processed. For each kth necessary resource of operation o_i the subset $M_i^k \subset M$ denotes the compatible resources. We denote with $Sg(i, k)$ the stages during which the kth resource of operation o_i is required. An example of the separation of an operation into stages is shown in Fig. 1.

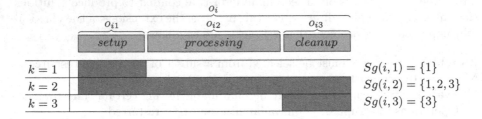

Fig. 1. Operation o_i with $m(i) = 3$ and $q(i) = 3$

Routing constraints are defined by $PR(i)$, the set of operations preceding operation o_i and $SR(i)$, the set of operations succeeding operation o_i. These notations are extended to stages such that $PR(i, s)$ is the set of stages preceding o_{is} and $SR(i, s)$ the set of stages succeeding o_{is}.

Resources may be unavailable during certain time periods. Let $U_r = \{[su_l^r, cu_l^r] \mid \forall l = 1..l_r\}$ be the set of unavailabilities for each resource m_r, where l_r is the number of unavailability intervals of resource r. An unavailability is a time interval $[su_l^r, cu_l^r]$ within which resource m_r is unavailable. Operations interrupted by an unavailability require to be restarted from their first stage.

The processing time of an operation may be divided according to its stages and assigned resources. We define pt_{irs} as the processing time of operation o_i on resource r during stage s. The real processing time of each stage correspond the maximum processing time of each resource required during that stage.

We will represent solutions of our problem by a solution graph [7] inspired by the disjunctive graph representation. A solution graph is a directed acyclic graph $G = (\mathcal{O}, \mathcal{R}, \mathcal{S})$. $\mathcal{O}$ is the set of all vertices, one for each stage of each operation. To maintain consistency, we will denote $\mathcal{O}_{is}$ as the vertex associated with stage o_{is}. Two fictitious vertices, the source $\mathcal{O}_0$ and the sink $\mathcal{O}_*$ are also added, representing the start and end of the whole schedule respectively. $\mathcal{R}$ is the set of routing edges corresponding to the routing constraints PR and SR. $\mathcal{S}$ is the set of all sequence edges, which correspond to the solution of our problem. An edge between two stages $\mathcal{O}_{is}$ and $\mathcal{O}_{i's'}$, either non consecutive or from different operations, exists only if o_{is} is the direct predecessors of $o_{i's'}$ on a shared resource. We will note $pi(i, k)$ (resp. $pk(i, k)$) the index of the operation (resp. the index of the corresponding necessary resource) that precedes operation o_i on its kth necessary resource. Respectively, we will note $si(i, k)$ and $sk(i, k)$ the index of the operation succeeding o_i on its kth necessary resource. Stages that do not have a predecessor on one of their necessary resources have an incoming edge from $\mathcal{O}_0$ and, respectively, stages that do not have a successor on one of their necessary resources have an outgoing edge to $\mathcal{O}_*$. All edges have a weight equal to the maximum processing time pt_{irs} of any resource r assigned to o_i during the stage o_{is} corresponding the source of the edge $\mathcal{O}_{is}$. Edges outgoing from the source $\mathcal{O}_0$ have a weight of 0.

Computing the starting and completion times of all operations can then be done by a traversal of each vertex in their topological order, with potential backtracking if an operation is interrupted by an unavailability.

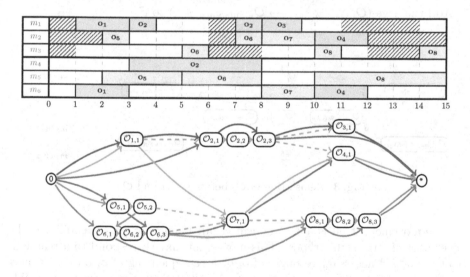

Fig. 2. Solution representations (Gantt chart - Graph representation)

An example of a solution for a problem with 8 operations (o_i) and 6 resources (m_r) is given as a Gantt chart and as its corresponding graph representation (Fig. 2), routing edges are grey and dotted and sequence edges are coloured according to their corresponding resource.

3   Neighbourhood-Based Resolution Approach

The partner company, HERAKLES, has stressed the importance of the resolution of this problem in a short time frame. As a first step, we have decided to develop a simulated annealing approach for its flexibility.

3.1   Neighbourhoods Based on the Graph Representation

We focus on a neighbourhood structure based on a movement adapted from and inspired by [4,5]. The basic idea of this movement is the removal and reinsertion of an operation in the solution graph. We have adapted the movement to our different graph structure to handle our constraints. The move is defined as follows (illustrated in Fig. 3):

1. Select an operation o_i and one of its necessary resources k
2. Remove the incoming edge $(\mathcal{O}_{u,\max Sg(u,k_u)} \longrightarrow \mathcal{O}_{i,\min Sg(i,k)})$ with $u = pi(i,k)$ and $k_u = pk(i,k)$
3. Remove the outgoing edge $(\mathcal{O}_{i,\max Sg(i,k)} \longrightarrow \mathcal{O}_{v,\min Sg(v,k_v)})$ with $v = si(i,k)$ and $k_v = sk(i,k)$
4. Add the edge $(\mathcal{O}_{u,\max Sg(u,k_u)} \longrightarrow \mathcal{O}_{v,\min Sg(v,k_v)})$ to $\mathcal{S}$
5. Select and remove an edge $(\mathcal{O}_{u',\max Sg(u',k_{u'})} \longrightarrow \mathcal{O}_{v',\min Sg(v',k_{v'})})$ from $\mathcal{S}$
6. Add the edge $(\mathcal{O}_{u',\max Sg(u',k_{u'})} \longrightarrow \mathcal{O}_{i,\min Sg(i,k)})$ to $\mathcal{S}$
7. Add the edge $(\mathcal{O}_{i,\max Sg(i,k)} \longrightarrow \mathcal{O}_{v',\min Sg(v',k_{v'})})$ to $\mathcal{S}$

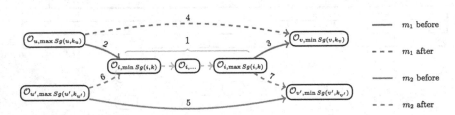

Fig. 3. Reinsertion of $\mathcal{O}_i$ between $\mathcal{O}_{u'}$ and $\mathcal{O}_{v'}$

The movement itself is not complicated but maintaining its feasibility is the clever part of our neighborhood. This movement may correspond to a reassignment of the kth necessary resource of o_i or a re-sequencing of o_i on its kth necessary resource depending on where it is reinserted. Such a movement is feasible if and only if the assignment constraints are respected and the resulting solution graph stays acyclic. This movement is designed to involve a single resource, however it may be required for some operations to do this movement on multiple necessary resources at once otherwise no feasible solution may be reached due to multiple parallel edges between two vertices. We thus extend the definition of our movement to take into account this case.

We define a neighbourhood $N1$ as the set of feasible solutions reachable from such a movement. Feasibility can be checked efficiently through a set of sufficient conditions involving the original solution graph. Unfortunately this neighbourhood may be rather large. We can reduce the size of this neighbourhood using the *critical path* theory by computing a critical path for each job finishing later than its due date. Any improving movements of $N1$ involve operations appearing on these paths. We can thus define a neighbourhood $N2$ as the set of feasible solutions reachable from the reinsertion of an operation appearing in at least one of the critical paths of each job.

There have been a few works on the efficient evaluation of a movement without any graph traversal, based on timings, heads, tails and critical path theory, thus providing significant gains with respect to computational time. Unfortunately, unavailability constraints invalidate most of these properties and make evaluation without graph traversal not precise enough to be useful. A proper evaluation of a movement requires graph traversal and thus a significant computation time. This is currently the main factor limiting our resolution approach.

3.2 Simulated Annealing

We have implemented a straight forward version of the simulated annealing [6] using our neighbourhood structure $N2$. A simulated annealing was chosen due to it's capability to escape local optimum without having to explore the neighbourhood structure fully. We start the simulated annealing from an initial solution generated using a constructive algorithm providing semi-active schedules. At each iteration, a random neighbour solution from $N2$ is selected and evaluated. It is accepted if either this new solution is better than the previous one, or based on a certain probability depending on the annealing schedule used.

The simulated annealing parameters heavily depend on the size of our instances and a set time limit and thus parameter tuning was required for each instance. The initial temperature T_0 is set such that most of the neighbours of the initial solutions are accepted. The cooling scheme is a geometric one, using an α such that $T_{i+1} = \alpha \times T_i$. α is set such that the solutions converge within a given time limit, i.e. our stopping criterion. The acceptance criterion is the same as the original SA formulation, i.e. $P_{accept} = exp(-\Delta(s', s)/T)$.

4   Experiments and Results

To our knowledge, there is no existing benchmark corresponding to our problem. Consequently, we generated our own test instances, extending instances from a well known existing benchmark set for the flexible job shop problem [1]. As requested by the partner company, we use a fixed time limit of 5 min as the stopping criterion, mimicking a more realistic industrial context, even if the computational time of an iteration can largely vary with the size of the instance.

Table 1 illustrates average results following the execution of the proposed approach 30 times on each of the 15 considered extended problem instances.

For each instance, we report its name, the number of resources, the number of operations, the total lateness of the initial solution (obtained by a heuristic not described here), the average total lateness and the best objective function value obtained over the 30 runs of the SA. The last column measures the improvement of the best solution over the initial solution.

Table 1. Average results on the extended problem instances

Instances	m	n	Init. sol.	Av. $\sum L_j$	Best $\sum L_j$	Imp. %
Mk^+1	9	55	179.7	132.71	99.52	44.62%
Mk^+2	9	58	176.6	69.14	35.33	79.99%
Mk^+3	12	150	2188.15	1689.8	1421.07	35.06%
Mk^+4	12	90	822.71	448.64	374.25	54.51%
Mk^+5	6	106	3265.68	2347.64	2213.98	32.2%
Mk^+6	15	150	281.25	148.35	72.7	74.15%
Mk^+7	8	100	3706.55	2731.98	2423.1	34.63%
Mk^+8	15	225	10042.28	9114.74	8762.24	12.75%
Mk^+9	15	240	6703.84	6268.78	6146.06	8.32%
Mk^+10	23	240	2967.8	2766.39	2638.26	11.1%
Mk^+11	8	179	25173.5	23479.62	22598.15	10.23%
Mk^+12	15	193	19029.67	18094.75	17201.3	9.61%
Mk^+13	15	231	17297.17	15695.58	15125.23	12.56%
Mk^+14	23	277	18306.55	17176.16	16807.75	8.19%
Mk^+15	23	284	12190.56	11460.81	10905.68	10.54%

For the small size instances, the initial solution does not seem to be good since it is easily improved by SA. As the size of the instances grows, the computational time for each iteration is larger (due to the neighborhood structure) and SA has not enough time to improve the initial solution. At this stage, we require more experiments and a fair comparison with other approaches to validate the computational experiments.

5   Conclusion

Dealing with an industrial problem is a real challenge, especially when the problem contains a set of difficult constraints such as multi-resources, unavailabilities, partially necessary resources or non-linear routing, all together. This paper presents a first metaheuristic, namely a simulated annealing based on a powerful neighborhood. This neighborhood manages to maintain the feasibility of the new solution. Our approach proposes also to reduce the size of the neighborhood to avoid unnecessary computations. Despite our efforts, the results still lack in performance and also in a fair comparison with competing methods.

We are currently working to improve the neighborhood structure in order to further reduce the required computational time. We are also considering a different metaheuristic framework in order to intensify the search procedure. A preliminary MILP model was also constructed for comparison.

References

1. Brandimarte, P.: Routing and scheduling in a flexible job shop by tabu search. Ann. Oper. Res. **41**(3), 157–183 (1993)
2. Chaudhry, I.A., Khan, A.A.: A research survey: review of flexible job shop scheduling techniques. Int. Trans. Oper. Res. **23**(3), 551–591 (2016)
3. Dauzère-Pérès, S., Paulli, J.: An integrated approach for modeling and solving the general multiprocessor job-shop scheduling problem using tabu search. Ann. Oper. Res. **70**, 281–306 (1997)
4. Dauzère-Pérès, S., Pavageau, C.: Extensions of an integrated approach for multi-resource shop scheduling. IEEE Trans. Syst. Man Cybern. Part C **33**(2), 207–213 (2003)
5. Dauzère-Pérès, S., Roux, W., Lasserre, J.B.: Multi-resource shop scheduling with resource flexibility. Eur. J. Oper. Res. **107**(2), 289–305 (1998)
6. Kirkpatrick, S., Gelatt, C.D., Vecchi, M.P.: Optimization by simulated annealing. Science **220**(4598), 671–680 (1983)
7. Mastrolilli, M., Gambardella, L.M.: Effective neighbourhood functions for the flexible job shop problem. J. Sched. **3**(1), 3–20 (2000)
8. Mauguiere, P., Billaut, J.C., Bouquard, J.L.: New single machine and job-shop scheduling problems with availability constraints. J. Sched. **8**(3), 211–231 (2005)
9. Pinedo, M.L.: Scheduling, vol. 29. Springer, New York (2012). https://doi.org/10.1007/978-1-4614-2361-4

A Matheuristic for Multi-Depot Multi-Trip Vehicle Routing Problems

Tiziana Calamoneri[1] [ID], Federico Corò[2(✉)] [ID], and Simona Mancini[3,4] [ID]

[1] Sapienza University of Rome, Rome, Italy
calamo@di.uniroma1.it
[2] Missouri University of Science and Technology, Rolla, USA
federico.coro@mst.edu
[3] University of Palermo, Palermo, Italy
simona.mancini@unipa.it
[4] University of Klagenfurt, Klagenfurt am Wörthersee, Austria
simona.mancini@aau.at

Abstract. Starting from a real-life application, in this short paper, we propose the original *Multi-Depot Multi-Trip Vehicle Routing Problem with Total Completion Times minimization (MDMT-VRP-TCT)*. For it, we propose a mathematical formulation as a MILP, design a matheuristic framework to quickly solve it, and experimentally test its performance.

It is worth noting that this problem is original as in the literature its characteristics (*i.e.*, multi-depot, multi-trip and total completion time) can be found separately, but never all together. Moreover, regardless of the application, our solution works in any case in which a multi-depot multi-trip vehicle routing problem must be solved.

Keywords: Matheuristic · Multi-trip · Multi-depot · Vehicle routing problems · Total completion time minimization

1 Introduction

In this short paper, we study the *Multi-Depot Multi-Trip Vehicle Routing Problem with Total Completion Times minimization (MDMT-VRP-TCT)*. This problem arises from a Search & Rescue application: immediately after a natural disaster, a fleet of unmanned aerial vehicles (UAVs) helps rescue teams to individuate people needing help inside an affected area. In this context, typically diverse civil defence rescue teams rush from the vicinity to the most affected area, so they give rise to multiple depots. Moreover, UAVs return to depots to substitute their batteries and leave for a new tour, so introducing a multi-trip scenario. Finally, to save as many lives as possible, the most important goal is to get the job done in the shortest possible time, so we aim at minimizing the total completion time.

The resulting optimization problem is original, as in the literature these three characteristics can be found separately, but never all together. Indeed, many problems having similarities with ours can be found, but they also have essential differences with respect to ours.

In this short paper we only refer to [2] as a survey paper on Multi-Trip Vehicle Routing Problems (MTVRP) (where there is a single depot), to [3] for the description of Multiple Traveling Repairperson Problem (mTRP) (that could appear similar to ours but the latency is minimized instead of the completion time), to [1] for the Multiple Traveling Salesperson Problem (mTSP) (where there are no battery constraints *i.e.*, a single trip for each vehicle), and to [4] for a description of the Rooted Min-Max Cycle Cover Problem (RMMCCP) (with single depot).

The rest of this short paper is organized as follows: in Sect. 2, we model MDMT-VRP-TCT as a MILP; in Sect. 3 we propose a matheuristic framework to face reasonably large instances and, in Sect. 4, we experimentally test it.

2 Mathematical Formulation

In this section, we present a mathematical formulation for MDMT-VRP-TCT.

Assume to have an area of interest (*e.g.* the one affected by a natural disaster) with a set I of *target nodes* to monitor (*e.g.* all the damaged buildings). Around the area, there is a set D of *depots* where a set U of vehicles start from (*e.g.* the places where some rescue teams settle down their bases, each one with a sub-fleet of UAVs); in general, each vehicle u is characterized by a different *budget* b_u and then it has to come back to the depot it is uniquely associated to (*e.g.* each UAV has a battery; when it runs down, it is necessary to substitute it with a charged one, and this can be done only at its own depot).

The *traveling distance* between each pair of nodes $i, j \in I \cup D$ is known. Each node $i \in I$, has an associated *service time* s_i (*e.g.* the needed time to overfly it).

A *sequence* is any ordered set k of target nodes; the *duration* of sequence k is computed as the sum of all traveling distances between consecutive target nodes in k plus the service times of all the target nodes in k.

The aim of our problem consists in assigning to each vehicle $u \in D$ an ordered set of sequences such that, from its depot, u is able to reach the first target of any of the sequences assigned to it, serve all its target nodes, come back to o_u, and start again. A sequence k with the addition of the depot of u is called a *trip* and its duration, t_{ku} is given by the duration of k plus the traveling distances between the depot associated to u and the two extremes of k.

A sequence k is *compatible* with a vehicle u if its duration is upper bounded by b_u. A compatibility index, Φ_{ku} is defined equal to 1 if sequence k is compatible with vehicle u and equal to 0 otherwise. Of course, k can be assigned to u only if it is compatible with it (i.e. if $\Phi_{ku} = 1$). A sequence k is considered *feasible* if it is compatible with at least one vehicle. Only feasible sequences are considered.

A *solution* for our problem consists in selecting a set of sequences K whose union covers I and in assigning them to compatible vehicles. We define the *total completion time of a solution* the maximum over all times required by each UAV to fly over all the trips assigned to it by that solution.

Then, we introduce the following decision variables:

- $X_k \in \{0,1\}$, $k \in K$, is a binary variable assuming value equal to 1 if sequence k is selected and 0 otherwise;
- $Y_{ku} \in \{0,1\}$, $k \in K$ and $u \in U$, is a binary variable assuming value equal to 1 if sequence k is executed by vehicle u;
- T_u is the completion time of vehicle u;
- τ is a non-negative variable representing the total completion time.

The mixed integer programming formulation is reported in the following

$$\min \; \tau \tag{of}$$

$$\sum_{k \in \Omega_i} X_k = 1 \quad \forall i \in I \tag{C1}$$

$$\sum_{u \in U} Y_{ku} = X_k \quad \forall k \in K \tag{C2}$$

$$Y_{ku} \le \Phi_{ku} \quad \forall k \in K, \forall u \in U \tag{C3}$$

$$T_u = \sum_{k \in K} t_{ku} Y_{ku} \quad \forall u \in U \tag{C4}$$

$$\tau \ge T_u \quad \forall u \in U \tag{C5}$$

The objective function consisting into the minimization of the total completion time, as reported in (of). Constraints (C1) ensure that each target is covered by exactly one sequence. If a sequence is selected, it must be assigned to exactly one vehicle, chosen among those compatible with it, (constraints (C2) and (C3)). The cumulative working time for each vehicle is computed by means of constraints (C3). The total completion time must be larger than the cumulative working time of each vehicle, as stated in (C4). This formulation distinguishes from the trip based standard one in the objective function.

We conclude this section highlighting that the novelty of our approach lies in considering sequences that can be assigned to vehicles located in different depots (in fact to all vehicles whose depot position makes them compatible with them) instead of constructing closed trips (as for example in [5]), that are inherently partitioned among the vehicles. Moreover, regardless of the application, our modeling approach works in any case in which a multi-depot multi-trip vehicle routing problem must be solved, whichever is the objective function to be optimized. Therefore, it can be applied also in cases in which the goal is to minimize the total covered distance, as common in logistics applications.

3 A Model Based Matheuristic Framework

The main idea under the above presented mathematical model consists into generating all possible feasible sequences and associating them to the set of their compatible vehicles. When the number of feasible trips is too large to be handled, the mathematical model becomes intractable. If, for instance, target nodes are very near among each other, or batteries are very large, so that several target nodes can be visited in a single sequence, even small instances may become difficult to handle exactly.

To overcome this issue, and be able to address larger instances, we derive from our model a heuristic approach, in which we generate only a subset of the feasible sequences, $\tilde{K}$, to be passed to the model. It is clear that the choice of the sequences can dramatically change the performance of the heuristic. Therefore, the problem of determining which sequences to generate assumes a crucial importance.

In the following, after giving some operative definitions, we describe how we generate promising sequences to be passed to the mathematical model.

A sequence k is *dominated (strictly dominated)* by another one, k', if they have the same extremes and contain exactly the same target nodes (possibly in a different order), but k' has a lower or equal (strictly lower) duration than k.

We initially generate all the sequences containing only one target node and directly insert them in the set of sequences to be passed to the model, $\tilde{K}$. We also insert them in a temporary set of sequences K^{tmp} which contains sequences to be expanded. All the sequences included in K^{tmp} are then processed. N_c child sequences are generated from each sequence k adding after the last target node in the sequence, l_k, the jth nearest node to l_k among those not already included in k, with j varying in $\{1, N_c\}$. For each child sequence k^c, we apply a first feasibility check: if the duration sequence k^c is larger than the maximum autonomy of a vehicle, $B_{max} = \max_{u \in U} b_u$, then the sequence is immediately discarded. Otherwise, we pass the corresponding trip to a second feasibility check, in order to verify that the sequence is neither strictly dominated by nor it strictly dominates another sequence already belonging to $\tilde{K}$. If a domination occurs, the dominated sequence is discarded otherwise it is kept in $\tilde{K}$. At this point, we set, for all the vehicles u compatible with k, $\Phi_{ku} = 1$, and for all the others $\Phi_{ku} = 0$.

Once all the child sequences of a sequence k have been analyzed, k is removed from K^{tmp}. The procedure terminates when K^{tmp} is empty or when a maximum allowed number of sequences, K_{max} have been added to $\tilde{k}$.

K_{max} is a parameter of the algorithm and it plays a crucial role in the performance of the algorithm. A larger value of K_{max} would yield to a better global solution but would increase the computational time required by the heuristic. To obtain an effective and efficient algorithm, this parameter must be carefully tuned in order to achieve a good balance between solutions quality and computational times. The maximum number of children generated by each trip, N_c, also plays an important role. The larger the value of N_c, the larger the number of sequences containing a specific number of target nodes. Note that keeping fixed the value of the maximum number of sequences to generate K_{max}, lower values of N_c allow us to generate sequences containing more target nodes, which could be promising; on the other hand, in those sequences only nodes which are very close to each others are visited sequentially, and this would imply that isolated targets would appear in very few sequences. Instead, with very large values of N_c, even targets which are not very close to each others can be visited, but in this case each sequence have several children, and so the maximum allowed number of sequence is reached already considering sequences containing a small number of targets, longer sequences are not generated, and this could negatively affect the solution quality.

After the sequence generation process is finished, the set of sequences $\tilde{K}$ is given in input to the mathematical model ((of)–(C4)).

4   Computational Results and Discussion

In this section, we study the performance of our matheuristic, comparing it with the exact model on small instances) and varying some parameters of the problem.

In this short paper we selected only some experiments, shown in Fig. 1.

In all charts, on the x axis, 3, 4, 5 and 6 represent the used values of N_c. The y coordinates of the dots correspond to an average computed on 20 random instances on the same number of nodes: every column of charts corresponds to a different value of n (increasing going from left to right). The red lines represent the benchmark values achieved by the model. It is worth to note that when n is small ($n \leq 30$), there are results corresponding to the model; when $n = 40$, the model terminates only in 11 cases out of 20, and computational times varied between 101.38 and 30559.5 s; probably the corresponding instances are particularly easy to solve (*e.g.*, without clustered target nodes) and for this reason we decided not to report the average value of the optimal solution. Instead, when $n = 50$, the model is not able to terminate in any case.

The experiments perfectly confirm the expectations. Indeed:

- The first three charts in the first row show the percentage gap from the optimum completion time, that is the main objective function of our problem; it is clear that it tends to 0 as N_c grows up and, when $N_c = 6$, it is very close to 0, showing that our matheuristic works very well. We can also observe that also with $N_c = 3$ the gap is very small for instances with 10 nodes, while it tends to increase for larger instances. The last two charts of the first row, instead, show the percentage gap w.r.t. the case $N_c = 3$; since large values of N_c lead to better solutions, clearly, these gaps are negative.
- The second row of charts corresponds to the number of trips generated in order to individuate the solution. As expected, the matheuristic dramatically decreases the number of trips, that is higher and higher when N_c increases but anyway reasonable. Just this reduction makes the matheuristic tractable even for large instances.
- The third row of charts corresponds to the time necessary for running the model and the matheuristic. Clearly, the computational time of the model is much higher and, for what concerns the matheuristic, it grows up when N_c is increased.

The novelty of the approach, consists into generating (open) sequences of nodes, that can be assigned to different vehicles at different costs, instead of generating complete routes including the depot. This way, the problem can be

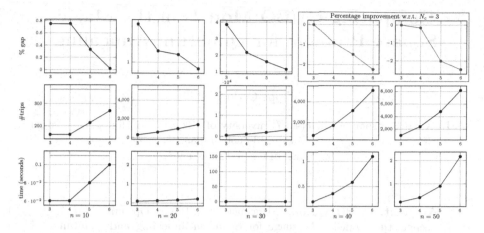

Fig. 1. Experimental results. On the x axis, 3,4,5,6 represent the used values of N_c; the red lines represent the benchmark values achieved by the model. (Color figure online)

modeled as a multiple choice knapsack, with knapsack-dependent items weight and maximum knapsack occupancy minimization. Such approach is not only valid for this specific problem, but can be used for a wide class of multi-depot multi-trip problems, including those having different objective functions, such as the *classical* total travel distance minimization, the minimization of the number of vehicles used, or the minimization of total cost given by vehicles purchasing costs plus travel costs. Furthermore, this matheuristic framework can be used whichever is the method exploited to generate promising sequences. In particular, it can be combined with well known solutions generation algorithms such as the Greedy randomized adaptive search (GRASP). However, we believe our method is more suitable for problems with heterogeneous fleet, since we generate sequences of increasing length, among which, the smaller ones are compatible also with vehicles with a limited autonomy (or capacity). Conversely, GRASP is designed for problems with homogeneous fleet and tend to generate sequences which exploit the whole capacity/autonomy of the vehicle.

References

1. Bektas, T.: The multiple traveling salesman problem: an overview of formulations and solution procedures. Omega **34**(3), 209–219 (2006)
2. Cattaruzza, D., Absi, N., Feillet, D.: Vehicle routing problems with multiple trips. 4OR **14**(3), 223–259 (2016). https://doi.org/10.1007/s10288-016-0306-2
3. Méndez-Díaz, I., Zabala, P., Lucena, A.: A new formulation for the traveling deliveryman problem. Discret. Appl. Math. **156**(17), 3223–3237 (2008)
4. Nagarajan, V., Ravi, R.: Approximation algorithms for distance constrained vehicle routing problems. Networks **59**, 209–214 (2012)
5. Paradiso, R., Roberti, R., Laganà, D., Dullaert, W.: An exact solution framework for multitrip vehicle-routing problems with time windows. Oper. Res. **68**(1), 180–198 (2020)

Comparing QUBO Models of the Magic Square Problem for Quantum Annealing

Philippe Codognet[(✉)]

JFLI, CNRS/Sorbonne University/University of Tokyo, 7-3-1 Hongo, Bunkyo-ku, Tokyo 113-8656, Japan
codognet@is.s.u-tokyo.ac.jp

Abstract. QUBO (Quadratic Unconstrained Binary Optimization) has become the modeling language for quantum annealing and quantum-inspired annealing solvers. We present different modeling in QUBO of the Magic Square problem, which can be modeled by linear equations and a permutation constraint over integer variables. Different ways of encoding integers by Booleans in QUBO amounts to models that have very different performance. Experiments performed on the Fixstars Amplify Annealer Engine, a quantum-inspired annealing solver, show that using unary encoding for integers performs much better than using the classical one-hot encoding.

1 Introduction

Quantum Annealing (QA) [8,10] has become in the recent years an interesting approach for solving combinatorial problems, in particular with the development of quantum hardware such as D-Wave computers [2] and quantum-inspired dedicated hardware such as Fujitsu's Digital Annealer Engine [1]. From a metaheuristic viewpoint, QA can be seen as a variant of simulated annealing where escape from local minima is done by the physical phenomenon of *quantum tunneling*, allowing to traverse energy barriers in the energy landscape as long as they are not too large. In the QA paradigm, combinatorial optimization problems can be described in the *Ising model*, a mathematical model of ferromagnetism in statistical mechanics, and converted to Ising Hamiltonians, the ground states of which correspond to the minimal solutions of the original problem, see for instance [12]. These ground states can be computed by quantum devices using adiabatic quantum evolution, an example of which being quantum annealing [15].

In the field of combinatorial optimization, Quadratic Unconstrained Binary Optimization (QUBO) has been proposed as a simple but powerful modeling language for a variety of problems, see [9,11] for a detailed presentation and examples. Interestingly, there is a straightforward transformation to convert Ising models to QUBO and vice-versa, and thus QUBO has become the standard input language for quantum and quantum-inspired annealing hardware.

L. Di Gaspero et al. (Eds.): MIC 2022, LNCS 13838, pp. 470–477, 2023.
https://doi.org/10.1007/978-3-031-26504-4_35

Consider n Boolean variables $x_1, ..., x_n$, a QUBO problem consists in minimizing an *objective function* defined by a quadratic expression over $x_1, ..., x_n$:

$$\sum_{i \leq j} q_{ij} x_i x_j$$

It is therefore usual to represent a QUBO problem by a vector x of n binary decision variables and a square $n \times n$ matrix Q with coefficients q_{ij}, and the problem can be written as: *minimize* $y = x^t Q x$, where x^t is the transpose of x.

Although the "U" in QUBO stands for "Unconstrained", constraint expressions can be introduced in QUBO models as *penalties* in the objective function to minimize, that is, as quadratic expressions whose value is minimal when the constraint is satisfied. An easy way to formulate such a penalty is to create a quadratic expression which has value 0 if the constraint is satisfied and a positive value if the constraint is not satisfied, representing somehow the degree of violation of the constraint, see [9] for a set of penalty expressions for basic constraints on a few Boolean variables.

This opens to the idea of modeling complex constraint satisfaction problems or constrained optimization problems in QUBO in order to solve them by quantum annealing. We would like in this paper to consider the *magic square problem* as such an example and detail its modeling in QUBO. We will consider two different QUBO models with different encoding of integers by Booleans and therefore different set of penalties to represent the constraints. The two models indeed have different performance, and we will report preliminary performance evaluation.

This paper is organized as follows. Section 2 describes how to model the magic square problem with constraints. Then Sect. 3 details the first QUBO encoding using one-hot encoding for integers, while Sect. 4 presents another QUBO encoding using unary/domain-wall encoding for integers. Experiments and performance evaluation on quantum-inspired hardware are then presented in Sect. 5. A short conclusion ends the paper.

2   The Magic Square Problem

The well-known magic square problem consists in placing on a $n \times n$ square all the numbers in $\{1, \dots, n^2\}$ such as the sum of the numbers in all rows, columns and the two main diagonals are the same. It is usually modeled by considering n^2 integer variables x_{ij} with domains $\{1, \dots, n^2\}$ and the following constraints:

- a permutation constraint stating that all variables have a different value: all-different $(x_{ij}, \forall i \in \{1, \dots, n\}, \forall j \in \{1, \dots, n\}))$
- $2n + 2$ linear equations for the sums on the rows, columns and the two main diagonals:
  - $\forall i \in \{1, \cdots, n\}, \sum_{j=1}^{n} x_{ij} = m,$
  - $\forall j \in \{1, \cdots, n\}, \sum_{i=1}^{n} x_{ij} = m,$
  - $\sum_{i=1}^{n} x_{ii} = m, \quad \sum_{i=1}^{n} x_{n+1-i,i} = m.$

The value m that should be the sum of all rows, columns and the two diagonals can be easily computed to be $\frac{n}{2}(n^2+1)$.

[7] presents different modeling of the magic square problem and compares its solving by different methods: Constraint Satisfaction Techniques (Microsoft Solver Foundation), linear programming (CPLEX), variable neighborhood local search, and genetic algorithms. As expected, metaheuristics methods are more efficient, in particular because they can implement the permutation constraint more efficiently. Large instances of the magic square problem (e.g. 100×100) can indeed be solved in a few seconds by iterated local search [3].

For modeling this problem in QUBO, we first need to encode integer variables by Booleans. There are mainly two schemes currently used in the QA community: the classical *one-hot* encoding and the *unary/domain-wall* encoding. We can encode each integer variable in $\{1, \ldots, n^2\}$ by n^2 Booleans with one-hot encoding and by $n^2 - 1$ Booleans with unary/domain-wall encoding. We then need to transform the permutation constraint and linear equations over integers into quadratic penalties over Booleans, as described previously, and the penalties corresponding to the constraints depend on the encoding, as will be detailed in the following sections. Observe that when several constraints c_i are to be modeled in the QUBO problem, the corresponding penalties should be added to the objective function with a penalty coefficient p_i for each penalty in order to make it compatible with the original objective function to optimize. The penalties coefficient corresponding to the constraint should be large enough to make such constraints "hard", whereas the objective function is considered "soft".

3   One-Hot Encoding

The natural way to encode integers with Booleans variables in QUBO is the so-called *one-hot encoding*: an integer variable $x \in \{1, \ldots, n\}$ can be represented by n Boolean variables x_k that have value 1 if the original variable x has value k and value 0 otherwise. Such a model has been proposed in [6] and we will briefly recall it in this section.

To model the Magic Square problem, we will need a total of n^4 Boolean variables $x_{ij}^k, \forall i \in \{1, \ldots, n\}, \forall j \in \{1, \ldots, n\}, \forall k \in \{1, \ldots, n^2\}$ for representing the integer variables x_{ij} corresponding to a cell (i, j). A Boolean variable x_{ij}^k has value 1 if the number in the cell (i, j) has value k and 0 otherwise.

To enforce that each variable as only one value we need the constraint $\sum_{k=1}^{n^2} x_{ij}^k = 1$. Remarking that $\sum_{k=1}^{n^2} x_{ij}^k = 1 \iff (\sum_{k=1}^{n^2} x_{ij}^k - 1)^2 = 0$, we can develop this quadratic expression and remove the constant term to obtain the *penalty* expression $P_{\text{one-hot}}(x_{ij})$ for each original integer variable x_{ij}, to be added to the QUBO objective function:

$$P_{\text{one-hot}}(x_{ij}) = -\sum_{k=1}^{n^2} x_{ij}^k + 2 \sum_{k<k'} x_{ij}^k x_{ij}^{k'}$$

For the Magic Square problem, we also need to enforce the fact that each value $k \in \{1, \ldots, n^2\}$ is assigned exactly once, therefore enforcing a *permutation constraint*, also known as *two-way one-hot* in the QA literature and used for instance to model in QUBO the TSP or the QAP [13]. We can encode such a permutation constraint in QUBO by $2 \times n^2$ pseudo-Boolean constraints representing one-hot constraints: one corresponding to each of the n^2 variables x_{ij} stating that it can have only one value k (i.e., one-hot encoding) and one for each of the n^2 values k stating that it can be assigned to only one variable x_{ij}^k. Adding all penalty expressions together and simplifying the quadratic expression gives the following penalty for the permutation constraint:

$$P_{\text{perm-1hot}} = \sum_{i=1,j=1}^{n} \sum_{\{k<k'\}} x_{ij}^k x_{ij}^{k'} + \sum_{k=1}^{n} \sum_{\{n*i+j<n*i'+j'\}} x_{ij}^k x_{i'j'}^k - \sum_{i=1,j=1,k=1}^{n} x_{ij}^k$$

Let us now define the penalties corresponding to the linear constraints on the rows, columns and two diagonals of the magic square. Consider m integer variables $y_1, \ldots, y_m$ in $\{1, \ldots, n\}$ and a linear constraint $\sum_{i=1}^{m} a_i y_i = b$, with each y_i being one-hot encoded by Boolean variables $y_{i,1}, \ldots, y_{i,n}$. As $y_i = \sum_{j=1}^{n} j * y_{ij}$, the linear constraint over variables y_i can be re-written:

$$\sum_{i=1}^{m} a_i \left(\sum_{j=1}^{n} j * x_{ij} \right) = b \iff \sum_{\substack{i \in \{1,\ldots,m\} \\ j \in \{1,\ldots,n\}}} a_i * j * y_{ij} = b$$

We now have a pseudo-Boolean linear constraint to which we can apply the transformation #1 from [9] in order to obtain the penalty $P_{\text{lin-1hot}}$ corresponding to the linear constraint over integer variables:

$$P_{\text{lin-1hot}} = \sum_{\substack{i \in \{1,\ldots,m\} \\ j \in \{1,\ldots,n\}}} a_i j (a_i j - 2b) y_{ij} + 2 \sum_{i*m+j < i'*m+j'} a_i a_{i'} j j' y_{ij} y_{i'j'}$$

In the Magic Square model, a constraint on column j, $\sum_{i=1}^{n} x_{ij} = \frac{n}{2}(n^2 + 1)$ is thus equivalent to $\sum_{i=1}^{n} (\sum_{k=1}^{n^2} k x_{ij}^k) = \frac{n}{2}(n^2 + 1)$, and will lead to a penalty:

$$P_{\text{column-j-1hot}} = \sum_{i} k(k - n(n^2 + 1)) x_{ij}^k + 2 \sum_{n*i+k < n*i'+k'} kk' x_{ij}^k x_{i'j}^{k'}$$

Linear constraints on lines and the two diagonals are treated similarly.

4   Unary/Domain-Wall Encoding

An alternative to the one-hot encoding, called *domain-wall encoding*, has been proposed in [4] and developed in [5] in an Ising model setting. When transposed to a Boolean setting, it amounts to the well-known *unary encoding* on a fixed number of bits: a number n is encoded by n bits set to 1, followed by zeros. Let us give a new definition for this format, adapted for defining penalties in QUBO.

A vector of $n-1$ booleans $x_0 \cdots x_{n-2}$ is a *unary/domain-wall encoding* of an integer $x \in \{0, \ldots, n-1\}$ if and only if the following properties holds:

1. $\forall i \in \{0, \ldots, n-3\}, \ x_i \geq x_{i+1}$
2. $x = \sum_{i=0}^{n-2} x_i$

This means that if $x_i = 1$ then, for all indexes $j < i$, $x_j = 1$, or equivalently if $x_i = 0$ then, for all indexes $j > i$, $x_j = 0$. For instance, 1100 is a valid unary/domain-wall encoding and represents the integer value 2, while 1101 and 0100 are not valid unary/domain-wall encodings. Observe that it is easier in this encoding to refer to integer values between $\{0, \ldots, n-1\}$ rather than between $\{1, \ldots, n\}$, and these values can be coded by using $n-1$ Booleans rather than n Booleans with one-hot encoding.

As described in [9], a Boolean constraint $x \geq y$ can be represented in QUBO by the penalty $y - xy$, therefore the constraint for a valid unary/domain-wall encoding corresponds to the penalty P_{unary}:

$$P_{\text{unary}} = \sum_{i=0}^{n-3} (x_{i+1} - x_i \, x_{i+1})$$

To model the Magic Square problem, we need a total of $n^2(n^2 - 1)$ Boolean variables x_{ij}^k, $\forall i \in \{1, \ldots, n\}, \forall j \in \{1, \ldots, n\}, \forall k \in \{1, \ldots, n^2 - 1\}$ for representing the integer variables x_{ij} corresponding to a cell (i, j), with $x_{ij}^0 \cdots x_{ij}^{n^2-1}$ representing the unary/domain-wall encoding of x_{ij} with corresponding penalty P_{unary}.

Let us now define the permutation constraint in its general form. Consider n integer variables $x_0, \ldots, x_{n-1}$ with values in the domain $\{0, \ldots, n-1\}$ and $n \times (n-1)$ Booleans x_{ij} such that $x_{i,0} \cdots x_{i,n-2}$ is the unary/domain-wall encoding of x_i. We can remark that $(x_0, \ldots, x_{n-1})$ is a permutation of $(0, \ldots, n-1)$ if and only if:

$$\forall j \in \{0, n-2\} \ \sum_{i=0}^{n-1} x_{ij} = (n-1) - j$$

We can now sum up all the penalties corresponding to these $n-1$ equations and obtain the penalty $P_{\text{perm-u}}$ corresponding to a permutation constraint on $x_0, \ldots, x_{n-1}$ with each x_i being unary/domain-wall encoded by $n-1$ Boolean x_{ij}:

$$P_{\text{perm-u}} = \sum_{j=0}^{n-2} \left(\, (2(j-n)+3) \sum_{i=0}^{n-1} x_{ij} + 2 \sum_{i<k} x_{ij} x_{kj} \, \right)$$

Let us now define the penalties corresponding to the linear equations over integers encoded with unary/domain-wall encoding.

Consider an equation $\sum_{i=1}^{m} a_i x_i = b$ with m integer variables $x_0, \ldots, x_{m-1}$ in $\{0, \ldots, n-1\}$, with each x_i unary/domain-wall encoded by Boolean variables $x_{i,0}, \ldots, x_{i,n-2}$. As $x_i = \sum_{j=1}^{n-1} x_{ij}$, the original linear constraint over integer variables x_i can be re-written as:

$$\sum_{i=0}^{m-1} a_i \Big(\sum_{j=0}^{n-2} x_{ij} \Big) = b \iff \sum_{\substack{i \in \{0,\ldots,m-1\} \\ j \in \{0,\ldots,n-2\}}} a_i x_{ij} = b$$

We can then apply as in Sect. 3 the transformation #1 from [9] to this pseudo-Boolean linear equation and obtain the penalty $P_{\text{lin-u}}$:

$$P_{\text{lin-u}} = \sum_{\substack{i \in \{0,\ldots,m-1\} \\ j \in \{0,\ldots,n-2\}}} a_i(a_i - 2b)x_{ij} + 2 \sum_{i*m+j < i'*m+j'} a_i a_{i'} x_{ij} x_{i'j'}$$

We thus have penalties that are slightly simpler than for one-hot encoding, and we can use this transformation for all the linear equations representing the constraints on the rows, columns and diagonals of the magic square.

5   Experiments with Fixstars Amplify Digital Annealer

Our original aim was to implement these two QUBO models on the D-Wave quantum computer. However, due to the architecture of the D-Wave Advantage system and in particular to the fact that the Pegasus architecture do not implement the complete connection graph between qubits, these machines are very limited in the size of the instances that can be solved. Indeed, we cannot find a solution satisfying all constraints for $n > 3$.

We therefore looked at *quantum-inspired annealing* (a.k.a. *digital annealing*) systems that use QUBO as an input language and simulate the behavior of quantum annealers on classical electronics, either on dedicated CMOS hardware such as the Fujitsu Digital Annealing Unit [1], or on clusters of parallel hardware, such as Fixstars Amplify Annealing Engine (AE) [14]. We implemented the QUBO models on the Fixstars Amplify AE, a digital annealer running on a cluster of Graphics Processing Units (GPUs), with a capacity of 65,536 bits connected by a complete graph. It can run more complex models than D-Wave because it has a complete connection graph topology between bits (i.e. QUBO variables), but it is based on classical (non-quantum) electronics, hence bits and not qubits.

The performance evaluation of both QUBO encodings on the Fixstars Amplify Annealer Engine is shown in Table 1 below (timings are in seconds). Unary/domain-wall encoding is clearly more efficient than one-hot encoding, being up to two orders of magnitude faster. Unary/Domain-wall encodings also have slightly less variables (e.g. 5,832 for a 9×9 magic square versus 6,561 for one-hot encoding) and the encoding of constraints is very different.

Table 1. Performance of different QUBO models on Magic Square

Size	One-hot		Unary/Domain-wall	
	Success rate (timeout = 1 mn)	Time to solution	Success rate	Time to solution (timeout = 1 mn)
4×4	100%	0.162	100%	0.064
5×5	100%	2.435	100%	0.087
6×6	100%	19.80	100%	0.209
7×7	40%	42.70	100%	0.598
8×8	0%	-	100%	1.775
9×9	0%	-	100%	25.13

6   Conclusion

We have detailed two QUBO models for the Magic Square problem as an experiment to implement complex combinatorial problems with various types of constraints in QUBO. Magic Square is modeled by linear equations and a permutation constraint over integer variables that are then encoded in QUBO with two different encodings for integers: one-hot encoding and unary/domain-wall. As quantum computers such as D-Wave systems cannot solve instance of this problem larger than $n = 3$, we have also implemented those QUBO models on a quantum-inspired annealer based on a cluster of GPUs, the Fixstars Amplify Annealer Engine. The performance evaluation shows that the unary/domain-wall encoding is more efficient than the classical one-hot encoding, being about two orders of magnitude faster. This shows the importance of investigating different types of modeling techniques in QUBO, including alternative codes for integers.

References

1. Aramon, M., Rosenberg, G., Valiante, E., Miyazawa, T., Tamura, H., Katzgraber, H.G.: Physics-inspired optimization for quadratic unconstrained problems using a digital annealer. Front. Phys. **7**, 48 (2019)
2. Bunyk, P.I., et al.: Architectural considerations in the design of a superconducting quantum annealing processor. IEEE Trans. Appl. Supercond. **24**(4), 1–10 (2014)
3. Caniou, Y., Codognet, P., Richoux, F., Diaz, D., Abreu, S.: Large-scale parallelism for constraint-based local search: the costas array case study. Constraints **20**(1), 30–56 (2016)
4. Chancellor, N.: Domain wall encoding of discrete variables for quantum annealing and QAOA. Quantum Sci. Technol. **4**, 045004 (2019)
5. Chen, J., Stollenwerk, T., Chancellor, N.: Performance of domain-wall encoding for quantum annealing (2021). arXiv:2102.12224v2 (quant-ph)
6. Codognet, P.: Constraint solving by quantum annealing. In: ICPP Workshops 2021: 50th International Conference on Parallel Processing, August 2021. ACM (2021)

7. Denic, A.: Application of exact and heuristic methods to magic square problem. Math. Balkanica **25**(5), 491–498 (2011)
8. Farhi, E., Goldstone, J., Gutmann, S., Lapan, J., Lundgren, A., Preda, D.: A quantum adiabatic evolution algorithm applied to random instances of an NP-complete problem. Science **292**(5516), 472–475 (2001)
9. Glover, F.W., Kochenberger, G.A., Du, Y.: Quantum bridge analytics I: a tutorial on formulating and using QUBO models. 4OR **17**(4), 335–371 (2019)
10. Kadowaki, T., Nishimori, H.: Quantum annealing in the transverse Ising model. Phys. Rev. E **58**, 5355–5363 (1998)
11. Kochenberger, G., et al.: The unconstrained binary quadratic programming problem: a survey. J. Comb. Optim. **28**(1), 58–81 (2014). https://doi.org/10.1007/s10878-014-9734-0
12. Lucas, A.: Ising formulations of many NP problems. Front. Phys. **2**, 5 (2014)
13. Matsubara, S., et al.: Digital annealer for high-speed solving of combinatorial optimization problems and its applications. In: 25th Asia and South Pacific Design Automation Conference (ASP-DAC), pp. 667–672 (2020)
14. Matsuda, Y.: Research and development of common software platform for ising machines. In: 2020 IEICE General Conference (2020). https://amplify.fixstars.com/docs/_static/paper.pdf
15. McGeoch, C.C.: Adiabatic Quantum Computation and Quantum Annealing: Theory and Practice. Morgan & Claypool, San Rafael (2014)

Self-adaptive Publish/Subscribe Network Design

Vittorio Maniezzo[1]([✉]), Marco A. Boschetti[2][iD], and Pietro Manzoni[3][iD]

[1] Department of Computer Science, Università di Bologna, Cesena, Italy
vittorio.maniezzo@unibo.it
[2] Department of Mathematics, Università di Bologna, Cesena, Italy
marco.boschetti@unibo.it
[3] Universitat Politécnica de Valéncia, Valencia, Spain
pmanzoni@disca.upv.es

Abstract. The *pub/sub pattern* is gaining momentum in IoT architectures, thanks to its robustness and since it offers many-to-many communication. Efficient network management is needed when only scarce and unreliable resources are available as network infrastructure. Moreover, any form of centralized control should be avoided so as not to limit the application potential. This *short paper* presents preliminary results of a research line casting pub/sub communication as a dynamic network design problem and supporting optimized adaptive routing via a fully distributed Lagrangian matheuristic applied to an extension of the integer multicommodity flow problem.

Keywords: Publish/subscribe pattern · Matheuristics · Network design

1 The Publish/Subscribe Pattern

The Internet of Things (IoT), renownedly a global network of connected devices, people, and processes, is expanding at an exponential pace. It transfers and shares collected data over the internet, having a huge impact on both the social and the digital world. The information generated by IoT devices is typically sent to servers hosted in the cloud that can be far away. The average round-trip time from various geographically distributed points to their hosts can be relatively high, and to this transfer time we should add the possible temporary connection failures, a major problem for time-critical applications. To face this challenge, "edge" and "fog" computing have been proposed to denote multilevel hierarchies of nodes spanning from the cloud to IoT devices, permitting to bring IoT solutions in areas where connectivity is scarce and, in general, resources are limited, for example, in rural or remote areas.

A transition from a centralized, cloud-based architecture to an interoperable and decentralized dynamic IoT architecture is in progress. A current research

© The Author(s), under exclusive license to Springer Nature Switzerland AG 2023
L. Di Gaspero et al. (Eds.): MIC 2022, LNCS 13838, pp. 478–484, 2023.
https://doi.org/10.1007/978-3-031-26504-4_36

challenge is how to include fog computing features in current infrastructures, ensuring the easy-to-use and high availability of network resources.

The *publish/subscribe pattern* (or simply *pub/sub*) provides an alternative to traditional client-server architectures along with this request. It decouples the client that sends a message (the *publisher*) from the clients, possibly more than one, that receive the messages (the *subscribers*). The publishers and subscribers never contact each other directly, in fact, they are not even aware that the other exists. The connection between them is handled by a third component (the *broker*), that filters all incoming messages and distributes them based on topic contents to relevant subscribers.

The *Message Queue Telemetry Transport* (MQTT), is an OASIS (Organization for the Advancement of Structured Information Standards) protocol for messaging between IoT devices that follows the Pub/Sub paradigm. MQTT permits to transport messages between devices requiring a small code footprint and limited network bandwidth. An MQTT client is any device, from a microcontroller to a server, that connects to an MQTT broker to exchange messages. The communication follows the topic-based publish/subscribe pattern with brokers acting as message dispatchers. The combination of the protocol simplicity at the client side and of the support for reliability and quality of service (QoS), makes MQTT effective for resource-constrained applications.

The centralized structure of MQTT, but more in general of Pub/Sub systems, has anyway drawbacks, and new solutions are currently being tested to deal with them. Choosing the proper infrastructure for broker messaging is crucial; otherwise, scaling can be hindered, and reliability issues may appear.

This work describes the first attempt of mathematical optimization of a distributed MQTT broker system. The problem is cast as a network design problem, whose goal is to find the network configuration that maximizes the number of messages relayed to subscribers, and is resilient to network topology variations. Moreover, we are focusing on the situation where links among brokers are bandwidth limited, for example, due to LPWAN technologies like LoRa used as network infrastructure [1].

2 The Lagrangian Matheuristic

We consider a scenario where several sensors are distributed in an area producing data tagged according to their content. End nodes behave either as *publishers* or *subscribers*, internal nodes are *brokers*. Clients accessing the network can connect to one or many of the actual brokers and, through them, publish and receive data. Data, as well as the associated tags (topics), are characterized by the *bandwidth* required for their transmission in the network.

All existing connections in the network have limited bandwidth, as is the case for example of LPWAN links. The application layer protocol used by the network components to communicate with each other is assumed to be the MQTT protocol. Clients are not limited to connecting to only one broker but can choose to connect to any of the visible ones, changing these links dynamically based on the

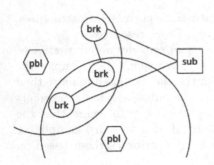

Fig. 1. A simple example showing the connections among publishers (hexagons) clients (squares) and brokers (circles)

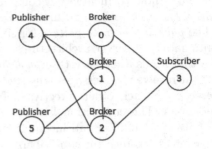

Fig. 2. Network model (named *tinyInstace*) of the case of Fig. 1

load of connections a broker is subjected to. A simple example is shown in Fig. 1, where two publishers (hexagons) can connect wirelessly to any broker (circle) within their range. Brokers are connected among themselves and with the single subscriber (square) by cables. If one of the brokers became saturated, a client might decide to switch the connection to another of the brokers it has access to. The network can also change dynamically, with the possibility of nodes disappearing or new ones appearing, just as connections between nodes can appear or disappear.

Brokers do not store data but only its tags. They are responsible for forwarding data compatible with the requests. Dynamic paths between data producers and consumers have to be identified.

Summing up, the problem is a network design problem, whose goal is to find the network configuration that minimizes the number of requests that are not satisfied by the network. We denote this problem as *Publish/subscribe network design Problem*, P/SP for short. P/SP is an extension of a standard max-flow min-cost *multicommodity flow* problem, whose graph can be described using the following elements:

- K is the set of *commodities* transmitted by the network, where the *commodities* correspond to the *topics* made available by the system.
- S is the set of source nodes, the clients that publish data. Each commodity k originates in a single client.
- T is the set of destination nodes, the subscribers to the different topics. Each commodity k can be requested by a set of clients $T^k \subseteq T$.
- B is the set of intermediate nodes, the brokers present in the network.
- A is the set of *directed* arcs present in the graph, edges are represented by pairs of arcs. A bandwidth constraints limits the sum of data flows in both directions, in case of pairs of opposing arcs.

The P/SP problem cannot be directly modeled as a multicommodity network design problem because the commodities generated by the sources could be requested by multiple destinations, and *duplicated* by the brokers met along the paths. Thus, the data flow exiting from the sources is not equal to the sum of the flows reaching the destinations, but possibly much smaller.

In our formulation, we further postulate that commodity flows cannot be split and recombined at destination, and that sources and sinks are not brokers for their respective commodity, though sinks can be brokers for commodities they do not request. Figure 2 presents the very simple graph model of the publish/subscribe network of Fig. 1 that will be optimized as the instance named *tinyInstance*. Based on the elements so far introduced, a first integer linear formulation for problem P/SP is the following one. We define:

- x_{ij}^k as the *integer* variable that denotes how many data paths connect the publisher of k and single subscribers going through arc (i, j).
- ξ_{ij}^k as the *binary* variable that takes value 1 if commodity $k \in K$ is transmitted along arc (i, j). We also introduced artificial variable to manage the case in which some subscribers cannot receive the commodity.
- cap_{ij} as the capacity of edge $\{i, j\}$, accounting both for arc (i, j) and (j, i).
- finally, c_{ij} as the penalty that is paid if the arc (i, j) is used.

The mathematical formulation, denoted as **F1** is as follows:

$$\textbf{min } z_{F1} = \sum_{k \in K} \sum_{i \in N} \sum_{j \in \Gamma_i^{-1}} c_{ji} x_{ji}^k \tag{1}$$

$$\textbf{subject to } \sum_{j \in \Gamma_t^{-1}} x_{jt}^k = 1 \qquad\qquad t \in T^k, k \in K \tag{2}$$

$$\sum_{j \in \Gamma_i^{-1}} x_{ji}^k = \sum_{j \in \Gamma_i} x_{ij}^k \qquad\qquad i \in B, k \in K \tag{3}$$

$$|T^k|\xi_{ij}^k \geq x_{ij}^k \qquad\qquad (i, j) \in A, k \in K \tag{4}$$

$$\sum_{k \in K} a^k (\xi_{ij}^k + \xi_{ji}^k) \leq cap_{ij} \qquad\qquad i \in N, j \in \Gamma_i \tag{5}$$

$$x_{ij}^k \in Z_0^+ \qquad\qquad (i, j) \in A, k \in K \tag{6}$$

$$\xi_{ij}^k \in \{0, 1\} \qquad\qquad (i, j) \in A, k \in K \tag{7}$$

where $N = S \cup B \cup T$.

Formulation F1 can be decomposed on the contribution of every single node and the whole problem can thus be optimized in a fully distributed fashion, provided that routing decisions can be enforced downpath. Unfortunately, problem P/SP and its distributed counterparts is NP-hard due to the arc capacity constraints, besides being based on the integer multicommodity flow problem, and we cannot, in general, expect to solve it within the characteristic time needed to operate real-world IoT networks. We resort therefore to a distributed heuristic, specifically a Lagrangian matheuristic [4], that leverages the mathematical formulation F1. In detail, we relax constraints 5 and we insert them in the objective function with non negative penalties λ_{ij}, $(i, j) \in A$. The resulting formulation LP is as follows:

Algorithm 1: Core Lagrangian heuristic

---
1 function LagrHeuristic();
   Input : Control parameters
   Output: A feasible solution $\mathbf{x}^*$ of value z*
2 Initialize the penalty vector $\boldsymbol{\lambda}$;
3 Identify an "easy" subproblem LR($\boldsymbol{\lambda}$);
4 **repeat**
5   solve subproblem LR($\boldsymbol{\lambda}$) obtaining the possibly infeasible solution $\mathbf{x}$;
6   check for unsatisfied constraints;
7   update penalties $\boldsymbol{\lambda}$;
8   construct problem solution $\mathbf{x}^h$ using $\mathbf{x}$ and $\boldsymbol{\lambda}$;
9   if $z(\mathbf{x}^h) < z^*$ then $\mathbf{x}^* = \mathbf{x}^h$;   // $z(\mathbf{x}^h), z^*$ are the costs of $\mathbf{x}^h, \mathbf{x}^*$
10 **until** *end_condition*;

$$\min z_{LP} = \sum_{k \in K} \sum_{i \in N} \sum_{j \in \Gamma_i^{-1}} (c_{ji} x_{ji}^k + \lambda_{ji} a^k (\xi_{ij}^k + \xi_{ij}^k)) - \sum_{(i,j) \in A} \lambda_{ij} cap_{ij} \tag{8}$$

$$\textbf{subject to } (2), (3), (4), (6), (7),$$

$$\lambda_{ij} \geq 0 \qquad (i,j) \in A \tag{9}$$

The constraint matrix now has a block structure and decomposes over the different commodities, the only linkage being the objective function. It is thus possible to solve separately for each commodity, identifying a Dijkstra tree rooted in the corresponding publisher and having leaves in the subscriber nodes.

Formulation F1 actually contains the standard integer programming formulation of the Single Source Shortest Path problem (SSSP), it keeps it as a subproblem, therefore the solution of the subproblems is readily made by means of any code for the SSSP. Unfortunately, the Dijkstra algorithm lends itself poorly to a distributed implementation. However, distributed alternatives to Dijkstra for standard SSSP instances have been studied.

Most Lagrangian heuristics share a common general structure, that constitutes a metaheuristic of its own. The structure is presented in Algorithm 1. In our case, the steps requiring most attention are Step 5 and Step 7.

Step 5 asks for solving problem LP, given the arc weights, and it decomposes into solving $|S|$ single source shortest path problems, one for each publisher. We implemented a fully parallel *Multiple Source Shortest Path* (MSSP) algorithm. The algorithm is based on a dynamic programming adoption of Dijkstra's cost update equation (details at the conference). The estimate at time t of the minimum cost of the path from source s to node i, $f_i(t)$, is

$$f_i(t) = \begin{cases} 0, & \text{if } i = s \\ \min_{j \in \Gamma_i^{-1}} f_j(t-1) + c(j,i), & \text{otherwise} \end{cases} \tag{10}$$

and the predecessor *pred(i)* is accordingly updated.

Step 7 requires special attention, because we need to distribute penalty updates over the nodes, but the standard subgradient update rule, Polyak rule, makes use of a normalization factor computed over all network subgradients. This was unacceptable for the application, but we could solve it by means of a rule similar to the one used in [2], called *quasi constant step size rule*. More details at the conference.

3   Preliminary Computational Results

We validated our approach first by coding in C# a fully distributed implementation of our approach, then by moving it in C++ to be run in the Omnet++ [3] network simulator. To validate our approach, we generated some simple P/SP instance ourselves and adapted to Pub/Sub a few well-known multicommodity network flows instances arising in the telecommunication industry, up to 71 nodes and 244 arcs. All instances are available from [5].

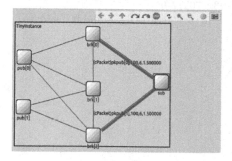

Fig. 3. Omnet++ ongoing simulation for TinyInstance

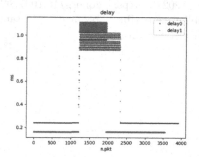

Fig. 4. Packet delays with operational and corrupted arc

Given constraint on short paper length, here we present only some illustrative obtained results. Figure 3 shows the Omnet++ rendering of the instance TinyInstance of Figs. 1 and 2, as taken during a simulation.

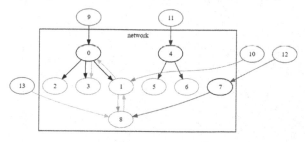

Fig. 5. Optimized data paths for instance res0

Figure 4 shows the final result of a simulation on TinyInstance, where the network was initially complete, then, at about 1/3 of the simulation, arc (0,3) was dropped and later reinstalled. There is a clear disruption of the level of service, that affects both transmissions as the network quickly reroutes all packets from 4 to 3 through brokers 1 and 2. When the arc is reinstalled, the original situation is recovered (but note the different adaptation times). Finally, Fig. 5 shows the optimized topology of a more complex case, resulting from the adaptation of one of the integer multicommodity flow instances.

References

1. Nakamura, K., et al.: A LoRa-based protocol for connecting IoT edge computing nodes to provide small-data-based services. Digit. Commun. Netw. **8**, 257–266 (2021)
2. Boschetti, M.A., Maniezzo, V., Roffilli, M.: A fully distributed lagrangean solution for a peer-to-peer overlay network design problem. INFORMS J. Comput. **23**(1), 90–104 (2011)
3. Varga, A.: OMNeT++. In: Wehrle, K., Güneş, M., Gross, J. (eds.) Modeling and Tools for Network Simulation, pp. 35–59. Springer, Berlin (2010)
4. Maniezzo, V., Boschetti, M.A., Stützle, T.: Matheuristics, Algorithms and Implementations. EURO Advanced Tutorials on Operational Research, Springer, Heidelberg (2021). https://doi.org/10.1007/978-3-030-70277-9
5. Publish/subscribe problem instances. http://astarte.csr.unibo.it/pspdata/PSPinstances.html

An Efficient Fixed Set Search
for the Covering Location
with Interconnected Facilities Problem

Isaac Lozano-Osorio[1]([✉]) [ID], Jesús Sánchez-Oro[1] [ID], Anna Martínez-Gavara[2] [ID],
Ana D. López-Sánchez[3] [ID], and Abraham Duarte[1] [ID]

[1] Universidad Rey Juan Carlos, Madrid, Spain
{isaac.lozano,jesus.sanchezoro,abraham.duarte}@urjc.es
[2] Universidad Pablo de Olavide, Sevilla, Spain
gavara@uv.es
[3] Universitat de València, Valencia, Spain
adlopsan@upo.es

Abstract. This paper studies the Coverage Location Problem with Interconnected Facilities (CPIF). It belongs to the family of Facility Location Problems, but being more realistic to nowadays situations as surveillance, or natural disaster control. This problem aims at locating a set of interconnected facilities to minimize the number of demand points that are not covered by the selected facilities. Two facilities are considered as interconnected if the distance between them is smaller than or equal to a predefined distance, while a facility covers a demand point if the distance to it is smaller than a certain threshold. The wide variety of real-world applications that fit into this model makes them attractive for designing an algorithm able to solve the problem efficiently. To this end, a metaheuristic algorithm based on the Fixed Set Search framework is implemented. The proposed algorithm will be able to provide high-quality solutions in short computational times, being competitive with the state-of-the-art.

Keywords: Combinatorial optimization · Covering location problem · Fixed set search

1 Introduction

The Facility Location Problem (FLP) [1] seeks to select the best location for a set of facilities optimizing a certain criterion, e.g., the cost of deployment, the total travel time for the clients to satisfy their demands, etc. There are many FLP variants in the literature, depending on the criterion used to place the facilities and the constraints to be considered [6]. Furthermore, practitioners are forced to introduce new constraints to model more realistic situations as stated in [2].

This paper addresses a variant of the p-Coverage problem [2] which considers that the selected facilities are interconnected, i.e., each pair of facilities are

L. Di Gaspero et al. (Eds.): MIC 2022, LNCS 13838, pp. 485–490, 2023.
https://doi.org/10.1007/978-3-031-26504-4_37

located within a maximum distance between them. This problem is named the Coverage Location Problem with Interconnected Facilities, from now on CPIF, where the main constraint indicates that the distance between two facilities cannot exceed a given distance r, and a facility covers a demand point if and only if the distance between them does not exceed a given threshold R. The original work [2] proposed three exact formulations for the problem and a metaheuristic method to solve it in a short computational time.

The interest in solving the CPIF problem is due to the fact that there are many realistic problems that fit into this model. For example, in the context of the Internet of Things, the deployment of a network of heterogeneous sensors such as alarms or motion detectors [8] or minimizing the impact of a natural disaster [2].

The paper is organized as follows: Sect. 2 defines the location problem addressed in this manuscript. Section 3 describes the proposed algorithm and the new strategies that we have implemented to solve it. Section 4 includes the computational results. Finally, the conclusions and future research are discussed in Sect. 5.

2   Problem Description

The Coverage Problem with Interconnected Facilities (CPIF) was recently introduced [2]. The problem is defined as a set of locations available to host a facility F and a set of demand points D that require the services of a facility. Let us define a distance function $d(i, j)$ which evaluates the shortest distance between two points (either facilities or demand points).

Without loss of generality, it can be assumed that $F = D$, i.e., every demand point is able to host a facility, instead of considering two disjoint sets, $F \cap D = \emptyset$. Notice that node 0 is denoted as the root node that always hosts a facility, as it is the site where the central facility is located [2].

Two facilities $i \in F$ and $j \in F$ are interconnected if and only if the distance between them is smaller than or equal to a given threshold r, that is, $d(i, j) \leq r$. Additionally, a demand point x is considered covered by a facility i if and only if $d(x, i) \leq R$, where R is the maximum distance in which a facility can satisfy the demand of a client. It is worth mentioning that in CPIF there are no capacity constraints, so each demand point is always assigned to its closest facility.

The objective of CPIF is to place a set of interconnected facilities, $S \subseteq F$, with $|S| = p$, where p is a fixed constraint a priori, in order to minimize the number of demand points that are not covered, i.e., the distance between the demand point and its closest facility is strictly larger than the given threshold R. Therefore, given a solution S, the objective function of CPIF is evaluated as follows:

$$CPIF(S) \leftarrow \left| \left\{ i \in D : \min_{j \in S} d(i, j) > R \right\} \right| \tag{1}$$

Then, CPIF aims to find a solution S^* with the minimum objective function value. More formally,

$$S^* \leftarrow \arg \min_{S \in \mathbb{SS}} CPIF(S) \tag{2}$$

where $\mathbb{SS}$ is the complete search space, which is conformed by all possible combinations of $p-1$ facilities (notice that node 0 always belong to the solution since it is the root node).

Figure 1 illustrates the CPIF with a example considering $p = 3$, $r = 8$, and $R = 10$. In the figure, candidate facilities are represented by a square, i.e., $F = \{1, 2, 3, 8\}$, being 0 the root node, while demand points are represented by a circle, i.e., $D = \{4, 5, 6, 7, 9, 10, 11\}$. In this example, a selected facility is represented by a gray gradient from black to white. When a demand point is covered by a facility, it is colored with the same gray gradient. Given the distance constraints, the only feasible solutions are those depicted in the figure, since any other combination of three facilities exceeds the maximum distance between the selected facilities r to consider that they are interconnected. In Fig. 1(a) facilities $S_1 = \{0, 1, 3\}$ are selected. Facilities 1 and 3 can be selected as facilities, since $d(0, 1) = 5 \leq r$ and $d(0, 3) = 5 \leq r$. Then, every demand point is assigned to its closest facility: 5 and 10 are assigned to 0, 4 and 11 are assigned to 3, 6 and 9 are assigned to 1, and finally, 7 cannot be assigned to any facility since none of them satisfies the distance constraint. In this case, the objective function value is $CPIF(S_1) = 1$ since there is only one demand point that is not covered.

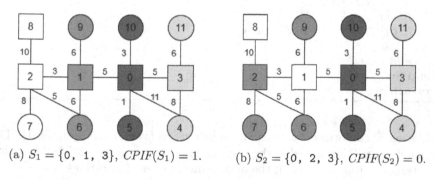

(a) $S_1 = \{0,\ 1,\ 3\}$, $CPIF(S_1) = 1.$ (b) $S_2 = \{0,\ 2,\ 3\}$, $CPIF(S_2) = 0.$

Fig. 1. Example of two possible solutions with 4 candidate facilities (square) and 7 demand points (circle).

Figure 1(b) shows another feasible solution, $S_2 = \{0, 2, 3\}$. In this solution, 5 and 10 are assigned to 0, 4 and 11 are assigned to 3, and finally, 6, 7 and 9 are assigned to 2. Notice that 9 cannot be assigned to 0 since $d(0, 9) > R$, but it can be assigned to 2 since $d(2, 9) = d(2, 1) + d(1, 9) = 9 \leq R$. In this case, the objective function value is $CPIF(S_2) = 0$ since all demand points are covered, being S_2 the optimal solution for this instance.

In this particular example, there are no additional feasible solutions $\mathbb{SS} = \{S_1, S_2\}$, since any other combination of selected facilities does not satisfy the requirement that they are interconnected. It is remarkable that fixing the root node at 0 limits the search, eventually resulting in solutions where the central facility 0 is not assigned to any demand point.

3   Fixed Set Search

Fixed Set Search (FSS) is a recently proposed population-based metaheuristic that includes a learning mechanism to the constructive procedure. It was originally introduced to solve the traveling salesman problem [3], and was later used to solve the Minimum Weighted Vertex Cover Problem [4] and minimizing makespan [5].

Algorithm 1 shows the pseudocode of the FSS. The algorithm has 4 input parameters, namely: δ, number of initial solutions generated by the constructive method; κ, number of selected solutions, ι number of maximum repetitions, and τ, starting size of the solution. The method starts by generating an initial

---

Algorithm 1 $FSS(\delta, \kappa, \iota, \tau)$

1: $P \leftarrow Populate(\delta)$
2: $S_b \leftarrow \arg\min_{S' \in P} CPIF(S')$
3: **for** $i \in 1 \ldots \iota$ **do**
4:     $P_\kappa \leftarrow RND(P, \kappa)$
5:     $S \leftarrow RND(P)$
6:     $S' \leftarrow Fix(S, P_\kappa, (1 - \tau) \cdot |S|)$
7:     $S'' \leftarrow RGF(S')$
8:     $P \leftarrow P \cup S'''$
9:     **if** $CPIF(S''') < CPIF(S_b)$ **then**   ▷ Improve
10:         $S_b \leftarrow S'''$
11:    **end if**
12: **end for**
13: **return** S_b

---

population P conformed by δ random solutions to favor diversity (step 1). The best solution in P is selected as the best current solution in step 2. The algorithm then iterates (steps 3-12) until the maximum number of iterations ι is reached.

In each iteration, a set of not repeated solutions P_κ is generated (step 4) as a random subset of κ solutions from P. Next, a single random solution is set to start the procedure (step 5). Then, a partial solution is generated by the Fix procedure (step 6). This solution is created by selecting the most frequent facilities in P_k, among those that satisfy the constraint that guarantees that the facilities are interconnected. The procedure RGF is a greedy procedure responsible for generating a complete solution S'' starting from the partial solution S' (step 7). In particular, RGF adds elements to S' selecting the interconnected facilities that produce the best quality based on the objective function value. After that, solution S'' is added to P in step 8. Finally, if solution S'' is better than the best solution found so far S_b (steps 9-11), it is updated (step 10). The procedure ends with the return of the best solution found during the search (step 13).

4   Computational Results

This section describes the computational experiments designed to evaluate the performance of the proposed algorithms and analyze the results obtained. All experiments have been performed in an Intel Core i7-9750H (2.6 GHz) with 16 GB RAM and the algorithms were implemented using Java 17 and the *Meta-heuristic Optimization framewoRK* (MORK) 10 [7]. All instances and source code of the proposed methods have also been made publicly available at https://grafo.etsii.urjc.es/CPIF-MIC. We would like to thank the authors of the previous work [2] for kindly sending us their source code. This has helped us to provide a fair comparison by executing all algorithms under the same hardware constraints. The testbed of instances used in this work is the same set considered in the previous work, which is derived from the well-known OR-Library http://people.brunel.ac.uk/~mastjjb/jeb/info.html, 555 instances are considered where $D = \{100, ..., 900\}$, $F = \{5, ..., 200\}$, $r = \{25, 50, 80, 100, 150\}$ and $R = \{8, 10, 12, 15, 20\}$.

The results compare the performance of our proposal with the state-of-the-art metaheuristic, based on Iterated Local Search (ILS) and the best exact method, which is a Mixed-Integer Linear Programming implemented in the commercial CPLEX solver (CPLEX) [2] in those instances in which the exact algorithm is able to provide the optimal solution.

The Fixed Set Search (FSS) parameters are experimentally set as follows $\iota = 20$, $\kappa = 10$, $\delta = 10$, $\tau = 1/2$. Table 1 contains the following performance metrics: the average objective function value, Avg.; the average execution time of the algorithm measured in seconds, Time (s); the average deviation with respect to the exact solution, Dev. (%); and, finally, the number of times that the algorithm is able to reach the best solution in the experiment (#Best).

Table 1. Results of the FSS algorithm versus the state-of the-art procedures.

Algorithm	Avg.	Time (s)	Dev. (%)	#Best
FSS	154.17	2.05	5.87%	266
ILS	155.04	2.05	8.09%	212
CPLEX	151.83	36.43	0.00%	555

Table 1 shows competitive results when comparing both heuristic approaches. In terms of deviation, ILS reports 8.09% versus 6.51% of the FSS in a similar computing time (2.05 s in the state of the art versus 2.11 s in the FSS). Analyzing the number of best solutions found, FSS is able to reach 246 out of the 555 available instances, while ILS obtains 212 best solutions.

5   Conclusion

In this paper, the Covering Location Problem with Intermediate Facilities is addressed with a Fixed Set Search algorithm. The results obtained are promising when comparing them with the state-of-the-art algorithm. In particular, FSS is able to outperform the results obtained by ILS and remains close to the optimal values reported by CPLEX.

As future research, it would be interesting to address an intelligent way to construct the initial population as stated in [3]. Additionally, an efficient local search method will be able to further improve the obtained solutions. Finally, a further study on the parameters of the Fixed Set Search will be performed and more challenging instances derived from real-life applications, where the exact methods are not able to obtain solutions, will be considered.

Acknowledgments. The authors acknowledge support from the Spanish Ministry of Ciencia, Innovación y Universidades under grant ref. PID2021-125709OA-C22 and PID2021-126605NB-I00, Comunidad de Madrid and Fondos Estructurales of the European Union with grant references S2018/TCS-4566, Y2018/EMT-5062, the Spanish Ministry of Economía, Industria y Competitividad through Projects PID2019-104263RB-C41, and from Junta de Andalucía, FEDER-UPO Research & Development Call, reference number UPO-1263769.

References

1. Balinski, M.L.: Integer programming: methods, uses, computations. Manag. Sci. **12**(3), 253–313 (1965)
2. Cherkesly, M., Landete, M., Laporte, G.: Median and covering location problems with interconnected facilities. Comput. Oper. Res. **107**, 1–18 (2019)
3. Jovanovic, R., Tuba, M., Voß, S.: Fixed set search applied to the traveling salesman problem. In: Blesa Aguilera, M.J., Blum, C., Gambini Santos, H., Pinacho-Davidson, P., Godoy del Campo, J. (eds.) HM 2019. LNCS, vol. 11299, pp. 63–77. Springer, Cham (2019). https://doi.org/10.1007/978-3-030-05983-5_5
4. Jovanovic, R., Voß, S.: Fixed Set Search Applied to the Minimum Weighted Vertex Cover Problem. In: Kotsireas, I., Pardalos, P., Parsopoulos, K.E., Souravlias, D., Tsokas, A. (eds.) SEA 2019. LNCS, vol. 11544, pp. 490–504. Springer, Cham (2019). https://doi.org/10.1007/978-3-030-34029-2_31
5. Jovanovic, R., Voß, S.: Fixed set search application for minimizing the makespan on unrelated parallel machines with sequence-dependent setup times. Appl. Soft Comput. **110**, 107521 (2021)
6. López-Sánchez, A.D., Sánchez-Oro, J., Laguna, M.: A new scatter search design for multiobjective combinatorial optimization with an application to facility location. INFORMS J. Comput. **33**(2), 629–642 (2020)
7. Martín, R., Cavero, S.: rmartinsanta/mork: v0.11 (2022)
8. Srinidhi, N.N., Kumar, S.M.D., Venugopal, K.R.: Network optimizations in the internet of things: a review. Eng. Sci. Technol. Int. J. **22**(1), 1–21 (2019)

Hybrid PSO/GA+solver Approaches for a Bilevel Optimization Model to Optimize Electricity Dynamic Tariffs

Maria João Alves[1,2](✉) , Carlos Henggeler Antunes[2] , and Inês Soares[2]

[1] CeBER and Faculty of Economics, University of Coimbra, Coimbra, Portugal
mjalves@fe.uc.pt
[2] INESC Coimbra, Department of Electrical and Computer Engineering, University of Coimbra, Coimbra, Portugal
{ch,inesgsoares}@deec.uc.pt

Abstract. Electricity retail markets are subject to competition and retailers generally work with thin commercialization margins. Thus, to increase their market share, these companies should offer attractive tariff options to consumers, including time-of-use pricing schemes, to maximize profits by exploiting the differences between buying energy in wholesale markets and selling it to consumers. In turn, consumers aim to minimize the electricity bill by making the most of time-differentiated prices. For this purpose, consumers may be assisted by an automated energy management system performing on their behalf the integrated optimization of appliance operation, charging/discharging of electric vehicle and stationary batteries, on-site generation, and exchanges with the grid. The retailer's problem considering the consumer's demand response can be formulated as a bilevel mixed-integer nonlinear programming model. The retailer is the leader acting first by setting the prices, and the consumer is the follower reacting to those prices. Two hybrid PSO+solver and GA+solver algorithms have been developed to cope with the complexity of this model. The PSO and the GA deal with the upper-level search determining the prices. The exact solver computes the solution to the lower-level problem for each price instantiation, which becomes a mixed-integer linear program to determine the corresponding optimal demand schedule. Results are presented for realistic data, comparing the two hybrid approaches. The GA+solver approach achieved slightly better results than the PSO+solver approach.

Keywords: Bi-level mixed-integer nonlinear programming · Hybrid meta-heuristic · Dynamic tariffs

1 Introduction

The liberalization of the electricity sector led to competitive retail markets, in which retailers compete for residential, commerce/services and industrial consumers. This paper focuses on residential consumers, although the models can be applied to other

L. Di Gaspero et al. (Eds.): MIC 2022, LNCS 13838, pp. 491–498, 2023.
https://doi.org/10.1007/978-3-031-26504-4_38

consumer types. Retailers buy electrical energy in different types of wholesale markets and sell it to consumers, aiming to maximize profits. Dynamic tariffs are interesting for retailers, enabling them to exploit differences in buying and selling prices, as well as for consumers, making the most of their flexibility in the scheduling of energy resources to minimize costs considering comfort needs. The consumers' decisions encompass appliance operation, charging/discharging of electric vehicle and stationary batteries, on-site generation, and exchanges with the grid (grid to home and home to grid). There is a hierarchical relation between the two decision makers pursuing distinct aims: the retailer is the leader setting prices, and the consumer is the follower reacting to those prices. This decision setting can be represented by a bilevel mixed-integer nonlinear programming model to assist the retailer to define dynamic tariffs considering the demand-side response. The design of retail tariffs integrating demand response has been addressed using bilevel models in a considerable number of studies in the literature [1]. However, most of them consider very simplified formulations of the appliance operation in the consumer's problem, which are not realistic.

In this work, we consider a consumer's model to optimize the operation of shiftable appliances, an electric water heater, an air conditioner, static and electric vehicle batteries and microgeneration, including buying and selling exchanges with the grid. Two approaches have been developed to cope with the complexity of the bilevel model, which use either a particle swarm optimization (PSO) algorithm or a genetic algorithm (GA) to guide the search for price solutions in the upper level (retailer's) problem, and a mixed-integer linear programming (MILP) solver to obtain an optimal solution to the lower level (consumer's) problem for each price setting. This approach follows the work in [2], where only the PSO was used, and a much more comprehensive consumer's model is considered herein.

This paper is organized as follows. In Sect. 2 the bilevel model is outlined. In Sect. 3, the PSO-solver and the GA-solver algorithms are described, and the numerical results comparing the two approaches for realistic data are presented. The main conclusions are drawn in Sect. 4.

2 A Bilevel Nonlinear Optimization Model to Optimize Electricity Dynamic Tariffs

Bilevel problems are difficult to solve because there is a lower-level optimization problem nested into another optimization problem, the upper-level problem. Thus, the leader must incorporate into his/her optimization process the follower's reaction because it affects the leader's objective value. Bilevel programming is known to be strongly NP-hard [3]. Recently, there has been an increasing interest in evolutionary techniques for bilevel problems due to their potential to tackle these problems and their application in real world problems [4].

In the bilevel problem addressed in this work, the retailer (*leader*) determines the electricity prices x_i (€/kWh) to be charged to the consumer (*follower*) in each predefined subperiod $P_i(i = 1, \ldots, I)$ of the planning period Γ, which is discretized in T time intervals $(t = 1, \ldots, T)$. The retailer's objective function is to maximize the profit.

Knowing the electricity prices, the consumer optimizes his/her energy uses aiming to minimize the electricity bill.

The consumer's model parameterized on the electricity prices is considered at the lower level of the bilevel problem, which includes the accurate modeling of different energy resources and exchanges with the grid:

- *Shiftable loads*, which are characterized by an operation cycle associated with each program and cannot be interrupted (examples are laundry machines, dishwashers and cloth driers). For these loads, the consumer provides the comfort time slots for load operation according to his/her preferences; the optimization determines the starting operation time and guarantees that the cycle is entirely executed in the due sequence within any of the comfort time slots.
- An *electric water heater*, whose operation is controlled by a thermostat. Its modelling requires several technical inputs (the power of the heating element, the ambient and inlet water temperatures, the tank characteristics, etc.) and data from the consumer (minimum/maximum allowed temperatures and water withdrawals); the optimization determines the on/off status of the heating element in each time, which defines the hot water temperature in the tank.
- An *air conditioner* system, which accounts for the behavior of the control thermostat with hysteresis. Technical inputs include the nominal power, performance of the system, outdoor temperatures and the thermal characteristics of the building envelope, while the consumer provides the minimum and the maximum comfort indoor temperatures; the optimization determines the on/off status of the air conditioner in each time, which defines the indoor temperature.
- A *static battery* and an *electric vehicle battery*, which provide energy exchanges. Technical inputs include charging/discharging efficiencies, minimum/maximum allowed battery charges, and data provided by the consumer mainly concerned with the utilization of the electric vehicle; the optimization determines the charging and discharging patterns of the static and electric vehicle batteries.
- A *base load* not deemed for control (examples are tv set and oven) and local microgeneration using a *photovoltaic* system are also considered in the model.

The consumer's MILP model requires a large number of binary variables and constraints. The optimization process yields the energy required from the grid (grid to home) and the energy sold to the grid (home to grid) in each time interval t: E_t^{G2H} and E_t^{H2G} (kWh), respectively. The set of all constraints of the consumer's model is represented below by $G(E_t^{G2H}, E_t^{H2G})$.

The bilevel model is outlined as follows:

$$\max_{x} F = \sum_{i=1}^{I} \sum_{t \in P_i} \left(x_i E_t^{G2H} \right) - \sum_{t=1}^{T} \left(\pi_t E_t^{G2H} \right) \tag{1}$$

$$\text{s.t.} \quad \underline{x}_i \leq x_i \leq \overline{x}_i \quad i = 1, \ldots, I \tag{2}$$

$$\frac{1}{T} \sum_{i=1}^{I} \left(P_{U_i} - P_{L_i} + 1 \right) x_i \leq x^{AVG} \tag{3}$$

$$\min_{E^{G2H}, E^{H2G}} f = \sum_{i=1}^{I} \sum_{t \in P_i} \left(x_i E_t^{G2H} \right) - \sum_{t=1}^{T} \left(c E_t^{H2G} \right) \tag{4}$$

$$\text{s.t.} \quad \text{Constraints } G(E_t^{G2H}, E_t^{H2G}) \tag{5}$$

The retailer's objective function (1) is to maximize profit (the difference between the revenue with the sale of energy to consumers and the acquisition cost in the wholesale market), where π_t (€/kWh) is the energy acquisition price incurred by the retailer in time t. Constraints (2) impose minimum ($\underline{x}_i$) and maximum ($\overline{x}_i$) values on x_i in each sub-period $P_i = \left[P_{L_i}, P_{U_i} \right] \subset \Gamma$. Constraint (3) imposes a maximum average price (x^{AVG}) for the whole planning horizon. These upper-level constraints, (2) and (3), aim to represent market competition of retailer prices as proposed in [5]. The lower-level objective function (4) is to minimize the consumer's net cost, where c (€/kWh) is the remuneration to the consumer by selling energy to the grid. Lower-level constraints (5) model all the above-mentioned consumer energy resources and exchanges with the grid.

Due to space limitations, the detailed model and the description of all variables and parameters are available at https://home.deec.uc.pt/~ch/consumermodel.

3　Hybrid PSO+solver and GA+solver Approaches

The hybrid PSO+solver and GA+solver approaches have a similar structure. The population consists of N individuals $x^n = \left(x_1^n, \ldots, x_I^n \right), n = 1, \ldots, N$, each one representing a price vector (leader's decision variables). The initial population is randomly generated, and then the population evolves in the search space according to the principles of PSO and GA in each approach. For each feasible x^n, the follower's problem (4–5) is solved by the MILP solver *Cplex*. The algorithms run during G iterations. The general scheme of these hybrid approaches is displayed in Fig. 1.

The PSO tries to iteratively improve the particles (price vectors) by moving them along the best directions. Each particle x^n is influenced by its best-known position $-x^{n^{best}}$, and the best position of the entire swarm $- g^{best}$. These positions are updated whenever better solutions to the bilevel problem are found, i.e., according to the leader's objective function F in (1). The movement in iteration q of each particle x^{n^q} is determined by its previous position $x^{n^{q-1}}$ and the velocity vector v^{n^q} as follows:

$$v_i^{n^q} = \eta v_i^{n^{q-1}} + r_1 C_1 \left(x_i^{n^{best}} - x_i^{n^{q-1}} \right) + r_2 C_2 \left(g_i^{best} - x_i^{n^{q-1}} \right), \forall i = 1, \ldots, I$$

$$x^{n^q} = x^{n^{q-1}} + v^{n^q}$$

where η is the inertia weight, C_1 and C_2 are the cognitive and social parameters, and $r_1, r_2 = rand\ (0,1)$. For simplicity reasons, we will omit the iteration index q in what follows.

A turbulence (mutation) operator is applied to x^n with probability p_m, by making $x_i^n \leftarrow x_i^n + \zeta$, with random $\zeta \in \left[-\delta(\overline{x}_i - \underline{x}_i), \delta(\overline{x}_i - \underline{x}_i)\right], \forall i = 1,\ldots,I$ (δ is a pre-defined constant). This operator aims to diversify the search. After the movement and turbulence, if x^n does not satisfy the upper-level constraints (2–3), the repairing routine described in [2] is called to fix it. Since, in practice, electricity prices have a fixed number of decimal places (4 with prices expressed in €/kWh), this issue is also taken into account in the repairing routine, ensuring that constraints are satisfied for x_i^n rounded to 4 decimal places.

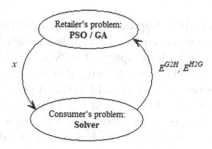

Fig. 1. General scheme of the PSO+solver and GA+solver approaches.

The lower-level problem (4–5) parameterized on x^n is solved by *Cplex*, for each $n = 1,\ldots,N$, yielding $\left(E^{G2H^n}, E^{H2G^n}\right)$. The solutions $\left(x^n, E^{G2H^n}, E^{H2G^n}\right)$ are then evaluated by F. The output of the algorithm is the solution that attains the highest F value (F^{best}) over the G iterations.

The GA starts each new generation by creating the offspring population of N individuals: for each mating, one parent is selected using a binary tournament (decided by the F value) and the other parent is picked randomly. Both individuals have the same probability of being the first or the second parent, and they are subject to one-point crossover to generate one offspring: if $x^{n1} = \left(x_1^{n1},\ldots,x_I^{n1}\right)$ and $x^{n2} = \left(x_1^{n2},\ldots,x_I^{n2}\right)$ are the first and the second parents, respectively, then the offspring is $x^c = \left(x_1^{n1},\ldots,x_{i_1}^{n1},x_{i_1+1}^{n2}\ldots,x_I^{n2}\right)$ where i_1 is the crossover point drawn at random in $[2, I-1]$. Then, a mutation operator (like the turbulence operator used in the PSO) is applied to x^c with probability p_m and, if the new x^c does not satisfy the constraints (2–3), the repairing routine is called. N children are generated in this process.

As in the PSO, the follower's problem (4–5) parameterized on x^c is solved by *Cplex* for each $c = 1,\ldots,N$; the solutions are then assessed by the leader's objective function F defined in (1), which is the fitness function. In the selection process that determines the population for the next generation, the best individual in the parent population and the best one in the offspring are first selected; the other $N-2$ individuals are chosen by binary tournaments without replacement between an individual of the current population and another one of the offspring, both randomly selected. The output of the algorithm is the solution with the highest F value (F^{best}) over the G iterations.

Other features were embedded in both algorithms (PSO and GA): 1) Adaptive mutation scheme: If F^{best} has no improvement during a predefined number G' of consecutive iterations (i.e., $\frac{F^{best q} - F^{best q-1}}{F^{best q}} < \tau$ for a given threshold τ), then the mutation probability p_m is increased to promote further exploration; 2) Adopting an *optimistic* perspective of the bilevel problem, i.e., when the follower has alternative optimal solutions for his/her objective function, he/she will choose the solution that most benefits the leader: at the end of each iteration, the lower-level problem with ties broken in favor of the leader is solved again for the best solution found in that iteration. The result of this problem is used for the possible update of F^{best}. Other perspectives (e.g., the pessimistic one) could be adopted in these hybrid algorithms without increasing the computational difficulty.

3.1 Results

The algorithms were coded in R language and run in a computer with an Intel Xeon Gold 6138 CPU@3.7 GHz processor. The *Cplex* solver is called from the R code. The following parameters were considered: specific for PSO, $\eta = 0.4$ and $C_1 = C_2 = 1$; for both algorithms, $p_m = 0.05$ and it increases to 0.1 if there is no improvement of F after $G' = 5$ consecutive iterations; $\tau = 0.001$ and $\delta = 0.2$. A planning horizon of 24 h has been discretized into intervals of 15 min ($T = 96$), and six periods ($I = 6$) are considered for the electricity prices charged to the consumers. The consumer's problem includes three shiftable loads (dishwasher, laundry machine and clothes dryer), an electric water heater, an air conditioner system, a stationary battery, and an electric vehicle battery. All the data are available at http://dx.doi.org/10.17632/j2vr7jgmcz.1.

For each instantiation of the electricity prices, the lower-level problem (4–5) comprises 933 binary variables, 1440 continuous variables and 2232 constraints. Although being MILP problems, the lower-level problems are difficult to solve to optimality. Thus, a predefined maximum computation time of 1 min was imposed to solve each lower-level problem by means of *Cplex* (the algorithms need to solve $G \times N$ lower-level problems in each run). For the determination of the best *optimistic* solution of each iteration (one resolution per iteration) a 5 min computation time was given. Ten independent runs were performed for each algorithm, considering $G = 20$ and $N = 10$.

The performance of both approaches was very similar in terms of solution quality, with a slight advantage of the GA+solver approach regarding the maximum and average F^{best} values but displaying a higher standard deviation than the PSO+solver approach. The prices in the solutions with the maximum F^{best} values are also alike. A summary of the results is displayed in Table 1. The prices in the solution with the highest retailer's profit (which was obtained by the GA+solver, $F = 4.0700$), as well as the corresponding grid to home and home to grid energy flows, are displayed in Fig. 2.

This solution of a dynamic tariff can be compared with a flat rate equal to the average price (0.1620 €/kWh). Considering a computation time of 5 min to solve the consumer's problem with this flat rate, the retailer's profit is $F = 3.8762$, which is worse than the best solutions found using both hybrid approaches. Moreover, the consumer's cost corresponding to the best solution given by the GA+solver approach is $f = 6.9661$, which compares with the cost $f = 7.3135$ considering the flat rate tariff. Therefore, both the retailer and the consumer can be better off in a time-differentiated tariff setting, which is expected to become a prevalent pricing scheme in smart grids.

Table 1. Summary of the results of 10 runs with PSO+solver and GA+solver.

In 10 runs:	maximum F^{best}	average F^{best}	minimum F^{best}	st. dev. of F^{best}	Average time of one run	Prices (€/kWh) in the best solution
PSO+solver	**3.9501**	3.8370	3.6451	0.1004	4h 01m	(0.0956, 0.1969, 0.2017, 0.1972, 0.2068, 0.1378)
GA+solver	**4.0700**	3.8777	3.6365	0.1402	4h 13m	(0.0996, 0.1917, 0.1928, 0.1944, 0.1933, 0.1620)

4 Conclusion

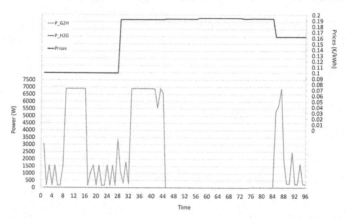

Fig. 2. Power flows in the best solution: grid to home (P_G2H) and home to grid (P_H2G).

Two hybrid approaches combining a meta-heuristic and a solver have been presented to tackle a bilevel mixed-integer nonlinear optimization model to determine the electricity dynamic tariffs that maximize the retailer's profit accounting for the consumers' reaction to minimize costs. The PSO and the GA deal with the upper-level search to find the prices, and the solver computes an optimal demand-side energy resource schedule in the lower-level problem for each price instantiation. The GA+solver approach achieved marginally better results than the PSO+solver approach. Moreover, it was shown that optimally-designed time of use tariffs can be profitable for the retailer and the consumer in comparison with flat rates.

Acknowledgments. This work has been funded through FCT – Fundação para a Ciência e a Tecnologia, I.P., within Projects UIDB/05037/2020, UIBD/00308/2020 and 3SQAIR-SOE4/P1/E10041.

References

1. Henggeler Antunes, C., Alves, M.J., Ecer, B.: Bilevel optimization to deal with demand response in power grids: models, methods and challenges. TOP **28**(3), 814–842 (2020). https://doi.org/10.1007/s11750-020-00573-y
2. Soares, I., Alves, M.J., Antunes, C.H.: Designing time-of-use tariffs in electricity retail markets using a bi-level model – estimating bounds when the lower level problem cannot be exactly solved. Omega **93**, 102027 (2020)
3. Hansen, P., Jaumard, B., Savard, G.: New branch-and-bound rules for linear bilevel programming. SIAM J. Sci. Statist. Comput. **13**(5), 1194–1217 (1992)
4. Sinha, A., Malo, P., Deb, K.: A review on bilevel optimization: from classical to evolutionary approaches and applications. IEEE Trans. Evol. Comput. **22**(2), 276–295 (2018)
5. Zugno, M., Morales, J.M., Pinson, P., Madsen, H.: A bilevel model for electricity retailers' participation in a demand response market environment. Energy Econ. **3**, 182–197 (2013)

An Agent-Based Model
of Follow-the-leader Search Using
Multiple Leaders

Martha Garzón, Lindsay Álvarez-Pomar, and Sergio Rojas-Galeano[✉]

Universidad Distrital Francisco José de Caldas, Bogotá, Colombia
srojas@udistrital.edu.co

Abstract. In this paper we study a swarm optimisation algorithm for real-valued bound-constraint cost functions whose search strategy operates on the basis of follow-the-leader intensification and random walk diversification. We studied the single-leader and multi-leader modes of operation. The simplicity of the search rules allows for a straightforward implementation of the algorithm as an agent-based model. In addition, various techniques were devised to prevent premature convergence to local minima and stagnation. We evaluate the efficacy/efficiency of the algorithm with empirical experiments on a testbed of well-known unconstrained real-valued cost functions using the NetLogo simulation environment. Our results indicate that the multi-leader configuration, with a small number of followers, proved to be advantageous both in accelerating convergence to the optimum on all testbed problems and in improving success rates compared to its single-leader version.

Keywords: Swarm intelligence · Real-valued unconstrained optimisation

1 Introduction

Despite the fact that many of the numerous swarm metaheuristics proposed in this century have taken inspiration from natural, biological or social metaphors [3,6], they essentially rely on an underlying algorithmic model that performs iterative shifts to candidate solutions derived from other representative solutions, an operation characterised as a *differential vector movement* [10]. When these representatives solutions improve their quality (fitness) as the algorithm progresses, the exploration of the swarm is guided towards increasingly more promising regions of the search space, resulting on a sort of "follow the leader" behaviour, the classic example being the *Particle Swarm Optimisation* (PSO) algorithm [9]. The ability of the leader to escape from local minima and the capacity of the followers to maintain diversity will determine the potential success of this type of algorithms.

In a recent review of 323 bio-inspired metaheuristics [10], a taxonomy of algorithms was outline from a behaviour perspective regardless of their inspirational metaphor. The largest category correspond to algorithms that perform

L. Di Gaspero et al. (Eds.): MIC 2022, LNCS 13838, pp. 499–505, 2023.
https://doi.org/10.1007/978-3-031-26504-4_39

differential vector movements of candidate solutions derived from a small group of representative solutions (169 out of the 323, or 53%), a nod to the efficacy that the follow-the-leader strategy has demonstrated in many natural occurring systems, and consequently, in the metaheuristics inspired in such behaviour.

Certainly many of the algorithms in such category will likely share some resemblance to the search rules we study in this paper; particularly we found that the *Leaders and Followers* (LaF) algorithm described in [5] is similar in spirit to our proposal. At the core of LaF, a subset of followers can follow any leader from the subset of leaders in each iteration (so the follower base is shared among leaders). In contrast, we investigate the behaviour of the search procedure when separate (non-overlapping) groups of followers are influenced by each leader. The details of the latter, along with and some additional operators intended to improve exploration and avoid stagnation as well as premature convergence, will be described in Sect. 2 where we focus our analysis on the algorithmic aspects of the resulting metaheuristic rather than on a metaphor-oriented rationale [11].

2 Methods and Materials

2.1 Algorithm Design

Let us define a real-valued optimisation problem by a cost function to be minimised, $f : \Re^d \to \Re$, where the goal is to search for the optimum solution $\mathbf{x}^* = \text{ARGMIN } f(\mathbf{x})$, subject to the bound constraints $a_k \le x_k \le b_k, k = 1, \ldots, d$, where $\mathbf{x} = [x_1, \ldots, x_d]$. The proposed algorithm maintains a set $S = \{\mathbf{x}_i \in \Re^d, 1 \le i \le n\}$ of candidate solutions to the problem f, that evolve based on the combination of two search rules: follow-the-leader intensification and random walk diversification. Thus, unlike other "follow-the-leader" type algorithms such as PSO or LaF, we incorporate the random walk operator in the hope of improving the exploration capability of the follow-the-leader exploitation strategy. For this purpose, each agent behaves as one of three possible roles: follower, leader or walker (hence we call it the *FLW* algorithm); the effect of each of these roles is discussed in Sect. 3. Below, a detailed description of the algorithm components is given.

Swarm Initialisation. The initial set S of n solutions is sampled using a uniform distribution across the search space defined by the bound constraints. This initial set is divided into three subsets $S = L \cup F \cup W$ as follows. Given a percentage η_W of walkers, a number $n_W = \eta_W * n$ of agents are randomly chosen and assigned to W; then, a number n_L of leader agents are randomly chosen and assigned to L. Lastly, given $n_F = n - n_L - n_W$, a number $\lfloor \frac{n_F}{n_L} \rfloor$ of agents is randomly chosen without replacement and assigned as followers to each of the leaders in L, resulting in non-overlapping leader/followers subgroups. The solution pool size n, walkers rate η_W and number of leaders n_L are the setup parameters of the algorithm; we will discuss the behaviour of several configuration scenarios for these parameters in Sect. 3.

Intensification Rules. Follow-the-leader is the core search rule of the algorithm. It is intended to drive followers to exploit local areas close to their corresponding leaders. Thus for an arbitrary follower $\mathbf{x}_i \in F$ we defined a simple update that shifts its location towards its leader $\mathbf{x}_\ell$, as shown in Eq. (1):

$$\mathbf{x}_i' = \mathbf{x}_i + \epsilon_i(\mathbf{x}_\ell - \mathbf{x}_i) + \mathbf{v}_i, \tag{1}$$

where $\epsilon_i \sim Uniform(0, 2)$ and $\mathbf{v}_i \sim N(\mathbf{0}, 1)$ is a small random perturbation.

Besides, each leader is also moved so as to conduct local search using a small perturbation $\mathbf{u}_i \sim N(\mathbf{0}, 1)$ as per Eq. (2):

$$\mathbf{x}_\ell' = \mathbf{x}_\ell + \mathbf{u}_i, \tag{2}$$

Once all the solutions in the F and L sets are updated, leaders roles are updated by choosing the best local solution from their follower base. The best performing leader represent the best solution ever found by the algorithm, i.e. $\mathbf{x}_\ell^* = \mathrm{ARGMIN}\, f(\mathbf{x}_\ell'), \mathbf{x}_\ell \in L$. Notice that the algorithm allows for a single-leader or multi-leaders modes of operation as controlled by the parameter n_L.

Diversification Rules. Random walk search was included into the algorithm in order to promote exploration of the search space. Thus, for each arbitrary walker $\mathbf{x}_i \in W$ we defined the simple update of Eq. (3):

$$\mathbf{x}_j' = \mathbf{x}_j + \mathbf{w}_j, \tag{3}$$

where $\mathbf{w}_i \sim N(\mathbf{0}, 20)$ is a large random perturbation. If a walker becomes a better solution than the current top leader, it moves to such location, $\mathbf{x}^* = \mathbf{x}_j'$, however they still remain as different elements in their respective sets, $\mathbf{x}_\ell^* \in L$ and $\mathbf{x}_j' \in W$.

On the other hand, it may be the case that two leaders end up exploring the same promising region, and even collapse jointly with their followers into the same location. If this occurs, in order to preserve diversity, the algorithm incorporate a clash safeguard that choose one of the colliding leaders and replace it with a new sample, dragging its followers along the way.

Furthermore, in order to avoid stagnation due to premature convergence to local minima, the algorithm perform a warm restart of the entire set S, where agents are uniform randomly dispersed to random locations under a uniform distribution across the search space. This restart is activated when the average cohesion of the leaders to their followers fall below a predefined threshold; alternatively, instead of cohesion an ageing mechanism was defined [2], where agents are allocated a preset lifespan (number of iterations) before being restarted.

Lastly, we equipped the algorithm with a straightforward learning rule that recalls the best solution ever found by moving a single walker to that location at the beginning of each iteration. This is reminiscent of the *elitism* property of other population-based metaheuristics, but also can be think of an *information-sharing* mechanism between subgroups of leaders and followers, one that prevents the swarm performing a new blind search after each warm restart.

We note in passing that the rules designed for this algorithm falls into the *differential vector movement* category defined in the metaheuristic behaviour taxonomy proposed in [10], and more specifically into *differential vector as a function of groups of solutions* sub-category. Besides, a distinctive feature of

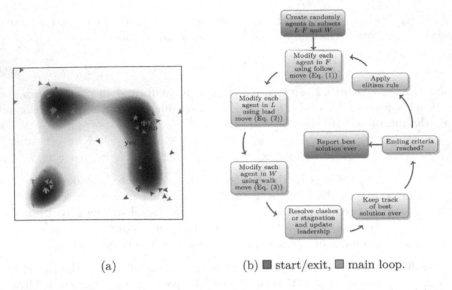

(a)   (b) ■ start/exit, ■ main loop.

Fig. 1. Illustrations of the FLW algorithm: (a) Agent-based simulation view. (b) Flow chart steps: L, F, W are the sets of leaders, followers and walkers.

this algorithm is that during its execution, the three subsets of non-overlapping agents L, F and W coexist, so the rules of diversification and intensification can be applied separately and simultaneously at each iteration (see Fig. 1b).

2.2   Model Implementation

Building upon our previous work on metaphor-based metaheuristics [1,4] we developed an agent-based model of the proposed algorithm using the Netlogo programming language and platform [12]. Since Netlogo operates over a discrete simulation environment, firstly the cost function of the optimisation problem is quantised and mapped onto the square grid of cells that make up the search space where the agents move. Each agent would be located on a cell with coordinates (x_1, x_2) in the resulting 2D landscape; the value of the cost function evaluated at that cell indicates its quality (or fitness) as a candidate solution for the chosen problem. Notice that the resolution level of the quantisation process affects the sampling rate of the problem cost function and thus modifies the size of the search space (the larger this size, the more difficult the search for the cell with the optimum value). A testbed of 14 benchmark problems were used in our empirical study to assess the performance of the model: ACKLEY, BEALE, BOHACHESVSKY, BOOTH, EASOM, EGGHOLDER, HIMMELBLAU, MICHALEWICZ, PARSOPOULOS, RASTRIGIN, ROSENBROCK, SPHERE, SPHERE-OFFSET, THREE-HUMP CAMEL (see [7] for function definitions and properties).

A snapshot of the model's simulated environment is shown in Fig. 1a ($n_L = 4, n_F = 28, n_W = 8$); there, agents are searching for promising solutions around the four global minima regions of the HIMMELBLAU problem (leaders are shown with a 🐚 symbol, best solution ever with ☆, followers with △ same color as their leaders, and walkers as grey △). In accordance with good laboratory practices [8] we have made the code publicly available in the Netlogo Modelling Commons repository, see: http://modelingcommons.org/browse/one_model/6978.

3   Empirical Study

In this section we aim to assess the behaviour of different configurations of the algorithm parameters, focusing on comparing the performance of the single vs. multi-leader modes of search, including the following scenarios:

- *Simultaneous Global Random Search (GRS)*: $n = 5, n_L = 1, \eta_W = 0.8$, no elitism. This is a configuration for a walkers-only swarm.
- *Simultaneous Local Random Search (LRS)*: $n = 5, n_L = 4, \eta_W = 0.2$, no elitism). This is a configuration for a swarm of singleton leaders.
- *Single-leader Swarm Search (SSS)*: $n = 5, n_L = 1, \eta_W = 0.2$, no elitism. This would be a swarm with an unique leader and its followers, plus a walker.
- *Multi-leader Swarm Search (MSS)*: $n = 20, n_L = 4, \eta_W = 0.2$, no elitism. This would be a swarm of agents with multiple leaders, each one with a different group of followers, plus a few walkers.
- *FLW search*: $n = 20, n_L = 4, \eta_W = 0.2$, elitism. Same as MMS plus the elitism enforcement. This correspond to the fully-equipped FLW algorithm.

The experiments were conducted using the BehaviorSpace tool included in NetLogo platform v6.1, on a 1.4 GHz Intel Core i5 running Mac OS X version 10.13.6; statistics were analysed using Python with libraries `pandas` and `seaborn`.

Fig. 2 summarises the empirical results of LRS, GRS, SSS, MSS and FLW configurations on the testbed problems, with different resolution levels for the problem landscapes, averaged over 30 repetitions. Regarding the success rates (Fig. 2a), the strategies based on random search, GRS and LRS, yielded values below 50% for all problems at all resolutions (except GRS at 100×100, which achieved slightly more than 50%, suggesting that such a simple strategy can be useful in small-size search spaces).

The effect of the combination of follow-the-leader and random-walk operators is noticeable in the other three scenarios (SSS, MSS and FLW); the single leader configuration (SSS), however, struggles for high success rates in 5 out of 14 problems, particularly on EGGHOLDER, RASTRIGIN and ROSENBROCK at finer resolutions; this is probably because the single leader and his followers converge to local minima in these types of valley-shaped or multimodal cost functions. The latter is alleviated in the multi-leader setting (MSS), which allows the swarm to escape local minima by spreading the multiple leaders across different regions, eventually discovering the optimal one; still, EGGHOLDER which is a multimodal problem with irregular and non-symmetric local minima remains difficult to solve, specially at higher resolutions. In addition, the incorporation of elitism in FLW, as a mechanism for information-sharing between multiple leaders, is

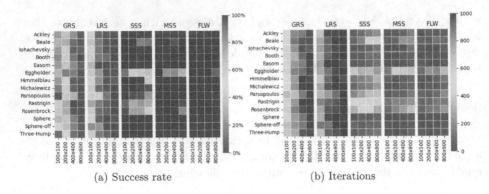

(a) Success rate   (b) Iterations

Fig. 2. Average empirical results for 30 runs of LRS, GRS, SSS, MSS and FLW configurations on the testbed problems, with different resolution levels for the search space. (a) Success rate: % of runs where optimum was found (warmer is better). (b) Number of iterations taken to termination criteria (cooler is better).

useful to improve performance, raising the success rate on all problems in all resolutions to nearly 100% efficacy.

Another interesting view of the algorithm performance is given by the efficiency in terms of function evaluations per execution (Fig. 2b), which hints at the benefit of combining the described intensification and diversification rules with the multiple leader mode, yielding a reduction of the number of iterations to converge to the optima in the latest MMS and FLW configurations; FLW in particular managed, on average, less than 300 iterations to solve all the problems.

4   Conclusions

Our study reports empirical evidence suggesting that the strategy of follow-the-leaders with non-overlapping groups of followers yields a more effective search behaviour than the single-leader strategy, particularly across a variety of problems with different optimisation properties.

The addition of random walk exploration, stagnation restarts, clash safeguards and elitism mechanisms improved success rate and efficiency, particularly on multimodal, valley-shaped, and non-symmetric cost function problems. Although the discrete nature of the implemented agent-based model can induce quantisation errors, the method was robust at different resolution levels. This behaviour suggests the suitability of the algorithm to also solve continuous-valued problems in larger dimensionalities, questions that we plan to address in our future work, including a comparative analysis with other state-of-the-art methods from an efficiency and algorithmic complexity viewpoints.

References

1. Blanco, A., Chaparro, N., Rojas-Galeano, S.: An urban pigeon-inspired optimiser for unconstrained continuous domains. In: Proceedings of 8th Brazilian Conference on Intelligent Systems (BRACIS 2019). IEEE Xplore Digital Library (2019)
2. Chen, W.N., et al.: Particle swarm optimization with an aging leader and challengers. IEEE Trans. Evol. Comput. **17**(2), 241–258 (2012)
3. Dokeroglu, T., Sevinc, E., Kucukyilmaz, T., Cosar, A.: A survey on new generation metaheuristic algorithms. Comput. Ind. Eng. **137**, 106040 (2019)
4. Garzón, M., Rojas-Galeano, S.: An agent-based model of urban pigeon swarm optimisation. In: 2019 IEEE Latin American Conference on Computational Intelligence, pp. 1–6. IEEE (2019)
5. Gonzalez-Fernandez, Y.: Leaders and followers-a new metaheuristic to avoid the bias of accumulated information. In: Congress on Evolutionary Computation, pp. 776–783. IEEE (2015)
6. Hussain, K., Mohd Salleh, M.N., Cheng, S., Shi, Y.: Metaheuristic research: a comprehensive survey. Artif. Intell. Rev. **52**(4), 2191–2233 (2019)
7. Jamil, M., Yang, X.S.: A literature survey of benchmark functions for global optimisation problems. Int. J. Math. Model. Numer. Optim. **4**(2), 150–194 (2013)
8. Kendall, G., et al.: Good laboratory practice for optimization research. J. Oper. Res. Soc. **67**(4), 676–689 (2016)
9. Kennedy, J., Eberhart, R.: Particle swarm optimization. In: Neural Networks, IEEE International Conference on, vol. 4 (1995)
10. Molina, D., Poyatos, J., Ser, J.D., García, S., Hussain, A., Herrera, F.: Comprehensive taxonomies of nature-and bio-inspired optimization: inspiration versus algorithmic behavior, critical analysis recommendations. Cogn. Comput. **12**(5), 897–939 (2020)
11. Sörensen, K.: Metaheuristics-the metaphor exposed. Int. Trans. Oper. Res. **22**(1), 3–18 (2015)
12. Tisue, S., Wilensky, U.: Netlogo: a simple environment for modeling complexity. In: International Conference on Complex Systems. New England Complex Systems Institute (2004)

A Scatter Search Approach
for the Parallel Row Ordering Problem

Raul Martín-Santamaría(ID), Jose Manuel Colmenar(✉)(ID),
and Abraham Duarte(ID)

Universidad Rey Juan Carlos, Calle Tulipán s/n, Móstoles, Madrid, Spain
josemanuel.colmenar@urjc.es

Abstract. In this work, we present a new approach for the Parallel Row Ordering Problem (PROP), based on the Scatter Search metaheuristic. The PROP focuses on minimizing the total weighted sum of all distances between each pair of facility centers in a linear layout. The proposed method is able to obtain all known optimal values in a fraction of the time required by the previous exact methods for the set of smaller instances, and it outperforms the current state of the art metaheuristic for the set of larger instances, spending a comparable computing time.

Keywords: Facility layout problems · Scatter Search · Combinatorial optimization

1 Introduction and Problem Description

Facility layout problems (FLPs) focus on determining the best arrangement of a set of facilities. Some examples of their practical applications are distributing rooms in a floor plan and organizing shelves to minimize material handling effort. The FLP family includes a large set of problems, due to the variety of layout constraints under consideration. We refer the reader to [1] for a detailed survey on FLPs.

The Parallel Row Ordering Problem (PROP), first proposed by [2], considers a two-row layout in which each facility $i \in F$ is assigned to one row. Therefore, it has to be placed only in the assigned row. Each facility i has an associated length l_i, and a vector of weights w, where w_{ij} corresponds with the weight between facilities $i, j \in F$. As in most of the FLPs, the objective function tries to minimize the weighted distances between each pair of facility centers, as defined in Eq. (1), where c_i represents the position of the center of facility i in the layout.

$$min \sum_{i,j \in F} w_{ij} * |c_i - c_j| \tag{1}$$

This work has been partially supported by the Spanish Ministerio de Ciencia, Innovación y Universidades (MCIU/AEI/FEDER, UE) under grant refs. PGC2018-095322-B-C22 and PID2021-126605NB-I00; and Comunidad de Madrid y Fondos Estructurales de la Unión Europea with grant ref. P2018/TCS-4566.

L. Di Gaspero et al. (Eds.): MIC 2022, LNCS 13838, pp. 506–512, 2023.
https://doi.org/10.1007/978-3-031-26504-4_40

In order to avoid overlaps between the set of facilities in the same row (F_r), the restriction shown in Eq. (2) must be observed.

$$|c_i - c_j| \geq \frac{l_i}{2} + \frac{l_j}{2} \;\; \forall i, j \in F_r, \;\; i \neq j \tag{2}$$

Because PROP is an NP-Hard problem [2,3], we tackle this problem from a metaheuristic point of view in order to solve the different proposed instances in reasonable computing times.

2   Algorithmic Proposal

The selected metaheuristic approach is Scatter Search. In this section, the main structure of the method will be described, as well as the details of the combination method, which is a critical step in this algorithm.

2.1   Scatter Search

Scatter Search (SS) is a population-based evolutionary procedure firstly proposed in [4], in the context of surrogate constrain relaxation procedures. It mainly works by constantly evolving and combining a set of solutions in the Reference Set (*RefSet*). In each iteration, a set of solutions is chosen from the *RefSet*, new solutions are created as combinations of each pair, and the *RefSet* is updated to include the new solutions if they are better or more diverse than the already existing ones. The stopping criteria is usually set to a maximum number of update iterations in the *RefSet*, or an iteration without a *RefSet* update, whichever happens first.

Although it may appear similar to other evolutionary algorithms, such as Genetic Algorithms (GA), it has some key differences. On the one hand, the *RefSet* size, which favors small values, is between 10 and 30, while the GA population size is usually one order of magnitude bigger. On the other hand, Scatter Search lacks randomization, which is usually provided by mutation operators in Genetic Algorithms [8].

A high-level view of the algorithm is presented in Algorithm 1, where $\pi = (\pi^0, \pi^1)$ represents an incumbent solution and π^k is the layout of row k. Step 1 of the algorithm shows how the current best solution π^* is initialized by random. Then, the multi-start loop begins in step 2 performing t_{max} iterations of the algorithm. The initial *RefSet*, denoted as R in the code, is initialized as a set of r_{size} solutions generated by randomly shuffling the facility order for each row and then applying a best improvement local search based on the displacement move, in steps 3 to 7. Each generated solution π' is added to the initial *RefSet* in step 7.

Then, in step 9, the main loop of SS executes. In the first step in this loop (line 10 of the code) a copy of the original *RefSet* is retained, in order to later stop the execution if the set has not changed. Then, each pair of existing solutions (π_i, π_j) in the *RefSet* is combined in step 11. Notice that the combination is performed individually for each row, since the facilities are assigned to one of the rows. The resulting solutions are stored in the candidate *RefSet*, denoted as R'. The original *RefSet* is then merged with the candidate *RefSet* in step 12 in order to include the r_{size} best solutions from both sets, and the iteration counter is increased (step 13). The main SS loop ends when either the *RefSet* is not updated in the current iteration or the maximum number of iterations l_{max} has been reached. After the loop ends, the best solution in the *RefSet* is obtained in step 15 and compared with the current best in step 16. In case of improvement, the current best is updated in step 17 and returned in step 18.

Algorithm 1: Multi-Start Scatter Search$(I, t_{max}, l_{max}, r_{size})$

---

1 $\pi^* \leftarrow \text{Shuffle}(I)$
2 **for** $t = 1$ **to** t_{max} **do**
3      $R \leftarrow \varnothing$
4   **for** $t = 0$ **to** r_{size} **do**
5          $\pi \leftarrow \text{Shuffle}(I)$
6          $\pi' \leftarrow BestImprovementLocalSearch(\pi)$
7          $R \leftarrow R \cup \{\pi'\}$
8      $l \leftarrow 0$
9   **do**
10         $R_{old} \leftarrow R$
11         $R' \leftarrow \{Combine(\pi_i^0, \pi_j^0), Combine(\pi_i^1, \pi_j^1)\}\ \forall \pi_i, \pi_j \in R,\ \pi_i \neq \pi_j$
12         $R \leftarrow MergeSet(R, R')$
13         $l \leftarrow l + 1$
14   **while** $l < l_{max} \wedge R \neq R_{old}$
15     $\pi' \leftarrow \arg\min \mathcal{F}(I, \pi)\ \forall \pi \in R$
16   **if** $\mathcal{F}(I, \pi') < \mathcal{F}(I, \pi\star)$ **then**
17         $\pi^* \leftarrow \pi'$

18 **return** π^*

2.2 Combination Method

The combination method proposed in this paper is detailed in Algorithm 2. As stated before, it receives two permutations of solutions corresponding to two individual row layouts, denoted as π^p and π^q. Then, for each position i of the layout (step 1) a random facility is chosen after a uniform distribution from either π^p (step 3) or π^q (step 5), depending on the generated random number. Here, it is possible to pick the same facility twice in different positions, but because π is a set, repeated facilities are not included multiple times. Due to this, π may have missing facilities, so the solution needs to be repaired in step 6 including possible missing facilities (step 7 and 8) and generating a complete new row π. Finally, a best improvement local search based on the displacement move is executed in step 9, and the improved row layout is returned in step 10.

Algorithm 2: Combine(π^p, π^q)

---

1 **for** $i = 0$ **to** $|\pi^p|$ **do**
2 | **if** $Random([0,1]) \leq 0.5$ **then**
3 | | $\pi[i] \leftarrow \pi^p[i]$
4 | **else**
5 | | $\pi[i] \leftarrow \pi^q[i]$

6 **for** $i = 0$ **to** $|\pi^p|$ **do**
7 | $\pi_r \leftarrow \pi^p \setminus \pi$
8 | $\pi \leftarrow \pi \cup \pi_r$

9 $\pi' \leftarrow BestImprovementLocalSearch(\pi)$
10 **return** π'

---

3 Experimental Experience

In this section, we first describe how the proposal has been tuned, and continue by comparing the results obtained against both an exact method and the state-of-the-art heuristic for the PROP.

Instances are divided in two sets: a smaller set, called *Set1*, for which the optimum values are known, formed by 31 instances; and a set of 25 more challenging instances, called *Set2*, for which the optimum values are not known. Instances are taken from [9] and [7] respectively. For each instance, the number of facilities in the first row is given by the following formula: $\lfloor |F|/c \rfloor$, for $c \in \{2,3,4,5\}$, while the facilities in the second row are the remaining ones in lexicographical order.

All experiments have been executed in a virtual machine with 32 cores and 16 GB of RAM, using Java 17 and our development framework. All the instances and code are available upon request.

3.1 Proposal Tuning

The proposed algorithm has been automatically tuned using the irace software package, which finds the most appropriate settings given a set of instances for any given optimization problem [6]. The preliminary instance set used during tuning is formed by a randomly chosen representative set of 11 instances, which corresponds with approximately 20% of the complete instance set size.

Parameters such as number of iterations or the *RefSet* size cannot be directly calibrated using irace, because if given the opportunity, irace will always try to maximize the computational effort used. However, by defining a computing budget b such as $b = t_{max} \cdot r_{size}$, the computational effort is distributed between the number of iterations and the *RefSet* size parameters.

Using a budget of $b = 600$, the chosen parameters by irace are $r_{size} = 20$, $t_{max} = 30$. The maximum number of *RefSet* update operations (l_{max}) is fixed to 100, in order to guarantee a maximum execution time.

3.2 Results

The results obtained by the proposal are compared to the current state-of-the-art methods: an exact model implemented in the commercial Gurobi solver [9], denoted as *Exact* in the results and a heuristic approach based on a population-based Simulated Annealing algorithm, denoted as *PSA* [5].

As the instance sets are different, the comparison will be divided in two different tables. Table 1 sums up the comparison between the exact model and our proposal. Instances have been grouped according to the number of elements c in each row, reporting the number of times the algorithm reaches the best solution (#Best), average deviation to best known value (%Dev), the average execution time in seconds (T(s)) and the standard deviation of the execution time in seconds (Std. T(s)). As it can be seen it the table, the Scatter Search proposal reaches the best values for all instance types in a fraction of the time required by the exact model.

Table 1. Summary comparison between the exact model and the Scatter Search proposal for instances in *Set1*. The number of facilities per row is $t = \lfloor |F|/c \rfloor$, for $c \in \{2, 3, 4, 5\}$.

c	Exact				Scatter Search			
	#Best	%Dev	T(s)	std T(s)	#Best	%Dev	T(s)	std T(s)
2	31	0.00	1342.32	1790.17	31	0.00	4.09	2.40
3	31	0.00	334.38	544.30	31	0.00	4.72	2.83
4	31	0.00	541.74	1300.40	31	0.00	5.45	3.31
5	31	0.00	1101.25	2043.17	31	0.00	5.94	3.57

Table 2 summarizes the results obtained by both the previous state-of-the-art PSA and our Scatter Search proposal. Instances are similarly grouped by the number of elements per row, and the reported elements are similar to the previous table. As seen, the Scatter Search proposal obtains the highest number of best results, with a total of 89 best results out of 100, while the PSA gets 57 out of 100, using comparable execution times. Specifically, with $c = 2$ both metaheuristics reach the same number, but as the c parameter increases the quality difference becomes stronger.

Extended result tables for each experiment detailing the results for each instance are available at Zenodo (https://dx.doi.org/10.5281/zenodo.6616412).

Table 2. Summary comparison between the previous PSA work, and the Scatter Search proposal for instances in *Set2*. The number of facilities per row is $l = \lfloor |F|/c \rfloor$, for $c \in \{2, 3, 4, 5\}$.

c	PSA				Scatter Search			
	#Best	%Dev	T(s)	std T(s)	#Best	%Dev	T(s)	std T(s)
2	17	0.03	18.3	14.3	17	0.03	30.3	14.3
3	13	0.02	22.1	17.6	23	0.00	35.5	19.7
4	14	0.01	24.9	19.4	24	0.00	41.5	25.2
5	13	0.01	31.7	25.2	25	0.00	45.5	28.0

4   Conclusions and Future Work

In this work, we introduce a new Scatter Search approach for the Parallel Row Ordering Problem (PROP). The proposal is based in the combination of solutions by means of randomly merging the content of each row of pairs of solutions, as well as on a best improvement local search. The comparison with the state-of-the-art manifests the effectiveness of the proposal, obtaining all known best values in a fraction of the time required by an exact model, and improving the results of the previous state-of-the-art PSA heuristic for the larger instances.

The promising obtained results suggest that proposing a bigger set of more challenging instances may increase the competitiveness of the algorithm. So, further improvements in the PROP family of problems are possible. In particular, more efficient local search neighborhoods and new different combination methods may be tested, in order to increase the efficiency of the proposal. Learning from this experience, future woks will involve similar problems from the facility layout problem family.

References

1. Ahmadi, A., Pishvaee, M.S., Jokar, M.R.A.: A survey on multi-floor facility layout problems. Comput. Ind. Eng. **107**, 158–170 (2017)
2. Amaral, A.R.: A parallel ordering problem in facilities layout. Comput. Oper. Res. **40**(12), 2930–2939 (2013)
3. Garey, M.R., Johnson, D.S.: Computers and Intractability, vol. 174. Freeman, San Francisco (1979)
4. Glover, F.: Heuristics for integer programming using surrogate constraints. Decis. Sci. **8**(1), 156–166 (1977)
5. Gong, J., Zhang, Z., Liu, J., Guan, C., Liu, S.: Hybrid algorithm of harmony search for dynamic parallel row ordering problem. J. Manuf. Syst. **58**, 159–175 (2021)
6. López-Ibáñez, M., Dubois-Lacoste, J., Pérez Cáceres, L., Stützle, T., Birattari, M.: The irace package: iterated racing for automatic algorithm configuration. Oper. Res. Perspect. **3**, 43–58 (2016). https://doi.org/10.1016/j.orp.2016.09.002
7. Maadi, M., Javidnia, M., Jamshidi, R.: Two strategies based on meta-heuristic algorithms for parallel row ordering problem (PROP). Iran. J. Manag. Stud. **10**(2), 467–498 (2017)

8. Martí, R., Corberán, Á., Peiró, J.: Scatter search. In: Martí, R., Pardalos, P., Resende, M. (eds.) Handbook of Heuristics, pp. 717–740. Springer, Cham (2018). https://doi.org/10.1007/978-3-319-07124-4_20
9. Yang, X., Cheng, W., Smith, A.E., Amaral, A.R.S.: An improved model for the parallel row ordering problem. J. Oper. Res. Soc. **71**(3), 475–490 (2020)

A Multi-Population BRKGA
for Energy-Efficient Job Shop Scheduling
with Speed Adjustable Machines

S. Mahdi Homayouni[1]([✉]) [iD], Dalila B. M. M. Fontes[1,3] [iD],
and Fernando A. C. C. Fontes[2,3] [iD]

[1] LIAAD- INESCTEC, Porto, Portugal
smh@inesctec.pt
[2] Universidade do Porto, Porto, Portugal
[3] SYSTEC-ISR-ARISE LA, Porto, Portugal
fontes@fep.up.pt, faf@fe.up.pt

Abstract. Energy-efficient scheduling has become a new trend in industry and academia, mainly due to extreme weather conditions, stricter environmental regulations, and volatile energy prices. This work addresses the energy-efficient Job shop Scheduling Problem with speed adjustable machines. Thus, in addition to determining the sequence of the operations for each machine, one also needs to decide on the processing speed of each operation. We propose a multi-population biased random key genetic algorithm that finds effective solutions to the problem efficiently and outperforms the state-of-the-art solution approaches.

Keywords: Job shop scheduling problem · Energy efficiency · Speed adjustable machines · Biased random key genetic algorithm

1 Introduction

The steady increase in energy consumption and public awareness of its environmental impacts bring new challenges for the industrial sector. Thus, practitioners are looking for ways of reducing energy consumption. An effective way to accomplish such a reduction is to implement an energy-efficient scheduling methodology which, additionally, needs no significant capital investments and is particularly relevant for small and medium-sized enterprises [1].

The energy-efficient job shop scheduling problem (EEJSP) attempts to lower energy consumption while providing the same level of service mainly by adjusting machines processing speed [9] and/or by switching machines into a power-saving mode when idle [8]. The former strategy balances energy consumption and production time, while the latter balances energy savings from resorting to the use of a power-saving mode and energy requirements to restart and warm up machines.

This work is financed by National Funds through the Portuguese funding agency, FCT - Fundação para a Ciência e a Tecnologia, within project LA/P/0063/2020.

L. Di Gaspero et al. (Eds.): MIC 2022, LNCS 13838, pp. 513–518, 2023.
https://doi.org/10.1007/978-3-031-26504-4_41

Furthermore, under the first strategy, non-processing energy consumption can be decreased as adjusting the processing speed may reduce machines' idle time.

Although only recently energy-efficient scheduling became the subject of systematic research, one of the first works on EEJSP with speed adjustable machines is that of He et al. [5]. They propose a Tabu search (TS) algorithm to find a sequence of operations for each machine as well as the machine processing speed for each operation such that both the makespan (C_{max}) and the total energy consumption $(\mathcal{E})$ are optimized. Later, [10] proposes a mixed-integer programming (MIP) model and a genetic-simulated annealing algorithm. Recently, [9,11] propose similar genetic algorithms (GAs). While [9] minimizes the weighted sum of the normalized C_{max} and the normalized $\mathcal{E}$, [11] also includes the normalized noise emissions in its weighted sum. More recently, hybrid metaheuristics have been proposed. For example, [12] embeds two local search heuristics into a GA to find solutions that minimize the total weighted tardiness (wT) and the $\mathcal{E}$. One of the heuristics attempts to improve the wT by swapping disjoint pairs of adjacent operations on each machine and the other attempts to improve the $\mathcal{E}$ by decreasing the processing speed of non-critical operations. Another example is the work in [7] that decrease energy consumption by decelerating the processing of non-critical operations as much as possible without affecting the completion time of any other operation. A more complete account of recent work on energy-efficient job shop scheduling problems can be found in [3].

This work proposes a multi-objective multi-population biased random key genetic algorithm (multi-objective mpBRKGA) that finds effective solutions to the EEJSP problem efficiently. The computational experiments performed on a set of 13 benchmark problem instances show that the mpBRKGA outperforms the state-of-the-art algorithms. We describe the problem being addressed in Sect. 2 and the proposed multi-objective mpBRKGA in Sect. 3. The computational experiments and results are reported in Sect. 4. Finally, Sect. 5 draws some conclusions and points out future research directions.

2   Problem Definition

The classical job shop scheduling problem (JSP) comprises a set J of independent jobs and a set of machines. Each job consists of a set $O_j = \{1, 2, \ldots, n_j\}, \forall j \in J$ of n_j ordered operations. Each operation must be processed on a given machine and each machine can only process one operation at a time.

The EEJSP with speed adjustable machines requires solving simultaneously two interdependent combinatorial problems: scheduling the operations on each machine (machine scheduling) and determining the speed at which each operation is processed (operation speed assignment).

A machine can process an operation at one of the various speeds requiring a predefined processing time and power consumption. (The higher the processing speed, the higher the power consumption required.) Idle machines are considered in "stand-by" mode and with negligible power consumption.

An operation can be started right after the completion of the previous operation of the same job (or immediately if it is the job's first operation) if the required machine is idle; otherwise, the job waits for the machine in a buffer. Whenever an operation is completed the machine can start processing the next operation immediately. Among all possible solutions, we are interested in those that minimize the makespan C_{max} and the energy consumed to process all production operations $\mathcal{E}$.

3 The Proposed Multi-objective MpBRKGA

Biased random key genetic algorithms (BRKGAs) have been successfully proposed and implemented to find good quality solutions in several application areas. The multi-objective mpBRKGA proposed in this work borrows its principles from [4] and evolves $\Omega + \Pi$ populations: each of the Ω populations considers one of the Ω objectives, while each of the Π populations considers simultaneously all Ω objectives. The initial populations are uniformly randomly generated. Nevertheless, 20% of the initial solutions of the single-objective population that minimizes C_{max} have the processing speeds set to the maximum value. Similarly, the processing speeds are set to the minimum value for 20% of the initial solutions of the single objective population that minimizes $\mathcal{E}$.

A solution is **encoded** as a two-part vector with $N = \sum_{j \in J} n_j$ elements each. Each vector element is a random key, i.e., a real number in the interval $[0, 1]$. The first part provides a sequence of operations and the second assigns a processing speed to each operation. (See Fig. 1.a for an example with three jobs, each with three operations that can be processed at one of the three available speeds.)

The **decoding** of the first N elements uses the smallest position value rule to construct a permutation representation of the RKs. Elements are sorted in ascending order, thus, providing a sequence of operations, see part (i) of Fig. 1.b). Then, the permutation is converted into a feasible sequence of operations by sorting the operations of each job in ascending order; this way ensuring precedence constraints, see part (i) of Fig. 1.c). Each RK in the second part of the chromosome is decoded into the processing speed of the corresponding operation. The processing speed of operation l is given by the smallest integer greater than or equal to $RK_l \times |\mathcal{P}_l|$, where $|\mathcal{P}_l|$ is the number of speed values available for operation l, see Fig. 1 - part (ii).

The **evolutionary strategy** is similar for all populations. At each generation, a new population is obtained by joining the set N_e of n_e elite solutions, the set N_o of n_o offspring solutions, and the set N_m of n_m mutant solutions. For each of the $\omega \in \Omega$ single objective populations, the set N_e^ω of elite solutions comprises the best n_e solutions of its current generation. Regarding each of the $\pi \in \Pi$ multi-objective populations, the set N_e^π of elite solutions are the best n_e solutions chosen from a pool of solutions consisting of the best n_e solutions of its current generation and the best n_e solutions of the current generation of each $\omega \in \Omega$ population, from which the repeated solutions are removed. Additionally, the Π multi-objective populations exchange their elite solutions after

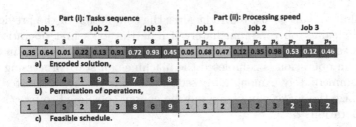

Fig. 1. Solution encoding and decoding procedures.

a pre-determined number of generations (g_{ex}). Thus, every g_{ex} generations, the pool of solutions also contains the best n_e solutions of each $\pi \in \Pi$ population. To identify the best solutions for the multi-objective populations, we employ a non-dominated ranking procedure [2] and the crowding distance [2] to sort solutions of the same rank.

Regardless of the population being evolved, the n_o offspring solutions are obtained by a biased parameterized uniform crossover [4,6] and the n_m mutants are randomly generated in the same manner as the initial population was.

The evolutionary process is repeated for a predetermined number of generations G_{max}. Then, a pool is created by joining the best solutions of the Ω populations and the non-dominated solutions of the Π populations and removing repeated and dominated solutions. The remaining solutions are Pareto solutions and provide an approximation to the Pareto optimal front.

4   Numerical Results

The multi-objective mpBRKGA was implemented in Python® 3.8 and all computational experiments were carried out on a 3.20 GHz Intel® Core™ i7-8700 PC with 24 GB RAM. We solve three problem instances proposed in [11] and 10 proposed in [9]. The former instances are designated as Yin01, Yin02, and Yin03 and have four, 10, and 20 jobs with 12, 40, and 60 operations, respectively. Instances Yin01 and Yin03 have five machines while instance Yin02 has six. Operations can be processed at two or three speeds. The instances proposed in [9] are designated as Sal01 $\sim$ 10 and each has three jobs with 25 operations each and three machines. Operations can be processed at three speeds.

Following on literature recommendations, the control parameters of the mpBRKGA were set as: population size $P = 500$, number of elite solutions $n_e = 0.2P$, number of mutants $n_m = 0.1P$, and inheritance probability $p_e = 0.7$ [4,6]. The remaining control parameters were set through empirical testing to $G_{max} = 300$, $\Pi = 2$, $\Omega = 2$, and $g_{ex} = 30$.

The benchmark instances were solved in [11] by a genetic algorithm (GA) and in [9] by a multi-objective GA (moGA) and constraint programming (CP). Both [11] and [9] employed a weighted summation approach and reported the C_{max} and $\mathcal{E}$ values under different weight assumptions. Table 1 reports the boundary solutions, that is, the minimum makespan C^*_{max} and associated energy consumption $\mathcal{E}$ and the minimum energy consumption $\mathcal{E}^*$ and associated makespan C_{max}, obtained by the mpBRKGA and by the GA [11] for the Yin instances. Similar results are reported for the Sal instances. However, the reported values are averages over the 10 problem instances since this is what is reported for the CP and the moGA approaches in [9]. As it can be seen, the mpBRKGA is capable of finding very good solutions and it outperforms the GA, the CP, and the moGA approaches. Although the mpBRKGA finds the same boundary solutions and PF for Yin01, it finds better ones for Yin02 and Yin03. Indeed, the mpBRKGA boundary solutions dominate those of [11]. Regarding instances Sal01 ~ 10, our averages are always better, except for the minimum energy that we have the same value.

Table 1. The makespan and energy consumption values at boundary solutions for the EEJSP benchmark problem instances.

	Solution approach	C^*_{max}	$\mathcal{E}$	C_{max}	$\mathcal{E}^*$
Yin01	mpBRKGA	22.0	5.81	35.0	4.82
	GA	22.0	5.81	35.0	4.82
Yin02	mpBRKGA	40.0	18.34	56.0	17.73
	GA	40.0	19.48	60.0	18.13
Yin03	mpBRKGA	62.0	22.94	111.0	20.96
	GA	64.0	23.68	105.0	21.73
Sal01 ~ 10	mpBRKGA	1673.3	6699.7	2786.4	3827.1
	CP	1673.4	6732.2	3160.0	3827.1
	moGA	1697.0	7088.5	2888.8	3827.1

Figure 2 depicts the Pareto fronts (PFs) obtained by the mpBRKGA and the ones obtained by the GA [11] for problem instances Yin01 ~ 03. As it can be seen, for instance Yin01, the (PFs) are the same for both methods. However, for Yin02 and Yin03, the mpBRKGA produces much better solutions. All the solutions found by the GA proposed in [11] are dominated by the ones found by the mpBRKGA. In contrast, none of the mpBRKGA solutions are dominated.

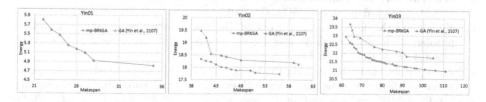

Fig. 2. Pareto fronts found by the mpBRKGA and the GA [11] for instances Yin01 ~ 03.

5   Conclusion

We address the EEJSP with speed adjustable machines. We propose a multi-objective mpBRKGA that is capable of finding good solutions efficiently and outperforms the state-of-the-art solution approaches. Experiments show that, our PFs are the better than those of [11] for instances Yin01 $\sim$ 03; and our average values for boundary solutions are better than those reported in [9] for instances Sal01 $\sim$ 10. This work can be extended by considering other energy factors such as peak power, which usually is a major concern if renewable energies are being used.

References

1. Bansch, K., et al.: Energy-aware decision support models in production environments: a systematic literature review. Comput. Ind. Eng. **159**, 107456 (2021)
2. Deb, K., Pratap, A., Agarwal, S., Meyarivan, T.: A fast and elitist multiobjective genetic algorithm: Nsga-ii. IEEE Trans. Evolut. Comput. **6**(2), 182–197 (2002)
3. Fernandes, J.M., Homayouni, S.M., Fontes, D.B.M.M.: Energy-efficient scheduling in job shop manufacturing systems: a literature review. Sustainability **14**(10), 6264 (2022)
4. Gonçalves, J.F., Resende, M.G.: A parallel multi-population biased random-key genetic algorithm for a container loading problem. Comput. Oper. Res. **39**(2), 179–190 (2012)
5. He, Y., Liu, F., Cao, H.J., Li, C.B.: A bi-objective model for job-shop scheduling problem to minimize both energy consumption and makespan. J. Cent. South Univ. Technol. **12**(2), 167–171 (2005)
6. Homayouni, S.M., Fontes, D.B.M.M., Gonçalves, J.F.: A multistart biased random key genetic algorithm for the flexible job shop scheduling problem with transportation. Int. Trans. Oper. Res. **30**(2), 688–716 (2023)
7. Lu, C., Zhang, B., Gao, L., Yi, J., Mou, J.: A knowledge-based multiobjective memetic algorithm for green job shop scheduling with variable machining speeds. IEEE Syst. J. **16**(1), 844–855 (2021)
8. Meng, L., Zhang, C., Shao, X., Ren, Y.: MILP models for energy-aware flexible job shop scheduling problem. J. Clean. Prod. **210**, 710–723 (2019)
9. Salido, M.A., Escamilla, J., Giret, A., Barber, F.: A genetic algorithm for energy-efficiency in job-shop scheduling. Int. J. Adv. Manuf. Technol. **85**(5), 1303–1314 (2016)
10. Tang, D., Dai, M.: Energy-efficient approach to minimizing the energy consumption in an extended job-shop scheduling problem. Chin. J. Mech. Eng. **28**(5), 1048–1055 (2015). https://doi.org/10.3901/CJME.2015.0617.082
11. Yin, L., Li, X., Gao, L., Lu, C., Zhang, Z.: Energy-efficient job shop scheduling problem with variable spindle speed using a novel multi-objective algorithm. Adv. Mech. Eng. **9**(4), 1–21 (2017)
12. Zhang, R., Chiong, R.: Solving the energy-efficient job shop scheduling problem: a multi-objective genetic algorithm with enhanced local search for minimizing the total weighted tardiness and total energy consumption. J. Clean. Prod. **112**, 3361–3375 (2016)

An Evolutionary Algorithm Applied to the Bi-Objective Travelling Salesman Problem

Luis Henrique Pauleti Mendes[1]([⊠]) [iD], Fábio Luiz Usberti[1] [iD],
and Mário César San Felice[2] [iD]

[1] Institute of Computing, State University of Campinas, Campinas, SP, Brazil
{luis.mendes,fusberti}@ic.unicamp.br
[2] Department of Computing, Federal University of São Carlos, São Carlos, SP, Brazil
felice@ufscar.br

Abstract. This paper presents an evolutionary algorithm for multi-objective optimization problems, based on the Biased Random-Key Genetic Algorithms and on the Elitist Non-dominated Sorting Genetic Algorithm. Computational experiments applied to the Bi-Objective Travelling Salesman Problem compared our algorithm with other well-known multi-objective evolutionary algorithms from the literature. The results show that our methodology consistently outperformed the other approaches with respect to the hypervolumes from the obtained non-dominated fronts.

Keywords: Multi-objective optimization · NSGA-II · BRKGA

1 Introduction

Multi-objective Optimization Problems (MOPs) are an important field of operations research applied to many real-world problems that require taking into account multiple conflicting points of view. A MOP can be stated as:

$$\min_{x \in \mathcal{X}} \mathbf{f}(x) = (f_1(x), \ldots, f_m(x))^T,$$

where $\mathcal{X} \subseteq \mathbb{R}^n$ is an n-dimensional decision space, $\mathbf{f} : \mathcal{X} \to \mathbb{R}^m$ consists of m real-valued objective functions, where each one needs to be minimized, and $\mathcal{Y} = \{\mathbf{f}(x) : x \in \mathcal{X}\} \subseteq \mathbb{R}^m$ is an attainable objective space. Improvement of one objective function may lead to the deterioration of another. Thus, generally no single solution can optimize all the objectives. Instead, the best trade-off solutions, called efficient solutions, are of interest to a decision maker.

Population-based methods such as Evolutionary Algorithms (EAs) have become increasingly popular for solving MOPs over the past decades [8]. These methods, known as Multi-Objective Evolutionary Algorithms (MOEAs), carry the advantage of finding a set of efficient solutions, because of their inherent parallel character and group strategy.

L. Di Gaspero et al. (Eds.): MIC 2022, LNCS 13838, pp. 519–524, 2023.
https://doi.org/10.1007/978-3-031-26504-4_42

Our contribution consists of a new MOEA applied to the Bi-Objective Travelling Salesman Problem (BOTSP), whose performance is compared with other MOEAs from the literature. Our methodology combines the benefits of Biased Random-Key Genetic Algorithms (BRKGAs) and of the Elitist Non-Dominated Sorting Genetic Algorithm (NSGA-II). We believe our methodology can be easily applied to MOPs in general.

2   Base Algorithms

2.1   Biased Random-Key Genetic Algorithms

Random-Key Genetic Algorithms (RKGAs) were first introduced by [2] for solving single-objective combinatorial optimization problems involving sequencing. In a RKGA, chromosomes are represented as vectors of randomly generated real numbers in the interval $[0, 1]$. A deterministic algorithm, called decoder, maps any individual to a solution of the optimization problem, for which a fitness value can be computed.

The initial population P_0 is made up of vectors of random-keys, where each allele is generated independently at random in the interval $[0, 1]$. After the fitness of each individual is computed by the decoder, the population P_0 is partitioned into a small set of elite individuals P_0^e, that is, those with the best fitness value, and a larger set of non-elite individuals $P_0^{\bar{e}} = P_0 \backslash P_0^e$, containing the remaining individuals. In each generation t, the elite set P_t^e is copied unchanged to generation $t + 1$ and a small set P_t^m of random solutions called mutants are introduced in the new population. Considering that $|P_t^e \cup P_t^m| \leq |P_t|$, the population is completed by offsprings, generated through the process of mating.

A BRKGA [6] differs from a RKGA in the way parents are selected for mating. In an RKGA [2], the two parents are selected at random from the entire population. Whereas in a BRKGA, each offspring is generated combining one individual from the elite set P_t^e and one individual from the non-elite set $P_t^{\bar{e}}$. Mating is done using parameterized uniform crossover, where a gene is taken from the elite individual with probability $\rho > 0.5$, or otherwise it is taken from the other individual.

The Multi-Parent Biased Random-Key Genetic Algorithm (BRKGA-MP) [9], a variant of BRKGA, selects as many parents as the parameter π determines, and rank them according to their fitness values. Among these, π^e are elite parents and $\pi^{\bar{e}} = \pi - \pi^e$ are non-elite parents, where $\pi^e \geq \pi^{\bar{e}}$. Then, a parameterized uniform crossover is performed such that each allele is taken from an individual chosen by the roulette method. Each parent has a probability of passing its alleles to the offspring, which is calculated using the bias of the parent. Parent bias is defined by a pre-determined, non-increasing weighting bias function $\Phi : \mathbb{N} \to \mathbb{R}_+^*$ over its rank r.

2.2   Elitist Non-dominated Sorting Genetic Algorithm

The NSGA-II [5], one of the most popular algorithms for finding multiple Pareto-optimal solutions for multi-objective optimization problems, has the following features: it uses an elitist principle, it uses an explicit diversity preserving mechanism, and it emphasizes non-dominated solutions.

Initially, a random parent population P_0 is created. The population is sorted based on the non-domination. Each solution receives a rank equal to its non-domination level. At first, the binary tournament selection, simulated binary crossover, and polynomial mutation are used to create an offspring population Q_0 of size $|P_0|$. Elitism is introduced by comparing current population with previously found best non-dominated solutions.

Next, we describe the t-th generation of the NSGA-II. First, a combined population $R_t = P_t \cup Q_t$, of size $2|P_t|$, is formed. Then, the fast non-dominated sorting is used to classify the entire population R_t. Although this requires more computation compared to performing a non-dominated sorting on Q_t alone, it allows a global non-domination check among offspring and parent solutions, ensuring elitism. Now, solutions belonging to the best non-dominated set $S_1^\star$ are the best solutions in R_t and must be emphasized. If the size of $S_1^\star$ is smaller than $|P_t|$, all members of $S_1^\star$ are chosen for the next population P_{t+1}. The remaining members of the population P_{t+1} are chosen from subsequent non-dominated set in order of their ranking. Thus, solutions from the set $S_2^\star$ are chosen next, followed by solutions from the set $S_3^\star$, and so forth. This procedure is continued until no more sets can be accommodated. When the last acceptable non-dominated set $S_l^\star$ is being considered, there may exist more solutions in the $S_l^\star$ than the remaining slots in the new population P_{t+1}. Instead of arbitrarily discarding some members of the $S_l^\star$, a niching strategy is used to choose the solutions that reside in the least crowded region of $S_l^\star$, increasing the diversity of the solutions chosen.

3   Non-dominated Sorting Biased Random-Key Genetic Algorithm

In this work, we extend the single-objective BRKGA-MP framework for solving MOPs. The fitness of each chromosome must be taken into account for all objective functions. Thus, the main question that arises is how to determine an order for the solutions to select the elite set. To that end, we used the fast-non-dominated sorting and the crowding distance assignment from NSGA-II [5]. Our proposed algorithm is named Non-dominated Sorting Biased Random-key Genetic Algorithm (NS-BRKGA). The t-th generation of the NS-BRKGA is presented in Algorithm 1.

The standard BRKGA have already been hybridized with NSGA-II to tackle multi-objective combinatorial optimization problems [3,4]. However, to the best of our knowledge, no other work from the literature have extended the BRKGA-MP for multi-objective optimization. Furthermore, we do not insert mutant individuals into the population. Instead, we apply polynomial mutation on the offsprings. Moreover, the NS-BRKGA uses crowding distance sorting to preserve diversity on the objective space. To enhance the diversity on the genotype space, we emphasize solutions with high diversity to be part of the elite set. Such mechanism is described in Algorithm 2.

Algorithm 1. NS-BRKGA

1: Sort P_t using fast-non-dominated-sort and crowding-distance-sort
2: $P_t^e \leftarrow$ select-elite-set$(P_t, S_1^\star, D, p_{min}^e, p_{max}^e)$
3: $P_{t+1} \leftarrow P_t^e$ ▷ Initialize next population
4: **while** $|P_{t+1}| < |P_t|$ **do**
5:      $c \leftarrow$ biased-multi-parent-crossover$(P_t, P_t^e, \pi, \pi^e, \Phi)$
6:      $c' \leftarrow$ polynomial-mutation(c)
7:      $P_{t+1} \leftarrow P_{t+1} \cup \{c'\}$ ▷ Add offspring to next population
8: $t \leftarrow t + 1$

Algorithm 2. Select elite set

1: $P_t^e \leftarrow P_t[1 : p_{min}^e]$
2: **if** $|P_t^e| < |S_1^\star|$ **then**
3:      $P_t^e \leftarrow S_1^\star$
4: **if** $|P_t^e| > p_{max}^e$ **then**
5:      $P_t^e \leftarrow P_t[1 : p_{max}^e]$
6: **for** $p \leftarrow |P_t^e| + 1$ **to** p_{max}^e **do**
7:      **if** $D(P_t^e) < D(P_t[1 : p])$ **then**
8:          $P_t^e \leftarrow P_t[1 : p]$

Line 1 initializes the elite set P_t^e with the first p_{min}^e elements of the current population P_t. Lines 2–3 make sure that P_t^e contains every non-dominated solutions $S_1^\star$ from P_t. Lines 4–5 ensure that P_t^e do not exceed the maximum size p_{min}^e. The loop of lines 6–8 makes sure that P_t^e consists of the prefix of P_t that maximizes the value of the diversity function D. Diversity is measured using the average distance to centroid [7]: $D_v(P) = \frac{1}{|P|} \sum_{i=1}^{|P|} \|P[i] - \bar{P}\|$, where $\bar{P} = \frac{1}{|P|} \sum_{i=1}^{|P|} P[i]$ is the centroid of the population.

4 Bi-Objective Travelling Salesman Problem

The Travelling Salesman Problem (TSP) is an $\mathcal{NP}$-hard permutation-based combinatorial optimization problem, which have been extensively studied in the literature [1]. Consider a complete weighted graph $G = (V, E, c)$, where V is the set of vertices, E is the set of edges, and $c : E \to \mathbb{R}$ is a function that assigns to each edge $e \in E$ a cost $c(e) \in \mathbb{R}$. The goal of the TSP is to find a minimal cost Hamiltonian cycle of G.

In the Bi-Objective TSP (BOTSP), an instance consists of a complete weighted graph $G = (V, E, \mathbf{c})$, where $\mathbf{c} : E \to \mathbb{R}^m$ is a function that assigns to each edge $e \in E$ a vector $(c_1(e), c_2(e))^T$ with 2 costs [12]. Many engineering problems, such as network structure design problems, machine scheduling problems and vehicle routing problems, can be formulated as BOTSPs. Consequently, the BOTSP is frequently employed as a benchmark problem to evaluate the performance of multi-objective optimization algorithms.

5   Computational Experiments

Instances. We consider symmetric instances of size ranging from 100 to 1000 [10]. The instances with at most 200 cities have been generated combining single-objective TSP instances of the TSPLIB. The instances of at least 300 cities, were generated with random coordinates. The costs between the cities are computed by calculating the Euclidean distance between each city.

Decoder. A chromosome that encodes a solution for an instance $G = (V, E, \mathbf{c})$ of the BOTSP is represented by a vector with $|V| - 1$ real numbers in the interval $[0, 1]$. One vertex $v \in V$ is fixed as the initial vertex of the Hamiltonian cycle of G, and the order in which the vertices in $V \setminus \{v\}$ are visited is given by the value of each of the $|V| - 1$ genes. Hence, the decoder sorts the genes of the chromosome and sum up the travel costs for each of the 2 objectives, which gives a computational time complexity of $\mathcal{O}(|V| \log |V|)$.

Benchmark. The proposed algorithm was compared with the following algorithms, and their implementations available at the PAGMO2 library[1]: Elitist Non-dominated Sorting Genetic Algorithm (NSGA-II) [5] and Multi-Objective Evolutionary Algorithm by Decomposition with Differential Evolution (MOEA/D-DE) [13]. We performed five independent runs for each algorithm and instance on a computer with an Intel Xeon CPU E5-2420 CPU of 1.90 GHz and 32 GB of RAM, with a time limit of one hour.

Evaluation Metric. The algorithms are compared by the hypervolume indicator I_{HV} [11]. Figure 1 presents the hypervolume obtained by each algorithm. Figure 1a shows the mean hypervolume obtained for each instance size, and Fig. 1b shows the mean hypervolume as a function of the execution time.

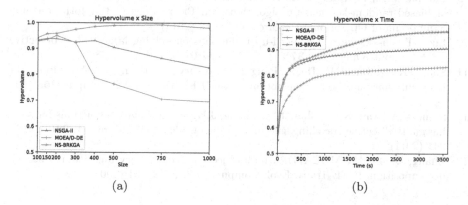

        (a)   (b)

Fig. 1. Hypervolume obtained by each algorithm.

---

[1] https://esa.github.io/pagmo2/.

Results. Analyzing Fig. 1a, we can conclude that the NS-BRKGA is more scalable, showing increasing margins of improvements for bigger instances. Moreover, on Fig. 1b, we can see that the NS-BRKGA converges more rapidly to the best hypervolumes compared to the other algorithms. To conclude, computational experiments show that the proposed methodology consistently outperformed the previous approaches, in both solution quality and execution time.

References

1. Applegate, D.L., Bixby, R.E., Chvatál, V., Cook, W.J.: The Traveling Salesman Problem: A Computational Study. Princeton University Press, Princeton (2007)
2. Bean, J.C.: Genetic algorithms and random keys for sequencing and optimization. ORSA J. Comput. **6**(2), 154–160 (1994)
3. Chagas, J.B., Blank, J., Wagner, M., Souza, M.J., Deb, K.: A non-dominated sorting based customized random-key genetic algorithm for the bi-objective traveling thief problem. J. Heurist. **27**(3), 267–301 (2021)
4. Damm, R.B., Ronconi, D.P.: A multi-objective biased random-key genetic algorithm for service technician routing and scheduling problem. In: Mes, M., Lalla-Ruiz, E., Voß, S. (eds.) ICCL 2021. LNCS, vol. 13004, pp. 471–486. Springer, Cham (2021). https://doi.org/10.1007/978-3-030-87672-2_31
5. Deb, K., Pratap, A., Agarwal, S., Meyarivan, T.: A fast and elitist multiobjective genetic algorithm: NSGA-II. IEEE Trans. Evol. Comput. **6**(2), 182–197 (2002)
6. Gonçalves, J.F., Resende, M.G.: Biased random-key genetic algorithms for combinatorial optimization. J. Heurist. **17**(5), 487–525 (2011)
7. Lacevic, B., Amaldi, E.: Ectropy of diversity measures for populations in Euclidean space. Inf. Sci. **181**(11), 2316–2339 (2011)
8. Li, X.: A non-dominated sorting particle swarm optimizer for multiobjective optimization. In: Cantú-Paz, E., et al. (eds.) GECCO 2003. LNCS, vol. 2723, pp. 37–48. Springer, Heidelberg (2003). https://doi.org/10.1007/3-540-45105-6_4
9. Lucena, M.L., Andrade, C.E., Resende, M.G., Miyazawa, F.K.: Some extensions of biased random-key genetic algorithms. In: Proceedings of the 46th Brazilian Symposium of Operational Research, pp. 1–12 (2014)
10. Lust, T., Jaszkiewicz, A.: Speed-up techniques for solving large-scale biobjective TSP. Comput. Oper. Res. **37**(3), 521–533 (2010)
11. Shang, K., Ishibuchi, H., He, L., Pang, L.M.: A survey on the hypervolume indicator in evolutionary multiobjective optimization. IEEE Trans. Evol. Comput. **25**(1), 1–20 (2021)
12. Shim, V.A., Tan, K.C., Chia, J.Y., Chong, J.K.: Evolutionary algorithms for solving multi-objective travelling salesman problem. Flex. Serv. Manuf. J. **23**(2), 207–241 (2011)
13. Zhang, Q., Li, H.: MOEA/D: a multiobjective evolutionary algorithm based on decomposition. IEEE Trans. Evol. Comput. **11**(6), 712–731 (2007)

Hybrid Metaheuristic Approaches for Makespan Minimization on a Batch Processing Machine

Juan Carlos Rivera$^{(\boxtimes)}$ (iD) and Ana María Cortes

Mathematical Modeling Research Group, Universidad EAFIT,
Carrera 49 # 7 sur - 50, Medellín, Colombia
{jrivera6,acortesz}@eafit.edu.co

Abstract. A batch processing machine (BPM) is characterized as being able to process multiple jobs simultaneously. This type of machines is common in industrial processes such as electrolytic coating, heat treatments and drying ovens. The BPM scheduling problem consists of grouping a set of jobs into batches to be processed in a single machine with a limited capacity, in such a way that the time necessary to manufacture all jobs (makespan) is minimized. The BPM scheduling problem can be formulated as a mixed integer linear program (MILP). Nevertheless, it is usually addressed through metaheuristic algorithms due to it belongs to NP-Hard class of problems. In this paper, techniques such as the savings methods, NEH algorithm, Large Neighborhood Search (LNS) metaheuristic and splitting algorithm (order-first cluster-second) are adapted to solve the BPM. The performance of the algorithms is evaluated using known instances from literature with up to 100 jobs. The proposed algorithms improve some of the best known solutions in the literature.

Keywords: Batch processing machine · Metaheuristic algorithms · Machines scheduling · Combinatorial optimization

1 Introduction

The Batch Processing Machine (BPM) scheduling is a combinatorial optimization problem in which a machine or processor is able to process multiple jobs simultaneously. This type of machine is commonly found in manufacturing and metallurgical industry in processes like thermal treatments, drying ovens and painting processes, among others.

The search for BPM solution methods is essential at a practical level, taking into account that generally these machines have a high cost; converting them in a restrictive and scarce resource, that is, an optimal BPM schedule impacts directly on the real production capacity and the efficiency of the processes.

The problem consists on schedule a set $J = \{1, ..., n\}$ of jobs in a BPM which has a limited capacity B. Each job $j \in J$ is characterized by a release time r_j, a

© The Author(s), under exclusive license to Springer Nature Switzerland AG 2023
L. Di Gaspero et al. (Eds.): MIC 2022, LNCS 13838, pp. 525–530, 2023.
https://doi.org/10.1007/978-3-031-26504-4_43

minimum processing time p_j and a size s_j. They can be grouped into batches such that the sum of the jobs sizes does not exceed the machine capacity. The release time and the processing time of a batch are computed as the maximum of the release times and the processing times of the jobs that compose it, respectively. Therefore, it is understood that the jobs can be processed in a batch with longer processing time without affecting the final product quality. The objective is to find a schedule T, i.e. start and finish time for each job $j \in J$, so that the time required to manufacture the n jobs (C_{max}) is minimum [1].

There are several variants of the problem, which depend mainly on the consideration (or not) of release times and arbitrary sizes of the jobs. This paper focuses on the case in which both, release times and arbitrary sizes, are taken into account for each one of the jobs. This particular case has been one of the least studied in the literature.

Uzsoy [12] demonstrates that the problem with arbitrary sizes is NP-hard, so several (meta)heuristics have been proposed to solve it. Melouk *et al.* [7] propose a Simulated Annealing (SA) and a mixed integer linear program (MILP). Chou *et al.* [3] present a two-phase metaheuristic. In the first phase, a Genetic Algorithm (GA) defines a sequence of jobs, while the second one creates batches with the heuristic *First-Fit Longest Processing Time* (FFLPT). Chou [2] proposes a hybrid method combining Dynamic Programming with a Genetic Algorithm (DP+GA). GA is used to generate several sequences of jobs while DP evaluates each chromosome. This method outperforms the one presented by Chou *et al.* [3]. Vélez-Gallego *et al.* [13] extend the SKP (*Succcessive Knapsack Problem* [5]) heuristic by considering time windows to represent ready times. Xu *et al.* [14] propose an Ant Colony Optimization (ACO) metaheuristic with candidate lists to restrict the number of reachable solutions and a way to build heuristic information based on reducing waste and idle space.

In the sequel this paper is organized as follows: The proposed solution approaches are presented in Sect. 2. Section 3 describes the computational experiments, while Sect. 4 introduces main concluding remarks.

2   Metaheuristic Approaches

This section introduces seven proposed solution approaches: a constructive heuristic based on savings concept, a splitting procedure, a NEH heuristic and four variants of Large Neigborhood Search (LNS) metaheuristic. Five destroying operators and four repairing operators are also introduced.

Savings Algorithm: Savings concept, first proposed by Clarke and Wright [4] to solve Vehicle Routing Problems (VRP), is adapted to deal BPM.

The procedure starts with n batches, each one composed by a single job. When two batches can be processed simultaneously, i.e. they did not exceed machine capacity when performed together, savings are computed as the difference between the total processing time when they are processed separately (one batch after the other) and when they are processed in the same batch. The pairs of batches with the largest savings are prioritized.

Splitting Algorithm: This algorithm adapts the one proposed by Prins [9] for the VRP. More detailed information can be found on Prins *et al.* [10]. For the BPM, the algorithm translates a sequence of jobs into a sequence of batches. That is, given a sequence of jobs $\sigma = (\sigma_0, \sigma_1, ..., \sigma_n)$, a *Shortest Path Problem* (SPP) is defined on a digraph $G = (J', E)$. The set of nodes (jobs) $J' = J \cup \{0\}$ contains an initial dummy node and n jobs. Each job $j \in J$ has a size s_j, a processing time p_j and a release date r_j. The set E contains arcs (i, j) such that $i < j$ and $\sum_{k=i+1}^{j} s_{\sigma_k} \leq B$. In this version, each arc $(i, j) \in E$ represents a possible batch composed by jobs $(\sigma_{i+1}, ..., \sigma_j)$.

Belman-Ford algorithm is adapted to solve the proposed SPP. The algorithm scans each node $i \in \sigma$, expands all feasible arcs $(i, j) \, \forall \, j \in \sigma$, $j > i$, and evaluates the cost to reach the node j as $V_j = \min\{V_j, R_{ij} + P_{ij}, V_i + P_{ij}\}$. The path with the least value V_n is the one that represents the set of batches with minimum makespan (C_{max}).

As this method requires an initial sequence of jobs to get a BPM schedule, it is tested with a savings based sequence and others proposed in following subsections.

NEH-Based Algorithm: NEH algorithm was proposed by Nawaz, Enscore and Ham [8] for the Flow-Shop Scheduling Problem (FSP) and adapted here for the BPM. This algorithm starts by sorting the jobs according to LPT (longest processing time) dispatching rule. Then, first two jobs are scheduled in such a way that makespan is minimized. For the remaining jobs, on each step a job from the set of unscheduled jobs is inserted in the position which produces the best possible makespan. In our implementation the *split* procedure is performed on each resulting sequence to create batches and evaluate the makespan. The procedure can be also performed in opposite order (shortest processing time or SPT dispatching rule).

Large Neighborhood Search (LNS): LNS, proposed by Shaw [11], improves an incumbent solution by iteratively applying destroy and repair operators. Destroy operators release some variables and fixes others at current values, while Repair operators rebuild the solution, i.e. recover a complete solution.

In our implementation, the initial solution is get by Savings algorithm. Five destroy operators are used with the same selection probability: a. remove the biggest job from a random chosen batch, b. remove the longest job from a random chosen batch, c. remove the job with largest volume from a random chosen batch where volume is defined as processing time multiplied by size, d. remove a random chosen job, and e. remove a random chosen batch. Similarly, three repair operators are used with the same probability: a. insert jobs in the batch with the smallest capacity slack, b. insert jobs in the first possible batch, and c. best insertion, i.e. test all possible positions and choose the best one.

LNS$^+$: LNS$^+$ represents a variant of LNS method in which the Split procedure is applied on each iteration to each rebuilt solution after a Repair operator.

Iterated Large Neighborhood Search (ILNS): ILNS modifies Iterative Local Search (ILS), proposed by Lourenço et al. [6], by replacing neighborhoods based search for a LNS search structure. Thus, the main difference between ILNS and LNS is that ILNS use a *Perturbation* operator. It allows to restart the search procedure from a new solution produced by applying random moves on an incumbent solution. Perturbation operator removes a number of random chosen jobs and reinserts them in random positions.

NEH-Based LNS: In this variant of LNS, a single destroy operator is performed: a number of random chosen jobs are removed from an incumbent solution. Them, the removed jobs are reinserted by a repair operator based on NEH algorithm: jobs are sorted and tested one by one on each sequence position to select the best possible. Makespan is computed by applying spliting algorithm on each resulting sequence of jobs.

3   Computational Experimentation

Computational experiments are conducted in order to compare the proposed methods. All tests have been performed on a Intel core-i7 processor (2.6 GHz) with 16 GB of RAM, and methods have been coded on Python 2.7. We use the instances proposed by Xu et al. [14], where $n \in \{10, 20, 50, 100\}$, $p_j \sim U(8, 48)$, $r_j \sim U(0, LB)$ and LB is a lower bound. Two kind of problems are considered, problems with small jobs where $s_j \sim U(1, 30)$ and with large jobs where $s_j \sim U(15, 35)$. Machine capacity is always the same ($B = 40$). For each type and size, 10 instances have been tested.

Table 1 summarizes the obtained results. First two columns indicates the number of jobs and the type of problem: with small jobs (1) and with large jobs (2). For each method, the average perceptual gap with respect to optimal solution and average computational time in seconds are shown. Stop criteria for LNS-like methods is set to 100 iterations.

It can be noticed that type 1 instances get greater gaps and computational times. Split procedure improves 1.3% Savings solutions on average without significantly increasing running times. NEH/Split algorithm outperforms previous methods by improving both, solution quality and running times. LNS methods are able to improve solutions for type 2 instances. They also get all optimal solutions for instances with $n = 10$. ILNS improves the average gap obtained by NEH/Split. LNS/NEH gets the best average gap with the second best running time. LNS/NEH improves most of the solutions gotten by ACO Xu et al. [14] and it reaches 29 optimal solutions out of 80.

Table 1. Summary of results

Instance		Savings		Savings+Split		NEH/Split		LNS		LNS$^+$		ILNS		LNS/NEH	
n	Type	Gap	Time	Gap	Time	Gap	Time	Gap	Time	Gap	Time	Gap	Time	Gap	Time
10	1	3.87	0.00	2.35	0.00	0.88	0.00	0.00	0.34	0.00	0.89	0.00	0.89	0.00	0.04
10	2	1.73	0.00	1.07	0.00	0.80	0.00	0.00	0.21	0.00	0.95	0.00	0.95	0.00	0.03
20	1	6.84	0.00	3.91	0.01	1.07	0.00	2.53	0.38	0.42	1.52	0.03	1.75	0.48	0.09
20	2	1.46	0.00	1.46	0.00	1.04	0.00	0.17	0.28	0.15	1.89	0.00	2.08	0.54	0.08
50	1	12.44	0.24	9.48	0.24	4.30	0.03	8.64	1.20	6.89	5.11	4.94	5.34	3.44	0.37
50	2	2.90	0.04	2.72	0.04	1.94	0.04	1.33	0.60	1.18	6.32	0.66	6.56	1.11	0.52
100	1	10.66	6.81	8.57	6.81	3.61	0.14	8.57	11.38	7.88	23.84	6.28	24.78	3.46	1.01
100	2	2.41	1.05	2.34	1.05	1.12	0.33	2.11	3.05	1.76	21.87	1.16	23.18	1.09	2.37
Average	1	8.45	1.76	6.08	1.77	2.47	0.04	4.93	3.33	3.80	7.84	2.81	8.19	1.84	0.38
	2	2.12	0.27	1.90	0.27	1.22	0.10	0.90	1.04	0.77	7.76	0.46	8.19	0.68	0.75
	Total	5.29	1.02	3.99	1.02	1.85	0.07	2.92	2.18	2.29	7.80	1.63	8.19	1.26	0.56

4 Conclusions

In this paper, new metaheuristic algorithms are proposed for BPM; some of them have been adapted from classical algorithms for vehicle routing problems. NEH/Split, which combine NEH strategy with a splitting algorithm, is the fastest heuristic and gets better solutions than savings-based strategies. The proposed LNS-like metaheuristics, based on the destroy and repair principle, are successful at improving solutions by randomly alternate between five destroy operators and three repair operators. LNS/NEH obtains the best performance with respect to gap and computing time.

It is important to remark that job sizes (type of instance) have a great impact on methods performance. Thus, instances composed by larger jobs (type 2) are harder, so their gaps are larger for all tested methods.

As future research direction, an interesting approach is to adapt the proposed strategies to other variants like the parallel batch processing machine scheduling.

References

1. Cheng, T., Ng, C., Yuan, J., Liu, Z.: Single machine scheduling to minimize total weighted tardiness. Eur. J. Oper. Res. **165**(2), 423–443 (2005)
2. Chou, F.D.: A joint GA+DP approach for single burn-in oven scheduling problems with makespan criterion. Int. J. Adv. Manuf. Technol. **35**(5–6), 587–595 (2007)
3. Chou, F.D., Chang, P.C., Wang, H.M.: A hybrid genetic algorithm to minimize makespan for the single batch machine dynamic scheduling problem. Int. J. Adv. Manuf. Technol. **31**(3–4), 350–359 (2006)
4. Clarke, G., Wright, J.W.: Scheduling of vehicles from a central depot to a number of delivery points. Oper. Res. **12**(4), 568–581 (1964)
5. Dupont, L., Ghazvini, F.: Minimizing makespan on a single batch processing machine with non-identical job sizes. Eur. J. Autom. **32**, 431–440 (1998)
6. Lourenço, H.R., Martin, O.C., Stützle, T.: Iterated local search: framework and applications. In: Gendreau, M., Potvin, J.Y. (eds.) Handbook of Metaheuristics. International Series in Operations Research & Management Science, vol. 146, pp. 363–397. Springer, Boston (2010). https://doi.org/10.1007/978-1-4419-1665-5_12

7. Melouk, S., Damodaran, P., Chang, P.Y.: Minimizing makespan for single machine batch processing with non-identical job sizes using simulated annealing. Int. J. Prod. Econ. **87**(2), 141–147 (2004)
8. Nawaz, M., Enscore, E.E., Ham, I.: A heuristic algorithm for the m-machine, n-job flow-shop sequencing problem. Omega **11**(1), 91–95 (1983)
9. Prins, C.: A simple and effective evolutionary algorithm for the vehicle routing problem. Comput. Oper. Res. **31**, 1985–2002 (2004)
10. Prins, C., Lacomme, P., Prodhon, C.: Order-first split-second methods for vehicle routing problems: a review. Transp. Res. Part C **40**, 179–200 (2014)
11. Shaw, P.: Using constraint programming and local search methods to solve vehicle routing problems. In: Maher, M., Puget, J.-F. (eds.) CP 1998. LNCS, vol. 1520, pp. 417–431. Springer, Heidelberg (1998). https://doi.org/10.1007/3-540-49481-2_30
12. Uzsoy, R.: Scheduling a batch processing machine with non-identical job sizes. Int. J. Prod. Res. **38**(10), 2173–2184 (2000)
13. Vélez-Gallego, M., Damodaran, P., Rodríguez, M.: Makespan minimization on a single batch processing machine with unequal job ready times. Int. J. Ind. Eng. Theory Appl. Pract. **18**(10), 536–546 (2011)
14. Xu, R., Chen, H., Li, X.: Makespan minimization on single batch-processing machine via ant colony optimization. Comput. Oper. Res. **39**(3), 582–593 (2012)

Variable Neighborhood Descent
for Software Quality Optimization

Javier Yuste[✉][ID], Eduardo G. Pardo[ID], and Abraham Duarte[ID]

Universidad Rey Juan Carlos, 28933 Móstoles, Madrid, Spain
{javier.yuste,eduardo.pardo,abraham.duarte}@urjc.es

Abstract. In the Software Development Life-Cycle, the maintenance phase is often the most costly stage due to the efforts devoted to understanding the system. The Software Module Clustering Problem (SMCP) is an optimization problem which objective is to find the most modular organization of software systems to ease their comprehension. In this problem, software projects are frequently modeled as graphs. Then, the objective of the SMCP, which is proved to be $\mathcal{NP}$-hard, is to group the vertices in modules such that the modularity of the graph is maximized. In this work, we present an algorithm based on Variable Neighborhood Descent for the SMCP and study a novel quality metric, the Function of Complexity Balance, which was recently proposed for the problem. Our proposal has been favorably evaluated over a dataset of 34 real software projects, outperforming the previous state-of-the-art method, a Hybrid Genetic Algorithm, in terms of both quality and computing time. Furthermore, the results are statistically significant according to the Wilcoxon's signed rank test.

Keywords: Search-based software engineering · Software module clustering · Heuristics · Function of complexity balance · Maintainability

1 Introduction

In the Software Development Life-Cycle, the maintenance phase has long been known to be the most costly stage due to the efforts devoted to understanding the code [3]. To ease this task, software systems are usually divided into different components (e.g., classes) that are then grouped in modules (e.g., packages), trying to achieve high cohesion (components in the same module are closely related) and low coupling (components from different modules are loosely connected) [4]. The Software Module Clustering Problem (SMCP) is an optimization problem that tries to find the most modular organization of a given software project. In the SMCP, software systems are usually modeled in a graph structure known as Module Dependency Graph (MDG). Therefore, the SMCP consists of partitioning the vertices of the MDG in clusters, maximizing the modularity of the graph.

L. Di Gaspero et al. (Eds.): MIC 2022, LNCS 13838, pp. 531–536, 2023.
https://doi.org/10.1007/978-3-031-26504-4_44

To the best of our knowledge, Mancoridis et al. [6] were the first ones to tackle the SMCP as an optimization problem, introducing an objective function named Modularization Quality (MQ), which has been extensively studied in the area [12,14]. Lately, some related literature has highlighted issues in the modularity paradigm and in the MQ metric [11]. As an alternative, Mu et al. [10] proposed the Function of Complexity Balance (FCB) to deal with the problem of over-cohesiveness and to reduce the number of isolated clusters. Since the SMCP has been proven to be $\mathcal{NP}$-hard [1], exact methods are less suitable for this task than approximate search-based algorithms. In this context, evolutionary approaches have been prominently studied for the SMCP. Although some authors in the area have proposed trajectory-based metaheuristics [9], recent literature has highlighted the need for a deeper and richer exploration of search-based alternatives to evolutionary strategies [13].

In this paper, we present an efficient Variable Neighborhood Descent (VND) [8] algorithm for the SMCP. We study the FCB objective function and compare the results obtained with the Hybrid Genetic Algorithm (HGA) introduced in [10] over a set of 34 real instances previously curated by the community [9], producing solutions of equal or better quality in 29 out of the 34 tested instances. Furthermore, our approach is three orders of magnitude faster than the previous best method in the state of the art.

2   Problem Definition

In the SMCP, the software projects are often modeled in an MDG, a graph structure $G = (V, E, W)$ where V is the set of vertices that represents components of the system, E is the set of edges that represents relations between components, and W is the set of weights that represents the strenght of the relations. Then, a solution for the SMCP consists of a set $M = \{m_1, m_2, ..., m_k\}$ of disjoint subsets of V, where k represents the number of modules $(1 \leq k \leq |V|)$ and each m_i, with $1 \leq i \leq k$, is a subset of V. To evaluate the quality of a solution, we study the FCB objective function [10]. Given a solution x for the SMCP, the objective function is calculated as follows:

$$FCB(x) = \frac{C + \max(d_i)}{T}, \tag{1}$$

where C is the coupling of the architecture (that is, the sum of the weights of edges with endpoints in different modules), d_i is the cohesion of the ith module (i.e., the sum of the weights of edges with both endpoints in m_i), and T is the sum of the weights of the entire architecture, which is a constant value independent of the partitioning of the solution.

Then, the objective of the SMCP is to find a solution $x^\star$ among the set of all possible solutions X that minimizes Eq. 1. More formally:

$$x^\star = \arg\min_{x \in X}(FCB(x)). \tag{2}$$

3   Algorithmic Proposal

The algorithm presented in this paper is a multi-start method based on the Variable Neighborhood Search (VNS) methodology. VNS was originally proposed by Mladenović and Hansen in 1997 [8] as a general framework for solving hard combinatorial and global optimization problems, which has been later combined with multi-start approaches [2]. In particular, the method is composed by a random constructive and an improvement phase based on Variable Neighborhood Descent (VND), a well-known variant of VNS. It receives an MDG (i.e., the instance) and the number of iterations. In each iteration, an initial solution is built by the random constructive and then improved in the VND phase. When the algorithm reaches the maximum number of iterations, it returns the best solution found during all iterations. The VND procedure receives two input parameters: the initial solution obtained from the random constructive and the set of neighborhoods to explore. Then, the algorithm explores each neighborhood until all the neighborhoods have been explored sequentially without improving the current best solution. If a better solution is found while exploring any of the neighborhoods, the new solution is saved as the current best solution and VND restarts the exploration of all the neighborhoods. This exploration is performed with a local search that follows a first improvement strategy. When all the neighborhoods have been explored sequentially without improving the solution, it means that none of the neighborhoods contains a solution better than the current one. Consequently, the VND procedure exits the loop and returns the best solution found.

In this work, we propose the use of three different neighborhoods for the SMCP that will be explored during the VND phase. The first neighborhood (N_1) is composed of the solutions that can be reached with an insertion movement, which consists of deleting a vertex from its current module and inserting it into another module in the solution. Given a graph with $|V|$ vertices and $|M|$ modules, the size of this neighborhood is $|V| \cdot (|M| - 1)$. The second neighborhood (N_2) is composed of the solutions that can be reached with a merge movement. A merge consists of joining two different existing modules into a single one. For a graph with $|M|$ modules, the size of this neighborhood is $|M| \cdot \frac{|M|-1}{2}$. Finally, the third neighborhood (N_3) is composed by the solutions that can be reached with a split movement, which consists of splitting an existing module in half. Given a graph with $|M|$ modules, the size of the neighborhood is $\sum_{i=1}^{|M|} \frac{|m_i|!}{\lfloor |m_i|/2 \rfloor!(|m_i| - \lfloor |m_i|/2 \rfloor)!}$.

4   Experimental Results

In this section, we present the experimental results obtained in this research. We compared our approach with the best previous method in the state of the art (i.e., the HGA proposed in [10]) over a set of 34 real instances made publicly available by the community [9]. For the sake of reproducibility, we make publicly available both the code and the data used in this work[1]. Unfortunately, the code for the HGA procedure was not available, and the results have been obtained with our own implementation of the ideas proposed in [10]. All experiments were run on an AMD EPYC 7282 @ 2795 MHz CPU, with 8 cores and 8 GB RAM. The Operating System used was Microsoft Windows 10 Pro 10.0.19042 x64. Our proposal was coded in Java 17.0.1 and using the Metaheuristic Optimization framewoRK (MORK) project [7]. The HGA proposed in [10] was implemented in Java and Matlab (R2021b Update 1). However, we noticed that the implementation in Matlab, as originally proposed by the authors, benefited from the efficient use of matrix operations, which are important for the design of the algorithm. Therefore, we used the results obtained by this second implementation for comparison purposes. The search parameters of our proposal (the number of iterations and the order of the neighborhoods) have been empirically determined over a reduced preliminary dataset of 10 instances using the parameter-tunning software *irace* [5]. In particular, our proposal has been set to perform a maximum of 100 iterations per instance and the VND approach has been configured to explore the neighborhoods in the following order: N_1, N_3, and N_2.

In Table 1, we present the results obtained by our proposal (VND) and the state-of-the-art procedure (HGA [10]). Both algorithms were run once per instance. We report the average quality of the best solutions found (Avg. score), the average deviation to the best solution found in the experiment (Dev. (%)), the number of best solutions found (# Best), and the average CPU time per instance (CPUt (s)). As it can be observed, the quality of the solutions reached by VND is higher than the quality of the solutions found by HGA. On average, VND was able to obtain solutions closer to the best one found in the experiment (with a deviation of 1.37 %) than HGA (with a deviation of 8.62 %). Additionally, VND was able to find the best solution in 29 out of 34 instances, while HGA reached the best solution only for 11 instances. Moreover, VND was, on average, three orders of magnitude faster than HGA. According to the Wilcoxon's signed rank test, the results are statistically significant with $p < 0.01$.

Table 1. Comparison of the results of the algorithm presented in this work (VND) and the best previous method in the state of the art (HGA [10]).

Method	Avg. score	Dev. (%)	# Best	CPUt (s)
VND	**0.6293**	**1.37%**	**29**	**85.08**
HGA [10]	0.6708	8.62%	11	35,804.40

Finally, to compare the performance of the algorithms when executed for the same amount of time, we run them once per instance with different time limits. Particularly, we performed 3 experiments, stopping the algorithms after

---

[1] https://github.com/JavierYuste/Variable-Neighborhood-Descent-for-software-quality-optimization.

Table 2. Detailed results obtained for all the tested instances.

Instance	$|V|$	$|E|$	HGA [10]		GVNS	
			FCB	CPUt (s)	FCB	CPUt (s)
squid	2	2	1.0000	109.31	1.0000	0.14
small	6	5	0.6000	112.66	0.6000	0.00
compiler	13	32	0.6875	134.96	0.6875	0.07
sharutils	19	36	0.6442	369.09	0.6250	0.05
spdb	21	17	0.5000	363.47	0.5000	0.02
ispell	24	103	0.7320	184.55	0.7113	0.20
ciald	26	64	0.6563	180.85	0.6563	0.13
crond	29	112	0.7085	287.81	0.6911	0.68
seemp	30	61	0.5200	364.55	0.4800	0.09
dhcpd-2	31	122	0.6544	331.84	0.6460	0.61
cyrus-sasl	32	100	0.6504	329.45	0.6104	0.86
star	36	89	0.5506	438.29	0.5506	0.20
cia	38	185	0.6117	416.18	0.5900	0.77
dot	42	255	0.7298	354.47	0.7419	1.13
screen	42	292	0.7364	793.49	0.7157	1.43
slang	45	242	0.7213	776.15	0.6778	1.44
slrn	45	323	0.7304	735.84	0.7056	1.93
hw	53	51	0.6093	512.97	0.4188	0.27
imapd-1	53	298	0.7201	517.52	0.6587	2.09
javaocr	58	155	0.5420	518.28	0.5496	0.53
dhcpd-1	59	571	0.7184	665.12	0.6445	11.39
icecast	60	650	0.7509	489.67	0.7060	6.71
servletapi	61	131	0.4803	1013.90	0.4567	0.65
bunch2	65	151	0.6250	675.35	0.5000	0.60
grappa	86	295	0.5952	722.76	0.6032	1.93
inn	90	624	0.7368	1174.01	0.6994	13.63
acqCIGNA	114	179	0.6862	6254.05	0.5585	1.75
cia++	124	369	0.7395	8027.85	0.6377	6.23
JavaGeom	171	1445	0.7422	27942.71	0.6408	77.79
incl	174	360	0.4472	12310.84	0.6111	5.09
dom4j	195	930	0.7406	34328.38	0.6246	53.52
bunchall	324	1339	0.7424	189325.97	0.5818	117.95
JACE	338	1524	0.7430	347810.33	0.5195	158.20
res_cobol	470	7163	0.7539	578777.00	0.7953	2424.60

1, 10, and 60 min, respectively. Considering the results of the 3 experiments, on average, our approach obtained the best solutions in 33 out of the 34 instances, with a score of 0.6237 and a deviation of 0.28%. In contrast, the state-of-the-art algorithm obtained the best solutions in only 4 of the 34 instances on average, with a score of 0.7249 and a deviation of 18.82% (Table 2).

Acknowledgements. This research has been partially funded by the Spanish government (Refs. PGC2018-095322-B-C22 and PID2021-125709OA-C22) and by the Comunidad de Madrid (Ref. P2018/TCS-4566), cofinanced by European Structural Funds ESF and FEDER. The opinions, findings, and conclusions or recommendations expressed are those of the authors and do not necessarily reflect those of any of the funders.

References

1. Brandes, U., et al.: On modularity clustering. IEEE Trans. Knowl. Data Eng. **20**(2), 172–188 (2007)
2. Cavero, S., Pardo, E.G., Duarte, A.: A general variable neighborhood search for the cyclic antibandwidth problem. Comput. Optim. Appl. **81**, 657–687 (2022)
3. Chen, C., Alfayez, R., Srisopha, K., Boehm, B., Shi, L.: Why is it important to measure maintainability and what are the best ways to do it? In: 2017 IEEE/ACM 39th International Conference on Software Engineering Companion (ICSE-C), pp. 377–378. IEEE (2017)
4. International Organization for Standardization: ISO/IEC/IEEE 24765:2017 Systems and software engineering - Vocabulary (2017)
5. López-Ibáñez, M., Dubois-Lacoste, J., Cáceres, L.P., Birattari, M., Stützle, T.: The irace package: iterated racing for automatic algorithm configuration. Oper. Res. Perspect. **3**, 43–58 (2016)
6. Mancoridis, S., Mitchell, B.S., Rorres, C., Chen, Y.F., Gansner, E.R.: Using automatic clustering to produce high-level system organizations of source code. In: 6th International Workshop on Program Comprehension (IWPC 1998), pp. 45–52. IEEE (1998)
7. Martín, R., Cavero, S.: MORK: Metaheuristic Optimization framewoRK. https://doi.org/10.5281/zenodo.6241738
8. Mladenović, N., Hansen, P.: Variable neighborhood search. Comput. Oper. Res. **24**(11), 1097–1100 (1997)
9. Monçores, M.C., Alvim, A.C.F., Barros, M.O.: Large neighborhood search applied to the software module clustering problem. Comput. Oper. Res. **91**, 92–111 (2018)
10. Mu, L., Sugumaran, V., Wang, F.: A hybrid genetic algorithm for software architecture re-modularization. Inf. Syst. Front. **22**(5), 1133–1161 (2020)
11. de Oliveira Barros, M., de Almeida Farzat, F., Travassos, G.H.: Learning from optimization: a case study with apache ant. Inf. Softw. Technol. **57**, 684–704 (2015)
12. Praditwong, K., Harman, M., Yao, X.: Software module clustering as a multi-objective search problem. IEEE Trans. Softw. Eng. **37**(2), 264–282 (2010)
13. Ramirez, A., Romero, J.R., Ventura, S.: A survey of many-objective optimisation in search-based software engineering. J. Syst. Softw. **149**, 382–395 (2019)
14. Yuste, J., Duarte, A., Pardo, E.G.: An efficient heuristic algorithm for software module clustering optimization. J. Syst. Softw. **190**, 111349 (2022)

Iterated Local Search with Genetic Algorithms for the Photo Slideshow Problem

Labeat Arbneshi and Kadri Sylejmani[✉]

Faculty of Electrical and Computer Engineering, University of Prishtina, Prishtina, Kosova
{labeat.arbneshi,kadri.sylejmani}@uni-pr.edu

Abstract. In this extended abstract, we present a two-stage approach for solving the photo slideshow problem as defined in the qualification round of the Google Hash Code 2019. In the first stage, we apply a Genetic Algorithm to produce a good-quality initial solution, whereas, in the second stage, we apply an Iterated Local Search metaheuristic to further optimize the solution. The presented computational study in four challenging test instances shows that our approach produces comparable results to the ones achieved in the competition, where, for two of the instances, new benchmark results are obtained.

Keywords: Iterated Local Search · Genetic Algorithms · Automated photo slideshow design

1 Introduction

The photo slideshow problem was originally formulated in the qualification round of Google Hash Code competition for the year 2019 [1]. This problem is about automated design of a slideshow by using a set of photos, which are described by a set of tags (see the example in Fig. 1).

| House, white hat, flag, eagle | Eagle, mountain | Mountain, snow, house, dog | House, lute, eagle, white hat |

Fig. 1. List of photos and their associated tags

The work on this paper was supported by the HERAS+ progam within the project entitled "Automated Examination Timetabling in the Faculty of Electrical and Computer Engineering - University of Prishtina".

The aim is to arrange the given set of photos into a slideshow so that the transition between consecutive slides is both smooth and 'interesting', in terms of having some similar and different tags when transiting between the consecutive slides of the slideshow. A slide can be composed of either a single horizontal photo or two vertical photos.

The photo slideshow problem is related to the Traveling Salesman Problem (TSP) [2] and the Orienteering Problem (OP) [3] in the aspect that it is a sequencing problem that requires arranging a set of objects (nodes) into an optimized sequence. It is different from TSP in the aspect that it does not require to have all the objects within the sequence, and it is a maximization problem, while it is different from OP, since it optimizes based on the edges (i.e., slide transitions) rather than objects. Further, it is different from both TSP and OP in the aspect that objects are not strictly unary but can also be binary (the case of the slide with two vertical photos).

The main contribution of this work is twofold. First, we hybridize Iterated Local Search and Genetic Algorithms for designing a competitive approach for the photo slideshow problem, and second, for the same problem, we also propose a unique neighborhood exploration mechanism that include a combination of mutation and crossover operators.

2 Mathematical Formulation

In the original formulation of the photo slideshow problem [1], a set of P photos is given, where P_h of them have a horizontal orientation, while the remaining P_v are vertically oriented. In addition, each photo k in set P is described by using a set of tags T_k, which describe its content. The goal is to arrange the photos into a slideshow that is as 'interesting' as possible based on the so-called *Transition Rule* ($T^r_{i,i+1}$) that considers the intersection and difference of tags of any two consecutive slides i and $i + 1$, as defined in Eq. (1). A given photo can only be used once in the entire slideshow, while a given slide can contain either a single horizontal photo or two vertical photos. In the first case, the tags of the photo are also the tags of the slide, whereas in the latter case the tags of the slide are the union of the tags of both photos.

$$T^r_{i,i+1} = \min(|T \cap T_{i+1}|, |T_i \setminus T_{i+1}|, |T_{i+1} \setminus T_i|) \qquad (1)$$

Based on the number of horizontal and vertical photos, the total number of slides is $N = P_h + N_v$, where $N_v = P_v(P_v - 1)/2$ represents the total number of theoretical slides that can be created by using each pair of vertical photos (if we relax the constraint that a single photo cannot be used more than once). For satisfying the constraint that each photo can be used at most once, as part of the pre-calculation, it is convenient to create a list of tuples L that contains pairs of slides (i, j), where $i, j \in N$, having the same photos.

The presented mathematical formulation uses a Boolean decision variable z_{ij}, which equals either 1 (if slide i is followed by slides j) or 0 (otherwise).

$$max \sum_{i=1}^{N} \sum_{\substack{j=1 \\ j \neq i}}^{N} z_{ij} T^r_{i,j} \qquad (2)$$

$$\sum_{\substack{j=1 \\ j \neq i}}^{N} z_{ij} \leq 1, \forall i = [1, \ldots, N] \tag{3}$$

$$\sum_{\substack{i=1 \\ i \neq j}}^{N} z_{ij} \leq 1, \forall j = [1, \ldots, N] \tag{4}$$

$$z_{ij} + z_{ji} \leq 1, \forall i = [1, \ldots, N], \\ \forall j = [1, \ldots, N], i \neq j \tag{5}$$

$$\sum_{a=1}^{N} \sum_{\substack{b=1 \\ b \neq a}}^{N} z_{ab} f_{ij} \leq 1, \forall (i, j) \in L \tag{6}$$

$$f_{ij} = \begin{cases} 1, & if\, (a = i) \bigvee (a = j) \bigvee (b = i) \bigvee (b = j) \\ 0, & otherwise \end{cases}$$

Equation (2) maximizes the sum of the transitions between consecutive slides in the slideshow. Equations (3) and (4) ensure the continuity of the slideshow by making sure that each slide is followed by exactly one another slide. Equation (5) makes sure that each slide is used at most once. Finally, Eq. (6) ensures that each photo is used at most once in the slideshow.

3 Solution Approach

3.1 Preprocessing, Search Space and Fitness Function

In the preprocessing phase, besides grouping horizontal and vertical photos into two distinct clusters, we also calculate the transitions between all slides that include horizontal photos and any two pairs of vertical photos.

A state in the search space is represented by a vector, whose size determines the number of slides in the slideshow. Further, a particular vector member might have either the index of a horizontal photo or indexes of two vertical photos.

The fitness function is implemented based on Eq. (2) that is presented in Sect. 2.

Algorithm 1 Iterated Local Search with Genetic Algorithms

```
1:    procedure Solve (instance, totalTime, popSize, crossoverRate, mu-
      tationRate, numMutants, slidingWindow, tournamentSize, hillClimb-
      Time, localSearchTime, homeBaseRate, perturbIntensity, pertur-
      bRate)
2:       population = initializePopulation(instance, popSize, slid-
         ingWindow)
3:      while 0.2*totalTime not elapsed
4:         numChildren = popSize/2 + numMutants
5:         for c from 1 to numChildren
6:            p = random(0,1)
7:            child = population[c]
8:            if p < crossoverRate
9:               child = swapNHorizontalPhotos(child, tournamentSize)
10:           if p < mutationRate
11:              mutateOperator = selectMutationOperator(instance)
12:              child = applyMutation (child, mutateOperator)
13:           population = extendPopulation(child)
14:        currentBest = selectBest(population)
15:        currentBest = hillClimb(currentBest, hillClimbTime)
16:        population = extendPopulation(bestChild)
17:        population = updatePopulation(population)
18:     best = home = current = selectBest(population)
19:     while 0.8*totalTime not elapsed
20:        while localSearchTime not elapsed
21:           mutateOperator = selectMutationOperator(instance)
22:           neighbor = applyMutation(current,  mutateOperator)
23:           if neighbor better than current
24:              current = neighbor
25:           if current better than best
26:              best = current
27:        home = updateHomeBase(home, current, homeBaseRate)
28:        current = perturbHomeBase(home, perturbIntensity, pertur-
           bRate)
29:     return best
```

3.2 Initial Solution and Neighborhood Structure

The initial solution is generated based on a combined randomness and greediness strategy. The addition of a particular slide into the slideshow (i.e., representation vector) is done by first selecting, at random, several unassigned photos (as defined by the parameter named *slidingWindow*) and then, based on the transition rule (as defined by Eq. (1)), picking, amongst the selected photos, the best horizontal photo, or the best vertical pair of photos to become the next slide.

The neighborhood structure consists of four operators, the first three being mutation operators, and the last one a crossover operator, as described below.

Swap 2 slides – Selects two slides at random from a given parent and exchanges their position in the slideshow.

Shuffle *n* slides – Selects a random position within the slideshow and shuffles the next *n* slides from a given parent (if *n* exceeds slideshow size, then all remaining slides, after the randomly selected position, are shuffled).

Swap 2 vertical photos – First, from a given parent, two slides containing vertical photos are selected at random, and then, one of the two photos (chosen at random) are interchanged between the selected slides.

Swap n horizontal photos – Interchanges the photos of several randomly selected slides (as defined by parameter n) between the two selected parents. This operator is only applied between slides that contain horizontal photos.

3.3 Iterated Local Search with Genetic Algorithms

Iterated Local Search (ILS) and Genetic Algorithms (GA) have been hybridized in various related optimization problems, such as location-routing problem [4], function optimization [5] and ring loading problem [6]. In our implementation, as presented in Algorithm 1, in the first phase, we use GA to produce a good-quality initial solution, whereas, in the second phase, we apply ILS as an improvement method.

In each generation of the GA, from a population of size l, we select parents using the tournament selection method and produce m children by selecting genetic operators partially in random basis and partially based on the instance characteristics (e.g., number of horizontal/vertical photos, number of average/minimum/maximum tags per photo, etc.). The population is updated based on the $l + m$ schema, meaning that the best l members from the collection of l existing parents and the m children are selected to produce the next population.

In the second phase, the best returned solution from GA is further improved by using the ILS metaheuristic, which explores the solution neighborhood based only on the presented mutation operators. The home base mechanism [7] is implemented in a Monte Carlo search strategy, implying that a new home base can be adopted even if it is not better than the current solution (subject to the parameter *homeBaseRate*). The perturbation mechanism, with equal probability, either mutates the new home base several times (as specified in *perturbIntensity*), by using the defined operators, or it first removes several slides (as specified in *perturbRate*) and then re-inserts the corresponding photos into the slideshow using the greedy heuristic that is applied when generating the initial solution (as discussed in Sect. 3.2).

4   Computational Experiments

The proposed algorithm has been developed using the C# programming language and the source code can be found in GitHub[1]. The computational experiments have been performed using a machine with Intel Core i5–3470 processor, 8 GB of RAM memory and Windows 10 operating system.

4.1   Data Set and Parameter Tuning

The dataset, as presented in Table 1, consists of four instances, which differ in terms of number of photos (horizontal and vertical) and number of tags. There are instances that

---

[1] https://github.com/labeatarbneshi/photo-slideshow.

contain either horizontal or vertical photos, or both. In addition, some instances have a larger number of tags, including unique ones and a higher average number of tags per photo. Parameter tuning has been done based on some preliminary experimentation with all four instances of the dataset, where the algorithm has been executed five times for each instance and for three different configurations of the main algorithm parameters that include *popSize, tournamentSize, slidingWindow, perturbRate.*

4.2   Comparison Results

Based on our initial and limited experimental design, in Table 2, we present the preliminary computation results of our approach against the results of the two teams that have participated in the Google Hash Code competition, namely AIM Tech (the winner) and Fatture (the best ranked Italian team, whose solutions are available in GitHub[2]). The presented outcomes show that our approach produces best new results for instances *memorable moments* and *pet pictures*, while it has a gap of 2.17% and 4.34% for instances *lovely landscapes* and *shiny selfies,* respectively. Based on the complexity of a particular problem instance, the computation time varies from 20 min for the *memorable moments* up to six hours for the *lovely landscapes.* Despite our best efforts, we were not able to acquire the computation time for the results achieved by the two referred approaches, hence we are not able to make a direct comparison in this aspect.

Table 1. Dataset characteristics.

Instance name	# Photos	# Horizontal photos	# Vertical photos	# Tags	# Unique tags	# Avg. tags per photo
Lovely landscapes	80,000	80,000	0	1,440,000	840,000	18.0
Memorable moments	1,000	500	500	9,476	2,166	9.4
Pet pictures	90,000	30,000	60,000	902,256	220	10.0
Shiny selfies	80,000	0	80,000	1,527,981	500	19.1

---

[2] https://github.com/danieleratti/hashcode-2019.

Table 2. Comparison against best Google Hash Code competition results.

Instance name	Best known	AIM Tech (best)	Fatture (best)	GA-ILS (avg)	GA-ILS (best)	ILS-GA gap from best known (%)
Lovely landscapes	**205,563**	N/A	205,563	196,362	201,094	2.17
Memorable moments	1,764	1,764	N/A	1,722	**1,791**	−1.53
Pet pictures	394,697	394,697	347,793	391,376	**401,835**	−1.81
Shiny selfies	**559,233**	559,233	537,349	488,265	534,976	4.34

5 Conclusion and Future Work

The proposed ILS-GA approach exhibits its ability to produce competitive results against the best-known approaches submitted in Google Hasch Code competition [1]. As part of the future work, we aim to integrate memories that will record information about the search space exploration experience, which in turn would be used for intensification and diversification purposes.

References

1. Google Hash Code Archive Homepage: https://storage.googleapis.com/coding-competitions. appspot.com/HC/2019/hashcode2019_qualification_task.pdf. Accessed 20 Apr 2022
2. Jünger, M., Reinelt, G., Rinaldi, G.: The traveling salesman problem. In: Handbooks in Operations Research and Management Science, vol. 7, pp. 225–330 (1995)
3. Vansteenwegen, P., Souffriau, W., Van Oudheusden, D.: The orienteering problem: a survey. Eur. J. Oper. Res. **209**(1), 1–10 (2011)
4. Derbel, H., Jarboui, B., Hanafi, S., Chabchoub, H.: Genetic algorithm with iterated local search for solving a location-routing problem. Expert Syst. Appl. **39**(3), 2865–2871 (2012)
5. Lima, C.F., Lobo, F.G.: Parameter-less optimization with the extended compact genetic algorithm and iterated local search. In: Deb, K. (ed.) GECCO 2004. LNCS, vol. 3102, pp. 1328–1339. Springer, Heidelberg (2004). https://doi.org/10.1007/978-3-540-24854-5_127
6. Bernardino, A.M., Bernardino, E.M., Sanchez-Perez, J.M., Gomez-Pulido, J.A., Vega-Rodriguez, M.A.: Genetic algorithms and iterated local search to solve the ring loading problem. In: 2008 50th International Symposium ELMAR, vol. 1, pp. 265–268. IEEE (2008)
7. Luke, S.: Essentials of Metaheuristics. A Set of Undergraduate Lecture Notes. Zeroth Edition. Online Version, 1 (2011)

A Tabu Search Matheuristic for the Generalized Quadratic Assignment Problem

Peter Greistorfer[1]([✉]), Rostislav Staněk[2], and Vittorio Maniezzo[3]

[1] Operations and Information Systems, Karl-Franzens-Universität Graz,
Graz, Austria
peter.greistorfer@uni-graz.at
[2] Applied Mathematics, Montanuniversität Leoben, Leoben, Austria
rostislav.stanek@unileoben.ac.at
[3] Computer Science, Università di Bologna, Bologna, Italy
vittorio.maniezzo@unibo.it

Abstract. This work treats the so-called Generalized Quadratic Assignment Problem. Solution methods are based on heuristic and partially LP-optimizing ideas. Base constructive results stem from a simple 1-pass heuristic and a tree-based branch-and-bound type approach. Then we use a combination of Tabu Search and Linear Programming for the improving phase. Hence, the overall approach constitutes a type of mat- and metaheuristic algorithm. We evaluate the different algorithmic designs and report computational results for a number of data sets, instances from literature as well as own ones. The overall algorithmic performance gives rise to the assumption that the existing framework is promising and worth to be examined in greater detail.

Keywords: Generalized Quadratic Assignment · Matheuristic ·
Metaheuristic · Linear Programming · Tabu Search

1 Introduction

The problem of interest is the so-called *Generalized Quadratic Assignment Problem* (GQAP). The GQAP has been used as a model for several relevant actual applications, including order picking and storage layout in warehouse management, relational database design or scheduling activities in semiconductor wafer processing. Technically, it originates from the *Linear Assignment Problem* (LAP), where a number of agents (equivalently, machines or supplies) have to be assigned to a number of jobs (tasks or demands) while minimizing the total cost of service and obeying assignment constraints, which secure that each job has to be serviced by exactly one agent and vice versa. The LAP in turn is a special case of the *Generalized Assignment Problem* (GAP), in which assignment constraints on the supply-side are replaced with upper bound constraints, which model an

L. Di Gaspero et al. (Eds.): MIC 2022, LNCS 13838, pp. 544–553, 2023.
https://doi.org/10.1007/978-3-031-26504-4_46

agent's capacity consumed by the assigned task weights. Another generalization of the LAP is the *Quadratic Assignment Problem* (QAP), which models multiplicative cost factors between agents and jobs, e.g. in locational analyzes distances times interaction frequencies. Finally, the GQAP can be thought of as a combination of the GAP and the QAP, where the objective function receives a quadratic component in addition to the linear one, the same way as it is done in the QAP and suppliers have an upper bound constraint as in the GAP.

It is well-known that there exist efficient polynomial algorithms for the LAP, e.g. the Hungarian method, but the GAP, the QAP, and the GQAP are $\mathcal{NP}$-hard (e.g. see [1]). Matheuristics (see [6,8]), a synthesis of classical, as a rule, (real-valued) linear and integer linear programming methods (LP, ILP) with conventional heuristic methods (e.g. local search) and/or modern metaheuristic methods (Tabu Search, GRASP, Scatter Search etc.) have become popular with the rise of recent powerful hardware and even more because of the success of solvers like CPLEX or Gurobi.

Our research focuses on both mathematical formulations of the GQAP as well as on the development of LP- and ILP-based matheuristics. In this paper we review a branch-and-bound type heuristic tree search and elaborate on the basic decomposition idea of the LP-component in the improvement algorithm. Afterwards, the focus lies on the presentation of a new Tabu Search (TS), the *TS-matheuristic GQAP* approach (TS-GQAP) is presented. We conclude with current computational results and a short outlook.

2   Modelling the GQAP

The GQAP can be described by means of the following quadratic integer program. We are given $m \in \mathbb{N}$, the number of agents, $n \in \mathbb{N}$, the number of jobs, with linear assignment costs and weights $p_{ij} \in \mathbb{R}$ and $w_{ij} \in \mathbb{R}_0^+$, where $1 \leq i \leq m$ and $1 \leq j \leq n$, respectively. The weights present resource amounts to be spent by agent i for processing job j, without exceeding an available capacity $a_i \in \mathbb{R}_0^+$. Quadratic costs are defined by the product of $d_{ir} \in \mathbb{R}$, the cost factor between the agents i and r, and $f_{js} \in \mathbb{R}$, $1 \leq r \leq m$, $1 \leq s \leq n$, the cost factor between the jobs j and s. Binary decision variables $x_{ij} \in \{0,1\}$ determine whether a job j is served by agent i, or not. Then the GQAP is defined by the following binary quadratic program (BQP):

$$\min \sum_{i=1}^{m}\sum_{j=1}^{n} p_{ij}x_{ij} + \sum_{i=1}^{m}\sum_{j=1}^{n}\sum_{r=1}^{m}\sum_{s=1}^{n} d_{ir}f_{js}x_{ij}x_{rs} \tag{1}$$

$$\text{s.t.} \sum_{j=1}^{n} w_{ij}x_{ij} \leq a_i \quad \forall\, 1 \leq i \leq m, \tag{2}$$

$$\sum_{i=1}^{m} x_{ij} = 1 \quad \forall\, 1 \leq j \leq n, \tag{3}$$

$$x_{ij} \in \{0,1\} \quad \forall\, 1 \leq i \leq m,\, \forall\, 1 \leq j \leq n. \tag{4}$$

3 Solving the GQAP with TS-GQAP

3.1 Basic Concepts

We start with a short summary of the constructive solution procedure, which was in its original form presented as *Guided Adaptive Relaxation Rounding Procedure* (GARRP) in [7]. GARRP starts with an empty (infeasible) solution and iteratively creates partial solutions by fixing exactly one agent-task-assignment $x'_{ij} := 1$ at each iteration, based upon the optimum solution of a relaxation of BQP, leaving a subset of remaining decision variables as free. These partial solutions, X', i.e. incomplete or partial assignments of jobs j to single specific agents i, are stored within a tree, in which each tree-node represents one assignment made. So for any (partial) solution all fixed assignments left can be found by pegging links back to the root, which stores the result of the first relaxation solved (compare [4]). At any node, relaxed, i.e. continuous values x^*_{ij} received from a GQAP-LP-solution, define choice-probabilities among the k-best possible succeeding fixations $x^*_{ij} \geq x_{LB}$ (k and x_{LB} being parameters), from which the most likely option is chosen. The whole tree is explored in a depth-first manner, if needed investigating all potential successor assignments according to a width-second sequence. If at some node all successor options are unsuccessfully investigated (in terms of feasibility defined by capacity constraints), a backtracing process starts until a node with a free successor, becoming investigated, is found or until all nodes failed because of overall infeasibility. Normally, this iterative process stops as soon as n assignments have been made with a complete and feasible solution, otherwise it fails.

The present work, focusing on improvement methods for the GQAP, has its roots in the *Magnifying Glass Heuristic* approach, which was already successfully applied to the *Quadratic Travelling Salesman Problem* in [10]. Then, for the first time, this approach was adopted as MG-GQAP [2] to improve (heuristic) GQAP solutions. As might be expected, while the results were quite satisfying, it turned out that there was still room for improvement. First, we considered the possibility to include the quadratic components into the linear part by compressing the quadratic coefficient matrix into a one-dimensional vector, by means of techniques borrowed from data dimensionality reduction approaches, like principal component analysis. Specifically, we computed when possible (it has always been) the eigenvalues of the coefficient matrix and used them in our heuristic. The resulting code was efficient in the GQAP part, but the computation of the eigenvalues was very demanding in case of big instances, moreover local search was used in the method and it appeared as if the method was apparently not sufficiently capable to escape from local optima. So, the idea was to dismiss linearizations and, sort of the other way round, to use linear programming as an improvement to a TS. This led to the design of the TS-GQAP, which combines two distinct parts, the TS-neighbourhood and the MG-GQAP.

For the TS and its neighbourhood the performance of two prominent standard operators was analyzed: either two jobs swap agents (*exchange*) or a job changes to another agent (*insert*). We contrasted using these two operators together ver-

sus using them specifically. And since exclusive use of insert-moves turned out
to work best, this setting was chosen. In the course of the exploration the TS
sequentially checks all potential re-assignments of tasks. In doing so, a neighbourhood of admissible TS_nh_size trial moves is built by using a short-term
recency tabu-memory. This memory is defined by the time, i.e. iteration, when
a job is assigned to a new agent. Using this information, a newly positioned
task is not allowed to be removed from its agent for a parametric number of
tenure iterations. As aspiration criterion, the best-improvement-rule is used, in
which case the tabu status is released from a given trial move if it improves the
incumbent best solution.

MG-GQAP can be seen as a variable fixing heuristic, which decomposes the
overall problem into a fixed and a free component. Thus, interpreting release as
destruction, the method shares similarities with a *Large Neighbourhood Search*,
where solutions are partially destroyed and repaired by corresponding operators;
a process, which is iteratively repeated until a local optimum is reached (see [9]).
For the GQAP the corresponding idea works like this: according to pre-defined
selection patterns, we consider K chosen columns (jobs, in constraints (3) of
BQP) and create a new auxiliary instance containing only these K columns
(and all rows). Firstly, linear costs p_{ij} are modified including the quadratic costs
caused by relations between the assignment of j to i and the already fixed assignments outside of the chosen columns. Then, after re-adjusting the total amounts
of resources a_i in order to reflect the remaining capacities, we solve the auxiliary problem optimally and get a new, possibly improved solution with changes
restricted to the K chosen columns. Note that alternatively also subsets of *rows*
(agents, constraints (2)) may be selected to define new auxiliary subproblems,
which establishes the differentiation between a column- and a row-oriented variant of MG-GQAP. The overall process ends after a total number of given iterations.
It should be noted that the idea behind MG-GQAP is even capable of serving as
a construction procedure. In that case it starts from an empty unfeasible solution, iteratively increasing the size of a partial solution as long as assignments
are found, which were not prevented by binding capacity restrictions, while ultimately and ideally constructing a full feasible solution with n assignments. We
give this simple 1-pass start heuristic the name MG-GQAP-C and report results in
the computational section.

In summary, within TS-GQAP, the TS utilizes the LP-part in order to optimize
a good local solution in terms of intensification. The role of diversification is
taken over by the occasional use of elite solutions collected in a *pool* by the TS.
Section 3.2 explains this component and covers the algorithmic details.

3.2   Implementation Details

It is well-known that metaheuristic developments have proved to be successful
especially in cases, where their fundamental concepts are complemented with
pool-oriented approaches, originally rooted in Genetic Algorithms or Scatter
Search and Path Relinking. These algorithms maintain a reference set of high

quality solutions, which are repeatedly used during the search in order to guarantee a fruitful balance between diversification and intensification (see, e.g. [3]). Therefore, Algorithm 1: TS-GQAP uses a pool structure as follows.

Require: a GQAP instance with a solution $X := [x_{ij}]^{m \times n} \in \{0,1\}^{m \times n}$
Ensure: new solution X_{new} with $c(X_{new}) \leq c(X)$
1: $X^* := X$, $c_{best} := c(X^*)$
2: **repeat**
3:    **repeat** // TS-phase:
4:        $X' := best\_admissible\big(TS\_neighbourhood(X)\big)$
5:        **if** $c(X') < c_{best}$ **then**
6:            update $X^* := X'$
7:            $c_{best} := c(X')$
8:        **end if**
9:        maintain a set *pool* of good and diverse solutions
10:    **until** TS_term
11:    **if** TS_failed **then**
12:        $X' := unchecked\_solution(pool)$
13:    **end if**
14:    **repeat** // MG-GQAP-phase:
15:        $Col := select\_variable\_columns(X')$
16:        $X_{LP} := LP\_from(Col)$
17:        $X^*_{LP} := solve(X_{LP})$ // CPLEX
18:        $X'' := combine\_solutions(X', X^*_{LP})$
19:        **if** $c(X'') < c_{best}$ **then**
20:            update $X^* := X''$
21:            $c_{best} := c(X'')$
22:        **end if**
23:    **until** MG_GQAP_term
24:    $X := X''$
25: **until** TS_GQAP_term
26: **return** $x_{new} := X^*$

Algorithm 1: TS-GQAP

As an improvement procedure, the TS-GQAP builds on a feasible solution X with objective function value $c(X)$. It consists of alternating TS- and MG-GQAP-phases (lines 3–10 and 14–23), i.e. it is a series of TS1, MG-GQAP1, TS2, MG-QAP2,... until an overall termination criterion, TS_GQAP_term, gets true. Each TS is capable of maintaining the pool of elite solutions. Such solutions are collected to be used in future iterations of the search process. Solution X' in line 4 is the actual and best admissible solution iteratively drawn from consecutive TS-neighbourhoods, which involves the maintenance and usage of the TS-memory as described above.

The pool maintained by the TS, in line 9, collects a maximum of *pool_size* good and diverse solutions. In doing so, a solution is deemed good at the moment, when it has just been improved by the TS, i.e. goodness means objective function value and diversity targets the structural difference mapped in the values of the decision variables, i.e. assignments. In the current version such diversity is—at least with high probability—ensured by allowing into the pool only members with different objective function values. The diversification step is done in lines 11–13 as soon as the TS was not able to improve the current solution, expressed by a value *true*, returned from function TS_failed. To diversify the search, function $unchecked\_solution(pool)$ extracts a/the cost-minimum solution from the pool, which has not yet been output from the pool to be processed by the MG-GQAP. In the case that all pool solutions are already processed, this procedure forwards the running, i.e. the best solution from the last TS-phase to the MG-GQAP. For function $select\_variable\_columns(X')$ in line 15, which stores free assignments to be optimized by an LP in set Col, we considered a number of possibilities. For the determination of these $K := |Col|$ columns, several selection mechanisms were tested. These include *random* selection, so-called *plane* selection of columns ((1, 2, 3), (4, 5, 6), ..., (n − 2, n − 1, n); (2, 3, 4), (5, 6, 7), ...) or *binomial* selection with columns ((1), (2), (3), (1, 2), (1, 3), (2, 3), (1, 2, 3); (2), (3), (4),...), all examples underlying $K = 3$. Trends in efficiency are non-random strategies $\prec$ purely random $\prec$ mixed-random options. Therefore, the chosen strategy is a half and a half mixture of *random* and *plane*.

After building the LP-subproblem by $LP\_from(Col)$ and solving it, in lines 16 and 17, respectively, procedure $combine\_solutions(X', X_{LP}^*)$ adds the optimized values $x_{ij_{LP}}^*$ to the starting solution X' and thus includes the LP-optimized assignments of X_{LP}^*. If necessary, this is followed by an update of the best solution, the same way as it is done for the TS in lines 5–8. At the very end, with TS_GQAP_term getting true, the overall process stops and the algorithm returns its best solution found as X_{new}.

4   Computational Results

For the evaluation of the computational results we used three test beds with a total of 64 instances. Abbreviated with LAM, CEAL and OWN, there are 27 instances with $m = 6 - 30$ and $n = 10 - 16$ from Lee and Ma [5], 21 instances with $m = 6 - 20$ and $n = 20 - 50$ from Cordeau et al. [1] and 16 own randomly generated instances with $m, n = 10 - 200$. All algorithms were implemented in AMPL script V.20220310, using the solver CPLEX V.20.1.0.0 (MS VC++ 10.0, 64-bit). All runs were performed on a ThinkPad X1 notebook with an Intel(R) Core(TM) i7-8550U CPU @1.80 GHz (Aug. 2017) under Windows 10 Pro with 8 GB RAM.

We aimed for a unified parameter setting, but eventually this was only possible to a certain extent due to the large structural differences in the data sets: symmetry of the data matrices D and F, constancy of weights in W over the agent set and, very basic, instance sizes, where the latter directly affect and determine calculable LP subproblems, regarding the value of parameter K. To profit from faster calculations, a relatively low value of $TS\_nh\_size = 3$ is chosen. Tabu time becomes $tenure \approx 0.3n$ and $pool\_size = 20$. More numeric parametric details are reported below. Evaluation is split into two parts, judgement of the general design of the construction and improvement procedures, thereafter in part two supplemented by the discussion of the specific results obtained for the three data sets.

In order to test and evaluate the general design of the construction procedures, we used all 3 test sets. Starting with the tree heuristic GARRP and LAM, parameter settings were $x_{LB} = 0.5$ and $k = 50$. The procedure needs $min|avg|max = 6|2375|18364$ tree nodes and $min|avg|max = 0.3|168.5|1598.4$ seconds (sec.). The second test set, CEAL, is solved with $x_{LB} = 0.85, k = 25$ and $min|avg|max = 21|1842|16103$ nodes in $min|avg|max = 1.1|391.6|\underline{3600}$ sec. Note that while feasible solutions for all instances of LAM can be built, the method fails in two cases (#11 and #15) for CEAL, i.e. those ones for which one hour of computing time was not enough. Here starting solutions can be provided by solving the standard (linear) GAP. Results for set OWN, solved with $x_{LB} = 0.9, k = 30$, are $min|avg|max = 11|30.6|201$ nodes and $min|avg|max = 0.3|65.3|707.1$ sec. Again, as in LAM, no infeasible solution had to be accepted. Not much can be derived from these numbers, but two statements appear to be meaningful: the number of nodes, equivalent to the number of solved LPs, is relatively large for specific instances, which indicates a high CPU load. Secondly, it is a pleasing fact that the success rate of the procedure is quite high (62 of 64 cases). An objective-oriented reasoning of GARRP goes along with the judgement of the magnifying glass approach used as a construction procedure, i.e. algorithm variant MG-GQAP-C. This time, exclusively based on CEAL, setting $K = 4$, three MG-GQAP-C-runs with $MG\_GQAP\_term = 500|1000|10000$ iterations are performed. The hypothesis is that with an increasing number of iterations, i.e. with higher computing time, the numerical quality of the overall best solution will increase too, though by an unknown and decreasing amount. These expectations are met as follows. Designating the result of the 500-run as a basis, the relative stepwise improvements obtained between the 500- and the 1000-result and between the 1000- and the 10000-result, averaged over the objective values of all 21 instances, are 2.7% and 0.8%. Moreover, contrasting the best MG-GQAP-C-result, the 10000-result with GARRP, it becomes obvious that the tree-procedure works even better with an averaged additional 3.9% performance gain. It is interesting to note that these numerical ratios apply not only for the construction process but also for an improvement process that builds exactly on the former starting solutions. Total running time (min.) naturally gradually increases: 14.2|33.0|80.6 (MG-GQAP-C) vs. 137.1 (GARRP).

Next, we appraise the design of the proposed TS improvement procedure TS-GQAP. Essentially, it consists of a TS component (TS-phase) and an LP component (MG-GQAP-phase). So it is obvious to isolate the individual components and to compare three test runs: (1) only MG-GQAP-phase, (2) only TS-phase and (3) both phases combined (=TS-GQAP). Like above, test set CEAL is used, input comes in all cases from the tree start heuristic GARRP and parametrization from a unified, single parameter set. The outcome for (1) is an average improvement of $\delta_{tree}^{avg} = 5.75\%$ (in 35.1 min., whole set), for (2) it is 2.5% (38.0 min.) and finally for (3), the TS-GQAP, it is 6.39% (44.1 min.). Thus, even if the overall concept gets sufficiently motivated, some captious comments seem appropriate. Again, as an logical implication, objective function improvement comes at the expense of CPU time. In the present context, however, the contribution of the TS-phase is smaller than that of the MG-GQAP-phase. However, this is not surprising since the optimizing component based on an optimum solution algorithm will generally generate more visible improvements than a heuristic one. One can also see that the neighbourhood structure of the TS, which is more complex in terms of implementation, has a significant impact. Nonetheless, the contribution of the TS is also a significant one, which is only underscored by the overall effectiveness of the method.

The second part of the computational results covers the evaluation of the specific results calculated for the three test sets. Moderately sized instances of LAM allow the use of a mixed strategy. The number of LP-columns as well as the number of LP-rows is set with $K = 4$ and the LP-build strategy, column- or row-oriented, is changed every 25 iterations. Termination parameters are set as $TS\_term = 150, MG\_GQAP\_term = 50$ and $TS\_GQAP\_term = 20$. The result for LAM is $\delta_{tree}^{avg} = 11.66\%$, which takes 2.4 min. averaged over all instances. It can be observed that the algorithm finds the best solutions quite early. With respect to quality it can be stated that the results appear to be good, but no optimality gaps were calculated since optimal solutions are not published.

The next object of observation is the test set CEAL. Again, $K = 4$ and the LP-component follows a column approach, exactly as it is described in Algorithm 1, while completion criteria are given by $TS\_term = 500, MG\_GQAP\_term = 100$ and $TS\_GQAP\_term = 10$. This leads to an improvement of $\delta_{tree}^{avg} = 6.4\%$ in an average of 2.2 min. Because competing objective function values are available, deviations from best-known objectives can be determined. They are given with δ_{CEAL} and amount $min|avg|max = 0.0\%|1.4\%|8.29\%$. Results' quality clearly correlates with instance-density ρ, the ratio of total capacity demanded over total capacity available (reasonably only calculable for server independent constant demands, as it is the case with CEAL and LAM). As already indicated, the two problem instances, namely #11 and #15, cannot be improved. It should also be stated in an exculpatory manner that average quality gets destroyed by only a few outliers. Compensating these, an acceptable $\delta_{CEAL}^{avg} = 0.9\%$ can be achieved.

The third and last set, OWN, contains the hardest instances: asymmetric, non-constant weights and sizes up to $m, n = 200$. It is solved with the same column-oriented LP-strategy as used for CEAL. We set $K = 3$, however, due to advanced dimensions, it was necessary to reduce the value of K to 2 for instances with a high $m = 200$. Termination variables are given by $TS\_term = 150, MG\_GQAP\_term = 25$ and $TS\_GQAP\_term = 40$. With these parameters we can achieve a $\delta_{tree}^{avg} = 34.46\%$ in average 11.0 min. In the OWN case running time utilization is more efficient since for some instances the best result is only achieved after 90% of the total running time. The majority of instances ($\approx 11, 12$ out of 16) cannot be solved optimally with the means at our disposal, hence no reasonable deviations can be calculated.

As a final remark it should be noted that for all test sets, LAM, CEAL and OWN, algorithm TS-GQAP constitutes an improvement over the old MG-GQAP as described in [2]. In terms of objective value amount this progress is not outstanding, but it is clearly visible. It comes either as an increase in CPU productivity, i.e. percentage improvement divided by CPU time used, or actually as an improvement of the (average) objective.

5   Summary, Criticism and Outlook

This work is a further step in an ongoing research project looking for heuristic solution procedures for the GQAP. We combine the well-known metaheuristic TS with the strengths of mathematical programming and introduce a new matheuristic referred to as TS-GQAP. The basic idea behind it was originally coined as Magnifying Glass Heuristic, itself a matheuristic and successfully used for a *Quadratic Travelling Salesman Problem*. The design of the new method turned out to be successful in terms of CPU usage and the ability to deal with larger problem sizes, while also specific results could be improved. Moreover, in terms of competition, the method proved to be able to keep up with algorithms from the literature and the best-known gaps could be reduced by another level.

Even if the new algorithmic design is promising, there exist clearly visible improvement opportunities. New challenges raised are those about the course of the interaction between the TS neighbourhood and the LP decomposition or about the implementation of an overarching memory structure. Very basically, a stumbling block on the way to outstanding performance is founded in the capacity restricted nature of the underlying problem, which logically is intricate to navigate. Fast metaheuristics are able to play off their superiority for problem classes, which are unrestricted or endowed with a large number of feasible solutions (dense solution space) and often benefit from easier objective function calculations. As observed, the explorable solution space, more accurately, the neighbourhood induced solution landscape for some GQAP instances investigated is quite sparse. It has already been put forward [6] that sparse solution spaces are indicative of cases where matheuristics are probably more effective than plain metaheuristics, usually relying on local search or simple constructive procedures at their core. Moreover, GQAP is a representative problem of nonlinear combinatorial optimization, an area that so far received much less attention from research than its linear counterpart, despite its obvious relevance for modelling and solving compelling real-world problems.

This opens sufficient room to set up and tune new neighbourhood mechanisms to be developed, an endeavor, which of course cannot be done apart from designing more efficient memory structures. It is precisely this problem area that must be examined and analyzed more closely in the future.

References

1. Cordeau, J.-F., Gaudioso, M., Laporte, G., Moccia, L.: A memetic heuristic for the generalized quadratic assignment problem. INFORMS J. Comput. **18**(4), 433–443 (2006). https://doi.org/10.1287/ijoc.1040.0128
2. Greistorfer, P., Staněk, R., Maniezzo, V.: The magnifying glass heuristic for the generalized quadratic assignment problem. In: Proceedings of the XIII Metaheuristics International Conference, MIC 2019, pp. 22–24. Universidad de los Andes Sede Caribe (2019)
3. Greistorfer, P., Voß, S.: Controlled pool maintenance for metaheuristics. In: Sharda, R., Voß, S., Rego, C., Alidaee, B. (eds.) Metaheuristic Optimization via Memory and Evolution. Operations Research/Computer Science Interfaces Series, vol. 30, pp. 387–424. Springer, Boston (2005). https://doi.org/10.1007/0-387-23667-8_18
4. Kaufman, L., Broeckx, F.C.: An algorithm for the quadratic assignment problem using Bender's decomposition. Eur. J. Oper. Res. **2**(3), 204–211 (1978)
5. Lee, C.-G., Ma, Z.: The generalized quadratic assignment problem. Research report, Department of Mechanical and Industrial Engineering, University of Toronto (2003)
6. Maniezzo, V., Boschetti, M.A., Stützle, T.: Matheuristics: Algorithms and Implementations. EURO Advanced Tutorials on Operational Research, Springer, Cham (2021). https://doi.org/10.1007/978-3-030-70277-9
7. Maniezzo, V., Greistorfer, P., Staněk, R.: Exponential neighborhood search for the generalized quadratic assignment problem. EURO/ALIO International Conference on Applied Combinatorial Optimization, 25–27 June 2018, Bologna, Italy (2018)
8. Maniezzo, V., Stützle, T., Voß, S.: Matheuristics: Hybridizing Metaheuristics and Mathematical Programming, 1st edn. Springer, New York (2009). https://doi.org/10.1007/978-1-4419-1306-7
9. Ropke, S., Pisinger, D.: An adaptive large neighborhood search heuristic for the pickup and delivery problem with time windows. Transp. Sci. **40**(4), 455–472 (2006)
10. Staněk, R., Greistorfer, P., Ladner, K., Pferschy, U.: Geometric and LP-based heuristics for angular travelling salesman problems in the plane. Comput. Oper. Res. **108**, 97–111 (2019)

A Fast Metaheuristic for Finding
the Minimum Dominating Set in Graphs

Alejandra Casado[1] , Sergio Bermudo[2] , Ana Dolores López-Sánchez[2] ,
and Jesús Sánchez-Oro[1]([⊠])

[1] Universidad Rey Juan Carlos, Mostoles, Spain
{alejandra.casado,jesus.sanchezoro}@urjc.es
[2] Universidad Pablo de Olavide, Seville, Spain
{sbernav,adlopsan}@upo.es

Abstract. Finding minimum dominating sets in graphs is a problem
that has been widely studied in the literature. However, due to the
increase in the size and complexity of networks, new algorithms with the
ability to provide high quality solutions in short computing times are
desirable. This work presents a Greedy Randomized Adaptive Search
Procedure for dealing with the Minimum Dominating Set Problem in
large networks. The algorithm is conformed by an efficient constructive procedure to generate promising initial solutions and a local search
designed to find a local optimum with respect to those initial solutions.
The experimental results show the competitiveness of the proposed algorithm when comparing it with the state-of-the-art methods.

Keywords: Minimum dominating set · Grasp · Metaheuristics

1   Introduction

Due to the development of large networks in several context such as road networks, social networks, electrical networks, communication networks, computer
networks or security networks, among others, graph theory has regained the
interest of researchers and practitioners. The study of problems related to graph
domination is becoming more and more relevant. Since its original proposal [2]
and the main definition of domination number [10], more than two thousand
research papers have been published on this topic (we refer the reader to [6–8].

Given a graph $G = (V, E)$, a dominating set of vertices D is a subset of
V in which every vertex $u \in V \backslash D$ is adjacent to, at least, one vertex in D,
i.e., $\exists v \in D : (u, v) \in E$. The objective function of the minimum dominating
set problem (MDSP) is evaluated as the number of vertices belonging to the
dominating set, i.e., $MDSP(G, D) = |D|$. Then, the MDSP consists of finding a
minimum dominating set $D^\star$ among all possible dominating sets of the graph
under evaluation G. More formally,

$$D^\star \leftarrow \arg\min_{D \in \mathcal{D}} MDSP(G, D)$$

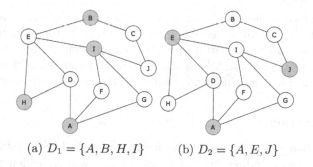

(a) $D_1 = \{A, B, H, I\}$ (b) $D_2 = \{A, E, J\}$

Fig. 1. Two feasible solutions D_1 and D_2 for a graph with 10 nodes and 12 edges.

where $\mathcal{D}$ represents the set of all possible dominating sets over the graph G.

Let us illustrate this problem with a graphical example, depicted in Fig. 1. In both examples, the vertices included in the dominating set are colored. Solution $D_1 = \{A, B, H, I\}$, depicted in Fig. 1(a), represents a dominating set for the example graph with an associated objective value of $|D_1| = 4$. If we now analyze solution $D_2 = \{A, E, J\}$, the objective function value is $|D_2| = 3$, being better than D_1. In fact, D_2 is the optimal solution for the considered graph (i.e., it is not possible to find a dominating set with less than three vertices).

This problem has several applications in different fields: social networks [9], radio stations [3], network surveillance [5], etc. The MDSP has been proven to be $\mathcal{NP}$-hard for arbitrary graphs [8], exact algorithms are not practical for large-sized graphs. Although there are only a few works focused on finding an algorithm to approximate the MDSP for general graphs [1,11], it has been extensively studied for particular graphs, proposing several bounds for the minimum size of the dominating set in general graphs also [6,8].

This work deals with the MDSP from a metaheuristic point of view for providing high-quality solutions in small computing times. As far as we know, the best approach for the MDSP is an order-based randomized local search (RLS) [1], which is shown to be better than previous approaches such as a greedy heuristic or an Ant Colony Optimization algorithm [11]. This algorithm use a permutation-based representation of the solution for MDSP which is transformed into dominating sets using a greedy approach. Then, a local search method based on randomized jump moves is applied.

The remaining of the paper is organized as follows: Sect. 2 presents the algorithmic approach, Sect. 3 details the computational experiments performed to test the quality of the proposal, and Sect. 4 draws some conclusions derived from the research.

2   Greedy Randomized Adaptive Search Procedure

The Greedy Randomized Adaptive Search Procedure (GRASP) is a multi-start metaheuristic originally introduced for solving the set covering problem [4]. It consists of two well-differenced phases. First, the construction phase is responsible for generating an initial feasible solution and, then, the local improvement stage finds a local optimum with respect to some predefined neighborhood. The distinguishing feature of GRASP is the inclusion of randomization in the construction phase, with the aim of increasing the diversity of the search.

2.1 Construction

The construction phase is one of the key parts of GRASP metaheuristic. It is responsible for including randomization in the search, which prevents the algorithm from stagnating in local optima. Algorithm 1 shows the pseudocode of the proposed constructive procedure, named SNF (Select Non-monitored First).

Algorithm 1. $SNF(G = (V, E), \alpha)$

1: $v \leftarrow Random(V)$
2: $D \leftarrow \{v\}$
3: $CL \leftarrow V \backslash \{v\}$
4: $M \leftarrow \{u \in V : (u, v) \in E\}$
5: **while** $|M| \leq |V \backslash D|$ **do**
6:     $g_{min} \leftarrow \min_{c \in CL} g(c)$
7:     $g_{max} \leftarrow \max_{c \in CL} g(c)$
8:     $\mu \leftarrow g_{max} - \alpha \cdot (g_{max} - g_{min})$
9:     $RCL \leftarrow \{c \in CL : g(c) \geq \mu\}$
10:     $d \leftarrow Random(RCL)$
11:     $D \leftarrow D \cup \{d\}$
12:     $CL \leftarrow CL \backslash \{d\}$
13:     $M \leftarrow M \cup \{u \in V : (u, d) \in E\}$
14: **end while**
15: **return** D

The method starts by randomly selecting the first vertex to be included in the solutions, as it is customary in GRASP to increase diversity (step 1). Then, the solution D is initialized with the selected vertex (step 2) and the candidate list CL is constructed with all the vertices except the selected one (step 3). Additionally, the set of monitored vertices is created with those vertices which are adjacent to the selected one, since those are now monitored (step 4). Then, the method iteratively adds a new vertex to the solution under construction until all the vertices are monitored (steps 5–14). In each iteration, a greedy function value g is considered to evaluate how promising a node is. In the context of MDSP, a vertex is promising if it is able to monitor a large number of non-previously monitored vertices. Therefore, the greedy function is evaluated as the number of adjacent vertices to the node under evaluation which are not monitored by any other node, i.e., $g(v) = |\{u \in V \backslash M : (u, v) \in E\}|$. The procedure then evaluates the minimum and maximum greedy function value (steps 6–7) to calculate a threshold μ (step 8). The threshold directly depends on the value of an input parameter named $\alpha \in [0,1]$, which controls the randomness/greediness of the method, i.e., if $\alpha = 0$, then $\mu = g_{\max}$ and the method is completely greedy, while $\alpha = 1$ results in $\mu = g_{\min}$, providing a completely random method. This threshold is used to construct a restricted candidate list RCL which contains all the candidate vertices whose greedy function value is better (larger) than μ (step 9). Then, the next vertex is selected at random from the RCL (step 10), including it in the solution under construction (step 11) and removing it from the CL (step 12). Additionally, the set of monitored nodes M is updated with all the adjacent vertices to the selected one (step 13). Finally, the method returns the constructed solution (step 15).

2.2   Local Improvement

The first element to be defined in order to propose a local search method is the move operator. In this work, we propose the use of swap moves, which removes a vertex from the solution and replaces it with a new one. In mathematical terms,

$$Swap(D, i, j) \leftarrow D\backslash\{i\} \cup \{j\}$$

Notice that this movement is not able to produce any improvement by itself, since the number of vertices in the dominating set will remain the same. However, the movement may result in a solution in which some vertices are redundant, i.e., they are covering vertices which are also covered by another vertex. All those redundant vertices are removed after each swap move, eventually leading to an improvement. Given this move operator, the neighborhood to be explored is defined as all the solutions that can be reached with a single swap move. More formally,

$$N(D) \leftarrow \{D' \leftarrow Swap(D, i, j) \ \forall i \in D \wedge \ \forall j \in V\backslash D\}$$

Having defined the move operator and the neighborhood explored, it is necessary to establish the strategy followed by the local search to explore the neighborhood. There are two main strategies to traverse the neighborhood: first improvement and best improvement. Since the problem has constraints related to computing time, we select first improvement as exploration strategy. Then, the first swap move that leads to an improvement is performed in each iteration of the search, avoiding exploring the complete neighborhood in each iteration, thus leading to reduce the computational effort without deteriorating the quality of the obtained solutions.

It is worth mentioning that when following a first improvement approach, the order in which the neighbors is explored is relevant. Although the neighborhood can be traversed at random, exploring first the most promising neighbors usually result in better solutions. In the context of MDSP, the nodes to be removed are selected in ascending order with respect to the number of nodes that remains non-dominated after its removal, with the aim of increasing the number of redundant nodes after the swap move. Additionally, the vertex which replaces it will be the one which dominates the maximum number of non-dominated nodes after the removal. This heuristic selection of the nodes to be removed and the ones to be added allows the local search procedure to find improvements in a small number of iterations, resulting in an efficient and effective local search method, as it can be seen in Sect. 3.

3 Computational Experiments

This section has two main objectives: 1) configure the parameters of the proposed algorithm and 2) perform a competitive testing against the state-of-the-art method, named RLS [1], previously described in Sect. 1.

The proposed algorithm have been implemented in Java 17 and all the experiments has been carried out in an Intel Core i7 2.7 GHz and 16 GB RAM, while the state of the art is executed in a similar computer but with 64 GB of RAM. In order to have a fair comparison, we have used the same testbed of instances as in the best previous work, conformed by 21 instances with nodes ranging from 34 to 16726, extracted from different applications of the problem. The results reports the following metrics: Time (s), average computing time required by the algorithm in seconds; Avg., average objective function value; Dev (%), average percentage deviation with respect to the best value found in the experiment; and #Best, number of times in which the best solution of the experiment is reached by the algorithm.

The preliminary experimentation is designed to select the best configuration of the proposed algorithm and, in order to avoid overfitting, a subset of 9 out of 21 representative instances are selected for this stage. The GRASP algorithm proposed has a single input parameter, which is the α parameter of the constructive procedure, which has to be tuned. In particular, the values tested are $\alpha = \{0.25, 0.50, 0.75, RND\}$, where RND indicates that it is selected at random in the range $[0, 1]$ in each construction. In all the cases, 100 constructions followed by its corresponding local improvement are performed. The best results are achieved when considering $\alpha = 0.25$, reaching 8 out of 9 best solutions with a deviation of 0.01%.

The competitive testing is performed against RLS to evaluate the performance of GRASP. In this experiment, the complete set of instances is considered. Table 1 shows the results obtained by both algorithms.

As it can be seen from the results, both algorithms provide similar performance in terms of quality. Although RLS is able to reach an additional best solution, the average deviation of GRASP is smaller than the one of RLS. Finally, the computing time required by GRASP is considerably smaller than the 10 min required by RLS. These results show that GRASP emerges as a competitive algorithm for finding a minimum dominating set in large graphs.

Table 1. Competitive testing between GRASP and RLS.

Algorithm	Time (s)	Avg.	Dev(%)	#Best
GRASP	33.66	603.19	0.40	19
RLS	600.00	603.05	0.68	20

4 Conclusion

The proposed GRASP algorithm is conformed by a greedy randomized adaptive constructive procedure and a local search method. The constructive procedure, named SNF, is able to generate high quality and diverse solutions in small computing times, providing promising starting points for the local search. The proposed local search is able to find a local optimum with respect to the initial point without requiring high computational efforts, resulting in an efficient and effective algorithm for tackling the MDSP. The results obtained show the competitiveness of GRASP when comparing it with the state of the art, providing similar results, being almost twenty times faster than RLS. As a future work, the testbed will be enlarged with more challenging instances derived from real-life networks to show the potential of GRASP.

References

1. Chalupa, D.: An order-based algorithm for minimum dominating set with application in graph mining. Inf. Sci. **426**, 101–116 (2018)
2. De Jaenisch, C.F.: Applications de l'Analuse mathematique an Jen des Echecs. Petrograd (1862)
3. Erwin, D.: Dominating broadcasts in graphs. Int. J. Comput. Eng. Technol. **42**, 89–105 (2004)
4. Feo, T.A., Resende, M.G.: A probabilistic heuristic for a computationally difficult set covering problem. Oper. Res. Lett. **8**(2), 67–71 (1989)
5. Haynes, T.W., Hedetniemi, S.M., Hedetniemi, S.T., Henning, M.A.: Domination in graphs applied to electric power networks. SIAM J. Discret. Math. **15**(4), 519–529 (2002)
6. Haynes, T.W., Hedetniemi, S., Slater, P.: Fundamentals of Domination in Graphs. CRC Press, Boca Raton (2013)
7. Haynes, T.W., Hedetniemi, S.T., Henning, M.A.: Structures of Domination in Graphs. Springer, Heidelberg (2021). https://doi.org/10.1007/978-3-030-58892-2
8. Haynes, T.: Domination in Graphs: Volume 2: Advanced Topics. Routledge (2017)
9. Lozano-Osorio, I., Sánchez-Oro, J., Duarte, A., Cordón, Ó.: A quick grasp-based method for influence maximization in social networks. J. Ambient Intell. Human. Comput. 1–13 (2021)
10. Ore, O.: Theory of Graphs. AMS, Providence (1962)
11. Potluri, A., Singh, A.: Hybrid metaheuristic algorithms for minimum weight dominating set. Appl. Soft Comput. **13**, 76–88 (2013)

Scheduling Jobs in Flexible Flow Shops
with s-batching Machines Using Metaheuristics

Jens Rocholl and Lars Mönch[✉]

Chair of Enterprise-Wide Software Systems, University of Hagen, Universitätsstraße 1,
58097 Hagen, Germany
{Jens.Rocholl,Lars.Moench}@fernuni-hagen.de

Abstract. A scheduling problem for a two-stage flexible flow shop with s-batching machines is considered. A batch is a group of jobs that are processed at the same time on a single machine. A maximum batch size is given. The jobs belong to incompatible families. Only jobs of the same family can be batched together. A setup time occurs between different batches. The processing time of a batch is the sum of the processing times of the jobs forming the batch, i.e., the jobs are processed in a serial manner. Batch availability is assumed. Each job has a weight, a due date, and a release date. The performance measure is the total weighted tardiness (TWT). An iterative decomposition approach (IDA) is proposed that uses a grouping genetic algorithm (GGA) or an iterated local search (ILS) scheme to solve the single-stage subproblems. Results of computational experiments based on randomly generated problem instances demonstrate that the IDA hybridized with ILS is able to determine high-quality solutions.

Keywords: Flow shop · s-batching · Decomposition · Iterated local search · Grouping genetic algorithm

1 Introduction

Flexible flow shop scheduling problems are important in many industrial domains such as semiconductor manufacturing, printing, food processing, or textile industries. Since flow shop scheduling problems are typically NP-hard, metaheuristics are applied to solve them. Solution approaches are often based on stage-based decomposition approaches. We discuss a two-stage flexible flow shop with s-batch processing machines on each stage. s-batch processing means that the jobs belonging to a batch are processed in a serial manner. We assume that the sum of the sizes of the jobs belonging to a batch does not exceed a maximum batch size. For instance, the maximum batch size can be seen as the capacity of a transportation cart or a cassette for transporting wafers in semiconductor manufacturing [7]. s-batching problems with a maximum batch size are not often discussed in the literature. We demonstrate by computational experiments that an IDA based on ILS outperforms the one that is based on GGAs.

The paper is organized as follows. In the next section, we describe the problem and discuss related work. A decomposition approach and two metaheuristic approaches to

© The Author(s), under exclusive license to Springer Nature Switzerland AG 2023
L. Di Gaspero et al. (Eds.): MIC 2022, LNCS 13838, pp. 560–568, 2023.
https://doi.org/10.1007/978-3-031-26504-4_48

tackle the resulting subproblems are described in Sect. 3. The results of computational experiments are presented in Sect. 4. Finally, conclusions and future research directions are discussed in Sect. 5.

2 Problem Setting and Related Work

2.1 Problem Description

We consider a two-stage flexible flow shop where identical s-batching machines are assumed on each stage, i.e., the processing time of a batch is the sum of the processing times of the jobs that belong to the batch. Incompatible job families are assumed, i.e., only jobs of the same family can belong to a batch. Each job $j = 1, \ldots, n$ has a size s_j, a ready time $r_j \geq 0$, a due date d_j, and a weight w_j to express the importance of job j. A constant setup time q_s is assumed between two batches at stage s. We consider batch availability, i.e., the jobs of a batch can only be processed further when all jobs of the batch are completed. We have a maximum batch size $B_s < n$ on stage s, i.e., the sum of the sizes of the jobs belonging to the batch does not exceed B_s. The performance measure is the $TWT := \sum_{j=1}^{n} w_j (C_j - d_j)^+$ where C_j is the completion time of job j, and we abbreviate $x^+ := \max(x, 0)$. Using the three-field notation the problem can be written as

$$FF2|F_s, s - batch, r_j, s_j, q_s|TWT, \tag{1}$$

where $FF2$ indicates a flexible flow shop, $s - batch$ s-batching machines, and F_s the incompatible families at stage s. A mixed integer linear programming (MILP) formulation for problem (1) is provided in the supplement belonging to this paper. It is shown in [1] that even the single-machine version of problem (1) with the total tardiness performance measure is NP-hard. Hence, we have to look for efficient heuristics.

2.2 Discussion of Previous Work

A survey for batching including s-batching is provided in [10]. A single-machine s-batching problem with a minimum and a maximum batch size and batch availability and the total completion time measure is studied in [8]. Efficient optimal solution procedures are designed. In [2], dynamic programming algorithms are proposed for a single-machine scheduling problem with s-batching, minimum and maximum batch sizes, and batch availability. The performance measure is the sum of the tardiness and the setup costs. A tabu search approach for a single-machine s-batching problem with maximum batch size and batch availability and the TWT measure is discussed in [13]. Polynomial-time approaches to find Pareto-optimal schedules are established. A parallel-machine scheduling problem with s-batching, batch availability, family sequence-dependent setup times, and the total weighted completion time performance measure is studied in [12]. Variable neighborhood search procedures that iterate between batch formation and sequencing are proposed. A single-machine s-batching problem with a minimum and a maximum batch size and batch availability where the maximum lateness and makespan are the performance measures is analyzed in [5]. All the discussed scheduling problems are for single- or parallel-machine environments, metaheuristic approaches are rarely used. Hence, to the best of our knowledge, scheduling problem (1) is not discussed in the literature so far. A stage-wise IDA for a two-stage flexible flow shop with p-batching and the TWT measure is designed in [14]. This IDA will be applied in the present paper to obtain parallel-machine subproblems for which metaheuristic approaches are proposed.

3 Metaheuristic Approaches

3.1 Iterative Decomposition Scheme

The IDA proposed in [14] is applied to problem (1). As a result, the problem is decomposed to obtain single-stage $Pm|F_s, s - batch, r_j, s_j, q_s|TWT$ subproblems, where Pm refers to identical parallel machines. The main idea of the IDA is to iteratively solve the subproblems of both stages for parameter values obtained in previous iterations and thus successively find improved solutions for problem (1). Reasonable parameter values for each subproblem are required to allow for computing schedules which are favorable with respect to the overall scheduling problem. We must set appropriate internal due dates for the subproblem of the first stage which consider the processing times at the second stage. Ready times of the jobs for the subproblem of the second stage depend on the solution of the former subproblem. Internal due dates for each job are calculated during the first iteration by adding half of its maximum slack to its earliest possible completion time at the first stage. Waiting times observed in the previous iteration are taken into account in the following iteration. A internal due date of job j in iteration l at the first stage is set according to

$$d_{j1}^{(l)} := (1 - \alpha)r_{j2}^{(l-1)} + \alpha\left(d_j - pb_{j2} - v_{j2}^{(l-1)}\right) + \beta v_{j2}^{(l-1)}, \tag{2}$$

where pb_{j2} is the processing time of the batch to which j belongs. Moreover, $r_{j2}^{(l-1)}$ and $v_{j2}^{(l-1)}$ are the internal ready time and the waiting time of j in the second stage of the previous iteration, respectively. The parameters α and β are set to control the influence of internal release dates and waiting times of preceding iterations. Due to space limitations, we refer to [14] for a detailed description of the parameter settings. Internal ready times for the subproblem of the second stage are set to the actual completion times taken from the schedule of the first stage of the same iteration.

3.2 GGA for Solving the Subproblems

GGAs are introduced in [3] by observing that conventional representation schemes can be too disruptive for grouping problems. A gene does not represent a job or its position but a batch in the GGA encoding scheme. We use a two-part random key to encode the batch-to-machine assignment and the batch sequence. A special crossover operator is designed to better conserve favorable grouping decisions. Preliminary experiments show that the performance of a GGA profits from hybridizing it with local search (LS). Therefore, for a certain portion of each generation, namely 10%, five neighborhood structures are used to insert a job into another batch, swap two jobs of different batches, insert a batch at another position, swap the positions of two batches, and split a batch. The LS is limited to ten moves for each neighborhood to make sure that enough time is allocated to the evolution.

3.3 ILS for Solving the Subproblems

An initial schedule is obtained by sorting the jobs with respect to the apparent tardiness cost (ATC) dispatching rule. The values of the look-ahead parameter κ are taken from a grid over [0.1, 5.0] with step size of 0.1. The schedule that leads to the smallest TWT value is taken. The jobs are assigned to machines using list scheduling. The LS is then applied to form batches until a local optimum is found. A randomized variable neighborhood descent similar to the one proposed in [11] is applied. The neighborhood structures used in the GGA are shuffled randomly. They are then searched one by one in a greedy manner until no better solution is found. If an improving schedule is found, then the neighborhood structures are shuffled again and the search starts over with the first one. Otherwise the search continues with the next neighborhood structure. The LS converges eventually to a local optimum and the solution is saved. To escape this local optimum and exploit different regions of the search space, perturbation moves randomly chosen from the neighborhood structures job insert, job swap, and batch split are performed. The LS and the perturbation phases are executed in an alternating manner until a termination criterion is met. A trade-off exists between the number of iterations and time provided for solving each individual subproblem. The GGA stops after five generations without improvement. The ILS terminates after 25 iterations of no improvement. After ten iterations without improvement the number of perturbation steps is doubled. These parameter values are justified by preliminary computational experiments based on a small set of problem instances.

4 Computational Experiments

The proposed heuristics are coded in the C++ programming language using the Galib framework [15]. All experiments are conducted on a computer with Intel Core I7-2600 CPU with 3.40 GHz and 16 GB RAM. A set of 12 small-sized problem instances with ten jobs of two incompatible families at the first stage is examined (see the design of experiments in the supplement). The MILPs for the problem instances are solved using IBM CPLEX. However, the solver cannot compute optimal solutions for all instances within a prescribed computing time limit of one hour per instance. The relative MIP gap is reported. The heuristic approaches are performed with a prescribed computing time limit of 60 s per instance. The relative performance $PR_H := TWT_H/TWT_{CPLEX}$ is shown for heuristic H in Table 1. Best comparable results are marked bold.

With the bottleneck at the second stage, optimality of the solutions found by CPLEX is guaranteed for four out of six instances. For these instances no heuristic computes a better solution which indicates that the algorithms are implemented correctly. All but one result found by the heuristics are within five percent of the best solutions. We use $\alpha = 0.66$ and $\beta = 1.66$ in the IDA based on preliminary experiments using a small set of problem instances.

Table 1. Results of experiments with small-sized problem instances

Inst.	Bottleneck stage	Range of releases	GGA	ILS	CPLEX	Rel. MIP gap
1	1	Tight	**1.000**	**1.000**	**1.000**	99.9%
2			**0.952**	**0.952**	1.000	99.7%
3			**1.000**	**1.000**	**1.000**	100.0%
4		Wide	**0.994**	1.012	1.000	100.0%
5			**1.000**	**1.000**	**1.000**	100.0%
6			**1.000**	**1.000**	**1.000**	100.0%
7	2	Tight	**1.000**	**1.000**	**1.000**	0.0%
8			1.009	1.046	**1.000**	0.0%
9			**0.994**	1.026	1.000	72.4%
10		Wide	1.002	**1.000**	**1.000**	0.0%
11			1.021	**1.000**	**1.000**	0.0%
12			1.003	1.006	**1.000**	33.7%

Table 2 shows the average results of five independent replications with a time limit of 180 s for the large-sized instances (see the supplement). Each row represents the TWT average of 24 instances. ILS provides the best results and serves as benchmark. We observe that while delivering results with 17% deterioration compared to the results obtained by the ILS for $n = 30$, the performance of the GGA quickly deteriorates with an increasing number of jobs. Only the ILS scheme can preserve a good performance even for larger instances.

Table 2. Results of experiments with large-sized problem instances

Bottleneck stage	Number of jobs	GGA	ILS
1	30	1.128	1.000
	45	1.292	1.000
	60	1.413	1.000
2	30	1.165	1.000
	45	1.222	1.000
	60	1.319	1.000

5 Conclusions and Future Research

We discussed a two-stage flexible flow shop scheduling problem with s-batching machines on each stage. We applied an IDA that allows to solve subproblems for each of the two stages. A GGA and an ILS scheme were designed for the subproblems. It was demonstrated by designed computational experiments that the IDA based on the ILS outperforms the one based on a GGA.

There are several directions for future research. It seems interesting to design an ILS scheme that works on a disjunctive graph representation of the scheduling problem [9]. We expect that it is possible to extend the batch-oblivious approach from [6] from a p-batch to an s-batch setting. Moreover, we believe that an extension to a flexible flow shop setting with more than two stages is interesting. Another fruitful direction is considering a machine environment where p-batching or s-batching can be found at the different stages. Such mixed settings exist in semiconductor manufacturing [4].

Appendix A: MILP Formulation

The problem can be formulated as a MILP. We have the following sets and indices:

$j = 1, ..., n$: set of jobs

$f = 1, ..., F_s$: set of families of stage $s \in \{1,2\}$

$b = 1, ..., b_{sm}$: set of batches on machine m of stage $s \in \{1,2\}$

$m = 1, ..., m_s$: set of machines at stage $s \in \{1,2\}$.

Moreover, the following parameters are used in the formulation:

B_s : maximum batch size of a machine on stage s

p_{fs} : processing time of family f at stage s

d_j : due date of job j

q_s : setup time at stage s

w_j : weight of job j

r_j : ready time of job j

s_j : size of job j

e_{jfs} : $\begin{cases} 1, & \text{if job } j \text{ belongs to family } f \text{ at stage } s \\ 0, & \text{otherwise} \end{cases}$

M : big number.

The following decision variables are applied in the model:

x_{jbsm}: $\begin{cases} 1, & \text{if job } j \text{ belongs to batch } b \text{ on machine } m \text{ at stage } s \\ 0, & \text{otherwise} \end{cases}$

y_{bfsm}: $\begin{cases} 1, & \text{if batch } b \text{ on machine } m \text{ at stage } s \text{ belongs to family } f \\ 0, & \text{otherwise} \end{cases}$

C_{js}:   completion time of job j at stage s

T_j: tardiness of job j

a_{bsm}:   : start time of batch b on machine m at stage s.

The model is formulated as follows:

$$\min \sum_{j=1}^{n} w_j T_j \tag{A1}$$

subject to

$$\sum_{b=1}^{b_{sm}} \sum_{m=1}^{m_s} x_{jbsm} = 1, j = 1, \dots, n, \ s \in \{1,2\}, \tag{A2}$$

$$\sum_{j=1}^{n} s_j x_{jbsm} \leq B_s, s \in \{1,2\}, \ m = 1, \dots, m_s, \ b = 1, \dots, b_{sm}, \tag{A3}$$

$$\sum_{f=1}^{F_s} y_{bfsm} = 1, s \in \{1,2\}, \ m = 1, \dots, m_s, \ b = 1, \dots, b_{sm}, \tag{A4}$$

$$e_{jfs} x_{jbsm} \leq y_{bfsm}, j = 1, \dots, n, s \in \{1,2\}, \ f = 1, \dots, F_s, \ m = 1, \dots, m_s, \\ b = 1, \dots, b_{sm} \tag{A5}$$

$$r_j x_{jb1m} \leq a_{b1m}, \ j = 1, \dots, n, \ b = 1, \dots, b_{1m}, \ m = 1, \dots, m_1, \tag{A6}$$

$$q_1 \leq a_{b1m}, \quad j = 1, \dots, n, \ b = 1, \dots, b_{1m}, \ m = 1, \dots, m_1, \tag{A7}$$

$$a_{bsm} + q_s + \sum_{j=1}^{n} \sum_{f=1}^{F_s} e_{jfs} p_{fs} s_j x_{jbsm} \leq a_{b+1,s,m}, s \in \{1,2\}, \\ b = 1, \dots, b_{sm}, \ m = 1, \dots, m_s, \tag{A8}$$

$$a_{bsm} + \sum_{k=1}^{n} \sum_{f=1}^{F_s} e_{kfs} p_{fs} s_k x_{kbsm} \leq C_{js} + M(1 - x_{jbsm}), \\ j = 1, \dots, n, s \in \{1,2\}, \ m = 1, \dots, m_s, \ b = 1, \dots, b_{sm}, \tag{A9}$$

$$C_{j1} \leq M(1 - x_{jb2m}) + a_{b2m}, j = 1, \dots, n, \ m = 1, \dots, m_s, b = 1, \dots, b_{sm}, \tag{A10}$$

$$C_{j2} - T_j \leq d_j, j = 1, \dots, n, \tag{A11}$$

$$x_{jbsm}, y_{bfms} \in \{0,1\}, C_{js}, T_j, a_{bsm} \geq 0, j = 1, \dots, n, s \in \{1,2\}, \\ f = 1, \dots, F_s, b = 1, \dots, b_{sm}, \ m = 1, \dots, m_s. \tag{A12}$$

The objective function (A1) to be minimized is the TWT. Constraint set (A2) ensures that each job on each stage is assigned to exactly one batch. The constraints (A3) enforce that the maximum batch size is respected for each formed batch. Constraint set (A4) makes sure that exactly one family is assigned to a batch. The family of the jobs belonging to a batch and the family assigned to a batch is the same. This is modeled by constraint set (A5). Constraint set (A6) relates the starting time of the first batch on a given machine to the ready time of the jobs that belong to the batch. The constraints (A7) make sure that setup times are considered at the first stage. The starting times of adjacent batches on a given machines are modeled by constraint set (A8). Constraint set (A9) computes the completion time of jobs based on the completion time of the related batch. The ready time of a batch on the second stage is not earlier than the completion time of any job of the batch on the first stage. This is expressed by the constraints (A10). Constraint set (A11) linearizes the tardiness. The domains of the decision variables are modeled by (A12).

Appendix B: Design of Experiments

Table B1. Design of experiments for small-sized problem instances

Factor	Level	Count
Number of jobs per family	$n_f := 10/F$	1
Number of families at first stage	$F_1 := 2$	1
Number of families at second stage	with probability 0.5 a family of the first stage is divided into two families at the second stage	1
Total number of machines	$m := 3$	1
Maximum batch size	$B_s := 10$	1
Setup time at stage s	$q_s := \left(\dfrac{s_j^{min} + s_j^{max}}{2} p_{js}^{max} \right)/2$	1
Sizes of the jobs	$s_j \sim DU[1,8]$	1
Weights of the jobs	$w_j \sim DU[1,5]$	1
Processing times of the jobs	5 with probability 0.2 10 with probability 0.3 15 with probability 0.3 20 with probability 0.2	1
Ready times of the jobs	$r_j \sim U\left(0, \alpha \left(\dfrac{1}{\overline{B}_1 m_1} \sum_{j=1}^{n} s_j p_{j1} + \dfrac{1}{\overline{B}_2 m_2} \sum_{j=1}^{n} s_j p_{j2} \right) \right),$ $\alpha = 0.25, 0.75$	2
Due date of the jobs	$d_j := r_j + 1.3(s_j p_{j1} + s_j p_{j2})$	1
Bottleneck machine configuration	stage 1, stage 2	2
Number of instances per factor combination	3	3
Total		12

The average batch size at stage i size is computed by dividing the maximum batch size at stage s by the average job size, i.e. $\overline{B}_s := nB_s/\sum_{j=1}^{n} s_j$.

A set of 144 large-sized instances is generated with 10, 15, and 20 jobs per family, three families at the first stage and overall six machines. The maximum batch size at each stage is 15 and 30. The remaining parameters are identical to the ones of the small-sized instances found in Table B1.

References

1. Cheng, T.C.E., Kovalyov, M.Y.: Single machine batch scheduling with sequential job processing. IIE Trans. **33**(5), 413–420 (2001)
2. Chrétienne, P., Hazır, Ö., Kedad-Sidhoum, S.: Integrated batch sizing and scheduling on a single machine. J. Sched. **14**, 541–555 (2011)
3. Falkenauer, E.: A hybrid grouping genetic algorithm for bin packing. J. Heuristics **2**(1), 5–30 (1996)
4. Fowler, J.W., Mönch, L.: A survey of scheduling with parallel batch (p-batch) processing. Eur. J. Oper. Res. **298**(1), 1–24 (2022)
5. He, C., Lin, H., Lin, Y.: Bounded serial-batching scheduling for minimizing maximum lateness and makespan. Discrete Optim. **16**, 70–75 (2015)
6. Knopp, S., Dauzère-Pérès, S., Yugma, C.: A batch-oblivious approach for complex job-shop scheduling problems. Eur. J. Oper. Res. **263**(1), 50–61 (2017)
7. Mönch, L., Fowler, J.W., Mason, S.J.: Production planning and control for semiconductor wafer fabrication facilities: modeling, analysis and systems. Springer, New York (2013). https://doi.org/10.1007/978-1-4614-4472-5
8. Mosheiov, G., Oron, D.: A single machine batch scheduling problem with bounded batch size. Eur. J. Oper. Res. **187**, 1069–1079 (2008)
9. Ovacik, I.M., Uzsoy, R.: Decomposition methods for complex factory scheduling problems. Springer, New York (1997). https://doi.org/10.1007/978-1-4615-6329-7
10. Potts, C.N., Kovalyov, M.Y.: Scheduling with batching: a review. Eur. J. Oper. Res. **120**(2), 228–249 (2000)
11. Queiroga, E., Pinheiro, R.G.S., Christ, Q., Subramanian, A., Pessoa, A.A.: Iterated local search for single machine total weighted tardiness batch scheduling. J. Heuristics **27**(3), 353–438 (2020). https://doi.org/10.1007/s10732-020-09461-x
12. Shen, L., Mönch, L., Buscher, U.: Simultaneous and iterative approach for parallel machine scheduling with sequence dependent family setups. J. Sched. **17**(5), 471–487 (2014)
13. Suppiah, S., Omar, M.K.: A hybrid tabu search for batching and sequencing decisions in a single machine environment. Comput. Ind. Eng. **78**, 135–147 (2014)
14. Tan, Y., Mönch, L., Fowler, J.W.: A hybrid scheduling approach for a two-stage flexible flow shop with batch processing machines. J. Sched. **21**(2), 209–226 (2017). https://doi.org/10.1007/s10951-017-0530-4
15. Wall M.: Galib: A C++ library of genetic algorithms components (2017). http://lancet.mit.edu/ga/

Correction to: Evaluating the Effects of Chaos in Variable Neighbourhood Search

Sergio Consoli and José Andrés Moreno Pérez

Correction to:
Chapter "Evaluating the Effects of Chaos in Variable Neighbourhood Search" in: L. Di Gaspero et al. (Eds.): *Metaheuristics*, LNCS 13838, https://doi.org/10.1007/978-3-031-26504-4_15

Chapter "Evaluating the Effects of Chaos in Variable Neighbourhood Search" was previously published non-open access. It has now been changed to open access under a CC BY 4.0 license and the copyright holder updated to 'The Author(s)'. The book has also been updated with this change.

The updated original version of this chapter can be found at
https://doi.org/10.1007/978-3-031-26504-4_15

© The Author(s) 2023
L. Di Gaspero et al. (Eds.): MIC 2022, LNCS 13838, p. C1, 2023.
https://doi.org/10.1007/978-3-031-26504-4_49

Author Index

L. Di Gaspero et al. (Eds.): MIC 2022, LNCS 13838, pp. 569–571, 2023.
https://doi.org/10.1007/978-3-031-26504-4

Printed in the United States
by Baker & Taylor Publisher Services

Printed in the United States
by Baker & Taylor Publisher Services